Birkhäuser

Progress in Mathematics

Volume 292

J.J. (Hans) Duistermaat (1942–2010)

Preface

Geometric concepts often play an essential role in obtaining a profound understanding of many areas of analysis and mechanics, for instance, in the theory of Fourier integral operators and in (semi)classical mechanics. This interaction between geometry and analysis or mechanics is a very dominant and also unifying theme in the publications of Hans Duistermaat. At the occasion of his 65th birthday, leading investigators convened at Utrecht University, in August 2007, to discuss recent developments along these lines and in other areas related to the scientific interests of Duistermaat. This volume contains refereed contributions from most of the speakers at this conference and, additionally, two from invited speakers who were unable to attend.

During the preparation of the conference proceedings, Hans Duistermaat passed away unexpectedly, on March 19, 2010. There is no doubt in our minds that Duistermaat would have wished the publication of these proceedings as planned. Accordingly, we decided to leave the format unchanged, but to add an overview of Duistermaat's scientific work as well as some reminiscences by V.W. Guillemin, A. Weinstein, G. Heckman, and R.H. Cushman, as friends and co-authors.

The thirteen research articles published in this volume cover grosso modo three different topics: pseudodifferential operators and (inverse) spectral problems, index theory and localization, and group actions.

Pseudodifferential operators and (inverse) spectral problems. A characterization of the local solvability for square systems of pseudodifferential operators is the topic of the paper of N. Dencker, while J. Sjöstrand describes results on eigenvalue distributions and Weyl laws for non-self-adjoint operators. F. Alberto Grünbaum discusses matrix-valued polynomials satisfying differential equations both with respect to the space and the spectral variables. There are three papers, by S. Vũ Ngọc, Y. Colin de Verdière and V.W. Guillemin, and Y. Colin de Verdière, respectively, on the question to what extent the semiclassical spectrum of an operator determines properties of the operator.

Index theory and localization. In his article, J.-M. Bismut explains the relations between refined versions of index theory on a manifold X and the localization formulas of Duistermaat–Heckman on LX, the associated loop space. P.-E. Paradan studies the local invariants associated to the Hamiltonian action of a compact torus and obtains wall-crossing formulas between invariants attached to adjacent connected components of regular values of the moment map. L. Boutet de Monvel, E. Leichtnam, X. Tang, and A. Weinstein use equivariant Toeplitz operator calculus in order to give a new proof of the Atiyah–Weinstein conjecture on the index of Fourier integral operators and the relative index of CR structures. L.C. Jeffrey and B. McLellan consider the analog of nonabelian localization results of Beasley and Witten when the gauge group G is the abelian group $G = \mathrm{U}(1)$. Finally, E. Meinrenken explains how to define the quantization of q-Hamiltonian SU(2)-spaces as push-forwards in twisted equivariant K-homology, and to prove the "quantization commutes with reduction" theorem for this setting.

Group actions. On a symplectic manifold equipped with a Hamiltonian torus action a real locus is defined to be a set of fixed points for an equivariant smooth anti-symplectic involution. J.-C. Hausmann and T. Holm observe that certain cohomological relations between such a real locus and the ambient manifold can be explained in terms of a purely topological structure, rather than a symplectic one. There is a close relationship between Mumford's geometric invariant theory (GIT) in algebraic geometry and the process of reduction in symplectic geometry. F. Kirwan's paper describes ways in which nonreductive compactified quotients, which cannot be treated by means of classical GIT, can be studied using symplectic techniques.

List of all speakers. Nalini Anantharaman (École Polytechnique, Palaiseau), Nicole Berline (École Polytechnique, Palaiseau), Jean-Michel Bismut (Université Paris-Sud), Yves Colin de Verdière (Université Grenoble), Richard Cushman (Utrecht University), Nils Dencker (Lund University), F. Alberto Grünbaum (University of California at Berkeley), Victor Guillemin (Massachusetts Institute of Technology), Tara Holm (University of Connecticut), Frances Kirwan (University of Oxford), Eugene Lerman (University of Illinois at Urbana-Champaign), Jiang-Hua Lu (University of Hong Kong), Eckhard Meinrenken (University of Toronto), Richard Melrose (Massachusetts Institute of Technology), Paul-Emile Paradan (Université Montpellier 2), Reyer Sjamaar (Cornell University), Gunther Uhlmann (University of Washington at Seattle), San Vũ Ngọc (Université Grenoble), Alan Weinstein (University of California at Berkeley).

Acknowledgements. We are grateful to Ann Kostant and Tom Grasso for expert editorial guidance and support. In addition, we thank all the authors for their fascinating and well-prepared lectures as well as their contributions, while the referees are also thanked for their valuable comments and suggestions. Main funding was provided by the Mathematical Institute and the Faculty of Science, both of Utrecht University. Additional financial support was provided by the Royal Netherlands Academy of Arts and Sciences (KNAW), the Netherlands Organisation for Scientific

Research (NWO), the Mathematical Research Institute (MRI), the Thomas Stieltjes Institute for Mathematics, and by the research clusters Geometry and Quantum Theory (GQT) and Nonlinear Dynamics of Natural Systems (NDNS+).

Utrecht
August 2010

Erik P. van den Ban
Johan A.C. Kolk

About J.J. Duistermaat

Ph.D. students of J.J. Duistermaat

1. J.A.C. Kolk, *The Selberg Trace Formula and Asymptotic Behaviour of Spectra*, 1977
2. G.J. Heckman, *Projections of Orbits and Asymptotic Behaviour of Multiplicities for Compact Lie Groups*, 1980
3. E.P. van den Ban, *Asymptotic Expansions and Integral Formulas for Eigenfunctions on a Semisimple Lie Group*, 1982
4. S.J. van Strien, *One Parameter Families of Vectorfields. Bifurcations near Saddle-connections*, 1982
5. H.E. Nusse, *Chaos, Yet No Chance to Get Lost*, 1983
6. J.C. van der Meer, *The Hamiltonian Hopf Bifurcation*, 1985
7. M. Poel, *Harmonic Analysis on* $\mathbf{SL}(n, \mathbf{R}) / \mathbf{GL}(n-1, \mathbf{R})$, 1986
8. P.J. Braam (University of Oxford), *Magnetic Monopoles and Hyperbolic Three-manifolds*, 1987
9. P.H.M. van Mouche, *Sur les Régions Interdites du Spectre de l'Opérateur Périodique et Discret de Mathieu*, 1988
10. R. Sjamaar, *Singular Orbit Spaces in Riemannian and Symplectic Geometry*, 1990
11. H. van der Ven, *Vector Valued Poisson Transforms on Riemannian Symmetric Spaces of Rank One*, 1993
12. J.B. Kalkman, *A BRST Model Applied to Symplectic Geometry*, 1993
13. J. Hermans, *Rolling Rigid Bodies With and Without Symmetries*, 1995
14. O. Berndt, *Semidirect Products and Commutative Banach Algebras*, 1996
15. E.A. Cator, *Two Topics in Infinite Dimensional Analysis. Convex Potential Theory on a Banach Space. Distributions on Locally Convex Spaces*, 1997
16. M.V. Ruzhansky, *Singular Fibrations with Affine Fibers, with Applications to the Regularity Theory of Fourier Integral Operators*, 1998
17. C.C. Stolk, *On the Modeling and Inversion of Seismic Data*, 2000

18. B.W. Rink, *Geometry and Dynamics in Hamiltonian Lattices, with Application to the Fermi–Pasta–Ulam Problem*, 2003
19. A.M.M. Manders, *Internal Wave Patterns in Enclosed Density-stratified and Rotating Fluids*, 2003
20. H. Lokvenec-Guleska, *Sum Formula for* $\mathbf{SL}_2$ *over Imaginary Quadratic Number Fields*, 2004
21. T. Gantumur, *Adaptive Wavelet Algorithms for Solving Operator Equations*, 2006
22. P.T. Eendebak, *Contact Structures of Partial Differential Equations*, 2007
23. A.Q. Vélez, *Equivalence of D-Brane Categories*, 2009

List of publications by J.J. Duistermaat

Articles

1. On contact-transformations. *Indag. Math.* **29** (1967), 111–114.
2. (thesis) Energy and entropy as real morphisms for addition and order. *Synthese* **18** (1968), 327–393.
3. Periodic solutions of periodic systems of ordinary differential equations containing a parameter. *Arch. Rational Mech. Anal.* **38** (1970), 59–80.
4. On periodic solutions near equilibrium points of conservative systems. *Arch. Rational Mech. Anal.* **45** (1972), 143–160.
5. (with L. Hörmander) Fourier integral operators II. *Acta Math.* **128** (1972), 183–269.
6. On first order elliptic equations for sections of complex line bundles. *Compositio Math.* **25** (1972), 237–243.
7. On Carleman estimates for pseudo-differential operators. *Invent. Math.* **17** (1972), 31–43.
8. On foliations induced by complex vectorfields. *Journées Complexes: Équations Différentiels et Fonctions de Plusieurs Variables (Metz, 1972)*, Exp. No. I, 10 pp. Inst. Recherche Math. Avancée, Strasbourg, 1973.
9. (with J. Sjöstrand) A global construction for pseudo-differential operators with non-involutive characteristics. *Invent. Math.* **20** (1973), 209–225.
10. Oscillatory integrals, Lagrange immersions and unfolding of singularities. *Comm. Pure Appl. Math.* **27** (1974), 207–281.
11. (with V.W. Guillemin) The spectrum of positive elliptic operators and periodic bicharacteristics. *Invent. Math.* **29** (1975), 39–79.
12. On the Morse index in variational calculus. *Advances in Math.* **21** (1976), 173–195.
13. (with I.M. Singer) Order-preserving isomorphisms between algebras of pseudo-differential operators. *Comm. Pure Appl. Math.* **29** (1976), 39–47.
14. (with R.H. Cushman) The behavior of the index of a periodic linear Hamiltonian system under iteration. *Advances in Math.* **23** (1977), 1–21.

15. (with J.A.C. Kolk, V.S. Varadarajan) Spectra of compact locally symmetric manifolds of negative curvature. *Invent. Math.* **52** (1979), 27–93. Erratum: *Invent. Math.* **54** (1979), 101.
16. On global action-angle coordinates. *Comm. Pure Appl. Math.* **33** (1980), 687–706.
17. On operators of trace class in $L^2(X, \mu)$. *Proc. Indian Acad. Sci.* **90** (1981), 29–32.
18. (with G.J. Heckman) On the variation in the cohomology of the symplectic form of the reduced phase space. *Invent. Math.* **69** (1982), 259–268.
19. (with G.J. Heckman) Addendum to "On the variation in the cohomology of the symplectic form of the reduced phase space." *Invent. Math.* **72** (1983), 153–158.
20. Convexity and tightness for restrictions of Hamiltonian functions to fixed point sets of an antisymplectic involution. *Trans. Amer. Math. Soc.* **275** (1983), 417–429.
21. (with J.A.C. Kolk, V.S. Varadarajan) Functions, flows and oscillatory integrals on flag manifolds and conjugacy classes in real semisimple Lie groups. *Compositio Math.* **49** (1983), 309–398.
22. Bifurcations of periodic solutions near equilibrium points of Hamiltonian systems. In *Bifurcation Theory and Applications (Montecatini, 1983)*. Lecture Notes in Mathematics 1057, Springer-Verlag, Berlin, 1984, 57–105.
23. On the similarity between the Iwasawa projection and the diagonal part. In *Analyse Harmonique sur les Groupes de Lie et les Espaces Symétriques (Kleebach, 1983)*. Mém. Soc. Math. France (N.S.) **15** (1984), 129–138.
24. Non-integrability of the 1:1:2-resonance. *Ergodic Theory Dynam. Systems* **4** (1984), 553–568.
25. (with F.A. Grünbaum) Differential equations in the spectral parameter. *Comm. Math. Phys.* **103** (1986), 177–240.
26. (with R.H. Cushman) The quantum mechanical spherical pendulum. *Bull. Amer. Math. Soc. (N.S.)* **19** (1988), 475–479.
27. Peculiarities in the development of the theory of Lie groups. In *Analysis, et cetera, Research Papers Published in Honor of Jürgen Moser's 60-th Birthday*. Academic Press, Boston, 1990, 263–282.
28. M. Riesz's families of operators. *Nieuw Arch. Wisk. (4)* **9** (1991), 93–101.
29. Selfsimilarity of "Riemann's nondifferentiable function." *Nieuw Arch. Wisk. (4)* **9** (1991), 303–337.
30. (with P.J. Braam) Normal forms of real symmetric systems with multiplicity. *Indag. Math. (N.S.)* **4** (1993), 407–421.
31. Equivariant cohomology and stationary phase. In *Symplectic Geometry and Its Applications (Sanda and Yokohama, 1993)*. Contemp. Math. **179** (1994), 45–62.
32. (with V. Guillemin, E. Meinrenken, S. Wu) Symplectic reduction and Riemann–Roch for circle actions. *Math. Res. Lett.* **2** (1995), 259–266.
33. (with R.H. Cushman) A characterization of the Ligon–Schaaf regularization map. *Comm. Pure Appl. Math.* **50** (1997), 773–787.
34. (with Wilberd van der Kallen) Constant terms in powers of a Laurent polynomial. *Indag. Math. (N.S.)* **9** (1998), 221–231.

35. The monodromy in the Hamiltonian Hopf bifurcation. *Z. Angew. Math. Phys.* **49** (1998), 156–161.
36. (with R.H. Cushman) Non-Hamiltonian monodromy. *J. Differential Equations* **172** (2001), 42–58.
37. On Hessian Riemannian structures. *Asian J. Math.* **5** (2001), 79–91.
38. On the boundary behaviour of the Riemannian structure of a self-concordant barrier function. *Asymptot. Anal.* **27** (2001), 9–46.
39. The universal barrier function of a convex polytope. In *Circumspice. Various Papers in around Mathematics in Honour of Arnoud van Rooij*. Subfaculteit Wiskunde, Katholieke Universiteit Nijmegen, 2001, 207–220.
40. (with M. Loog, L.M.J. Florack) On the behavior of spatial critical points under Gaussian blurring. A folklore theorem and scale-space constraints. In *Scale-Space and Morphology in Computer Vision: Third Internat. Conf. (Vancouver, 2001)*. Lecture Notes in Computer Science 2106, Springer-Verlag, Berlin, 2001, 183–192.
41. Second order contact of minimal surfaces. *Indag. Math. (N.S.)* **13** (2002), 313–350.
42. (with A.M.M. Manders, L.R.M. Maas) Wave attractors in a smooth convex enclosed geometry. *Phys. D* **186** (2003), 109–132.
43. (with Y.J. Lee) Toeplitz operators on the Dirichlet space. *J. Math. Anal. Appl.* **300** (2004), 54–67.
44. (with P. Hoyng) Geomagnetic reversals and the stochastic exit problem. *Europhys. Lett.* **68** (2004), 177–183.
45. (with A.E. Kyprianou, K. van Schaik) Finite expiry Russian options. *Stochastic Process. Appl.* **115** (2005), 609–638.
46. A mathematical model for geomagnetic reversals. In *The Breadth of Symplectic and Poisson Geometry. Festschrift in Honor of Alan Weinstein*. Progr. Math., 232, Birkhäuser, Boston, 2005, 59–73.
47. (with R.H. Cushman) Nearly flat falling motions of the rolling disk. *Regul. Chaotic Dyn.* **11** (2006), 31–60.
48. (with A. Pelayo) Symplectic torus actions with coisotropic principal orbits. *Ann. Inst. Fourier (Grenoble)* **57** (2007), 2239–2327.
49. (with A. Pelayo) Reduced phase space and toric variety coordinatizations of Delzant spaces. *Math. Proc. Cambridge Philos. Soc.* **146** (2009), 695–718.
50. (with A. Pelayo) Topology of symplectic torus actions with symplectic orbits. *Revista Matemática Complutense.* **24** (2011), 59–81.
51. (with A. Pelayo) Complex structures on four-manifolds with symplectic two-torus actions. *Internat. J. Math.* **22** (2011), 449–463.

Not (yet) published preprints

52. Integral Surfaces of Second Order Partial Differential Equations.
53. Chaplygin's sphere. http://arXiv.org/abs/math.DS/0409019.
54. (with N. Joshi) Okamoto's space for the first Painlevé equation in Boutroux coordinates. To appear in *Arch. Ration. Mech. Anal.*
55. Okamoto's space for the first Painlevé equation. Submitted for publication.

Books

1. *Fourier Integral Operators*. Courant Institute of Mathematical Sciences, New York University, New York, 1973. Reprinted in: Progress in Mathematics, Vol. 130. Birkhäuser, Boston, 1996. x+142 pp. ISBN: 0-8176-3821-0. Reissued in Softcover in: Modern Birkhäuser Classics. Birkhäuser, Boston, 2011. ISBN: 978-0-8176-8107-4.
2. *Classical Aspects of Partial Differential Equations*. Comm. Math. Inst., Vol. 13, Rijksuniversiteit Utrecht, Utrecht, 1980. 150 pp.
3. (with W. Eckhaus) *Analyse van Gewone Differentiaalvergelijkingen*. Epsilon Uitgaven, Utrecht, 1995. vi+233 pp. ISBN: 90-5041-039-1.
4. *The Heat Kernel Lefschetz Fixed Point Formula for the Spin-c Dirac Operator*. Progress in Nonlinear Differential Equations and Their Applications. Birkhäuser, Boston, 1996. viii+247 pp. ISBN: 0-8176-3865-2.
5. *Pseudodifferential Operators and Elliptic Boundary Problems*. Comm. Math. Inst, Vol. 20, Utrecht University, Utrecht, 1998.
6. (with J.A.C. Kolk) *Lie Groups*. Universitext, Springer-Verlag, Berlin, 2000. viii+344 pp. ISBN: 3-540-15293-8.
7. (with J.A.C. Kolk) *Multidimensional Real Analysis I: Differentiation*. Cambridge Studies in Advanced Mathematics, Vol. 86, Cambridge University Press, Cambridge, 2004. xvii+422pp. ISBN: 0-521-55114-5. Reprint edition for the People's Republic of China: Cambridge University Press and Beijing World Publishing Corporation, 2009. ISBN: 978-7-5100-0452-0.
8. (with J.A.C. Kolk) *Multidimensional Real Analysis II: Integration*. Cambridge Studies in Advanced Mathematics, Vol. 87, Cambridge University Press, Cambridge, 2004. xviii+376 pp. ISBN: 0-521-82925-9. Reprint edition for the People's Republic of China: Cambridge University Press and Beijing World Publishing Corporation, 2009. ISBN: 978-7-5100-0518-3.
9. (with R.H. Cushman, J. Śniatycki) *Geometry of Nonholonomically Constrained Systems*. Advanced Series in Nonlinear Dynamics, Vol. 26. World Scientific Publishing Co. Pte. Ltd, Hackensack, NJ, 2010. xviii+404 pp. ISBN-13: 978-981-4289-48-1. ISBN-10: 981-4289-48-5.
10. *Discrete Integrable Systems. QRT Maps and Elliptic Surfaces*. Springer Monographs in Mathematics, Vol. 304. Springer Science+Business Media, Berlin, 2010. xxi+620 pp. ISBN: 978-1-4419-7116-6.
11. (with J.A.C. Kolk) *Distributions: Theory and Applications*. Cornerstones in Mathematics. Birkhäuser, Boston, 2010. ISBN: 978-0-8176-4672-1.

Published lectures

1. Applications of Fourier integral operators. In *Séminaire Goulaouic–Schwartz 1971–1972: Équations aux Dérivées Partielles et Analyse Fonctionelle*. Exp. 27, 24 pp. (unpaged errata following Exp. 30.) Centre Math., École Polytech., Paris, 1972.

2. Fourier integrals II. In *R.C.P. 25 (Strasbourg, 1973)*. Inst. Recherche Math. Avancée, Université Louis Pasteur, Strasbourg, 1973.
3. On Carleman estimates for pseudo-differential operators. In *Colloque International CNRS sur les Équations aux Dérivées Partielles Linéaires (Orsay, 1972)*. Asterisque, 2 et 3, Soc. Math. France, Paris, 1973, 143–151.
4. (with J. Sjöstrand) Sur une classe d'opérateurs de type sous-elliptique. In *Séminaire Goulaouic–Schwartz 1972–1973: Équations aux Dérivées Partielles et Analyse Fonctionelle*. Exp. 7, 13 pp. Centre Math., École Polytech., Paris, 1973.
5. L'indice de Morse dans le calcul variationnel. In *Séminaire Goulaouic–Schwartz 1973–1974: Équations aux Dérivées Partielles et Analyse Fonctionelle*. Exp. 25, 13 pp. Centre Math., École Polytech., Paris, 1974.
6. (with V.W. Guillemin) The spectrum of positive elliptic operators and periodic geodesics. In *Differential Geometry (Stanford, 1973)*. Proc. Sympos. Pure Math., Vol. XXVII, Part 2, Amer. Math. Soc., Providence, 1975, 205–209.
7. On the spectrum of positive elliptic operators and periodic bicharacteristics. In *Fourier Integral Operators and Partial Differential Equations, Colloque International (Nice, 1974)*. Lecture Notes in Mathematics, 459, Springer-Verlag, Berlin, 1975, 15–22.
8. Applications of Fourier integral operators. In *Proceedings of the International Congress of Mathematicians (Vancouver, 1974)*. Vol. 2, Canad. Math. Congress, Montreal, Que., 1975, 263–268.
9. (with I.M. Singer) Isomorphismes entre algèbres d'opérateurs pseudo-différentiels. In *Séminaire Goulaouic–Lions–Schwartz 1974–75: Équations aux Dérivées Partielles Linéaires et Non Linéaires*. Exp. 24, 8 pp. Centre Math., École Polytech., Paris, 1975.
10. La formule de Kolk et Varadarajan. In *Séminaire Goulaouic–Schwartz 1976–77: Équations aux Dérivées Partielles et Analyse Fonctionelle*, Exp. 16, 10 pp. Centre Math., École Polytechnique, Palaiseau, 1977.
11. The light in the neighborhood of a caustic. In *Séminaire Bourbaki 1976/77, 29e Année (1976/77)*. Lecture Notes in Mathematics 677, Springer-Verlag, Berlin, 1978, 19–29.
12. Gauss on surfaces. In *Carl Friedrich Gauss 1777–1855. Four Lectures on His Life and Work*. Comm. Math. Inst., Vol. 7, Rijksuniversiteit Utrecht, Utrecht, 1978, 32–41.
13. Harmonic analysis on semi-simple Lie groups. In *Proceedings, Bicentennial Congress Wiskundig Genootschap (Amsterdam, 1978)*, Part I. Math. Centre Tracts, 100, Math. Centrum, Amsterdam, 1979, 45–58.
14. Asymptotics of elementary spherical functions. In *Analytical and Numerical Approaches to Asymptotic Problems in Analysis (Nijmegen, 1980)*. North-Holland Publ. Co., Amsterdam, 1981, 35–43.
15. Asymptotics of elementary spherical functions. 6 pp. In *Mathematische Arbeitstagung 1981*. Universität Bonn, 1981.
16. Asymptotics of elementary spherical functions. In *Differential Geometric Methods in Mathematical Physics (Clausthal, 1980)*. Lecture Notes in Mathematics 905, Springer-Verlag, Berlin, 1982, 100–107.

17. On the momentum map. In *Proceedings of the IUTAM–ISIMM Symposium on Modern Developments in Analytical Mechanics (Torino, 1982)*, Vol. I. Atti Accad. Sci. Torino Cl. Fis. Mat. Natur. **117** (1983), suppl. 1, 97–107.
18. Niet-integreerbaarheid van de 1:1:2-resonantie. In *Nederl. Akad. Wetensch. Verslag Afd. Natuurk.* **93** (1984), 17–19.
19. Bifurcations of periodic solutions of differential equations. In *Delft Progr. Rep.* **9** (1984), 19–27.
20. Differentiaalvergelijkingen in de eigenwaarde parameter. In *Nederl. Akad. Wetensch. Verslag Afd. Natuurk.* **95** (1986), 39–40.
21. The momentum map. In *Topics in Differential Geometry (Debrecen, 1984)*, Vol. I. Colloq. Math. Soc. János Bolyai, 46, North-Holland, Amsterdam, 1988, 347–392.
22. Multiplicités de représentations de groupes et l'espace de phase réduit. In *Actions Hamiltoniennes de Groupes. Troisième Théorème de Lie (Lyon, 1986)*. Travaux en Cours, 27, Hermann, Paris, 1988, 71–81.
23. The quantum spherical pendulum. In *Quantum Theories and Geometry (Les Treilles, 1987)*. Math. Phys. Stud., 10, Kluwer Acad. Publ., Dordrecht, 1988, 53–64.
24. Transformatiegroepen en een plaatje van SL(2, **R**). In *Afscheidssymposium Prof. dr. F. van der Blij (Utrecht, 1988)*. Nieuwe Wiskrant **7** (1988), 45–52.
25. Huygens' principle for linear partial differential equations. In *Huygens' Principle 1690–1990, Theory and Applications (The Hague and Scheveningen, 1990)*. Stud. Math. Phys., 3, North-Holland, Amsterdam, 1992, 273–297.
26. (with P.J. Braam) Normal forms of real symmetric systems with multiplicity. In *Panoramas of Mathematics (Warsaw, 1993/1994)*. Banach Center Publ., 34, Polish Acad. Sci., Warsaw, 1995, 29–35.
27. The heat kernel Lefschetz fixed point formula for the spin-c-Dirac operator. In *Partial Differential Equations and Mathematical Physics. (The Danish–Swedish Analysis Seminar, 1995)*. Birkhäuser, Boston, 1996, 369–370.
28. Zelfgelijkvormigheid van Riemann's niet-differentieerbare functie. In *Nederl. Akad. Wetensch. Verslag Afd. Natuurk.* **105** (1996), 30–33.
29. A short introduction to Riemannian geometry. In *Fourth Workshop on High Performance Optimization Techniques (Rotterdam, 1999)*. Erasmus University, Rotterdam, 1999.
30. (with A. Doelman, J. Grasman, A. van Harten) In memoriam Wiktor Eckhaus (1930–2000). In *Nieuw Arch. Wisk.* (5) **2** (2001), 18–20.
31. Niet-holonome inperkingen. In *Nederl. Akad. Wetensch. Verslag Afd. Natuurk.* **109** (2000).
32. Abel prize winner 2006; Lennart Carleson: achievements until now. *Nieuw Arch. Wisk.* (5) **8** (2007), 175–177.

Book reviews

1. Pseudodifferential Operators, by M.E. Taylor. *Bull. Amer. Math. Soc.* (N.S.) **7** (1982), 277–279.

2. Symplectic Techniques in Physics, by V. Guillemin and S. Sternberg. *Bull. Amer. Math. Soc.* (N.S.) **18** (1988), 97–100.
3. Lagrangian Manifolds and the Maslov Operator, by A.S. Mishchenko, V.E. Shatalov, B. Yu. Sternin. *Bull. Amer. Math. Soc.* (N.S.) **27** (1992), 174–178.
4. Propagation and Interaction of Singularities in Nonlinear Hyperbolic Problems, by M. Beals. *CWI Quarterly* **4** (1991), 73–74.

Hans Duistermaat (1942–2010)

Erik P. van den Ban* and Johan A.C. Kolk*

On March 19, 2010, mathematics lost one of its leading geometric analysts, Johannes Jisse Duistermaat. At age 67 he passed away, after a short illness following a renewed bout of lymphoma the doctors thought they had controlled. "Hans," as Duistermaat was universally known among friends and colleagues, was not only a brilliant research mathematician and inspiring teacher, but also an accomplished chess player, very fond of several physical sports, and a devoted husband and (grand)father. The remembrances and surveys that follow are from some of his many colleagues, students, and friends. We hope that they adequately convey the impressive breadth of Hans's life and work.

Hans Duistermaat was born December 20, 1942, in The Hague. After the end of World War II his parents moved to the Netherlands East Indies (Indonesia nowadays), where he spent a happy youth. Hans was a student at Utrecht University, where he wrote his Ph.D. thesis on mathematical structures in thermodynamics. The famous geometer Hans Freudenthal is listed as his advisor, but the topic was suggested and the thesis directed by Günther K. Braun, professor in applied mathematics, who tragically died one year before the defense of the thesis, in 1968.

Hans dropped the subject of thermodynamics, because the thesis had led to dissent between mathematicians and physicists at Utrecht University. Nevertheless, this topic exerted a decisive influence on his further development: in its study, Hans had encountered contact transformations. These he studied thoroughly by reading S. Lie, who had initiated their theory. In 1969–1970 he spent one year in Lund, where L. Hörmander was developing the theory of Fourier integral operators (FIOs); these are far-reaching generalizations of partial differential operators. Hans's knowledge of the work of Lie turned out to be an important factor in the formulation of this theory. Hans's mathematical reputation was firmly established by a long joint article with Hörmander concerning applications of the theory to linear partial differential equations. In 1972 Duistermaat was appointed full professor at the

Mathematical Institute, Utrecht University, The Netherlands
e-mail: e.p.vandenban@uu.nl; j.a.c.kolk@uu.nl

*Reprinted with the kind permission of the Nieuw Archief voor Wiskunde.

Catholic University of Nijmegen, and in 1974 at Utrecht University, as the successor to Freudenthal.

In these years, he continued to work on FIOs. At the Courant Institute in New York he wrote a paper on *Oscillatory integrals, Lagrange immersions and unfolding of singularities*, a survey of the subjects in the title that sets the agenda for the study of singularities of smooth functions and their applications to distribution theory. In some sense it is complementary to FIOs and parallel to work of V.I. Arnol'd. Furthermore, together with V.W. Guillemin he composed an article about application of FIOs to the asymptotic behavior of spectra of elliptic operators, and its relation to periodic bicharacteristics; see the article by Guillemin for more details. In these works one clearly discerns the red thread connecting most of Hans's achievements: on the basis of a complete clarification of the underlying geometry deep and powerful results are obtained in the area of geometric analysis.

It is characteristic for Hans's work that after a period of intense concentration on a particular topic, he would move to a different area of mathematics, bringing thereby acquired insights quite often to new fruition. Usually, this change was triggered by a question of a colleague, but more frequently of one of his Ph.D. students. Hans went to great efforts to accommodate the special needs of his students and help them develop in their own way, not in his way. In particular, in several cases Hans was willing and also able to guide students working on topics initiated by themselves. Examples are the theses of P.H.M. van Mouche and M.V. Ruzhansky.

It was by questions of J.A.C. Kolk and G.J. Heckman that Hans became interested in the theory of semisimple Lie groups. With Kolk and V.S. Varadarajan he published fundamental papers on harmonic analysis and the geometry of flag manifolds, with the method of stationary phase as the underlying theme. This work also provided an impetus for the ground-breaking work with Heckman that culminated in the Duistermaat–Heckman formula, which will be discussed separately by Heckman.

In the thesis of E.P. van den Ban one finds the novel idea, suggested by Hans, of taking the integrals representing the spherical eigenfunctions on a semisimple Lie group, which are integrals over a real flag manifold, into integrals on real cycles inside the complex flag manifold. This allowed application of the method of steepest descent in order to study their asymptotics, generalizing the approach known in the theory of hypergeometric functions.

One of Hans's basic mathematical interests, to which he returned throughout his life, was classical mechanics and its relations with differential equations. In this case too, it was often through the work of his students S.J. van Strien, H.E. Nusse, J.C. van der Meer, J. Hermans, B.W. Rink, and A.A.M. Manders that this topic was taken up again. His activities in this area will be further discussed by his colleague and co-author R.H. Cushman.

F.A. Grünbaum posed a problem that led to the joint article *Differential equations in the spectral parameter*. It classifies second-order ordinary differential operators of which the eigenfunctions also satisfy a differential equation in the spectral parameter. The classification is in terms of rational solutions of the Korteweg–de Vries equation.

Writing a review of the book *Lie's Structural Approach to PDE Systems* by O. Stormark led Hans to further study of that circle of ideas. The result was a paper on the contact geometry of minimal surfaces as well as the thesis of P.T. Eendebak.

Together with A. Pelayo he wrote several papers about symplectic differential geometry; furthermore, he directed the thesis of R. Sjamaar. In this part of mathematics Hans was a very influential figure: witness his frequent contacts with other leading investigators, such as Guillemin and A. Weinstein.

In the later part of his life, Hans had an intense interest in applications of mathematics elsewhere in society. For instance, he was a consultant to Royal Dutch Shell, which led to the thesis of C.C. Stolk on the inversion of seismic data. Interaction with mathematical economists during a conference at Erasmus University in Rotterdam, where Hans had been invited to give an introduction to Riemannian geometry, sparked his interest in barrier functions, used in convex programming. He also collaborated with the geophysicist P. Hoyng in modeling the polarity reversals of the earth's magnetic field. The lengths of the time intervals between the subsequent reversals form an irregular sequence with a large variation, which make the reversals look like a (Poisson) stochastic process. Within a short period of time he mastered the nontrivial stochastics needed in this problem.

The bibliography of Hans's work contains eleven books. *Fourier Integral Operators* gives an exposition of seminal results in the area of microlocal analysis. *The Heat Kernel Lefschetz Fixed Point Formula for the Spin-c Dirac Operator* is concerned with a direct analytic proof of the index theorem of Atiyah–Singer in a special case of interest for symplectic differential geometry. *Lie Groups*, jointly with Kolk, contains a new proof of Lie's third theorem on the existence of a Lie group associated to any Lie algebra. The construction of the group as the quotient of a path space in the Lie algebra was the model for many important generalizations, including the integration of Lie groupoids by M. Crainic and R.L. Fernandes.

Analysis of Ordinary Differential Equations (in Dutch), jointly with W. Eckhaus, grew out of a set of lecture notes. Similarly, together with Kolk he authored *Multidimensional Real Analysis I: Differentiation* and *II: Integration* (also published in a China edition), and *Distributions: Theory and Applications*. The last book contains a novel proof of the kernel theorem of L. Schwartz, which in turn is used to efficiently derive numerous important results, and a treatment of theories of integration and of distributions from a unified point of view. The last four books together form a veritable "cours d'analyse mathématique."

In the book *Discrete Integrable Systems: QRT Maps and Elliptic Surfaces*, QRT (= Quispel, Roberts, and Thompson) maps are analyzed using the full strength of Kodaira's theory of elliptic surfaces. A complete and self-contained exposition is given of the latter theory, including all the proofs. Many examples of QRT maps from the literature are analyzed in detail, with explicit formulas and computer pictures. The interest in QRT maps was triggered by interaction with J.M. Tuwankotta. Hans had the idea to use the technique of blowing up, which he had previously encountered in the article *Constant terms in powers of a Laurent polynomial* jointly with Wilberd van der Kallen.

While Hans clearly exerted a substantial influence on mathematics through his own research and that of his many Ph.D. students, the books written by him alone or jointly traverse a wide spectrum of mathematical exposition, both in topic or level of sophistication. But in this case again, there is a common characteristic: every result, how hackneyed it may be, had to be fully understood and explained in its proper context. In addition to this, when writing, he insisted that the original works of the masters be studied. Frequently he expressed his admiration for the depth of their treatment, but he could also be quite upset about incomplete proofs that had survived decades of careless inspection. The last project that he was involved in exemplifies this: in joint work with Nalini Joshi reliable proofs are provided of old but also many new results concerning Painlevé functions.

The mode of writing preferred by Hans was top-down exposition: starting from the general, descending to the more concrete. Yet, hidden under the façade of a polished and sometimes quite abstract exposition, there usually was a detailed knowledge of explicit and representative examples. Many of the notebooks he left are filled with intricate calculations, which he performed with great precision and unflagging concentration. Not surprisingly, he greeted the advent of formula manipulation programs like *Mathematica* with great enthusiasm. Furthermore, Hans put a high value on correct illustrations; in private, he could express annoyance about misleading or ugly pictures. In the days of the programming language Pascal and matrix printers, he spent a substantial amount of time in order to put a dot exactly at the position he wanted: one of his favorite techniques for creating complicated illustrations was by printing just a huge number of dots.

In addition to his patience and powers of concentration, he was capable of grasping the essence of a problem and its solution with lightning speed. When this happened during someone's lecture, he usually mentioned this not critically, but kindly and supportively.

As a teacher, Hans was quite aware that not every student was as gifted as he. Despite the fact that he could ignore all restrictions of time and demanded serious work from the students, he was very popular among them. Repeatedly he gave unscheduled courses on their request. He was an honorary member of A-Eskwadraat, the Utrecht Science Students' Society. He shared this honor with Nobel laureate G. 't Hooft and with J.C. Terlouw, a nuclear physicist who pursued a successful career in Dutch politics.

As an administrator, however, he was less successful. Although he served our institute, the mathematical community, and the Royal Netherlands Academy of Arts and Sciences in many different capacities, he was at his best with concrete issues that could be solved rationally, not with situations that required intricate political maneuvering. For instance, he was very actively involved with the Scientific Programme Indonesia–Netherlands, which was an initiative of the academy, aimed at the selection and training of new researchers, the improvement of the supervising infrastructure at Indonesian institutes, and the conduct of joint research activities. In addition, the task of refereeing manuscripts was taken very seriously by Hans: many authors greatly benefited from his long e-mails. He was a member of a substantial number of selection committees, devoting considerable energy to evaluating the candidates' achievements and potential.

In 2004, Hans was honored with a special professorship at Utrecht University endowed by the Royal Netherlands Academy of Arts and Sciences. This position allowed him to focus exclusively on his research, without being distracted by administrative obligations. The five years that followed were a happy period in which his mathematics blossomed. Hans demonstrated by the breadth and depth of his accomplishments that his chair was aptly named "pure and applied mathematics."

His mood was almost invariably one of equanimity; even in difficult situations, he always tended to look for positive aspects. Immense concentration on a topic of momentary interest was natural for him. In fact, on several occasions he confessed that he had a "one-track mind," which made it necessary to mentally exclude disturbances. At times, however, this trait of character could be infuriating for his colleagues.

Very remarkably, Hans had no personal vanity, neither in human nor in professional relations. About his own work he once expressed that he considered himself lucky for having become well known for results he considered to be relatively simple. Most of his more difficult work, which had been far more difficult to achieve, had not received similar recognition. Honors did not mean much to Hans, although he was at first surprised and then gratified by them. He gave himself without any reservation to his friends and colleagues, always illuminating whatever was under discussion with characteristic insights based on his wide knowledge of mathematical and other topics.

In mathematics, Hans's life was a search for exhaustive solutions to important problems. This quest he pursued with impressive single-mindedness, persistence, power, and success. We know that this is a very sketchy attempt to bring him to life. In our minds, however, he is very vivid, one of the most striking among the mathematicians we have met. We deeply mourn his loss; yet we can take comfort in memories of many years of true and inspiring friendship.

Recollections of Hans Duistermaat

Victor W. Guillemin*

The two paragraphs below are a few brief recollections of mine from the period 1973–1974, the two years in which Hans Duistermaat and I worked together on our article "The spectrum of positive elliptic operators and periodic bicharacteristics" (and for me the two most memorable and exciting years of that decade). In the summer of '73, Hans and I met for the first time at an AMS-sponsored conference on differential geometry at Stanford and began to formulate the ideas that became the wave trace part of our paper. Then in the fall of 1974 he made a long visit to MIT, during which we firmed up these ideas and also proved the periodic bicharacteristic results that became the second main part of our paper.

A little prehistory: In the early 1970s, Bob Seeley, David Schaeffer, Shlomo Sternberg, and I ran a seminar at Harvard which was largely devoted to Hörmander's papers [1] and [2] and Hörmander–Duistermaat [3]. In particular, we spent a lot of time going through [3], which was the first systematic application of micro-local techniques to the problem of propagation of singularities. (Like analysts the world over, we were amazed at how simple this subject becomes when viewed from the perspective of the cotangent bundle.) Therefore, when I met Hans that summer I was well primed to discuss with him the contents of these papers. However, what initiated our collaboration was another memorable event from that conference: the announcement by Marcel Berger of Yves Colin de Verdière's result on the spectral determinability of the period spectrum of a Riemannian manifold. I vividly remember sitting next to Hans at Berger's lecture and our exchanging whispered comments as it became more and more evident that what Yves had done was intimately related to the things the two of us were currently thinking about. By the time the conference ended we had formulated a trace theorem for FIOs which asserted that the singularities of the wave trace are supported on the period spectrum of P (and hence that the wave trace gives one a simple means of accessing these data). As I mentioned above, this result became the first part of our paper. The second part was based on

Department of Mathematics, MIT, Cambridge, MA, U.S.A.
e-mail: vwg@mit.edu

*Reprinted with the kind permission of the Nieuw Archief voor Wiskunde.

an observation that Hans and I had made (each independently) that arose apropos of a result of Hörmander's in [1]. One of the most quoted results of Hörmander's paper is a generalization of a theorem of Avakumovic in which he obtains an "optimal" error term in the Weyl law for an elliptic pseudodifferential operator P and shows that this error term is indeed optimal by showing that this is the case if P is the Laplace operator on the standard round sphere. I noticed that this can be related to the fact that for the n-sphere the bicharacteristic flow associated with P is periodic. (More explicitly, I noticed that if the bicharacteristic flow of an elliptic operator P is periodic (i.e., P is Zoll) there has to be a clustering of eigenvalues about a lattice which prevents a sharpening of the Weyl law and vice versa.) In proving this result I made essential use of techniques developed in [3], so it was not surprising that when I described it to Hans at Stanford, I found that he had been thinking along similar lines. Moreover, it slowly began to dawn on us that the Hörmander example was just the tip of the iceberg. Among other things we noticed that his optimal error term could be replaced by a slightly better optimal error term (a big "O" could be converted into a little "o") if P was not Zoll, and also noticed that in this case the Weyl law could be differentiated to give an equidistribution result for eigenvalues. We also obtained a much sharper version of my clustering result: we showed that the clusters are clearly demarcated eigenbands of fixed width. Subsequently Alan Weinstein and Yves Colin de Verdière added a further dimension to this story by discovering that when Zoll phenomena are present, these clusters satisfy their own beautiful distribution law. Furthermore, Bill Helton discovered an extremely clean and economical version of our result: Let A be set of numbers obtained by taking all differences of pairs of eigenvalues, and let B be the cluster set of A. Then if the bicharacteristic flow is periodic, B is an integer lattice and if not, it is the whole real line.

At any rate, to conclude these reminiscences, by the spring of 1974, most of the conjectures we had made at the Stanford meeting had been supplied with rigorous proofs, although Hans continued, as was his wont, to tinker with them for the next several months just to make sure that they were "best possible." (No one was going to be able to achieve instant immortality by slightly improving them.) When Hans visited me in the fall the only unfinished piece of business was the Zoll part of the paper, and that consumed all our energies for four intense weeks. (One typical "Hans" memory from that time: One evening I return home late in the evening exhausted in mind and spirit following a frustrating day in which the two of us struggled without success to settle a delicate point about how large sets of periodic bicharacteristics have to be for clustering to occur. At 2 o'clock in the morning I get jarred awake by a phone call from Hans letting me know that he'd settled it.) I remember the aftermath of Hans's visit as a period of a slow, painful decompression. Never before had I worked so intensely and so single-mindedly on a project (and, for better or for worse, was destined never to do so again).

References

1. L. Hörmander, The spectral function of an elliptic operator, *Acta Math.* **121** (1968), 193–218.
2. L. Hörmander, Fourier integral operators I, *Acta Math.* **127** (1971), 79–183.
3. J.J. Duistermaat and L. Hörmander, Fourier integral operators II, *Acta Math.* **128** (1972), 183–269.

Recollections of Hans Duistermaat

Alan Weinstein

I first met Hans in the fall of 1972; my notes from his lecture in Princeton on "Non-involutive operators" continue to make good reading. His work on ordinary differential equations was already well known to me; his study of periodic orbits for the spring pendulum was the inspiration for the thesis of my first PhD student, Jair Koiller.

Hans and his familty then stayed in Berkeley in the summer of 1973 while we were attending the AMS Summer Institute in differential geometry at Stanford. I still have a picture in my mind of our daughters, aged about 2 at the time, playing in the sandbox in our backyard. As for mathematics, that was the time when contact with Hans deepened my interest in Fourier integral operators. Although Hans was not a co-author of Part 1 of the illustrious pair of papers by Hörmander, his influence is clear (and he is the only person thanked by Hörmander in that article).

We met again at a 1974 meeting in Nice, and then spent a lot of time together at the 1975 Nordic Summer School in Grebbestad, Sweden. Here I was totally immersed in the world of microlocal analysis (and Hans was also immersed in the nearby sea, which was too cold for anyone but him and the Finns in our group). A Google search for Duistermaat and Grebbestad turns up exactly two results—links to Hans's famous paper on global action-angle coordinates and my own rather obscure one on the order and symbol of a distribution.

This was just the beginning of Hans's influence on me through his papers and our frequent meetings. Other important influences were the 1972 NYU Lecture Notes on Fourier Integral Operators, which I was very pleased to incorporate later on into the Progress in Mathematics book series, where it remains one of the best places to learn about this subject. Nowhere else is the symplectic geometry of this subject, including the geometry of the Maslov class, so beautifully and concisely explained. The book is also notable for a section at the end linking the homogeneous and "asymptotic" theories of microlocal analysis.

Department of Mathematics, University of California, Berkeley, CA, U.S.A.
e-mail: alanw@math.berkeley.edu

The wonderful 1975 work with Guillemin on spectrum and periodic bicharacteristics extended the work of Colin de Verdière and Chazarain and set the paradigm for how such matters should be treated. During a visit by Hans to Berkeley, we tried to understand how the Birkhoff normal form of a periodic orbit might be encoded in the spectrum, but that project unfortunately remained unfinished.

A series of papers with Kolk and Varadarajan (1979, 1983) treated harmonic analysis on noncompact semisimple Lie groups. Hans's interest in Lie groups also led to the book with Kolk (2000), based on a course which Hans taught at Berkeley, among other places, and for which a manuscript circulated for many years before publication. Their beautiful proof of Lie's third theorem, constructing a Lie group as a quotient of the paths in the Lie algebra, suggested to me that there should be a similar construction going from Lie algebroids to groupoids. This was carried out in fundamental work of Crainic and Fernandes whose influence continues to this day.

I last saw Hans in the summer of 2009, when I was in Utrecht for a PhD thesis defense. I have a nice photo of the two of us in academic garb in the garden of the cloister at the university. The photo was taken by Marius Crainic, but the setting was carefully managed by Hans. Hans told me that he would be having a surgery later that summer but did not sound particularly concerned; he was obviously hiding the worst from me.

The remarks above cover only a tiny part of Hans's life and work, but they show that both his personal influence and "higher-order effects" have left a lasting mark on mathematicians and mathematics. It was a shock to lose him so suddenly, and his presence in our world will be sorely missed.

Recollections of Hans Duistermaat

Gert Heckman*

I would like to share with you some recollections of Hans Duistermaat from the period 1978–1981, during which he played a crucial role in my mathematical development. In 1976 I had started my dissertation work under the guidance of Gerrit van Dijk. In his thesis of 1962, the Russian mathematician Alexander Kirillov had developed a very elegant geometric method, the so-called orbit method, for understanding the representation theory of connected nilpotent Lie groups. In this method the branching rule for understanding how an irreducible representation decomposes under restriction to a subgroup has a very simple and elegant answer.

Gerrit suggested to me that I try to understand to what extent this orbit method could shed new light on the representation theory of semisimple Lie groups, in particular for the discrete series representations. In my first short paper from the summer of 1978 I worked out a particular example for compact Lie groups. According to the customs of those days I sent it around to several potentially interested people, and in return quickly received a reaction from Hans. My main result turned out to be already in the literature, and in addition Hans sketched an alternative and more elegant geometric proof. Aware of the fact that his letter might be intimidating for me he wrote at the end: "It is maybe superfluous to emphasize that I do not write you this proof out of pedantry, but rather as a sign of interest for your work, and I do hope that it leads to a still better understanding of the whole situation."

So my first little paper went into the wastebasket, but I was to receive something more valuable in return. I visited Hans regularly in Utrecht, and in June 1980 I defended my dissertation in Leiden with both Gerrit and Hans as thesis advisors. I realize now how lucky I was to have these two complementary teachers: Gerrit with his extensive knowledge of the work of Harish-Chandra, and Hans as the eminent analyst and geometer.

In August 1980 I went to Boston, to spend two years as a postdoc at MIT, and in September I lectured in the Lie groups seminar about my thesis work: how the

Department of Mathematics, Radboud University Nijmegen, The Netherlands
e-mail: G.Heckman@math.ru.nl

*Reprinted with the kind permission of the Nieuw Archief voor Wiskunde.

orbit method for compact Lie groups describes the branching rules in an asymptotic way, and how this leads to a convex polytope in which the multiplicities of the branching rule have their support [9]. The talk was received well, most notably by Victor Guillemin. Victor knew Hans well, and had great admiration for him. In 1975 they had written a beautiful article on the spectrum of elliptic operators on compact manifolds [5]. Looking back at my time at MIT, I realize again how lucky I was to be there during that period with Victor around.

That fall a number of new insights were unveiled regarding continuous symmetry reduction in symplectic geometry through the work of Guillemin–Sternberg, Atiyah–Bott, and Mumford. In these at first sight rather different contexts, namely quantum mechanics, quantum field theory, and algebraic geometry, there was a single fundamental underlying concept for the description of symmetry, namely that of the geometry of the moment map (or momentum map as Hans preferred to call it). I quote from a survey article by Bott from 1988 [4]:

> In fact, it is quite depressing to see how long it is taking us collectively to truly sort out symplectic geometry. I became aware of this especially when one fine afternoon in 1980, Michael Atiyah and I were trying to work in my office at Harvard. I say trying, because the noise in the neighboring office made by Sternberg and Guillemin made it difficult. So we went next door to arrange a truce and in the process discovered that we were grosso modo doing the same thing. Later Mumford joined us, and before the afternoon was over we saw how Mumford's stability theory fitted with the Morse theory. The important link here is the concept of a moment map, which in turn is the mathematical expression of the relation between symmetries of Lagrangians and conserved quantities; in short, what the physicists call Noether's theorem and which is one of their great paradigms.

In this quotation Bott refers to the results of fundamental publications by Guillemin–Sternberg [7], [8], Mumford [12], Ness [13], Atiyah–Bott [1], and Kirwan [10]. Since then, symplectic geometry has become a truly independent field in its own right.

In the spring of 1981 Victor gave a course on symplectic geometry, with special emphasis on the geometry of the moment map, and I learned the subject well. During the month of August I went back to the Netherlands to visit family and friends. The day before my return I was doing some last-minute work at MIT, when it occurred to me that the rather complicated locally polynomial formulas for the multiplicities could be explained by a linear variation of the symplectic form in the cohomology of the reduced phase space, at least over the generic fiber. A nice idea, but I had no clue how to prove it. A few days after my return I visited Hans, and we spent a whole afternoon talking about symplectic geometry. I told him about my question, and he listened attentively. That same evening he called me up at my parents' house, and with a piece of scratch paper on my lap I got an exposition of what later would become our joint paper [6].

Our work was well received. Independently of one other, Berline–Vergne [3] and Atiyah–Bott [2] placed it in the more general framework of equivariant cohomology. Our article was used later by Ed Witten in his work on two-dimensional Yang–Mills theory [14]. More recently our theorem was used again by Mariyam Mirzakhani in her computation of the Weil–Petersson volumes of the moduli space of curves [11].

In September 1982 I obtained a permanent position in Leiden as an assistant to Gerrit van Dijk: a solid base from which to pursue mathematical work professionally. I now appreciate very well the important role played by Hans during the early stages of my career. It is not inconceivable that without him I would have become a high-school teacher rather than a university professor of mathematics.

After this period of intensive contact from 1978 to 1981 our mathematical roads diverged. Our personal relationship remained, however, and I cherish the memories of the parties held for his 60th birthday and on the occasion of his royal decoration.

The sudden passing of Hans leaves behind a great emptiness, in the first place for his wife Saskia, his daughters Kim and Maaike and his relatives, but also for the many mathematicians with whom he collaborated. During the cremation ceremony many affecting words were spoken about Hans. His sister Dineke told the story of how, when she asked him as a student why he had chosen mathematics, Hans replied that he had no other option, because his talent for mathematics was such a godsend. I realize how very lucky I am that Hans shared this talent with me so generously.

References

1. M.F. Atiyah, R. Bott: The Yang–Mills equations over Riemann surfaces. *Philos. Trans. R. Soc. London* **308** (1983), 523–615.
2. M.F. Atiyah, R. Bott: The moment map and equivariant cohomology. *Topology* **23** (1984), 1–28.
3. N. Berline, M. Vergne: Zéros d'un champ de vecteurs et classes caractéristiques équivariantes. *Duke Math. J.* **50** (1983), 539–549.
4. R. Bott: Morse theory indomitable. *Publ. Math. IHES* **68** (1988), 99–114.
5. J.J. Duistermaat, V.W. Guillemin: The spectrum of positive elliptic operators and periodic bicharacteristics. *Invent. Math.* **29** (1975), 39–79.
6. J.J. Duistermaat, G.J. Heckman: On the variation in the cohomology of the symplectic form of the reduced phase space. *Invent. Math.* **69** (1982), 259–268.
7. V.W. Guillemin, S. Sternberg: Convexity properties of the moment mapping. *Invent. Math.* **67** (1982), 491–513.
8. V.W. Guillemin, S. Sternberg: Geometric quantization and multiplicities of group representations. *Invent. Math.* **67** (1982), 515–538.
9. G.J. Heckman: Projections of orbits and asymptotic behavior of multiplicities for compact connected Lie groups. *Invent. Math.* **67** (1982), 333–356.
10. F.C. Kirwan: Convexity properties of the moment mapping III. *Invent. Math.* **77** (1984), 547–552.
11. M. Mirzakhani: Weil–Petersson volumes and intersection theory on the moduli space of curves, *J. Amer. Math. Soc.* **20** (2007), 1–23.
12. D. Mumford: Appendix: Proof of the convexity theorem. *Amer. J. Math.* **106** (1984), 1326–1329.
13. L. Ness: A stratification of the null cone via the moment map. *Amer. J. Math.* **106** (1984), 1281–1325.
14. E. Witten: Two dimensional gauge theories revisited. *J. Geom. Phys.* **9** (1992), 303–368.

Classical mechanics and Hans Duistermaat

Richard H. Cushman*

One of Hans's favorite subjects was classical mechanics. As can be seen from his list of publications, his interest in this area was wide-ranging. In order to describe what he did, I will organize these papers, somewhat arbitrarily, into three classes: (i) periodic solutions near an equilibrium point; (ii) monodromy in integrable systems; (iii) other topics.

In [1] Hans studied the persistence of periodic solutions near an equilibrium point of a two-degrees-of-freedom Hamiltonian system which is in $1 : 2$ resonance. This is the simplest situation in which a well-known theorem of Lyapunov on the persistence of periodic solutions fails. Years later Hans returned to this subject in the almost forgotten paper [4]. Here, using the theory of singularities of mappings which are invariant under a circle action that fixes the origin, he proved a stability result for the set of short-period periodic orbits near an equilibrium point of a resonant Hamiltonian system of two degrees of freedom. In particular, he showed that this set of periodic orbits is diffeomorphic to the set of critical points of rank one of the energy-momentum mapping. Here the energy is the Hamiltonian of the Birkhoff normal form of the original resonant Hamiltonian truncated at some finite order. The momentum of the circle action is the quadratic terms of this normal form. As far as I am concerned, this result is the definitive generalization of Lyapunov's theorem.

Hans's most important contribution to the geometric study of Hamiltonian systems is his discovery of the phenomenon of monodromy in [3]. To describe what monodromy is we look at a two-degrees-of-freedom Hamiltonian system on four-dimensional phase space, which we assume is Euclidean space. We suppose that this Hamiltonian system has another function, which is an integral, that is, is constant on the motions of the original Hamiltonian system. Such a Hamiltonian system is said to be completely integrable with integral map given by assigning to each point in phase space the value of the Hamiltonian and the extra integral. If we assume that the integral map is proper and each preimage of a point is connected, then

Department of Mathematics and Statistics, University of Calgary, Alberta, Canada
e-mail: r.h.cushman@gmail.com

*Reprinted with the kind permission of the Nieuw Archief voor Wiskunde.

the action-angle theorem shows that the preimage of a suitably small open 2-disk in the set of regular values of the integral map is symplectically diffeomorphic to a product of a 2-torus and 2-disk. Hans showed that this local theorem need not hold globally. In particular, if we have a smooth closed, nonintersecting curve in the set of regular values of the integral map, then the preimage of this curve in phase space under the integral map is the total space of a 2-torus bundle, which need not be diffeomorphic to a product of the closed curve and a 2-torus. To understand what this global twisting is, we note that a 2-torus bundle over a circle may be looked at as a product bundle over a closed interval with a typical fiber a 2-torus. Here each of its two end 2-tori, which are Euclidean 2-space modulo the lattice of points with integer coordinates, are glued together by an integer 2×2 matrix with determinant 1. The monodromy of this 2-torus bundle is just this integer matrix. If the monodromy is not the identity matrix, then the 2-torus bundle is not a product bundle. In [3] Hans gave a list of geometric and analytic obstructions for local action-angle coordinates to be global. Monodromy is just the simplest obstruction. Monodromy would not be interesting if there were no two-degrees-of-freedom integrable Hamiltonian systems having it. When Hans was starting to write [3], he asked me to find an example of such a system. The next day I told him that the spherical pendulum, which was studied by Christiaan Huygens in 1612, has monodromy and gave him a proof. When writing up the paper Hans found a much simpler geometric argument to show that the spherical pendulum has monodromy. In [6] Hans and I discovered that monodromy appears in the joint spectrum of the energy and angular momentum operators of the quantized spherical pendulum. This discovery has now been recognized as fundamental by chemists who study the spectra of molecules and has led to a very active area of scientific research. In the early days, showing that a particular integrable system had monodromy was not easy. In [8] Hans did this for the Hamiltonian Hopf bifurcation.

In the middle 1990s Hans became interested in nonholonomically constrained systems such as the disk or a dynamically symmetric sphere with its center of mass not at its geometric center. Both are assumed to be rolling without slipping on a horizontal plane under the influence of a constant vertical gravitational force. This interest gave rise to [9]. In this paper Hans gave a simple geometric criterion for a not necessarily Hamiltonian system to have monodromy. He showed that an oblate ellipsoid of revolution rolling without slipping on a horizontal plane under the influence of a constant vertical gravitational force has a cycle of heteroclinic hyperbolic equilibria whose local monodromies add up to the identity. This shows that it cannot be made into a Hamiltonian system. The book [11] clearly indicates Hans's contributions to the geometric study of nonholonomically constrained systems. Especially, it contains a complete qualitative study of the motion of the rolling disk, some of which was published in [10].

His remaining publications range from removing the incompleteness of the flow of the Kepler problem for all negative energies at the same time, see [7], to showing that the $1 : 1 : 2$ resonance is nonintegrable by looking at its fourth-order normal form. In two degrees of freedom, integrability cannot be decided by any finite-order normal form. In the remaining paper on periodic linear Hamiltonian systems [2]

Hans answered an old question of Bott's about the Morse index of iterates of a periodic geodesic. Bott showed that this index is the sum of the index of the periodic geodesic and invariants of the real symplectic conjugacy class of the linear Poincaré map. Hans gave an explicit formula for the Morse index.

Working with Hans and collaborating on our joint publications is at the core of my mathematical career. It is hard for me to realize that he cannot answer my questions anymore.

References

1. On periodic solutions near equilibrium points of conservative systems. *Arch. Rational Mech. Anal.* **45** (1972), 143–160.
2. (with R.H. Cushman) The behavior of the index of a periodic linear Hamiltonian system under iteration. *Advances in Math.* **23** (1977), 1–21.
3. On global action-angle coordinates. *Comm. Pure Appl. Math.* **33** (1980), 687–706.
4. Bifurcations of periodic solutions near equilibrium points of Hamiltonian systems. In *Bifurcation Theory and Applications (Montecatini, 1983)*. Lecture Notes in Mathematics 1057, Springer Verlag, Berlin, 1984, 57–105.
5. Nonintegrability of the 1:1:2 resonance. *Ergodic Theory Dynam. Systems* **4** (1984), 553–568.
6. (with R.H. Cushman) The quantum mechanical spherical pendulum. *Bull. Amer. Math. Soc. (N.S)* **19** (1988), 475–479.
7. (with R.H. Cushman) A characterization of the Ligon–Schaaf regularization map. *Comm. Pure Appl. Math.* **50** (1997), 773–787.
8. The monodromy in the Hamiltonian Hopf bifurcation. *Z. Angew. Math. Phys.* **49** (1998), 156–161.
9. (with R.H. Cushman) Non-Hamiltonian monodromy. *J. Differential Equations* **172** (2001), 42–58.
10. (with R.H. Cushman) Nearly flat falling motions of the rolling disk. *Regul. Chaotic Dyn.* **11** (2006), 31–60.
11. (with R.H. Cushman, J. Śniatycki) *Geometry of Nonholonomically Constrained Systems.* Advanced Series in Nonlinear Dynamics, **26**. World Scientific Publishing Co, Hackensack, 2010.

Contents

Duistermaat–Heckman formulas and index theory

Jean-Michel Bismut

To the memory of Hans Duistermaat

Le maître: *Eh bien! Jacques, l'histoire de tes amours?*
Jacques: *Je ne sais où j'en étais. J'ai été si souvent interrompu que je ferais tout aussi bien de recommencer.*

Diderot, Jacques le fataliste

Abstract The purpose of this paper is to explain the relations between refined versions of index theory on a manifold X and the localization formulas of Duistermaat–Heckman on the loop space LX. Starting from Atiyah's remark exhibiting such a connection for the Atiyah–Singer index formula for the Dirac operator acting on spinors, we ultimately explain the geometric counterpart on LX to η-invariants, $\widetilde{\eta}$-forms and holomorphic torsion.

Key words: Index theory and related fixed point theorems

Mathematics Subject Classification (2010): 58J20

Introduction

The purpose of this paper is to explain the deep connections between the localization formulas of Duistermaat–Heckman and index theory.

Let us first give the proper background to the subject. In [A85], Atiyah and Witten discovered the remarkable fact that when translating the McKean–Singer formula [MS67] for the index of the Dirac operator over spinors into the integral on the loop space of a differential form, this form is equivariantly closed with respect to the vector field generating the obvious action of S^1 on the loop space. A formal application of the localization formula of Duistermaat–Heckman [DH82, DH83]

Département de Mathématique, Université Paris-Sud, Bâtiment 425, 91405 Orsay, France
e-mail: Jean-Michel.Bismut@math.u-psud.fr

leads to the index theorem of Atiyah–Singer for this specific operator, essentially without analysis. This application remains formal, because integration of differential forms on an infinite dimensional manifold is not well-defined.

The merit of this approach is that it gives a geometric formula for the index which preexists the index formula. This remark led us first to extend Atiyah's remark to general Dirac operators [B85], and also to start the project of building up a dictionary between these two theories, so as to use whatever could be gained from one theory to the other. It is this project that we will try to present here in some detail.

Let us immediately mention that the specific approach which will be taken here gives a partial and somewhat paradoxical view of the subject. Also we will be led to overemphasize our own work. Nevertheless we felt such a perspective could still be useful in a field which may look arduous and overly technical to the outsider.

Let us give a few more details. Let X be a Riemannian manifold. It is well known that the theory of the heat equation on X is equivalent to the theory of Brownian motion over X. In particular the trace of the heat operator on X can be expressed as the integral of a well-defined measure over the continuous loop space L^0X, which is also called a path integral. Then S^1 acts naturally on LX, and the measure integrated over L^0X is S^1-invariant. Passing from the trace of an operator to the integral of a measure on the loop space is already passing from analysis to geometry. An important formula in this context is the Feynman–Kac formula. Above all, Itô's stochastic calculus gave a tremendous input to a better understanding of the correspondence between second order differential operators and path integrals.

From the point of view of physics, this correspondence can also be described as passing from a Hamiltonian, or operator theoretic perspective, to a Lagrangian or path integral formalism. As we shall see in the paper, this correspondence also has some features of a nonlinear Fourier transform.

There are two new inputs in Atiyah's point of view. On the operator side, there is the Dirac operator D^X, and its square, whose corresponding heat operator is used in the McKean–Singer formula for the index $\mathrm{Ind}(D^X_+)$, valid for $t > 0$,

$$\mathrm{Ind}(D^X_+) = \mathrm{Tr_s}[\exp(-tD^{X,2})]. \tag{0.1}$$

In (0.1), $\mathrm{Tr_s}$ is our notation for the supertrace. On the geometric side, measures are replaced by differential forms. Atiyah's ideas give a cohomological perspective to the Hamiltonian–Lagrangian correspondence. The main drawback is that this correspondence is not well defined.

A well-known method to prove the index theorem for Dirac operators is to make $t \to 0$ in (0.1). Taking the local expansion of the heat kernel and its supertrace leads to a proof of the local index theorem of Gilkey and Atiyah–Bott–Patodi [Gil84, ABP73], via the mechanism of the 'fantastic cancellations' anticipated by McKean–Singer [MS67]. Of course the local index theorem implies the index theorem, but it is stronger. In the index formula for Dirac operators with boundary, Atiyah–Patodi–Singer [APS75a, APS75b] exploited the local index theorem to establish their index theorem. On the other hand, many of the known proofs of the Duistermaat–Heckman formulas rely on explicit choices of differential forms, and this explicit choice has also a 'local' character.

In [B86b], we found that the heat equation method in index theory is by itself the infinite-dimensional version of an easy proof of the localization formula in finite dimensions, and that the 'fantastic cancellations' have a universal counterpart in the finite-dimensional proof of the localization formula. This made not only equivariant cohomology in infinite dimensions relevant in predicting the index theorem, but also allowed refined local versions of the index formula and of the localization formula to be put in one-to-one correspondence.

That such a correspondence possibly exists is interesting by itself. But one can well imagine that since it relates objects of a fundamentally different nature, whatever information one can gain from one side can be passed to the other side, with the important proviso that since integrals of differential forms on LX are not well defined, any information one could get on that side would have to be translated back in the real operator-theoretic world to make sense. A difficulty in using such a point of view was that at the time the objects to be constructed did not exist on either side of the correspondence.

The ultimate purpose of this paper is to describe this correspondence as a loop space functor, mapping ordinary K-theory into the equivariant cohomology of the loop space, and Hermitian K-theory into the secondary theory of currents naturally associated with the localization formulas. Objects like η-invariants, holomorphic torsion are shown to correspond to Chern–Simons or Bott–Chern currents on the loop space. Also by transforming K-theoretic or analytic objects into geometric objects, any statement related to the families index theorem or Riemann–Roch–Grothendieck acquires a purely geometric flavour, which makes it accessible to a geometric treatment; this in turn can be used to construct the proper analytic objects, and also to anticipate how they should behave.

A powerful motivation for this study has been the program by Gillet–Soulé, which led them to the proof of an arithmetic Riemann–Roch theorem [GSo92]. It turns out that the loop space functor provides a useful understanding of the analytic objects that are natural in that theory.

Needless to say, the fact that the same object can be understood and analysed from several points of view makes the object more interesting. What is offered here is not an alternative or better view of objects which are otherwise known, but it is another view, whose main drawback is that some of the objects are not always well defined.

This paper is organized as follows. In Section 1, we review the Duistermaat–Heckman localization formula and the construction of associated secondary currents.

In Section 2, we summarize simple results which connect the heat equation on a manifold to measures on its loop space.

In Section 3, we develop Atiyah's remarks, and also the connection between the 'fantastic cancellations' and the localization formula.

Finally, in Section 4, we relate Hermitian K-theory to the theory of secondary currents associated with the localization formula.

The author is indebted to a referee for carefully reading the manuscript.

1 The Duistermaat–Heckman formula

The purpose of this section is to present the Duistermaat–Heckman localization formula, and to give a proof of the formula using explicit universal differential forms. Secondary currents of Chern–Simons or Bott–Chern type are also attached to the localization formula.

This section is organized as follows. In Subsection 1.1, we prove the localization formula, using a form α_t which will reappear in the whole paper.

In Subsection 1.2, we discuss the functorial aspects of the formula.

In Subsection 1.3, we give a transgressed version of the formula, and we construct a current ε of Chern–Simons type, whose existence implies the localization formula.

In Subsection 1.4, we give an integration along the fibre version of the formula.

In Subsection 1.5, we refine the integration along the fibre formula using the currents ε.

In Subsection 1.6, we relate the question of the compatibility of the currents ε to composition of projections via adiabatic limits.

In Subsection 1.7, we discuss the localization formula for complex manifolds. In particular we obtain a current σ of Bott–Chern type, whose existence implies the localization formula, and from which the current ε can also be obtained.

Finally, in Subsection 1.8, we discuss integration along the fibre for complex manifolds.

1.1 The localization formula

Let X be a compact connected oriented manifold. Let T be a torus acting smoothly on X, and let $\mathfrak{t}$ be its Lie algebra. Let g^{TX} be a T-invariant metric on TX. If $f \in \mathfrak{t}$, let f^X be the corresponding smooth vector field on X. If $f \in \mathfrak{t}$, then f^X is a Killing vector field.

Now we fix a $K \in \mathfrak{t}$. Let $\Omega^{\cdot}(X)$ be the vector space of smooth forms on X. Let L_{K^X} denote the associated Lie derivative operator acting on the de Rham complex $(\Omega^{\cdot}(X), d^X)$. The Cartan formula asserts that

$$L_{K^X} = [d^X, i_{K^X}]. \tag{1.1}$$

In (1.1), i_{K^X} denotes interior multiplication by K^X. Since d^X and i_{K^X} are both odd operators, the supercommutator appearing in (1.1) is actually an anticommutator.

Put

$$d_K^X = d^X + i_{K^X}. \tag{1.2}$$

Then (1.1) can be rewritten in the form

$$d_K^{X,2} = L_{K^X}. \tag{1.3}$$

Let $\Omega_K^{\cdot}(X)$ be the vector space of smooth K^X-invariant forms on X, i.e., of the forms α which are such that

$$L_{K^X}\alpha = 0. \tag{1.4}$$

By (1.3), (1.4), we find that when acting on $\Omega_K^{\cdot}(X)$,

$$d_K^{X,2} = 0. \tag{1.5}$$

The vector space $\Omega_K^{\cdot}(X)$ is naturally $\mathbf{Z}_2$-graded, and d_K^X acts as an odd operator on $\Omega_K^{\cdot}(X)$. Set

$$H_K(X) = \ker d_K^X / d_K^X \Omega_K^{\cdot}(X). \tag{1.6}$$

Then $H_K(X)$ is a $\mathbf{Z}_2$-graded vector space, which is called the equivariant cohomology associated with K. Forms vanishing under d_K^X will be called equivariantly closed forms.

Remark 1.1. Let $S^{\cdot}(\mathfrak{t}^*)$ be the algebra of polynomials on $\mathfrak{t}$. If K is allowed to vary in $\mathfrak{t}$, then d_K acts naturally on $\Omega^{\cdot}(X) \otimes S^{\cdot}(\mathfrak{t}^*)$. By giving forms their classical grading and polynomials in $S^{\cdot}(\mathfrak{t}^*)$ twice their degree, $\Omega^{\cdot}(X) \otimes S^{\cdot}(\mathfrak{t}^*)$ acquires a $\mathbf{Z}$-grading, and d_K^X is a differential, i.e., it increases the total degree by 1. The cohomology of the complex $(\Omega^{\cdot}(X) \otimes S^{\cdot}(\mathfrak{t}^*), d_K^X)$ is the equivariant cohomology of X in the sense of Cartan [Ca51a, Ca51b, GuSt99]. We will mostly disregard this point of view, which is nevertheless intimately related to what is done here.

Let N be the number operator of $\Lambda^{\cdot}(T^*X)$. Then N defined the $\mathbf{Z}$-grading of $\Omega^{\cdot}(X)$. Note that if $s \in \mathbf{R}^*$,

$$s^N d_K^X s^{-N} = s d_{K/s^2}^X. \tag{1.7}$$

It follows that $H_K(X)$ is unchanged when replacing K by $tK, t > 0$.

Let ∇^{TX} be the Levi-Civita connection on TX, and let R^{TX} be its curvature. Since K^X is a Killing vector field, $\nabla_{\cdot}^{TX} K^X$ is an antisymmetric section of $\mathrm{End}(TX)$. Since K^X preserves ∇^{TX}, we get easily

$$\nabla_{\cdot}^{TX} \nabla_{\cdot}^{TX} K^X + i_{K^X} R^{TX} = 0. \tag{1.8}$$

Let $X_K \subset X$ be the zero set of K^X. Then X_K is a totally geodesic submanifold of X. Let i be the embedding of X_K into X. Let $N_{X_K/X}$ be the orthogonal bundle to TX_K in $TX|_{X_K}$. Then $i^* : (\Omega_K^{\cdot}(X), d_K^X) \to (\Omega^{\cdot}(X_K), d^{X_K})$ is a morphism of $\mathbf{Z}_2$-graded complexes. Witten [W82] has shown that this is a quasiisomorphism, i.e., the map $i^* : H_K(X) \to H^{\cdot}(X_K)$ is an isomorphism of $\mathbf{Z}_2$-graded vector spaces.

By (1.8), we find that if $U \in TX_K$,

$$\nabla_U^{TX} \nabla^{TX} K^X = 0. \tag{1.9}$$

Let $\nabla^{N_{X_K/X}}$ be the connection on $N_{X_K/X}$, which is the restriction of ∇^{TX} to $N_{X_K/X}$, and let $R^{N_{X_K/X}}$ be the curvature of the connection $\nabla^{N_{X_K/X}}$. We denote by J_K the restriction of $\nabla_\cdot^{TX} K$ to $N_{X_K/X}$. By (1.9), J_K is an antisymmetric parallel endomorphism of $N_{X_K/X}$, which is nondegenerate, so that $N_{X_K/X}$ is of even dimension.

If V is an oriented Euclidean vector space of even dimension n, if $A \in \mathrm{End}(V)$ is antisymmetric, let ω_A be the 2-form on V such that if $a, b \in V$,

$$\omega_A(a, b) = \langle a, Ab \rangle. \tag{1.10}$$

By definition if η is the unit volume form defining the orientation of V, the Pfaffian $\mathrm{Pf}[A] \in \mathbf{R}$ is such that

$$\frac{\omega_A^{n/2}}{(n/2)!} = \mathrm{Pf}[A]\eta. \tag{1.11}$$

We can rewrite (1.11) in the form

$$[\exp(\omega_A)]^{\max} = \mathrm{Pf}[A]\eta. \tag{1.12}$$

A basic property of the Pfaffian is that it is a square root of the determinant, i.e.,

$$\det[A] = \mathrm{Pf}^2[A]. \tag{1.13}$$

Then $N_{X_K/X}$ is naturally oriented by J_K, the orientation being such that

$$\mathrm{Pf}[J_K] > 0. \tag{1.14}$$

The orientation of $N_{X_K/X}$ induces a corresponding orientation of X_K.

Set

$$e_K(N_{X_K/X}, \nabla^{N_{X_K/X}}) = \mathrm{Pf}\left[\frac{J_K + R^{N_{X_K/X}}}{2\pi}\right]. \tag{1.15}$$

Then $e_K(N_{X_K/X}, \nabla^{N_{X_K/X}})$ is a closed even form on X_K, whose cohomology class $e_K(N_{X_K/X})$ does not depend on the connection $\nabla^{N_{X_K/X}}$. This class is called the equivariant Euler class of $N_{X_K/X}$. Because of (1.14), the closed form $e_K^{-1}(N_{X_K/X}, \nabla^{N_{X_K/X}})$ is well defined. We denote by $e_K^{-1}(N_{X_K/X})$ the corresponding even cohomology class on X_K.

Of course the real vector bundle $N_{X_K/X}$ splits according to the eigenvalues of J_K. The form $e_K(N_{X_K/X}, \nabla^{N_{X_K/X}})$ and the class $e_K(N_{X_K/X})$ can then be expressed as a finite product over the eigenvalues of J_K of the Euler forms or the Euler classes of the corresponding components of $N_{X_K/X}$.

Let $K^{X\prime}$ be the 1-form on X which is dual to K^X by the metric g^{TX}. Since K^X is Killing,

$$L_{K^X} K^{X\prime} = 0, \tag{1.16}$$

which by (1.3) can be rewritten as

$$d_K^X [d_K^X K^{X\prime}] = 0. \tag{1.17}$$

The K^X-invariant form $d_K^X K^{X\prime}$ is d_K^X-closed and d_K^X exact. Of course

$$d_K^X K^{X\prime} = d^X K^{X\prime} + |K^X|^2, \tag{1.18}$$

i.e., (1.18) is the sum of a 0-form and of a 2-form.

If $U, V \in TX$, we have the obvious formula,

$$d^X K^{X\prime}(U, V) = 2\langle \nabla_U^{TX} K, V \rangle. \tag{1.19}$$

Definition 1.2. For $t > 0$, set

$$\alpha_t = \exp(-d_K^X K^{X\prime}/2t). \tag{1.20}$$

By (1.17),

$$d_K^X \alpha_t = 0. \tag{1.21}$$

The even form α_t will play an essential role in the whole paper.

Now we state the localization formula of Duistermaat–Heckman [DH82, DH83] and Berline–Vergne [BeV83].

Theorem 1.3. *For any $\mu \in H_K(X)$, the following identity holds:*

$$\int_X \mu = \int_{X_K} \frac{\mu}{e_K(N_{X_K/X})}. \tag{1.22}$$

Proof. We will not reproduce the original proofs, although we will comment later on the proof by Berline–Vergne. Here we will give our proof in [B86b, Theorem 1.3].

If β is a smooth form on X,

$$\int_X d^X \beta = 0. \tag{1.23}$$

Moreover, since $i_{K^X}\beta$ cannot be of top degree, we also get

$$\int_X i_{K^X} \beta = 0. \tag{1.24}$$

By (1.23), (1.24), we obtain

$$\int_X d_K^X \beta = 0. \tag{1.25}$$

We still denote by μ a smooth form on X which is d_K^X closed and represents the corresponding cohomology class. We claim that for any $t > 0$,

$$\int_X \mu = \int_X \alpha_t \mu. \tag{1.26}$$

Indeed

$$\alpha_\infty = 1, \tag{1.27}$$

so that (1.23) holds at $t = +\infty$. Moreover, since $d_K^X \alpha_t = 0, d_K^X \mu = 0$,

$$\frac{\partial}{\partial t} \int_X \alpha_t \mu = \frac{1}{2t^2} \int_X \alpha_t [d_K^X K^{X\prime}] \mu = \frac{1}{2t^2} \int_X d_K^X [\alpha_t K^{X\prime} \mu] = 0. \tag{1.28}$$

Therefore we have established (1.26).

Now we will make $t \to 0$ in (1.26). By (1.18), it is clear that for $\epsilon > 0$, if U_ϵ is an ϵ tubular neighbourhood of X_K in X, then α_t converges to 0 uniformly on $X \setminus U_\epsilon$. Let π be the projection $N_{X_K/X} \to X_K$, and let Y be the tautological section of $\pi^* N_{X_K/X}$ on the total space $\mathcal{N}_{X_K/X}$ of $N_{X_K/X}$. Then $J_K Y$ is a fibrewise Killing vector field on $\mathcal{N}_{X_K/X}$. Let $J_K Y'$ be the corresponding fibrewise 1-form on the fibres $N_{X_K/X}$. Using the connection $\nabla^{N_{X_K/X}}$, we may consider $J_K Y'$ as a 1-form on $\mathcal{N}_{X_K/X}$, which vanishes horizontally. Let ω_{J_K} be the fibrewise 2-form associated with J_K as in (1.10). We can consider ω_{J_K} as a 2-form on $\mathcal{N}_{X_K/X}$ which vanishes horizontally. Then one verifies easily that

$$d^{\mathcal{N}_{X_K/X}} J_K Y' = -2\omega_{J_K} + \langle R^{N_{X_K/X}} Y, J_K Y \rangle. \tag{1.29}$$

Equivalently,

$$d^{\mathcal{N}_{X_K/X}} J_K Y' = -2\omega_{J_K} - \langle J_K R^{N_{X_K/X}} Y, Y \rangle. \tag{1.30}$$

Also $R^{N_{X_K/X}} J_K$ is a 2 form on X_K with values in symmetric endomorphisms of $N_{X_K/X}$. By (1.29), we get

$$-\frac{1}{2} d_{J_K Y}^{\mathcal{N}_{X_K/X}} J_K Y' = \frac{1}{2} \langle J_K (J_K + R^{N_{X_K/X}}) Y, Y \rangle + \omega_{J_K}. \tag{1.31}$$

We now take a geodesic coordinate system on the tubular neighbourhood U_ϵ in the directions normal to X_K. This way we identify U_ϵ to the ϵ-neighbourhood of X_K in $\mathcal{N}_{X_K/X}$. For $s > 0$, let k_s be the dilation of $\mathcal{N}_{X_K/X}$, which is given by $Y \to sY$. Then one verifies easily that

$$k_{\sqrt{t}}^* \frac{K^{X\prime}}{t} = J_K Y', \tag{1.32}$$

so that as $t \to 0$,

$$k_{\sqrt{t}}^* \frac{d_K^X K^{X\prime}}{t} \to d_{J_K Y}^{\mathcal{N}_{X_K/X}} J_K Y'. \tag{1.33}$$

From (1.33), it is not difficult to deduce the convergence of currents on X as $t \to 0$,

$$\alpha_t \to \alpha_0 = \pi_* [\exp(-d_{J_K Y}^{\mathcal{N}_{X_K/X}} J_K Y'/2)] \delta_{X_K}. \tag{1.34}$$

By combining (1.26) and (1.34), we finally obtain

$$\int_X \mu = \int_X \alpha_0 \mu. \tag{1.35}$$

Let us now evaluate α_0. By (1.31), $\pi_*[\exp(-d_{J_K Y} J_K Y'/2)]$ is just a fibrewise Gaussian integral, which produces the inverse of a square root of a determinant. A simple computation leads to the formula

$$\pi_*[\exp(-d_{J_K Y}^{\mathcal{N}_{X_K/X}} J_K Y'/2)] = \frac{1}{e_K(N_{X_K/X}, \nabla^{N_{X_K/X}})}, \tag{1.36}$$

so that

$$\alpha_0 = \frac{\delta_{X_K}}{e_K(N_{X_K/X}, \nabla^{N_{X_K/X}})}. \tag{1.37}$$

By (1.35) and (1.37), we get (1.22). The proof of our theorem is completed.

Remark 1.4. By (1.34), α_0 is the integral along the fibre of a $d_{J_K Y}^{\mathcal{N}_{X_K/X}}$-closed form on $\mathcal{N}_{X_K/X}$, which is just the form α_1 on $\mathcal{N}_{X_K/X}$ which is associated with the fibrewise Killing vector field $J_K Y$. This reduces the proof of the localization formula to the computation of a similar formula with integration replaced by integration along the fibre, albeit of a very simple form. In Subsection 1.4, we will study an integration along the fibre version of the localization formula. Still it is important to keep in mind that the proof of the ordinary localization formula already incorporates some version of integration along the fibre.

We will often refer to the convergence of currents in (1.34) to be a local Duistermaat–Heckman formula.

Let J be the canonical complex structure on $\mathbf{C} \simeq \mathbf{R}^2$. Let $K = -JY$ be the obvious Killing vector field on $\mathbf{R}^2$. If ω is the canonical orientation form on $\mathbf{R}^2$, then

$$\exp(-d_K^{\mathbf{R}^2} K'/2) = \exp(-|Y|^2/2)(1+\omega). \tag{1.38}$$

A tautological application of (1.22) leads to the Gaussian integration formula

$$\int_{\mathbf{R}^2} \frac{1}{2\pi} \exp(-d_K^{\mathbf{R}^2} K'/2) = 1. \tag{1.39}$$

Incidentally note that (1.39) has already been used in (1.36). Equation (1.39) will reappear later in an infinite-dimensional context.

1.2 Functoriality of the Duistermaat–Heckman formula

Here, we will describe the compatibility of equation (1.22) to submersions. Indeed let $\pi : M \to S$ be a T-equivariant submersion of smooth compact oriented

T-manifolds, with compact oriented fibre X. If $K \in \mathfrak{t}$, let K^M, K^S be the corresponding vector fields on M, S, so that $\pi_* K^M = K^S$. We use otherwise the notation of Subsection 1.1.

Let π_* denote integration of smooth forms along the fibre X. Note that

$$\pi_* d_K^M = d_K^S \pi_*. \tag{1.40}$$

Let $\mu \in \Omega^{\cdot}_K(M)$ be such that $d_K^M \mu = 0$. By (1.40), we get

$$d_K^S \pi_* \mu = 0. \tag{1.41}$$

Clearly,

$$\int_M \mu = \int_S \pi_* \mu. \tag{1.42}$$

Moreover, by Theorem 1.3, we get

$$\int_M \mu = \int_{M_K} \frac{\mu}{e_K(N_{M_K/M})}, \qquad \int_S \pi_* \mu = \int_{S_K} \frac{\pi_* \mu}{e_K(N_{S_K/S})}. \tag{1.43}$$

We will reconcile (1.42) and (1.43). Let $M^K = \pi^{-1} S_K$. Then M^K is a T-invariant submanifold of M which fibres on S_K, and the restriction of K^M to M^K is tangent to the fibres X. Also M_K is a submanifold of M^K. By applying Theorem 1.3 on M^K, we get

$$\int_{S_K} \frac{\pi_* \mu}{e_K(N_{S_K/S})} = \int_{M^K} \frac{\mu}{\pi^* e_K(N_{S_K/S})} = \int_{M_K} \frac{\mu}{\pi^* e_K(N_{S_K/S}) e_K(N_{M_K/M^K})}. \tag{1.44}$$

Observe that the Euler class is multiplicative, so that

$$e_K(N_{M_K/M}) = \pi^*[e_K(N_{S_K/S})] e_K(N_{M_K/M^K}). \tag{1.45}$$

Equation (1.45) ultimately explains the compatibility of (1.42) and (1.43).

There is a similar compatibility result to embeddings, which is more difficult to explain. For more details, we refer to [B92a].

1.3 Transgression currents and localization

We make the same assumptions as in Subsection 1.1. Recall that the form α_t on X was defined in (1.20). Put

$$\beta_t = \frac{K^{X\prime}}{2t^2} \alpha_t. \tag{1.46}$$

Let $\| \|_{C^1}$ be a norm on $\Omega^{\cdot}(X)$ associated with the uniform convergence of forms on X together with their first derivatives.

Theorem 1.5. *The following identity holds:*

$$\frac{\partial}{\partial t}\alpha_t = d_K^X \beta_t. \tag{1.47}$$

There exists $C > 0$ such that for $\mu \in \Omega^{\cdot}(X)$, $t \in]0,1]$,

$$\left|\int_X (\alpha_t - \alpha_0)\mu\right| \leq C\sqrt{t}\|\mu\|_{C^1}, \tag{1.48}$$
$$\left|\int_X t\beta_t\mu\right| \leq C\sqrt{t}\|\mu\|_{C^1}.$$

Proof. Equation (1.47) follows from (1.20) and (1.21). The first equation in (1.48) follows easily from the arguments in the proof of Theorem 1.3. Moreover, observe that as $t \to 0$,

$$k^*_{\sqrt{t}}\frac{K^{X\prime}}{t} \to J_K Y'. \tag{1.49}$$

By proceeding as in the proof of Theorem 1.3, we get easily the second identity in (1.48).

Remark 1.6. For more refined results involving microlocal convergence, we refer to [B86b, Theorem 1.3] and to [B92a, Theorem 2.5].

Definition 1.7. Let ε be the current on X,

$$\varepsilon = -\int_0^{+\infty} \beta_t dt. \tag{1.50}$$

When $t \to +\infty$, the integral in (1.50) converges trivially, and as $t \to 0$, Theorem 1.5 should be used to make sense of the integral.

Theorem 1.8. *The odd current ε is such that*

$$d_K^X \varepsilon = \frac{\delta_{X_K}}{e_K(N_{X_K/X}, \nabla^{N_{X_K/X}})} - 1. \tag{1.51}$$

Moreover, the wave-front set of ε is included in $N^*_{X_K/X}$. On $X \setminus X_K$, we have the identity,

$$\varepsilon = -\frac{K^{X\prime}}{d_K^X K^{X\prime}}. \tag{1.52}$$

Proof. Equation (1.51) follows from Theorem 1.5 and from (1.50). The fact that the wave-front set of ε is included in $N^*_{X_K/X}$ follows from the arguments in [B86b, B92a]. Finally, (1.52) is a consequence of (1.46) and (1.50).

Remark 1.9. If $\mu \in \Omega^{\cdot}(X)$, then

$$\int_X d_K^X(\varepsilon\mu) = 0. \tag{1.53}$$

If $d_K^X \mu = 0$, then (1.53) takes the form

$$\int_X (d_K^X \varepsilon)\mu = 0. \tag{1.54}$$

By (1.51), (1.54), we get (1.22). The existence of ε is stronger than the localization formula. The current ε is a current of Chern–Simons type.

Also observe that by (1.52), we have the obvious identity,

$$d_K^X \varepsilon = -1 \text{ on } X \setminus X_K. \tag{1.55}$$

Now equation (1.55) is an obvious consequence of (1.51). Still (1.51) is stronger since it extends ε as a current on X. In fact from equation (1.52), one deduces that ε is a L_1 current on X.

Inspection of (1.52) shows that the restriction of the form ε to $X \setminus X_K$ is closely related to a form used in Berline–Vergne [BeV83] in their proof of the localization formulas. Indeed Berline–Vergne consider a K^X-invariant 1-form a on $X \setminus X_K$ such that $i_{K^X} a = 1$ on $X \setminus X_K$, and define a form b on $X \setminus X_K$ by the formula

$$b = -\frac{a}{d_K^X a}. \tag{1.56}$$

Of course we still have

$$d_K^X b = -1 \text{ on } X \setminus X_K. \tag{1.57}$$

By using Stokes formula on $X \setminus X_K$, Berline–Vergne [BeV83] ultimately prove the localization formulas.

1.4 Localization formulas and integration along the fibre

Let $\pi : M \to S$ be a submersion of smooth manifolds, with compact oriented fibre X. Let T be a torus acting smoothly on M, and preserving the fibres X. If $f \in \mathfrak{t}$, f^M is now a smooth section of $TX = TM/S$, i.e., a vector field along the fibres X. This is why we will use the notation f^X instead of f^M.

We now fix $K \in \mathfrak{t}$, and we use the notation of the previous subsections. It follows from (1.40) that π_* gives a map $H_K^{\cdot}(M) \to H^{\cdot}(S)$.

Let $T^H M$ be a T-invariant horizontal subbundle of TM, so that $TM = T^H M \oplus TX$. Let g^{TX} be a T-invariant metric on TX. Let $K^{X\prime} \in T^*X$ be the fibrewise 1-form associated with K^X via the metric g^{TX}. We will consider $K^{X\prime}$ as a 1-form on M which vanishes on $T^H M$. As in (1.16), we have

$$L_{K^X} K^{X\prime} = 0. \tag{1.58}$$

It is important to observe that if g^{TM} is a T-invariant metric on M that restricts to g^{TX} on TX, and is such that $T^H M$ is orthogonal to TX with respect to g^{TM}, then

$K^{X\prime}$ is exactly an object of the type we already met in the proof of Theorem 1.3, by simply replacing X by M and g^{TX} by g^{TM} in that proof.

As before, M_K denotes the zero set of K^X. Then M_K is a submanifold of M, which fibres on S with compact fibre X_K which embeds in the fibre X. Let $N_{X_K/X}$ be the orthogonal bundle to TX_K in $TX|_{M_K}$ with respect to g^{TX}. We can identify $N_{X_K/X}$ to the normal bundle $N_{M_K/M}$.

By a construction given in [B86a], $(T^H M, g^{TX})$ determine a unique Euclidean connection on TX. This connection restricts to the Levi-Civita connection along the fibre X. If g^{TM} is a metric on M such that g^{TM} restricts to g^{TX} on TX, and moreover $T^H M$ and TX are orthogonal in TM with respect to g^{TM}, then ∇^{TX} is the projection of the Levi-Civita connection ∇^{TM} on TM with respect to the splitting $TM = T^H M \oplus TX$. Of course ∇^{TX} is T-invariant. It is easy to see that ∇^{TX} induces connections $\nabla^{TX_K}, \nabla^{N_{X_K/X}}$ on $TX_K, N_{X_K/X}$.

We still define the closed form $e_K(N_{X_K/X}, \nabla^{N_{X_K/X}})$ on M_K as in (1.15).

The form α_t is still defined as in (1.20), the only difference being that $K^{X\prime}$ is now defined as indicated at the beginning of the present subsection. The form α_t is a special case of the corresponding form in (1.20), when replacing X by M and g^{TX} by g^{TM}.

In the sequel, we will denote by $\int_X$ the integration along the fibre, which was previously denoted by π_*.

Let $\mu \in \Omega^{\cdot}(M)$ be such that $d_K^M \mu = 0$. By (1.41), $\int_X \mu$ is a closed form on S, and its cohomology class only depends on the d_K^M cohomology class of μ.

Now we state an extension of Theorem 1.3 which was established in [B86b, Theorem 1.9].

Theorem 1.10. *The following identity holds:*

$$\int_X \mu = \int_{X_K} \frac{\mu}{e_K(N_{X_K/X}, \nabla^{N_{X_K/X}})} \bmod d^S \Omega^{\cdot}(S). \tag{1.59}$$

Proof. The proof is formally exactly the same as the proof of Theorem 1.3.

By proceeding formally as in (1.25) and using instead (1.40), we find that for any $t > 0$,

$$\int_X \mu = \int_X \alpha_t \mu \bmod d^S \Omega^{\cdot}(S). \tag{1.60}$$

We define the current α_0 as in (1.34), (1.37), i.e.,

$$\alpha_0 = \frac{\delta_{M_K}}{e_K(N_{X_K/X}, \nabla^{N_{X_K/X}})}. \tag{1.61}$$

As we saw in (1.34), when $t \to 0$, we have the convergence of currents on M, $\alpha_t \to \alpha_0$. Therefore when $t \to 0$, we have the convergence of currents on S,

$$\int_X \alpha_t \mu \to \int_{X_K} \frac{\mu}{e_K(N_{X_K/X}, \nabla^{N_{X_K/X}})}. \tag{1.62}$$

By (1.60), (1.62), we get (1.59).

Remark 1.11. By proceeding as in [B90c, Theorems 5.1 and 5.4] and in [BGS90b, Theorem 3.12], one can prove that as $t \to 0$, $\alpha_t \to \alpha_0$ microlocally with respect to the topology of currents whose wave front set is included in $N^*_{X_K/X} = N^*_{M_K/M}$. By [Ho85, Theorem 8.2.12], the convergence in (1.62) is a uniform convergence of smooth forms and their derivatives on compact sets of S.

1.5 Transgression formulas and integration along the fibre

We make the same assumptions as in Subsection 1.4. Over M, we define the smooth form β_t as in (1.46). Let μ be a smooth d^M_K-closed form on M. By equation (1.47), we get

$$\frac{\partial}{\partial t}\int_X \alpha_t \mu = d^S \int_X \beta_t \mu. \tag{1.63}$$

As we saw in Remark 1.11, the estimates in (1.48) can be made microlocal in the class of currents whose wave-front set is included in $N^*_{X_K/X}$. By using again the results in [Ho85], we find that for $t \in]0,1]$,

$$\left|\int_X (\alpha_t - \alpha_0)\mu\right| \leq C\sqrt{t}, \tag{1.64}$$

$$\left|\int_X t\beta_t \mu\right| \leq C\sqrt{t}.$$

In (1.64), the estimate is taken with respect to the sup norm of compact sets of smooth forms and their derivatives of finite order. Also the constant C in (1.64) depends explicitly on μ.

We still define the current ε on M by a formula similar to (1.50). Set

$$\vartheta = \int_X \varepsilon\mu. \tag{1.65}$$

Then ϑ is a current on S.

Theorem 1.12. *The current ϑ is a smooth form on S, which is such that*

$$d^S\vartheta = \int_{X_K} \frac{\mu}{e_K(N_{X_K/X}, \nabla^{N_{X_K/X}})} - \int_X \mu. \tag{1.66}$$

Proof. The fact that ϑ is smooth follows from the microlocal estimates which were described above. Equation (1.66) is a consequence of (1.40) and (1.51).

1.6 Equivariant projections and adiabatic limits

We make again the same assumptions as in Subsection 1.2 and we use the corresponding notation.

Let g^{TM} be a T-invariant metric on TM, and let g^{TS} be a T-invariant metric on TS. Then g^{TM} restricts to a T-invariant metric g^{TM^K} on TM^K. Let g^{TX} be the restriction of g^{TM} to TX. Then the metric $g^{TX|_{M^K}}$ on $TX|_{M^K}$ is also T-invariant. The orthogonal bundle T^HM^K to $TX|_{M^K}$ in TM^K is also T-invariant.

The projection $\pi : M^K \to S_K$ verifies exactly the assumptions of Subsection 1.4. In particular the torus T now acts along the fibres of this projection.

As we saw in Subsection 1.2, the formulas of Duistermaat–Heckman are compatible to functorial operations. Given our choice of metrics, there are associated currents ε^M on M, ε^S on S and ε^{M^K} on M^K. One can then ask to what extent these currents are compatible. Analogues of natural identities relating similar currents were established in [BGS90b]. In connection with Remark 1.4, for $\epsilon > 0$, it is natural to introduce the T-invariant metric g_ϵ^{TM} on TM given by

$$g_\epsilon^{TM} = g^{TM} + \frac{1}{\epsilon}\pi^* g^{TS}. \tag{1.67}$$

and for $t > 0$, to replace g_ϵ^{TM} by g_ϵ^{TM}/t. Indeed by playing with the parameters ϵ, t, we can obtain in this way a whole range of localization formulas. They are associated with the smooth forms $\alpha_{\epsilon,t}$ on M which are attached to the metric g_ϵ^{TM}/t. When $\epsilon \to 0$, two phenomena occur. The first is that localization is forced on M^K. But a second related phenomenon is that the fibres of $\pi : M^K \to S_K$ are further and further apart.

Making $\epsilon \to 0$ is also called passing to the adiabatic limit.

We will not discuss refined identities on the currents ε much more in this paper. Holomorphic analogues for their complex analogues are discussed in much detail in the context of embeddings in [B92a].

1.7 Localization formulas and complex manifolds

We make the same assumptions as in Subsections 1.1 and 1.3. Also we assume that X is a complex manifold, and that the torus T acts holomorphically on X. In the sequel, TX denotes the holomorphic tangent bundle, and $T_{\mathbf{R}}X$ is the real tangent bundle. Let J be the complex structure of $T_{\mathbf{R}}X$, so that $TX, \overline{TX}$ are the eigenspaces J that are associated with the eigenvalues $i, -i$. The de Rham operator d^X splits as

$$d^X = \overline{\partial}^X + \partial^X. \tag{1.68}$$

We fix $K \in \mathfrak{t}$. Let $K^{X(1,0)}, K^{X(0,1)}$ be the components of K^X in $TX, \overline{TX}$, so that

$$K^X = K^{X(1,0)} + K^{X(0,1)}. \tag{1.69}$$

Put

$$\overline{\partial}_K^X = \overline{\partial}^X + i_{K^{X(1,0)}}, \qquad \partial_K^X = \partial^X + i_{K^{X(0,1)}}. \tag{1.70}$$

Clearly,

$$d_K^X = \overline{\partial}_K^X + \partial_K^X. \tag{1.71}$$

Moreover, since $K^{X(1,0)}$ is a holomorphic section of TX, using (1.3), we get

$$\overline{\partial}_K^{X,2} = 0, \qquad \partial_K^{X,2} = 0, \qquad [\overline{\partial}_K^X, \partial_K^X] = L_{K^X}. \tag{1.72}$$

By (1.72), we deduce that when acting on $\Omega^{\cdot}_K(X)$,

$$[\overline{\partial}_K^X, \partial_K^X] = 0. \tag{1.73}$$

Let g^{TX} be a T-invariant Hermitian metric on TX, let $g^{T_{\mathbf{R}}X}$ be the corresponding Riemannian metric on TX, and let $\langle , \rangle$ be the corresponding scalar product. Let ω^X be the associated Kähler form on X, i.e., if $U, V \in T_{\mathbf{R}}X$,

$$\omega^X(U, V) = \langle U, JV \rangle. \tag{1.74}$$

Then ω^X is a $(1, 1)$ form on X. Clearly,

$$L_{K^X}\omega^X = 0, \tag{1.75}$$

which by (1.72) can also be written as

$$\overline{\partial}_K^X \partial_K^X \omega^X = -\partial_K^X \overline{\partial}_K^X \omega^X. \tag{1.76}$$

Also observe that

$$K^{X\prime} = (i_{K^{X(1,0)}} - i_{K^{X(0,1)}}) i\omega^X. \tag{1.77}$$

Now we assume that the metric g^{TX} is Kähler, i.e., the form ω^X is closed, so that

$$\overline{\partial}^X \omega^X = 0, \qquad \partial^X \omega^X = 0. \tag{1.78}$$

By (1.77), (1.78), we get

$$K^{X\prime} = (\overline{\partial}_K^X - \partial_K^X) i\omega^X. \tag{1.79}$$

By (1.71), (1.76), and (1.79), we obtain,

$$d_K^X K^{X\prime} = 2\partial_K^X \overline{\partial}_K^X i\omega^X = -2\overline{\partial}_K^X \partial_K^X i\omega^X. \tag{1.80}$$

From (1.20), (1.80), we get the formula in [B90a, eqs. (14) and (15)],

$$\alpha_t = \exp(\overline{\partial}_K^X \partial_K^X i\omega^X / t). \tag{1.81}$$

Recall that β_t was defined in (1.46). Set

$$\gamma_t = \frac{\omega^X}{t} \exp(\overline{\partial}_K^X \partial_K^X i\omega^X/t). \tag{1.82}$$

Now we have the result in [B90a, Proposition 5] and [B92a, Theorem 2.3].

Theorem 1.13. *The following identities hold:*

$$\overline{\partial}_K^X \alpha_t = 0, \qquad \partial_K^X \alpha_t = 0, \tag{1.83}$$

$$\beta_t = \frac{\partial_K^X - \overline{\partial}_K^X}{2i}\frac{\gamma_t}{t}, \qquad \frac{\partial}{\partial t}\alpha_t = d_K^X \beta_t = \frac{\overline{\partial}_K^X \partial_K^X}{i}\frac{\gamma_t}{t}.$$

Proof. The first two identities follow from (1.21) and (1.71), or from (1.76) and (1.81). By (1.46), (1.77), (1.79), (1.81), we obtain the third identity. The fourth identity follows from (1.47) and from our third identity, or more directly from (1.81).

Note that X_K is a complex submanifold of X. Also $N_{X_K/X}$ is a holomorphic Hermitian vector bundle on X_K, and $\nabla^{N_{X_K/X}}$ is the corresponding holomorphic Hermitian connection.

By [B90a, eq. (40)] and [B92a, Theorem 2.7], as $t \to 0$, the current γ_t has an asymptotic expansion of the type,

$$\gamma_t = \frac{\lambda_{-1}}{t} + \lambda_0 + \mathcal{O}(t). \tag{1.84}$$

By Theorem 1.5,

$$\lambda_{-1} = \frac{\omega^X}{e_K(N_{X_K/X}, \nabla^{N_{X_K/X}})}\delta_{X_K}. \tag{1.85}$$

For $s \in \mathbf{C}$, set

$$F(s) = \frac{1}{\Gamma(s)}\int_0^{+\infty} t^{s-1}\gamma_t dt. \tag{1.86}$$

Using the above results, one can easily prove that the current can be defined as a meromorphic function of $s \in \mathbf{C}$, $\mathrm{Re}(s) < 1$, which is holomorphic at $s = 0$.

Set

$$\sigma = \frac{\partial}{\partial s}F(0). \tag{1.87}$$

The following result was established in [B90a, Theorem 6] and in [B92a, Theorem 2.12].

Theorem 1.14. *The current σ is a sum of currents of type (p, p), whose wave-front set is included in $N^*_{X_K/X,\mathbf{R}}$. Moreover, it verifies the equations of currents on X,*

$$\frac{\overline{\partial}_K^X \partial_K^X}{i}\sigma = 1 - \frac{\delta_{X_K}}{e_K(N_{X_K/X}, \nabla^{N_{X_K/X}})}, \qquad \frac{\overline{\partial}_K^X - \partial_K^X}{2i}\sigma = \varepsilon. \tag{1.88}$$

Proof. The first equation in (1.88) follows from (1.48) and from the last identity in (1.83). The second equation is a consequence of the third equation in (1.83).

Remark 1.15. Equation (1.88) indicates that the current σ is a current of Bott–Chern type [BoC65]. For a general theory of Bott–Chern currents, we refer to Gillet–Soulé [GSo90] and also to Bismut–Gillet–Soulé [BGS90b, BGS90a]. The paper [B92a] is devoted to the study of the functorial behaviour of the current σ under complex embeddings.

1.8 Complex manifolds and integration along the fibre

Here we make the same assumptions as in Subsections 1.4 and 1.5, while also assuming M and S to be complex manifolds, and $\pi : M \to S$ to be holomorphic, and also that the action of the torus T on M is holomorphic. Here $TM, TX = TM/S, TS$ will denote the obvious holomorphic tangent bundles.

One can try to adapt the arguments we gave in Subsection 1.4 to obtain secondary Bott–Chern forms on S by integration along the fibre of the currents σ of Subsection 1.7.

It is not difficult to see that what is needed is a closed real 2 form ω^M of type $(1,1)$, whose restriction ω^X to the fibres X induces a Kähler metric g^{TX} along the fibres X. Let $T^HM \subset TM$ be the orthogonal bundle to TX with respect to ω^X. If $\omega^{M,H}$ is the restriction of ω^M to $T^H_{\mathbf{R}}M$, we can write ω^M in the form

$$\omega^M = \omega^{M,H} + \omega^X. \tag{1.89}$$

Equations (1.81), (1.82) for α_t, γ_t are now replaced by

$$\alpha_t = \exp(\overline{\partial}^M_K \partial^M_K i\omega^M/t), \qquad \gamma_t = \frac{2\pi\,\omega^M}{t}\exp(\overline{\partial}^M_K \partial^M_K i\omega^M/t). \tag{1.90}$$

Note the form α_t in (1.90) is a special case of the form α_t which was considered in Subsection 1.4.

Let μ be a smooth form on M which is a sum of forms of type (p,p), and such that $d^M_K\mu = 0$, which is equivalent to $\overline{\partial}^M_K\mu = 0, \partial^M_K\mu = 0$. Instead of (1.63), we now have

$$\int_X \beta_t\mu = \frac{\partial^S - \overline{\partial}^S}{2i}\int_X \frac{\gamma_t}{t}\mu, \qquad \frac{\partial}{\partial t}\int_X \alpha_t\mu = \frac{\overline{\partial}^S\partial^S}{i}\int_X \frac{\gamma_t}{t}\mu. \tag{1.91}$$

For $s \in \mathbf{C}$, set

$$\Phi(s) = \frac{1}{\Gamma(s)}\int_0^{+\infty} t^{s-1}\int_X \gamma_t\mu. \tag{1.92}$$

Then for $s \in \mathbf{C}, \Re(s) < 1$, $\Phi(s)$ is a smooth form on S which depends holomorphically on S. Of course, if we define the form $F(s)$ on M as in (1.86), then

$$\Phi(s) = \int_X F(s)\mu. \tag{1.93}$$

Set

$$\tau = \frac{\partial}{\partial s}\Phi(0). \tag{1.94}$$

If we define the current σ on M as in (1.87), then

$$\tau = \int_X \sigma\mu. \tag{1.95}$$

Also $\nabla^N_{X_K/X}$ is exactly the holomorphic Hermitian connection on $N_{X_K/X}$. Moreover, τ is a smooth form on S which is a sum of (p,p) forms, and which is such that

$$\frac{\overline{\partial}^S\partial^S}{i}\tau = \int_X \mu - \int_{X_K}\frac{\mu}{e_K(N_{X_K/X}, N_{X_K/X})}. \tag{1.96}$$

Finally, if we define ϑ as in (1.65), by (1.88), (1.95), we get

$$\frac{\overline{\partial}^S - \partial^S}{2i}\tau = \vartheta. \tag{1.97}$$

2 Heat equation and measures on the loop space

The purpose of this section is to review classical results connecting the trace of the heat kernel on a Riemannian manifold to integrals over the continuous loop space of the manifold of well-defined measures.

This section is organized as follows. In Subsection 2.1, we consider first the case of finite-dimensional traces and corresponding integrals over a discrete loop space.

In Subsection 2.2, we relate the heat kernel on a Riemannian manifold to path integrals.

Finally, in Subsection 2.3, we state the Feynman–Kac formula.

2.1 Finite-dimensional traces

Let V be a finite-dimensional real vector space of dimension n, let $e_1, \ldots, e_n$ be a basis of V. Let $u \in \mathrm{End}(V)$, and let $A = (a_i^j)$, $1 \leq i, j \leq n$ be the matrix of u with respect to this basis. Then the trace $\mathrm{Tr}[u]$ is given by

$$\mathrm{Tr}[u] = \sum_{i=1}^n a_i^i. \tag{2.1}$$

More generally, for $q \in \mathbf{N}$,

$$\mathrm{Tr}[u^{q+1}] = \sum_{1 \le i_0, \dots i_q \le n} a_{i_0}^{i_1} a_{i_1}^{i_2} \dots a_{i_q}^{i_0}. \tag{2.2}$$

Set

$$X = \{1, 2, \dots, n\}. \tag{2.3}$$

Let μ_{q+1}^{A} be the measure on X^{q+1} such that

$$\mu_{q+1}^{A}(i_0, \dots, i_q) = a_{i_0}^{i_1} a_{i_1}^{i_2} \dots a_{i_q}^{i_0}. \tag{2.4}$$

Observe that X^{q+1} should be thought of as a discrete version of a loop space. Indeed $(i_0, \dots, i_q)$ should be considered as the vertices of a graph, whose edges are virtual lines connecting $(i_0, i_1), (i_1, i_2), \dots, (i_q, i_0)$. The weight $\mu_{q+1}^{A}(i_0, \dots, i_q)$ in (2.4) is the product of the weights of the ordered edges of this graph.

Then we can rewrite (2.3) in the form

$$\mathrm{Tr}[u^{q+1}] = \int_{X^{q+1}} d\mu_{q+1}^{A}. \tag{2.5}$$

Observe that identity (2.5) is just a tautological reformulation of (2.2).

Put $\mathbf{F}_{q+1} = \mathbf{Z}/(q+1)\mathbf{Z}$. It will be convenient to view $0, \dots, q$ as describing the elements of $\mathbf{F}_{q+1}$. The set $X^{q+1} = \{1, \dots, n\}^{q+1}$ can be identified with the set of maps from $\mathbf{F}_{q+1}$ into X. The group $\mathbf{F}_{q+1}$ acts on X^{q+1}, so that if $j \in \mathbf{F}_{q+1}$,

$$j(i_0, \dots, i_q) = (i_{0+j}, \dots, i_{q+j}). \tag{2.6}$$

Then the measure μ_{q+1}^{A} is obviously $\mathbf{F}_{q+1}$-invariant.

Assume the a_i^j to be nonzero. If $v \in \mathrm{End}(V)$, if $B = (b_i^j)$, $1 \le i, j \le n$ is the associated matrix, then the measure μ_{q+1}^{B} has a density $d_{A,q+1}^{B}$ with respect to μ_{q+1}^{A} which is given by

$$d_{A,q+1}^{B}(i_0, \dots, i_q) = \frac{b_{i_0}^{i_1} b_{i_1}^{i_2} \dots b_{i_q}^{i_0}}{a_{i_0}^{i_1} a_{i_1}^{i_2} \dots a_{i_q}^{i_0}}. \tag{2.7}$$

By (2.5), we get

$$\mathrm{Tr}[v^{q+1}] = \int_{X^{q+1}} d_{A,q+1}^{B} \, d\mu_{q+1}^{A}. \tag{2.8}$$

2.2 The heat kernel on a compact Riemannian manifold

Let (X, g^{TX}) be a compact Riemannian manifold, with volume dx. Let Δ^X be the Laplace–Beltrami operator. For $t > 0$, let $p_t(x, y)$ be the smooth heat kernel on X which is associated with the operator $\exp(t\Delta^X/2)$.

Let L_2 be the Hilbert space of square-integrable real functions on X. Then $\exp(t\Delta^X/2) \in \mathrm{End}(L_2)$. We will consider $p_t(x, y)$ as an infinite-dimensional matrix, indexed by couples $(x, y), x, y \in X$.

Clearly,

$$\mathrm{Tr}[\exp(t\Delta^X/2)] = \int_X p_t(x, x)dx. \tag{2.9}$$

For $q \in \mathbf{N}$,

$$\exp(t\Delta^X/2) = [\exp(t\Delta^X/2(q+1))]^{q+1}. \tag{2.10}$$

Equation (2.9) is equivalent to the identity

$$p_t(x, y) = \int_{X^q} p_{t/(q+1)}(x, x_1) \dots p_{t/(q+1)}(x_q, y)dx_1 \dots dx_q. \tag{2.11}$$

By (2.9), (2.11), we obtain

$$\mathrm{Tr}[\exp(t\Delta^X/2)] = \int_{X^{q+1}} p_{t/(q+1)}(x_0, x_1) \dots p_{t/(q+1)}(x_q, x_0)dx_0 \dots dx_q. \tag{2.12}$$

Let $d\mu_{t,q+1}$ be the positive measure on X^{q+1},

$$d\mu_{t,q+1} = p_{t/(q+1)}(x_0, x_1) \dots p_{t/(q+1)}(x_q, x_0)dx_0 \dots dx_q. \tag{2.13}$$

Then we can rewrite (2.12) in the form

$$\mathrm{Tr}[\exp(t\Delta^X/2)] = \int_{X^{q+1}} d\mu_{t,q+1}. \tag{2.14}$$

Equation (2.14) is a continuous analogue of (2.5).

Note that (2.14) does not depend on q. The underlying property of the measures μ_{q+1} is that in an algebraic sense they form a compatible projective system of measures. We now will make this idea more precise.

Indeed as in Subsection 2.1, we can replace X^{q+1} by the associated set of virtual graphs. However, we want to think of the edges connecting the pairs (x_i, x_{i+1}) as being paths in X. An obvious possibility would pick a minimizing geodesic. However, except when these points are close to each other, the geodesic is in general not unique. This can be partly compensated by the fact that as $t \to 0$, $p_t(x, y)dy$ converges to the Dirac mass at x.

If $S^1 = \mathbf{R}/\mathbf{Z}$, let L^0X be the space of continuous functions from S^1 into X. Note that S^1 acts on L^0X, so that if $s \in S^1$,

$$k_s x_\cdot = x_{s+\cdot}. \tag{2.15}$$

Given $q \in \mathbf{N}$, let $\pi_q : L^0X \to X^{q+1}$ be the map

$$\pi_q x_\cdot = (x_0, x_{1/(q+1)}, x_{2/(q+1)}, \dots, x_{q/(q+1)}). \tag{2.16}$$

In view of the above, it is natural to raise the question of the existence of an S^1-invariant positive measure μ_t on L^0X whose image by the maps π_q would just be the $\mu_{t,q+1}$.

The fact that this question has an obviously unique answer is a fundamental fact which is produced from the theory of Brownian motion and of the stochastic differential equations. The corresponding measure μ_t is the Brownian loop measure of parameter t. Brownian motion is a very complex and intriguing object of which we will say very little, except that μ_t a.e., its paths are nowhere differentiable. Brownian motion has locally a very erratic behaviour, whose physical manifestation is given by the well-known observations by Brown of the motion of pollen in water.

A consequence of equation (2.14) is that

$$\operatorname{Tr}[\exp(t\Delta^X/2)] = \int_{L^0X} d\mu_t. \tag{2.17}$$

Equation (2.17) should be thought of as being the continuous analogue of the discrete equation (2.5).

We have not yet explained the role of the metric g^{TX} and of the operator Δ^X in the construction of μ_t. This is what we will do now.

Let C be the infinite constant

$$C = \prod_1^{+\infty} 2k^2\pi. \tag{2.18}$$

By equation (2.13), using the gaussian approximation for $p_t(x, y)$ for t small and $d(x, y) \sim \sqrt{t}$, one arrives easily at the formal formula

$$d\mu_t = \frac{C^n}{(2\pi)^{n/2}} \exp\left(-\int_{S^1} |\dot{x}|^2 ds/2t\right) \frac{\mathcal{D}x}{t^{\infty/2}}. \tag{2.19}$$

In (2.19), $\mathcal{D}x$ denotes formally the Lebesgue measure on L^0X. The normalizing constant C^n in (2.19) can be found by inspection of the case where X is a torus.

There are several difficulties with (2.19). The first difficulty is that as we saw before

$$\int_{S^1} |\dot{x}|^2 ds = +\infty \ \mu_t \text{ a.e.}. \tag{2.20}$$

The second difficulty is that there is no Lebesgue measure $\mathcal{D}x$ on L^0X. The third related difficulty is the denominator $t^{\infty/2}$, where ∞ should be understood as $\dim LX$.

Such difficulties are unavoidable. Indeed it is well known that a Gaussian measure on an infinite-dimensional Hilbert space gives 0 measure to this Hilbert space. Although (2.19) does not mean anything, many of its consequences are correct. We will use (2.19) repeatedly in the sequel.

Let LX be the set of smooth maps $s \in S^1 \to x_s \in X$. By (2.17), (2.19), we get

$$\operatorname{Tr}[\exp(t\Delta^X/2)] = \frac{C^n}{(2\pi)^{n/2}} \int_{LX} \exp\left(-\int_{S^1} |\dot{x}|^2 ds/2t\right) \frac{\mathcal{D}x}{t^{\infty/2}}. \tag{2.21}$$

Incidentally, note that without fear of a contradiction, we have replaced L^0X by LX in (2.21).

In (2.20), we may replace S^1 by $S^1_t = \mathbf{R}/t\mathbf{Z}$. The effect is to make the factors t disappear on the right-hand side of (2.19).

In physics terminology, integrals like the ones that appear in the right-hand side of (2.17) and (2.21) are often called functional integrals, or path integrals. An equality like (2.21) relates a quantity in a Hamiltonian form in the left-hand side to another one in Lagrangian form on the right-hand side.

There is a Fourier transform quality to (2.21), as should be clear from the change of t on the left-hand side to $1/t$ on the right-hand side. This is related to the classical correspondence in quantization $p \to i\frac{\partial}{\partial x}$, which is here only made pointwise at every $x \in X$. There is indeed an implicit very strong relation between the pseudodifferential calculus and the functional integral. Indeed the pseudodifferential calculus enlarges the commutative algebra of smooth functions on X to a noncommutative algebra of operators, the functional integral enlarges the space X to the loop space LX. That these two enlargements turn out to be equivalent is more or less expressed in (2.21).

In this context, one could ask what difference there is, if any, between classical pseudodifferential calculus and stochastic calculus, once one takes for granted (2.17) and its formal version (2.21). The link is again Fourier transform, which maps the operator-theoretic (or Hamiltonian) picture to the path integral (or Lagrangian) picture. The main advantage of the Lagrangian picture is that it restores a geometric quality to the formulas, even if this geometry is infinite dimensional, in which the classical geometric intuition remains valid to a certain extent. As should be clear later, this is even more true in the context of index theory.

The Fourier transform quality of (2.19) is not related to the fact one expresses a trace as an integral over the loop space, which could be taken for granted for algebraic reasons, but because the formal expression (2.19) for $d\mu_t$ reflects Fourier transform, in the same way as the standard expression for the heat kernel on $\mathbf{R}$.

In equation (2.21), as $t \to +\infty$, $\operatorname{Tr}[\exp(t\Delta^X/2)]$ converges to 1 exponentially fast, while from the right-hand side, we can easily formally derive the behaviour of $\operatorname{Tr}[\exp(t\Delta^X/2)]$ as $t \to 0$. This is indeed typical of what happens with a Fourier transform. It is vain to expect from the right-hand side of (2.21) anything explicit concerning its behaviour as $t \to \infty$, in particular because the ergodic phenomena associated with long Brownian paths are difficult to describe in this formalism. In a formula like (2.21), there is a dichotomy which makes that both sides give a description of the same quantity which is relevant in different range of values of the parameter t.

2.3 The Feynman–Kac formula

Let $V : X \to \mathbf{R}$ be a smooth function. The Trotter formula asserts that

$$\begin{aligned}&\operatorname{Tr}[\exp(t(\Delta^X/2 - V))]\\&\quad = \lim_{q\to+\infty} \operatorname{Tr}[(\exp(t\Delta^X/2(q+1))\exp(-tV/(q+1)))^{q+1}].\end{aligned} \tag{2.22}$$

By proceeding as in (2.17) and using (2.22), we arrive at a version of the Feynman–Kac formula, which asserts that

$$\operatorname{Tr}[\exp(t(\Delta^X/2 - V))] = \int_{L^0X} \exp\left(-t\int_{S^1} V(x_s)ds\right) d\mu_t. \tag{2.23}$$

Equation (2.23) should be thought of as an infinite-dimensional version of (2.8). By (2.19), we can rewrite formally (2.23) in the form

$$\begin{aligned}&\operatorname{Tr}[\exp(t(\Delta^X/2 - V))]\\&\quad = \frac{C^n}{(2\pi)^{n/2}} \int_{LX} \exp\left(-\int_{S^1} |\dot{x}|^2/2t - t\int_{S^1} V(x_s)ds\right) \frac{\mathcal{D}x}{t^{\infty/2}}.\end{aligned} \tag{2.24}$$

The considerations we made after equation (2.21) remain also valid for (2.24).

The conclusion is that we have been able to express traces of heat operators as integrals on L^0X, such integrals being also called functional integrals or path integrals. The loop space L^0X is the space in which the computations occur. The measures on L^0X which appear naturally are S^1-invariant.

3 Index theory and differential forms on the loop space

In this section, we exhibit the formal connections between the index theorem for the Dirac operator and the localization formula over the loop space.

This section is organized as follows. In Subsection 3.1, we briefly review Clifford algebras.

In Subsection 3.2, we introduce the Dirac operator.

In Subsection 3.3, we give the McKean–Singer heat equation formula for the index of the Dirac operator, which depends on the time parameter $t > 0$.

In Subsection 3.4, we describe the 'fantastic cancellations' for the small time asymptotics of the local supertrace of the heat kernel.

In Subsection 3.5, we give a formula for the heat kernel associated with the square of the Dirac operator in terms of the scalar heat kernel.

In Subsection 3.6, via the McKean–Singer formula, we express the index as the integral over the continuous loop space L^0X as a S^1-invariant measure ν_t.

In Subsection 3.7, along the lines of Atiyah [A85], we describe the action of S^1 on the smooth loop space LX, and its generating vector field K.

In Subsection 3.8, we recall the fundamental remark of Atiyah and Witten [A85] expressing the index of the Dirac operator acting on untwisted spinors as a formal integral over LX of a d_K^{LX}-closed form, which turns out to be the form α_t, and we show that a formal application of the localization formula leads directly to the index formula of Atiyah–Singer.

In Subsection 3.9, we extend this remark to the case of general Dirac operators.

In Subsection 3.10, the fantastic cancellations in index theory are related to the proof of the localization formula which was given in Subsection 1.1.

In Subsection 3.11, we show that making sense of equivariant localization in infinite dimensions is as difficult as making sense of measures on infinite-dimensional Hilbert spaces.

Finally, in Subsection 3.12, we sketch a Hamiltonian–Lagrangian correspondence in index theory, which will be further elaborated in Section 4.

3.1 Clifford algebras

If $\mathcal{A} = \mathcal{A}_+ \oplus \mathcal{A}_-$ is a $\mathbf{Z}_2$-graded algebra, if $\alpha, \beta \in \mathcal{A}$, the supercommutator $[\alpha, \beta] \in \mathcal{A}$ is bilinear in α, β and such that if $\alpha, \beta \in \mathcal{A}_\pm$, then

$$[\alpha, \beta] = \alpha\beta - (-1)^{\deg(\alpha)\deg(\beta)}\beta\alpha. \tag{3.1}$$

In (3.1), deg is 0 on $\mathcal{A}_+$ and 1 on $\mathcal{A}_-$.

If $W = W_+ \oplus W_-$ is a $\mathbf{Z}_2$-graded vector space, let $\tau = \pm 1$ on $W_\pm$ be the endomorphism defining the $\mathbf{Z}_2$-grading. The algebra $\mathrm{End}(W)$ is naturally $\mathbf{Z}_2$-graded, the even (resp. odd) elements of $\mathrm{End}(W)$ commuting (resp. anticommuting) with τ. If $A \in \mathrm{End}(W)$, we define its supertrace $\mathrm{Tr_s}[A]$ by the formula

$$\mathrm{Tr_s}[A] = \mathrm{Tr}[\tau A]. \tag{3.2}$$

Then $\mathrm{Tr_s}$ vanishes on supercommutators.

Let V be a real vector space of dimension n, which is equipped with a bilinear symmetric form g^V. The Clifford algebra $c(V)$ attached to (V, g^V) is the real algebra generated over $\mathbf{R}$ by $1, e \in V$, and the commutation relations for $e, e' \in V$,

$$ee' + e'e = -2g^V(e, e'). \tag{3.3}$$

If $g^V = 0$, then $c(V)$ is just the exterior algebra $\Lambda^{\cdot}(V)$. The Clifford algebra $c(V)$ is equipped with an increasing filtration

$$\mathbf{R} = F^0c(V) \subset F^1c(V) \cdots \subset F^nc(V) = c(V). \tag{3.4}$$

Let $\mathrm{Gr}^{\cdot}$ be such that $\mathrm{Gr}^i = F^i c(V)/F^{i-1}c(V)$. Then

$$\mathrm{Gr}^{\cdot} \simeq \Lambda^{\cdot}(V). \tag{3.5}$$

From the above, it follows that $c(V)$ is a $\mathbf{Z}_2$-graded algebra.

Assume g^V to be a scalar product. If n is even, let $S^V = S^V_+ \oplus S^V_-$ be the $\mathbf{Z}_2$-graded vector space of (V, g^V) spinors. Then S^V is a $c(V)$ Clifford module. More precisely we have the identification of $\mathbf{Z}_2$-graded algebras,

$$c(V) \otimes_{\mathbf{R}} \mathbf{C} \simeq \mathrm{End}(S^V). \tag{3.6}$$

3.2 Spin manifolds and the Dirac operator

Let X be a compact oriented manifold of dimension n, and let g^{TX} be a Riemannian metric on TX. Let ∇^{TX} be the Levi-Civita connection on TX, and let R^{TX} be its curvature.

Let $c(TX)$ be the Clifford bundle of algebras associated with (TX, g^{TX}).

We will assume X to be even dimensional and spin. Let $S^{TX} = S^{TX}_+ \oplus S^{TX}_-$ be the $\mathbf{Z}_2$-graded vector bundle on (TX, g^{TX}) spinors. Let $\nabla^{S^{TX}} = \nabla^{S^{TX}_+} \oplus \nabla^{S^{TX}_-}$ be the corresponding unitary connection on $S^{TX} = S^{TX}_+ \oplus S^{TX}_-$.

Let (E, g^E, ∇^E) be a Hermitian vector bundle with unitary connection, and let R^E be the curvature of ∇^E. Let $\nabla^{S^{TX}\otimes E}$ be the unitary connection on $S^{TX} \otimes E$ which is induced by $\nabla^{S^{TX}}, \nabla^E$. Note that $S^{TX} \otimes E$ is a $c(TX)$ Clifford module.

Let $e_1, \ldots, e_n$ be an orthonormal basis of TX. The Dirac operator D^X is given by the formula

$$D^X = \sum_{i=1}^n c(e_i)\nabla^{S^{TX}\otimes E}_{e_i}. \tag{3.7}$$

Then D^X acts as an odd operator on $C^\infty(X, S^{TX} \otimes E)$. We can write D^X as a matrix with respect to the splitting $C^\infty(X, S^{TX} \otimes E) = C^\infty(X, S^{TX}_+ \otimes E) \oplus C^\infty(X, S^{TX}_- \otimes E)$,

$$D^X = \begin{bmatrix} 0 & D^X_- \\ D^X_+ & 0 \end{bmatrix}. \tag{3.8}$$

Put

$$c(R^E) = \frac{1}{2} \sum_{1\le i,j\le n} c(e_i)c(e_j)R^E(e_i, e_j). \tag{3.9}$$

The Lichnerowicz formula asserts that

$$D^{X,2} = -\Delta^H + \frac{S}{4} + c(R^E). \tag{3.10}$$

In (3.10), Δ^H is the Bochner horizontal Laplacian, which is the obvious version of a Laplacian acting on sections of a vector bundle, in which derivatives are replaced by covariant derivatives, and S is the scalar curvature. The remarkable fact is that there are no terms of length 4 in the Clifford bundle of algebras $c(TX)$, their contribution vanishes because of the circular identity on R^{TX}.

3.3 The index of D_+^X and the McKean–Singer formula

Since D_+^X is an elliptic operator, it is Fredholm. Also since D^X is formally self-adjoint, D_-^X is the formal L_2-adjoint of D_+^X. In particular the index of D_+^X is given by

$$\mathrm{Ind}(D_+^X) = \dim\ker(D_+^X) - \dim\ker(D_-^X). \tag{3.11}$$

As is well known, the index D_+^X is a homotopy invariant. In particular it does not depend on the choice of metrics or connections which was made above.

Since $D^{X,2}$ is elliptic of order 2, for $t > 0$, there is a heat kernel $P_t(x, y)$ corresponding to the operator $\exp(-tD^{X,2}/2)$. In particular $\exp(-tD^{X,2}/2)$ is trace class.

Also we have the Bianchi identity,

$$[D^X, D^{X,2}] = 0. \tag{3.12}$$

Now we state the McKean–Singer formula [MS67].

Theorem 3.1. *For any $t > 0$,*

$$\mathrm{Ind}(D_+^X) = \mathrm{Tr_s}[\exp(-tD^{X,2})]. \tag{3.13}$$

Proof. Note that if P is the orthogonal projection operator on $\ker D^X$, as $t \to +\infty$,

$$\exp(-tD^{X,2}) \to P, \tag{3.14}$$

where the convergence is taken in every possible sense. From (3.14), we find that as $t \to +\infty$,

$$\mathrm{Tr_s}[\exp(-tD^{X,2})] \to \mathrm{Ind}(D_+^X). \tag{3.15}$$

It is enough to prove that the right-hand side of (3.13) does not depend on $t > 0$. We will write the equations showing that the right-hand side of (3.13) does not depend on $t > 0$. Note that since D^X is odd,

$$D^{X,2} = \frac{1}{2}[D^X, D^X]. \tag{3.16}$$

Using (3.16), we have

$$\frac{\partial}{\partial t}\mathrm{Tr_s}[\exp(-tD^{X,2})] = -\mathrm{Tr_s}\left[\frac{1}{2}[D^X, D^X]\exp(-tD^{X,2})\right]$$
$$= -\frac{1}{2}\mathrm{Tr_s}[[D^X, D^X \exp(-tD^{X,2})]] = 0. \tag{3.17}$$

To get the last equality in (3.17), we used the fact that $\mathrm{Tr_s}$ vanishes on supercommutators. This concludes the proof of (3.13).

Remark 3.2. The usual proof of (3.13) is to use the spectral decomposition of the self-adjoint operator $D^{X,2}$. However, (3.17) makes clear that self-adjointness has little to do with (3.13). Indeed it is still true for any odd operator D^X having the same principal symbol as D^X. Proving the analogue of (3.2) in finite dimensions is also interesting.

Clearly,

$$\mathrm{Tr_s}[\exp(-tD^{X,2}/2)] = \int_X \mathrm{Tr_s}[P_t(x,x)]dx. \tag{3.18}$$

By (3.13), (3.18), we obtain

$$\mathrm{Ind}(D_+^X) = \int_X \mathrm{Tr_s}[P_t(x,x)]dx. \tag{3.19}$$

3.4 The fantastic cancellations

Since $D^{X,2}$ is elliptic of order 2, general considerations show that for $x \in X$, as $t \to 0$, $P_t(x,x) \in \mathrm{End}(S^{TX} \otimes E)_x$ has an asymptotic expansion of the form

$$P_t(x,x) = \frac{a_{-n/2}(x)}{t^{n/2}} + \frac{a_{-n/2+1}(x)}{t^{n/2-1}} + \cdots + a_0(x) + a_1(x)t + \cdots + a_k(x)t^k + o_x(t^k), \tag{3.20}$$

the $a_k(x)$ depend only locally on the metrics and connections, and the expansion is uniform in x. From (3.20), we get the asymptotic expansion as $t \to 0$,

$$\mathrm{Tr_s}[P_t(x,x)]$$
$$= \frac{b_{-n/2}(x)}{t^{n/2}} + \frac{b_{-n/2+1}(x)}{t^{n/2-1}} + \cdots + b_0(x) + b_1(x)t + \cdots + b_k(x)t^k + o_x(t^k), \tag{3.21}$$

and the b_k are also local. From (3.19) and (3.21), we get

$$\mathrm{Ind}(D_+^X) = \int_X b_0(x)dx, \qquad \int_X b_k(x)dx = 0 \text{ for } k \neq 0. \tag{3.22}$$

Equation (3.22) indicates that the global $\mathrm{Ind}(D_+^X)$ can be expressed as the integral of a local expression.

Set

$$\widehat{A}(x) = \frac{x/2}{\sinh(x/2)}. \tag{3.23}$$

We identify $\widehat{A}(x)$ with the corresponding multiplicative genus. We denote by $\widehat{A}(TX, \nabla^{TX})$ the Chern–Weil representative of the characteristic class $\widehat{A}(TX)$ which is associated with the connection ∇^{TX}. Similarly $\mathrm{ch}(E, \nabla^E)$ is the Chern character form for (E, ∇^E).

In [MS67], McKean and Singer conjectured that fantastic cancellations would occur so that (3.22) could be refined to be

$$b_k(x) = 0 \text{ for } k < 0, \qquad b_0(x) = [\widehat{A}(TX, \nabla^{TX})\mathrm{ch}(E, \nabla^E)]^{\max}. \tag{3.24}$$

Note that the expansion (3.20) always remain singular as $t \to 0$. The term 'fantastic cancellations' refers to the fact that the local vanishing of the singular terms occurs only when taking the supertrace in (3.20).

The McKean–Singer conjecture was proved by Gilkey [Gil84] and Atiyah–Bott–Patodi [ABP73]. From (3.22), (3.24), we get the Atiyah–Singer index theorem,

$$\mathrm{Ind}(D_+^X) = \int_X \widehat{A}(TX)\mathrm{ch}(E). \tag{3.25}$$

A statement like (3.24) is called a local index theorem, because of the local character of the expansion in (3.4).

The local index theorem was given a new impulse by the work by physicists [Al83] relating the index theorem to supersymmetry. Getzler [Ge86] gave a proof of the local index theorem using a fruitful rescaling technique on the Clifford algebra, a probabilistic proof was given in [B84a, B84b]. We refer to the book by Berline–Getzler–Vergne [BeGeV92] for more details.

3.5 *The heat kernel for $D^{X,2}$ as a path integral*

We will now use Feynman–Kac's formula and the Itô calculus to express $P_t(x, y)$ in terms of $p_t(x, y)$.

A first point is that although the Brownian paths are nowhere differentiable, still, it is possible to define parallel transport with respect to $\nabla^{S^{TX} \otimes E}$ along them. To do this, one can use Itô's stochastic differential equations, and/or use the approximation of Brownian motion by broken geodesics as outlined in Subsection 2.2, and show that parallel transport passes to the proper limit.

Let $x_\cdot$ be such a Brownian path, with x_0. Consider the differential equation on $V_\cdot \in \mathrm{End}(S^{TX} \otimes E)_{x_0}$,

$$\frac{dV}{ds} = -V\tau_0^s c(R^E_{x_s}/2)\tau_s^0, \qquad V_0 = 1. \tag{3.26}$$

In (3.26), τ_s^0 denotes parallel transport from x_0 to x_s, and τ_0^s is its inverse. Let $P^t_{x,y}$ be the probability law of the Brownian motion starting at x and conditioned to be y at time t. An application of the results mentioned above shows that

$$P_t(x,y) = p_t(x,y)E^{P^t_{x,y}}\left[\exp\left(-\int_0^t S(x_s)ds/8\right)V_t\tau_0^t\right]. \tag{3.27}$$

In (3.27), $E^{P^t_{x,y}}$ is just expectation with respect to $P^t_{x,y}$.

3.6 The index as a well-defined path integral

For $t > 0$, let V^t_s be the solution of (3.26), in which R^E is replaced by tR^E. Recall that the positive measure μ_t on L^0X was defined in Subsection 2.2. Let ν_t be the measure on L^0X,

$$d\nu_t = \exp\left(-\int_0^1 tS(x_s)ds/8\right)\mathrm{Tr_s}[V_1^t\tau_0^1]d\mu_t. \tag{3.28}$$

Proposition 3.3. *For any $t > 0$,*

$$\mathrm{Ind}(D^X_+) = \mathrm{Tr_s}[\exp(-tD^{X,2}/2)] = \int_{L^0X} d\nu_t. \tag{3.29}$$

Proof. Equation (3.29) follows from (3.13), (3.19) and (3.27).

Observe that by combining (2.19), (3.28) and (3.29), we get the formal equality,

$$\mathrm{Tr_s}[\exp(-tD^{X,2}/2)]$$
$$= \frac{C^n}{(2\pi)^{n/2}}\int_{LX}\exp\left(-\int_{S^1}|\dot{x}|^2/2t\right)\exp\left(-\int_0^1 tS(x_s)ds/8\right)\mathrm{Tr_s}[V_1^t\tau_0^1]\frac{\mathcal{D}x}{t^{\infty/2}}. \tag{3.30}$$

The remarkable fact about (3.29) is that the mass of ν_t is an integer, which remains constant with $t > 0$, and more generally does not depend on the metrics or connections that were used in its definition.

From a measure-theoretic point of view, the natural example one can give of a measure whose mass remains constant is precisely the heat kernel $p_t(x,y)$, which is such that

$$\int_X p_t(x,y)dy = 1. \tag{3.31}$$

One could think that (3.29) could just be a consequence of a heat equation on L^0X, with dv_t representing some sort of heat kernel on L^0X. Such a point of view is untenable for many reasons on which we will not elaborate.

The remark of Atiyah and Witten in [A85] is that the constancy of (3.29), (3.30) can be given a cohomological interpretation, at least at a formal level. This is what we will explain next.

3.7 The loop space and the action of S^1

Recall that LX is the smooth loop space of X. We will view LX as a smooth manifold, disregarding the technicalities. If $x. \in LX$, the tangent space $T_{x.}LX$ can be identified with smooth periodic sections of TX along $x.$. If $U, V \in T_{x.}LX$, set

$$\langle U, V\rangle = \int_{S^1} \langle U_s, V_s\rangle ds. \tag{3.32}$$

Then (3.32) defines a S^1-invariant metric g^{TLX} on TLX.

The smooth action of S^1 on LX is generated by the vector field K, which is given by

$$K(x.) = \dot{x}.. \tag{3.33}$$

The vector field K is a Killing vector field. Its zero set LX_K, which is the fixed point set of the action of S^1, is just given by

$$LX_K = X. \tag{3.34}$$

We will now use in this infinite-dimensional context the notation of Section 1. Note that K', the 1-form dual to K, is such that if $U \in TLX$,

$$K'(U) = \int_{S^1} \langle U, dx\rangle. \tag{3.35}$$

Let $\frac{D}{Ds}$ denote the covariant derivative operator acting on TLX. This operator is antisymmetric. One verifies easily that if $U, V \in TLX$,

$$d^{LX}K'(U, V) = 2\int_{S^1} \left\langle \frac{D}{Ds}U, V\right\rangle ds. \tag{3.36}$$

Using the conventions in (1.10), we find that

$$\omega_{\frac{D}{Ds}} = -\frac{d^{LX}K'}{2}. \tag{3.37}$$

Observe that

$$|K|^2 = \int_{S^1} |\dot{x}|^2 ds. \tag{3.38}$$

If V is an even-dimensional Euclidean oriented vector space, if $A \in \mathrm{End}(V)$ is antisymmetric, the Pfaffian $\mathrm{Pf}[A]$ was defined in (1.11). By (1.12), and by (1.13), $\mathrm{Pf}[A]$ is a square root of $\det[A]$.

Since X is oriented, the parallel transport τ_0^1 on TX along a loop $x.$ is an oriented isometry. Therefore its eigenvalues are given by $e^{\pm i\theta_j}$, $1 \leq j \leq n/2$, with $0 \leq \theta_j \leq \pi$. The eigenvalues of $\frac{D}{Ds}$ are given by

$$2i\pi k \pm i\theta_j,\ k \in \mathbf{Z}, 1 \leq j \leq n/2. \tag{3.39}$$

Let C' be the infinite constant,

$$C' = \prod_{k=1}^{+\infty} (4k^2\pi^2). \tag{3.40}$$

By (3.39), we can write formally,

$$\det\left[\frac{D}{Ds}\right] = \left[\prod_{j=1}^{n/2} C'\theta_j \prod_{k=1}^{+\infty}\left(1 - \frac{\theta_j^2}{4k^2\pi^2}\right)\right]^2. \tag{3.41}$$

Equation (3.41) can also be rewritten in the form

$$\det\left[\frac{D}{Ds}\right] = C'^n \left[\prod_{j=1}^{n/2} 2\sin(\theta_j/2)\right]^2. \tag{3.42}$$

Now we try to make sense of the Pfaffian $\mathrm{Pf}\left[\frac{D}{Ds}\right]$. For this Pfaffian to be defined, first we need LX to be even dimensional, at least formally. However, the nonzero eigenvalues of $\frac{D}{Ds}$ come by conjugate pairs. As to the zero eigenvalue, it corresponds to the eigenvalue 1 of the parallel transport operator τ_1^0 acting on TX. Since X is oriented, this space is even dimensional, which ultimately proves that LX is indeed formally even dimensional. If X was instead odd dimensional, a similar argument would show LX to be formally odd dimensional.

To define $\mathrm{Pf}\left[\frac{D}{Ds}\right]$, a second requirement is that LX should be an oriented manifold. As explained by Atiyah in [A85], it is equivalent to require $\det\left[\frac{D}{Ds}\right]$ to have a smooth square root, which will then be precisely the Pfaffian $\mathrm{Pf}\left[\frac{D}{Ds}\right]$. In any case, if $\mathrm{Pf}\left[\frac{D}{Ds}\right]$ can we defined, from (3.42), we will get

$$\mathrm{Pf}\left[\frac{D}{Ds}\right] = \pm C'^{n/2} \prod_{j=1}^{n/2} 2\sin(\theta_j/2). \tag{3.43}$$

Let $g \in \mathrm{SO}(V)$, let $\pm\theta_j$, $1 \leq j \leq n/2$ be the angles of g. Then g has two lifts in $\mathrm{Spin}(V)$, which differ by a sign. Let $g_\dagger$ be one of these lifts. Then $g_\dagger$ acts on $S^V = S_+^V \oplus S_-^V$. We have the identity,

$$\mathrm{Tr_s}^{S^V}[g_\dagger^{-1}] = \pm \prod_{i=1}^{n/2} (e^{i\theta_j/2} - e^{-i\theta_j/2}) = \pm i^{n/2} \prod_{j=1}^{n/2} 2\sin(\theta_j/2). \tag{3.44}$$

The sign in (3.44) is fixed by the fact that if $A \in \mathrm{so}(V)$ has angles $\pm\theta_j$ with respect to an oriented orthonormal basis of V, if g exponentiates A in $\mathrm{SO}(V)$, and $g_\dagger$ exponentiates A in $\mathrm{Spin}(V)$, the sign in (3.44) is $+$.

By comparing (3.42) and (3.44), and keeping in mind that $\mathrm{Pf}\left[\frac{D}{Ds}\right]$ should be a square root of $\det\left[\frac{D}{Ds}\right]$, we reach the conclusion that

$$\mathrm{Tr_s}^{S^{TX}}[\tau_0^1] = C'^{-n/2} i^{n/2} \mathrm{Pf}\left[\frac{D}{Ds}\right]. \tag{3.45}$$

Equation (3.45) fixes the choice of the smooth square root $\mathrm{Pf}\left[\frac{D}{Ds}\right]$ of $\det\left[\frac{D}{Ds}\right]$, and proves the orientability of LX when X is spin.

There is nothing truly exotic in (3.42) or (3.45). Indeed $\det\left[\frac{D}{Ds}\right]$ can be defined by a classical zeta function technique. Equation (3.42) is just a version of the Cheeger–Müller theorem [Ch79, M78] establishing the equality of analytic and Ray–Singer torsion.

3.8 The remark of Atiyah and Witten for the Dirac operator on spinors

We still follow [A85]. We take E to be the trivial bundle $\mathbf{R}$. Also we decide to ignore the scalar curvature in equation (3.30).

Finally, we will use equation (1.12) in infinite dimensions. Namely, we will write

$$\mathrm{Pf}\left[\frac{D}{Ds}\right] \frac{\mathcal{D}x}{t^{\infty/2}} = [\exp(\omega_{\frac{D}{Ds}}/t)]^{\max}. \tag{3.46}$$

Identity (3.46) is questionable. Indeed (1.12) just evaluates the component of $\exp(\omega_A)$ of top finite degree. In (3.30), the expansion (3.46) is infinite, the component of infinite degree is not there. It is only by a renormalization procedure of the type outlined in (3.39)–(3.45) that we can make sense of (3.46).

In view of (3.37), we can rewrite (3.46) in the form

$$\mathrm{Pf}\left[\frac{D}{Ds}\right] \frac{\mathcal{D}x}{t^{\infty/2}} = [\exp(-d^{LX}K'/2t)]^{\max}. \tag{3.47}$$

Now we take equation (3.30) in which we ignore the term containing the scalar curvature, and in which we make $V_1^t = 1$, because E is trivial. Using (3.38), (3.45), and (3.47), and comparing with the definition of α_t given in (1.20), we can rewrite (3.30) in the form

$$\operatorname{Tr_s}[\exp(-tD^{X,2}/2)] = \left(\frac{C}{\sqrt{C'}}\right)^n (-2i\pi)^{-n/2} \int_{LX} \alpha_t. \tag{3.48}$$

Note that

$$C/\sqrt{C'} = \prod_{k=1}^{+\infty} k. \tag{3.49}$$

Also by (1.21),

$$d_K^{LX}\alpha_t = 0. \tag{3.50}$$

We now proceed as Atiyah [A85]. Namely, we will use the localization formula of Duistermaat–Heckman in Theorem 1.3, forgetting about the fact that LX is infinite dimensional and noncompact. Since $LX_K = X$, by (1.22) and (3.48), we should get

$$\operatorname{Tr_s}[\exp(-tD^{X,2}/2)] = (-2i\pi)^{-n/2}(C/\sqrt{C'})^n \int_X \frac{1}{e_K(N_{X/LX})}. \tag{3.51}$$

We will evaluate the right-hand side of (3.51). Indeed $N_{X/LX}$ is exactly the set of fibre maps $s \in S^1 \to f_s \in TX$ such that $\int_{S^1} f ds = 0$. The action of J_K on $N_{X/LX}$ is just $\frac{d}{ds}$. The Fourier series decomposition of f corresponds to the eigenvalue decomposition of f with respect to J_K. If R^{TX} still denotes the curvature of ∇^{TX}, we find easily that

$$\begin{aligned} e_K(N_{X/LX}, \nabla^{N_{X/LX}}) &= \left(\prod_{k=1}^{+\infty} k\right)^n \prod_{k=1}^{+\infty} \det{}^{1/2}\left[1 + \frac{R^{TX,2}}{4k^2\pi^2}\right] \\ &= \left(\prod_{k=1}^{+\infty} k\right)^n \widehat{A}^{-1}[R^{TX}]. \end{aligned} \tag{3.52}$$

By (3.48)–(3.52), we get

$$\operatorname{Tr_s}[\exp(-tD^{X,2}/2)] = \int_X \widehat{A}(TX, \nabla^{TX}). \tag{3.53}$$

Comparing with (3.13) and (3.25), we obtain the stunning fact that the above formal arguments predict the Atiyah–Singer index theorem without analysis.

3.9 An extension to general Dirac operators

In [B85], we extended the remark of Atiyah to the case of a nontrivial twisting bundle (E, g^E, ∇^E). Without elaborating on the method, let us just say that (3.48) takes now the form

$$\mathrm{Tr_s}[\exp(-tD^{X,2}/2)] = \left(\frac{C}{\sqrt{C'}}\right)^n (-2i\pi)^{-n/2} \int_{LX} \alpha_t \wedge c. \tag{3.54}$$

We will briefly describe the construction of c.

Set $m = \dim E$. Then E is associated with a $\mathrm{U}(m)$ principal fibre bundle P. Let LE be the loop space of E over LX. More precisely given $x \in LX$, the fibre of LE is the vector space of smooth sections of E parametrized by S^1 over the given loop x. Then LE is associated with a $L\mathrm{U}(m)$ fibre bundle LP over LX. The metric g^E and connection ∇^E induce an obvious L_2 metric on LE, and a unitary connection ∇^{LE}.

The action of S^1 on LX lifts to LE, and this action preserves the metric and the connection. However, it does not induce a bundle automorphism of the fibre bundle LP. Indeed for $p \in P$, $g \in LU(m)$, if $s \in S^1$, we have the identity

$$k_s(pg) = k_s p k_s g. \tag{3.55}$$

Having a bundle automorphism would require that $k_s g \in L\mathrm{U}(m)$ would simply be g. This defect can be easily be cured by introducing instead the semidirect product $\widehat{L}\mathrm{U}(m) = L\mathrm{U}(m) \ltimes S^1$. The Lie algebra of $\widehat{L}\mathrm{U}(m)$ is the Lie algebra of first order differential operators on S^1 of the type $a\frac{d}{ds} + B$, $B \in L\mathrm{u}(m)$. When $a = 1$, these are the $\mathrm{u}(m)$ connection forms over the trivial $\mathrm{U}(m)$-bundle.

The point is now to construct a d_K^{LX}-closed characteristic form on LX, which is associated with the vector bundle (LE, ∇^{LE}). Let ω^E be the connection form on P which is associated with ∇^E, and let Ω^E be its curvature. Given $p \in LP$, one verifies easily that if Ω_K^{LE} is the K-equivariant curvature of ∇^{LE}, then

$$\Omega_K^{LE} = \frac{d}{ds} + \omega^E\left(\frac{dp}{ds}\right) + \Omega^E. \tag{3.56}$$

Then Ω_K^{LE} is the sum of a 0-form and of a 2-form with values in $\widehat{L}\mathrm{u}(m)$.

To obtain a characteristic form, one needs to pick a gauge invariant analytic function. Now given an operator of the type $\frac{d}{ds} + B$, $B \in L\mathrm{u}(m)$, an obvious gauge invariant analytic function is the trace of the inverse of its monodromy. Namely, one solves the differential equation over S^1,

$$\left(\frac{d}{ds} + B\right) g = 0, \qquad g_0 = 1. \tag{3.57}$$

Put

$$c = \mathrm{Tr}[g_1^{-1}]. \tag{3.58}$$

Then c is gauge invariant function. Equivalently, we may consider the equation

$$\frac{d}{ds}h - hB = 0, \qquad h_0 = 1. \tag{3.59}$$

Then

$$c = \mathrm{Tr}[h]. \tag{3.60}$$

In view of (3.56) and (3.57), we should consider the differential equation

$$\left(\frac{d}{ds} + \omega^E\left(\frac{dp}{ds}\right) + \Omega^E\right) g = 0, \qquad g_0 = 1. \tag{3.61}$$

Put

$$c = \mathrm{Tr}[g_1^{-1}]. \tag{3.62}$$

Let τ_s^0 denote parallel transport on E from x_0 to x_s along the path $x.$ with respect to the connection ∇^E, and let τ_0^s be its inverse. If one prefers the description of c in (3.59), (3.60), one should instead consider the differential equation

$$\frac{d}{ds} W = W \tau_0^s R^E \tau_s^0, \qquad W_0 = 1. \tag{3.63}$$

Then

$$c = \mathrm{Tr}[W_1 \tau_0^1]. \tag{3.64}$$

It is this c which appears in the formal formula (3.54) for $\mathrm{Tr_s}[\exp(-tD^{X,2}/2)]$.

By construction,

$$d_K^{LX} c = 0. \tag{3.65}$$

By (3.50) and (3.65), we get

$$d_K^{LX}(\alpha_t \wedge c) = 0. \tag{3.66}$$

Let i be the embedding $X \to LX$. Note that by (3.63), (3.64), we get

$$i^* c = \mathrm{Tr}[\exp(R^E)], \tag{3.67}$$

i.e., up to normalization, i^*c coincides with $\mathrm{ch}(E, \nabla^E)$.

A formal application of the Duistermaat–Heckman localization formula to (3.54) leads to the identity,

$$\mathrm{Tr_s}[\exp(-tD^{X,2}/2)] = \int_X \widehat{A}(TX)\mathrm{ch}(E), \tag{3.68}$$

which is again correct.

3.10 The fantastic cancellations and the localization formulas

What is truly remarkable in the above formal arguments is that they lead to the Atiyah–Singer index formula without taking $t \to 0$, as if algebra was taking over analysis.

From this point of view, we tried very hard to import in infinite dimensions any of the proofs existing at the time of the localization formulas, until we realized that the form α_t was by itself a perfectly legitimate way of proving the localization formulas also in finite dimensions. The formal argument of Atiyah was good enough not only to predict in a formal way the right index formula, but also was producing the remarkable form α_t, which is the key to the proof of the localization formula given in Theorem 1.3.

But the consequences are much vaster. Indeed, as we saw in the proof of Theorems 1.3 and 1.5, in a finite-dimensional context, as $t \to 0$, the current α_t on X converges to an explicit current α_0 localized on X_K, the existence of the limit not being entirely trivial.

From [B86b] and from the above arguments, one finds that the fantastic cancellations anticipated by McKean–Singer, which were described in Subsection 3.4, are just the infinite dimensional manifestation of the fact that also in infinite dimensions α_t should converge as a current to α_0. Understanding why this formal convergence should imply the fantastic cancellations is an easy matter left to the inspired reader.

From the above, we find that the localization formulas have provided a geometric explanation to an important analytic fact, the fantastic cancellations of McKean–Singer. Our project of importing to infinite dimensions any known proof of the localization formulas has been fulfilled tautologically, by exporting instead the heat equation method to finite dimensions. We will explore the consequences in the next section.

3.11 Formal versus rigorous arguments

Applying the localization formulas to the right-hand side of (3.54) is questionable, since we do not even know what form is ultimately integrated on LX. This form should again be the component of infinite degree of the wedge product of α_t and c, which are themselves series of well-defined forms of finite degree.

However, let us point that part of the difficulty is intrinsically related to infinite dimensions. Indeed let us assume that $X_k, k \in \mathbf{N}$ is a family of manifolds having exactly the same properties as X in Subsection 1.1. In particular they are equipped with an action of a torus T. Set

$$\mathcal{X} = \prod_1^{+\infty} X_k. \tag{3.69}$$

Then $\mathcal{X}$ is also equipped with an action of the torus T. Let $p_k : \mathcal{X} \to X_k$ be the obvious projection. For $k \in \mathbf{N}$, let μ_k be a $d_K^{X_k}$-closed form on X_k. Set

$$\mu = \prod_1^{+\infty} p_k^* \mu_k. \tag{3.70}$$

Note that the product (3.70) converges only if the product of the components of degree 0 of the μ_k converges.

For simplicity assume that for any $k \in \mathbf{N}$, $\mu_k^{\max}$ is nonnegative, and also that by an application of Theorem 1.3, we would get

$$\int_{X_k} \mu_k = 1. \tag{3.71}$$

Note that $m = \prod_1^{+\infty} \mu_k^{\max}$ is a well-defined positive measure on $\mathcal{X}$, and that

$$\int_{\mathcal{X}} dm = 1. \tag{3.72}$$

One can apply the localization formula of Theorem 1.3 to each μ_k. However, using this formula on the integral of μ over $\mathcal{X}$ is very difficult. Indeed if $\alpha_{k,t}$ is the form α_t on X_k, then

$$\alpha_t = \prod_{k=1}^{+\infty} p_k^* \alpha_{k,t} \tag{3.73}$$

is in general not well defined, precisely because the product of the components of degree 0 may well be 0, which forces the vanishing of the form α_t.

As an example of the above situation, consider the case where $X_k = \mathbf{C} \simeq \mathbf{R}^2$ as in Remark 1.4, and that the action of S^1 is the one specified there. Take

$$\mu_k = \frac{1}{2\pi} \exp(-d_K^{\mathbf{R}^2} K'/2). \tag{3.74}$$

By (1.39),

$$\int_{\mathbf{R}^2} \mu_k = 1. \tag{3.75}$$

The product μ in (3.70) does not exist for two reasons. The first is the presence of the factor $1/2\pi$. But the second more fundamental reason is that the Gaussian measure on an infinite-dimensional Hilbert space gives measure 0 to this Hilbert space. Of course, this is another manifestation of the fact that there is no Lebesgue measure in infinite dimensions, so that measures tend to be mutually singular.

The above provides an elementary version of the difficulties we have of making sense of the right-hand side (3.54). If X is a torus, the difficulty is exactly the one we had in defining μ from the μ_k in (3.70). For an arbitrary X, the problems are much more severe, because one cannot give a natural finite-dimensional approximation of the situation. Ultimately, to make sense of the right-hand side of (3.54), we need the left-hand side.

3.12 Hamiltonian–Lagrangian correspondence and index theory

As we saw in (2.17) and in its formal counterpart (2.21), the trace of a scalar heat kernel can be represented as the integral of a S^1-invariant measure over L^0X or LX.

In equation (3.30), we saw how to represent the supertrace of $\exp(-tD^{X,2}/2)$ as the integral of a S^1-invariant measure on L^0X. In (3.54), we also gave a formal expression of the same quantity as the integral of a d_K^{LX}-closed differential form over LX, of which the previous measure is supposed to represent the term of top degree. Needless to say, having a d_K^{LX}-closed form is much stronger than having a S^1-invariant measure. The price one has paid is that the formula one obtains remains formal.

Still one can establish a dictionary between index theory and integration of differential forms on loop spaces. Indeed $D^{X,2}$ is itself the square of D^X. In equation (3.30), this fact has been forgotten. One of the points of (3.54) is to resurrect this fact in the path integral itself, that is in the Lagrangian, or Fourier transformed picture of the Hamiltonian left-hand side.

Indeed we will start writing some elements of a dictionary between quantities involving supertraces of heat operators of the type $\exp(-tD^{X,2}/2)$ and integrals over LX of differential forms.

Table 3.1 The Hamiltonian–Lagrangian correspondence

Supertraces of heat operators	Integrals of differential forms on LX
$\mathrm{Tr_s}$	$\int_{LX}$
D^X	d_K^{LX}
$D^{X,2}$	L_K
$\mathrm{Tr_s}[[D^X, A]] = 0$	$\int_{LX} d_K^{LX}\mu = 0$
Local index theorem	Local Duistermaat–Heckman formula

The fact that equations (3.17) and (1.25) correspond to each other under the above dictionary reflects the amazing formal analogies in the proofs of the corresponding results.

A final question we would like to address is whether the fact that the formula (3.54) remains formal should necessitate further work to make integration of forms on loop spaces and localization formulas on such spaces rigorous. From the perspective of the present paper, the answer is left to the reader[1].

4 From localization formulas to Hermitian K-theory

In Section 1, and more specifically in Subsections 1.2, 1.4, and 1.6, we showed that the localization formulas are compatible to natural functorial operations.

[1] The author's point of view on this question is somewhat ambiguous. Indeed one could consider that the fact that many of the consequences of the formal theory have ultimately proved to be correct, and that new rigorous objects have been constructed as a consequence, represents the only way of making the formal theory rigorous.

In Section 3, we exhibited a correspondence that maps the index theory for the Dirac operator on a manifold to equivariant integration on the loop space LX. We also called this correspondence a Hamiltonian–Lagrangian correspondence, which we assimilated to a Fourier transform.

The index theorem is part of a more general theory involving the K-theory functor, the case of the index being the simplest case where the target space is just a point.

We will now ask whether the above correspondence can be extended to the possible functorial operations in both theories. Given the work which has been done in the field, the answer is certainly positive. At the time, the main difficulty was essentially that the objects involved in the correspondence were not even defined. Still the possibility that such a correspondence could possibly exist was one motivation for the discovery of the rigorous objects on the K-theory side, which would correspond to their natural counterparts on the localization side. Also if one accepts the fact that this is some sort of Fourier correspondence, part of the behaviour of the rigorous objects to be was already encoded in their Fourier counterpart.

We will give here mostly facts, and little justification.

This section is organized as follows. In Subsection 4.1, we show that in the Hamilton–Lagrangian correspondence, the families index theorem corresponds to equivariant integration along the fibre. Quillen's superconnections are needed in the Hamiltonian side to make sense of the correspondence.

In Subsection 4.2, we refine the Hamiltonian–Lagrangian correspondence to the local forms of the corresponding formulas. In this context, we show that the Getzler operator [Ge86] in local index theory is the curvature of a natural superconnection.

In Subsection 4.3, we recall the construction of the $\widetilde{\eta}$-forms in local families index theory, and we show that they are formally related to the currents ε and ϑ of Subsections 1.3 and 1.5.

In Subsection 4.4, we introduce the holomorphic torsion forms T, which themselves refine on the $\widetilde{\eta}$-forms, and we exhibit their relation to the forms τ of Subsection 1.8.

In Subsection 4.5, we discuss the extension of the above formalism to Lefschetz fixed point formulas.

Finally, in Subsection 4.6, we briefly consider the hypoelliptic Laplacian of [B05, B08] from the point of view of the present paper.

4.1 Families index theorem and equivariant integration along the fibre

In Subsection 1.4, we gave an integration along the fibre version of localization formulas.

Clearly, the K-theoretic version of integration along the fibre should be the families index theorem of Atiyah–Singer [AS71].

Let $\pi : M \to S$ be a submersion of smooth manifolds with compact oriented fibre X of even dimension n. We assume the vector bundle $TX = TM/S$ to be spin. Let g^{TX} be a metric on TX. Let (E, g^E, ∇^E) be a Hermitian vector bundle on M with unitary connection.

Each fibre X carries a fibrewise Dirac operator D^X. Set

$$\mathrm{Ind}(D_+^X) = \ker D_+^X - \ker D_-^X. \tag{4.1}$$

Note that $E \in K(M)$, and that the left-hand side of (4.1) is an element of $K(S)$. The right-hand side is a well-defined element of $K(S)$ only if the dimensions of $\ker D_+^X$ remains constant, in which case (4.1) is a definition of the left-hand side. In the general case, one can deform the operator D^X so that the assumptions we just made are correct.

Note here that by definition, the map $\pi_! : E \in K(M) \to \mathrm{Ind}(D_+^X) \in K(S)$ is part of the K-theory functor which we mentioned before.

The Chern character maps K-theory to rational even cohomology. The Atiyah–Singer families index theorem [AS71] asserts that

$$\mathrm{ch}(\pi_!(E)) = \pi_*[\widehat{A}(TX)\mathrm{ch}(E)] \text{ in } H^{\mathrm{even}}(S). \tag{4.2}$$

Equation (4.2) is compatible with the obvious functorial operations.

If the base is a point, then (4.2) is just the index formula in (3.25). So it is legitimate to ask whether the considerations of Section 3 can be extended to this more general situation.

We will use the loop space functor $M \to LM$. The map $\pi : M \to S$ induces an equivariant map $\pi : LM \to LS$. This new projection verifies precisely the assumptions in Subsection 1.2, where M, S should be replaced by LM, LS. Note that $LS_K = S$, and that $LM^K = \pi^{-1}S$ is the set of smooth loops in M which project to points in S, and the fibres are LX, the loop space of a given fibre X.

One can anticipate that (4.2) should be related to the integration along the fibre result of Theorem 1.10, when applied to the fibration $\pi : LM^K \to S$. This is especially true in light of the considerations of Subsection 1.6, where functoriality results for localization formulas under projections have been considered.

Let T^HM be a horizontal vector bundle on M, so that $TM = T^HM \oplus TX$. Then T^HM lifts to a horizontal vector bundle T^HLM^K on LM^K. Moreover, the metric g^{TX} induces a S^1-invariant metric g^{TLX} on TLX. Then we are precisely under the assumptions of Subsection 1.4. Let α_t be the $d_K^{LM^K}$-closed form on LM^K which one obtains as in that subsection. More precisely, let K' be the 1-form on LM^K which vanishes on T^HLM^K and coincides with the dual form to K along TLX. The form α_t is still given by (1.20), i.e.,

$$\alpha_t = \exp(-d_K^{LM^K} K'/2t). \tag{4.3}$$

Also over LM, we define the form c attached to (E, g^E, ∇^E) as in Subsection 3.9.

Let φ be the endomorphism of $\Omega^{\mathrm{even}}(S)$ given by $\alpha \to (-2i\pi)^{\deg(\alpha)/2}\alpha$. For $t > 0$, consider the form a_t on S,

$$a_t = \varphi(-2i\pi)^{-n/2}(C/\sqrt{C'})^n \int_{LX} \alpha_t \wedge c. \tag{4.4}$$

The integral in (4.4) is still formal. However, in degree 0, it is just the one in (3.54). Still let us pretend that it makes sense. If we could use (1.60), (1.62) as well as the considerations in Remark 1.11 and in (3.52) , then we would know the form a_t is closed on S, that its cohomology class does not depend on $t > 0$, and moreover that as $t \to 0$, a_t converges smoothly to a_0 given by

$$a_0 = \int_X \widehat{A}(TX, \nabla^{TX})\mathrm{ch}(E, \nabla^E). \tag{4.5}$$

In (4.5), ∇^{TX} is a Euclidean connection on TX, which is canonically attached to $(T^H M, g^{TX})$. However, a_0 is precisely a canonical representative of the right-hand side of (4.2). In degree 0, as we saw in (3.15), the behaviour of a_t (which is constant...) as $t \to +\infty$ is very easy to determine in the Hamiltonian picture. In the Lagrangian picture (4.4), whatever information one can get as $t \to +\infty$ is misleading for reasons already outlined in Subsection 2.2.

If we could determine the forms a_t rigorously in the operator-theoretic (or Hamiltonian) picture, there is a chance that these forms would provide also a proof of (4.2), simply because they would interpolate between the form a_0 in (4.5) as $t \to 0$, which is a natural representative of the right-hand side of (4.2), and whatever natural representative there is of the left-hand side as $t \to +\infty$, while remaining constant in cohomology.

The construction of the corresponding operator-theoretic object is obtained via Quillen's theory of superconnections [Q85a]. We will motivate the introduction of superconnections by discussing again adiabatic limits along the lines of Subsection 1.6. We will adopt the terminology of that subsection, except that of course M there should be LM, M_K should be M... Let g^{TM}, g^{TS} be Riemannian metrics on TM, TS. For $\epsilon > 0$, set

$$g_\epsilon^{TM} = g^{TM} + \frac{1}{\epsilon}\pi^* g^{TS}. \tag{4.6}$$

Here we will take $T^H M$ to be the orthogonal bundle to TX with respect to g^{TM}. Note that $T^H M$ is still the orthogonal bundle to TX with respect to g_ϵ^{TM}. Making $\epsilon \to 0$ is still called passing to the adiabatic limit.

As we saw in Subsection 3.7, the metrics g^{TM}, g_ϵ^{TM} induce corresponding S^1-invariant metrics on TLM, and g^{TS} a S^1-invariant metric on TLS. By (4.6), we get

$$g_\epsilon^{TLM} = g^{TLM} + \frac{1}{\epsilon} g^{TLS}, \tag{4.7}$$

which is just (1.67) in the present context.

In Subsection 1.6, we explained how integration along the fibre localization formulas could be related to adiabatic limits of ordinary localization formulas.

To guess what should be the Hamiltonian counterpart to the form a_t, we will also try adiabatic limits in the Hamiltonian formulation.

The metric g^{TX} induces a corresponding symmetric bilinear form on TM whose kernel is $T^H M$. To make our arguments simpler, instead of (4.7), we will now set

$$g_\epsilon^{TLM} = g^{TX} + \frac{1}{\epsilon}\pi^* g^{TS}. \tag{4.8}$$

The point about the metric g_ϵ^{TM} is that π is a Riemannian submersion with respect to the metric g_ϵ^{TM}.

We will also assume that TS is oriented, spin and even dimensional. Let $S^{TS} = S_+^{TS} \oplus S_-^{TS}$ be the associated bundle of (TS, g^{TS}) spinors. Set

$$S^{TM} = \pi^* S^{TS} \widehat{\otimes} S^{TX}. \tag{4.9}$$

Then S^{TM} can be identified with the vector bundle of (TM, g_ϵ^{TM}) spinors.

Let D_ϵ^M be the Dirac operator acting on $C^\infty(M, S^{TM} \otimes E)$, which is associated with g_ϵ^{TM} and ∇^E. First we will give an explicit formula for D_ϵ^M.

Let $\nabla^{S^{TX}}$ be the connection on S^{TX} induced by the canonical connection ∇^{TX}. Let ∇^{TS} be the Levi-Civita connection on (TS, g^{TS}), and let $\nabla^{S^{TS}}$ be the corresponding connection on S^{TS}. Let $\nabla^{S^{TM}}$ be the connection on S^{TM} which is induced by these two connections.

Let T^H be the curvature of the Diff(X) connection associated with $T^H M$. Let P^{TX} be the projection $TM = T^H M \oplus TX \to TX$. If $U \in TS$, let $U^H \in T^H M$ be the horizontal lift of U. If $U, V \in TS$,

$$T^H(U, V) = -P^{TX}[U^H, V^H]. \tag{4.10}$$

Let $f_1, \dots, f_m$ be an orthonormal basis of TS, and let $f^1, \dots, f^m$ be the corresponding dual basis of T^*S. Set

$$D^H = \sum_{i=1}^m c(e_i) \nabla_{e_i^H}^{S^{TM} \otimes E, u}, \tag{4.11}$$

The upper index u on the right-hand side of indicates that we have taken into account the variation of the volume form of X with respect to horizontal differentiation, so as to make the operator D^H self-adjoint. Then D^H is a horizontal Dirac operator. Put

$$c(T^H) = \frac{1}{2} \sum_{1 \le \alpha, \beta \le m} c(f_\alpha) c(f_\beta) c(T^H(f_\alpha^H, f_\beta^H)). \tag{4.12}$$

By [BC89, eq. (4.26)], D_ϵ^M can be written in the form

$$D_\epsilon^M = \sqrt{\epsilon} D^H + D^X - \epsilon \frac{c(T^H)}{4}. \tag{4.13}$$

The structure of (4.13) is interesting. In particular the third term on the right-hand side of (4.13) is intimately related with Leray's spectral sequence.

As $\epsilon \to 0$, from a simple minded point of view, the operator D_{ϵ}^{M} converges to D^{X}. However, in the case where M is product, i.e., $M = S \times X$, one is immediately tempted to do index theory over the base S.

The most direct way of what should be done is to use a Getzler rescaling [Ge86] on (4.13) as $\epsilon \to 0$. However, let us give here a shortcut. The reader will have noticed that to obtain the contribution of E to the index formula, in Lichnerowicz formula for $tD^{X,2}$ in (3.10), we just have to replace the Clifford variables $\sqrt{t}c(e_i)$ by the exterior variables $e^i\wedge$. So let us just do this in (4.13), while replacing t by ϵ.

Let Ω be the $\mathbf{Z}_2$-graded vector bundle on S of smooth sections of $S^{TX} \otimes E$ along the fibre X. If s is a smooth section of Ω on S, if $U \in TS$, put

$$\nabla_U^{\Omega} s = \nabla_{U^H}^{S^{TX}\otimes E,u} s. \tag{4.14}$$

Then equation (4.14) defines a unitary connection on Ω, whose curvature is a 2-form on S with values in first order differential operators along the fibre X.

We take $f_1, \ldots, f_m$ as before except that we do not assume any more this basis to be orthogonal. Then

$$\nabla^{\Omega} = \sum_{1\leq\alpha\leq m} f^{\alpha}\nabla_{f_{\alpha}}^{\Omega}. \tag{4.15}$$

Set

$$c^X(T^H) = \frac{1}{2}\sum_{1\leq\alpha,\beta\leq m} f^{\alpha} f^{\beta} c(T^H(f_{\alpha}, f_{\beta})). \tag{4.16}$$

The object we obtain from D_{ϵ}^{M} by the above shortcut is given by the formula

$$A = \nabla^{\Omega} + D^X - \frac{c^X(T^H)}{4}. \tag{4.17}$$

This object does not depend any more on the metric g^{TS}. It turns out that A is precisely a superconnection on the $\mathbf{Z}_2$-graded vector bundle Ω in the sense of Quillen [Q85a], and is called the Levi-Civita superconnection [B86a]. When replacing g^{TX} by g^{TX}/t, the corresponding object A_t is given by

$$A_t = \nabla^{\Omega} + \sqrt{t}D^X - \frac{c^X(T^H)}{4\sqrt{t}}. \tag{4.18}$$

We will not explain in detail Quillen's theory of superconnections. Let us just mention it is an extension of Chern–Weil theory. In the same way as the curvature of a connection can be viewed as the square of the connection, the curvature of the superconnection A_t is its square A_t^2. Here it is a second order elliptic operator acting along the fibre on $\Lambda^{\cdot}(T^*S)\widehat{\otimes}\Omega$.

Put

$$a_t = \varphi \mathrm{Tr_s}[\exp(-A_{t/2}^2)]. \tag{4.19}$$

We have deliberately kept the same notation in (4.4) and (4.19), since a_t in (4.19) is just the rigorous Hamiltonian version of (4.4). Then a_t is a smooth even form on S.

Let a_0 be the real even closed form on S given by

$$a_0 = \int_X \widehat{A}(TX, \nabla^{TX})\mathrm{ch}(E, \nabla^E), \tag{4.20}$$

The following result was established in [B86a], and is called the local families index theorem.

Theorem 4.1. *The even forms a_t are real, closed, and their cohomology class does not depend on $t > 0$. More precisely,*

$$[a_t] = \mathrm{ch}(\pi_! E). \tag{4.21}$$

As $t \to 0$,

$$a_t = a_0 + \mathcal{O}(t). \tag{4.22}$$

From Theorem 4.1, we recover the families index theorem of Atiyah–Singer [AS71] in the form given in (4.2).

Let us note here that the local index theorem in (3.24) is a special case of the local families index theorem, or more precisely of its proof. However, as explained in [B98], the local families theorem can be viewed as the adiabatic limit of the local index theorem.

Assume that the dimension of $\ker D^X_{\pm}$ is locally constant, so that $\ker D^X$ is now a smooth $\mathbf{Z}_2$-graded vector bundle on S. Let $\nabla^{\ker D^X}$ be the orthogonal projection of the connection ∇^{Ω} on $\ker D^X$. Put

$$a_\infty = \mathrm{ch}(\ker D^X, \nabla^{\ker D^X}). \tag{4.23}$$

The following result was established by Berline and Vergne in [BeGeV92].

Theorem 4.2. *As $t \to +\infty$,*

$$a_t = a_\infty + \mathcal{O}(1/\sqrt{t}). \tag{4.24}$$

Ultimately, we have established the following extension of (3.54),

$$a_t = \varphi(-2i\pi)^{-n/2}(C/\sqrt{C'})^n \int_{LX} \alpha_t \wedge c. \tag{4.25}$$

Let us mention again that while the right-hand side of (4.25) predicts the behaviour of (4.25) as $t \to 0$, this right-hand side does not say anything on the behaviour of (4.25) as $t \to +\infty$. Again this is natural from a Fourier perspective.

However, we will disregard this inconsistency, and extend (4.25) formally to $t = +\infty$, so that

$$a_\infty = \varphi(-2i\pi)^{-n/2}(C/\sqrt{C'})^n \int_{LX} c. \tag{4.26}$$

Equation (4.26) has to be taken with care, even from a formal point of view. If the bundle E was trivial, it would say that $a_\infty = 0$, which is not the case in general. Again the discrepancy comes from the fact that LX is not compact. We have actually warned enough the reader on this question.

4.2 The local Getzler operator and superconnections

In Remark 1.4, we pointed out that in our proof of the localization formula in Theorem 1.3, equation (1.36) for $e_K^{-1}(N_{X_K/X}, \nabla^{N_{X_K/X}})$ is obtained as the integral along the fibre of a $d_{J_K Y}^{\mathcal{N}_{X_K/X}}$-closed form, demonstrating this way that some version of the integral along the fibre is already present in the proof of the standard localization formula.

If one admits that in infinite dimensions, such integrals along the fibre correspond to superconnections, this indicates that there should be a superconnection version of the local index theorem.

This we will briefly demonstrate, along the lines of [B90b]. We work temporarily under the assumptions of Subsection 3.2. Let $\mathcal{X}$ be the total space of TX, and let Y be the tautological section of the fibre TX on $\mathcal{X}$. Let $\mathcal{L}$ be the operator along the fibres of TX,

$$\mathcal{L} = -\frac{1}{2}\sum_{i=1}^{n}\left(\nabla_{e_i} + \frac{1}{2}\langle R^{TX}Y, e_i\rangle\right)^2. \tag{4.27}$$

In (4.27), the curvature R^{TX} is still viewed as a 2-form on X. The operator $\mathcal{L}$ was obtained by Getzler in [Ge86] in his proof of the local index theorem. Let $q(Y, Y')$ be the smooth kernel associated with the heat operator $\exp(-\mathcal{L})$. The critical fact is that $q(Y, Y)$ does not depend on Y, and also that

$$q(0,0) = \widehat{A}(R^{TX}), \tag{4.28}$$

which is the key formula leading to the proof of the local index theorem. Equation (4.28) follows from Mehler's formula. It is a version of Lévy's stochastic area formula [L51]. Indeed Paul Lévy proved that if $(x_\cdot, y_\cdot)$ is a 2-dimensional Brownian bridge starting at $(0, 0)$ at time 0 and ending at $(0, 0)$ at time 1, for $a \in \mathbf{R}$,

$$E\left[\exp\left(\frac{ia}{2}\int_0^1 (x dy - y dx)\right)\right] = \widehat{A}(a), \tag{4.29}$$

from which (4.28) easily follows.

The reader should not have any difficulty in proving that (4.28) is exactly the proper infinite-dimensional version of (1.36).

But more should be true. Indeed $\mathcal{L}$ should be the curvature of the Levi-Civita superconnection $\mathcal{A}_{1/2}$ associated with the projection $\pi : \mathcal{X} \to X$, the metric along the fibres g^{TX} and the horizontal vector bundle determined by the connection ∇^{TX}.

The fact that this turns out to be exactly the case was established in [B90b], by an easy explicit computation. We can then write

$$\widehat{A}(TX, \nabla^{TX}) = \exp(-\mathcal{A}_{1/2}^2)(0,0). \tag{4.30}$$

One of the merits of (4.30) is that it expresses $\widehat{A}(TX, \nabla^{TX})$ as a sort of Chern character form. Pursuing along these lines would take us too far.

4.3 $\widetilde{\eta}$ forms and the currents ε

The results of Subsection 1.5 gives us a method to transgress the forms a_t in the Lagrangian formalism. More precisely, we define the form β_t on LM^K as in (1.46). Set

$$b_t = \varphi \frac{1}{\sqrt{2i\pi}}(-2i\pi)^{-n/2}(C/\sqrt{C'})^n \int_{LX} \beta_t \wedge c. \tag{4.31}$$

Then the analogue of equation (1.63) is given by

$$\frac{\partial}{\partial t} a_t = d^S b_t. \tag{4.32}$$

In the same way, one constructs the current ε on LM^K as in (1.50). By extending (1.66) in infinite dimensions, we should get

$$d^S \frac{1}{\sqrt{2i\pi}} \varphi(-2i\pi)^{-n/2}(C/\sqrt{C'})^n \int_{LX} \varepsilon \wedge c = a_0 - a_\infty. \tag{4.33}$$

Of course (4.31) is still formal. On the Hamiltonian rigorous side, one should recall that Quillen's theory is an extension of Chern–Weil's theory. Therefore from the rigorous formula form a_t in (4.9), one can construct a rigorous odd form b_t such that (4.32) holds by an extension of the Chern–Simons transgression mechanism. A tautological argument shows that the rigorous and non rigorous versions of b_t correspond.

Now we introduce the $\widetilde{\eta}$ forms of Bismut–Cheeger [BC89]. Put

$$\widetilde{\eta} = -\int_0^{+\infty} b_t dt. \tag{4.34}$$

As the notation indicates, the integral converges on the right-hand side of (4.34). Indeed one can derive the convergence from the proper application of Theorems 4.1 and 4.2.

By (4.32), we get

$$d^S \widetilde{\eta} = a_0 - a_\infty. \tag{4.35}$$

By the above, we have the formal identity,

$$\widetilde{\eta} = \varphi \frac{1}{\sqrt{2i\pi}} (-2i\pi)^{-n/2} (C/\sqrt{C'})^n \int_{LX} \varepsilon \wedge \beta. \tag{4.36}$$

If we define the form ϑ as in (1.65) with μ replaced by c, then

$$\widetilde{\eta} = \varphi \frac{1}{\sqrt{2i\pi}} (-2i\pi)^{-n/2} (C/\sqrt{C'})^n \vartheta. \tag{4.37}$$

We gave all the proper warnings concerning the interpretation of (4.36). However, the consequences are quite interesting.

The forms $\widetilde{\eta}$ are secondary objects from the point of view of K-theory. They are related with differential characters of Cheeger–Simons [ChS85]. On the other hand, the right-hand side is expressed in terms of integrals of currents on the loop space.

The forms $\widetilde{\eta}$ have found many uses. In particular the component of degree 1 (constructed prior to the construction of the forms $\widetilde{\eta}$) of $\widetilde{\eta}$ was identified in [BF86a, BF86b] as a connection form on the determinant bundle $\det(\ker D^X)$. Equation (4.35) in degree 2 is exactly the curvature theorem of [BF86b] for such determinant bundles.

In the case where the fibres X are instead odd-dimensional, there is a corresponding families index theorem of Atiyah–Singer, where the family of self-adjoint operator D^X is viewed as defining an element of $K^1(S)$. Equation (4.2) is still formally true. If $\ker D^X$ is of locally constant dimension, one can define the forms $\widetilde{\eta}$, which are now even forms, as before, and they verify an analogue of (4.35), i.e.,

$$d^S \widetilde{\eta} = a_0. \tag{4.38}$$

The component of degree 0 of $\widetilde{\eta}$ is nothing else than the η-invariant of the fibres X in the sense of Atiyah–Patodi–Singer [APS75a, APS76], which is a spectral invariant of the operator D^X. Equation (4.38) reduces to the variation formula of [APS76] for the η-invariant.

On the other hand, as explained in Subsection 3.7, the fibres LX can now be viewed as odd-dimensional. Ultimately, an analogue of identity (4.36) still holds in this case. In particular the η-invariant of [APS76] can be expressed formally as the integral of a secondary current on the corresponding loop space.

The importance of a formula like (4.36) should not be underestimated. Indeed it expresses a 'natural' object from the point of view of K-theory as the integral of another 'natural' object in equivariant integration theory, that is as a geometric object.

In Subsection 1.6, we mentioned that the secondary currents ε are compatible to natural functorial operations. When transferred to their infinite dimensional version, the $\widetilde{\eta}$-forms, these are exactly the results obtained in Bismut–Cheeger [BC89].

Similarly there are compatibility results of the currents ε to embeddings. For corresponding results for η-invariants, we refer to [BZ93].

4.4 Holomorphic torsion forms and Bott–Chern currents

Let us now assume that M and S are complex manifolds, and that $\pi : M \to S$ is holomorphic with compact fibre X of real dimension n. Again $TM, TX = TM/S, TS$ denote the corresponding holomorphic tangent bundles. Let (E, g^E) be a holomorphic Hermitian vector bundle on M, and let ∇^E be the associated holomorphic Hermitian connection. Let $\mathrm{ch}(E, g^E)$ be the Chern character form of E which is associated with ∇^E.

Let $(\Omega^{(0,\cdot)}(X, E|_X), \overline{\partial}^X)$ be the Dolbeault complex along the fibres X. Let $\overline{\partial}^{X*}$ be the fibrewise adjoint of $\overline{\partial}^X$.

We will explain the construction of the analytic torsion forms in Bismut–Gillet–Soulé [BGSo88a] and Bismut–Köhler [BK92]. We will inspire ourselves from Subsections 1.7 and 1.8.

First we try to reproduce formally the geometric situation of Subsection 1.8, where of course we deal here with loop spaces.

Let ω^M be a real closed $(1,1)$ form on M, whose restriction ω^X to the compact fibres X induces a Kähler metric g^{TX} along the fibres. Let ∇^{TX} be the holomorphic Hermitian connection on TX which is associated with g^{TX}. Let $\mathrm{Td}(TX, g^{TX})$ be the Todd form of TX associated with∇^{TX}.

Let $T^H M \subset TM$ be the orthogonal bundle to TX with respect to ω^M. Let $\omega^{M,H}$ be the restriction of ω^M to $T^H_{\mathbf{R}} M$. We can write ω^M in the form

$$\omega^M = \omega^{M,H} + \omega^X. \tag{4.39}$$

In such a geometric situation, one can construct the strict analogue B_t of the superconnection $A_{t/2}$ in (4.18). The superconnection B_t on $\Omega^{(0,\cdot)}(X, E|_X)$ can be written in the form

$$B_t = \nabla^{\Omega^{(0,\cdot)}(X,E|_X)} + \sqrt{t}(\overline{\partial}^X + \partial^{X*}) - \frac{c^V(T^H)}{2\sqrt{2t}}. \tag{4.40}$$

Underlying the definition of B_t is the fact that the antiholomorphic exterior algebra $\Lambda^{\cdot}(\overline{T^*X})$ is a $c(T_{\mathbf{R}}X)$ Clifford module, but we take this for granted.

Put

$$a_t = \varphi \mathrm{Tr_s}[\exp(-B_t^2)]. \tag{4.41}$$

In [BGSo88a], it is shown that a_t is a real closed form on S which is a sum of (p,p) forms. Let a_0 be th form on S,

$$a_0 = \int_X \mathrm{Td}(TX, g^{TX})\mathrm{ch}(E, g^E). \tag{4.42}$$

Let $R^{\cdot}\pi_* E \in K(S)$ be the direct image of E by π. The obvious analogue of Theorem 4.1 holds, simply because the above situation is a special case of the situation considered in Theorem 4.1. Of course $\pi_! E$ should be replaced here by $R^{\cdot}\pi_* E$.

By [Hi74], X is spin if and only if the line bundle $\det(TX)$ has a square root. In the sequel, the reader may assume X to be spin. If this is not the case, LX is no longer formally orientable in the sense of Subsection 3.7. Still the path integrals which appear in the sequel can easily be made sense of.

Let c be the form on LM^K which is associated with the vector bundle $E \otimes \det^{1/2}(TX)$ and its corresponding connection. Then the analogue of (4.25) holds. The form ω^M induces a S^1-invariant closed $(1,1)$ form ω^{LM^K} on LM^K. The vector bundle $T^H M$ lifts to a vector bundle $T^H LM^K \subset TLM^K$, which is exactly the orthogonal vector bundle to TLX in TLM^K. Also the restriction ω^{LX} of ω^{LM^K} to TLX is the Kähler form of the S^1-invariant Kähler metric on TLX which is induced by g^{TX}. If $\omega^{LM^K,H}$ is the restriction of ω^{LM^K} to $T^H_{\mathbf{R}} LM^K$, we have the identity,

$$\omega^{LM^K} = \omega^{LM^K,H} + \omega^{LX}. \tag{4.43}$$

The question is now to find an analogue of the double transgression equation (1.91) for the forms a_t. In view of (1.90), we need to find the operator-theoretic counterpart to the form γ_t in (1.90).

Let N^V be the number operator on $\Omega^{(0,\cdot)}(X, E|_X)$, which acts by multiplication by k on $\Omega^{(0,k)}(X, E|_X)$. For $t > 0$, set

$$N_t = N^V + i\frac{\omega^{M,H}}{t} - n/4. \tag{4.44}$$

In (4.44), $\omega^{M,H}$ is viewed as a section of $\pi^* \Lambda^{(1,1)}(T^*_{\mathbf{R}} S)$. Using the dictionary summarized in Table 3.1, it is not difficult to see that N_t lifts to the form ω^{LM^K} on LM^K, and that the decomposition (4.44) of N_t just reflects the splitting of ω^{LM^K} in (4.43).

Set

$$c_t = -\varphi \operatorname{Tr_s}[N_t \exp(-B_t^2)]. \tag{4.45}$$

In [BGSo88a, Theorem 2.9], it is shown that

$$b_t = \frac{(\partial^S - \overline{\partial}^S)}{4i\pi} c_t, \qquad \frac{\partial}{\partial t} a_t = \frac{\overline{\partial}^S \partial^S}{2i\pi} \frac{c_t}{t}. \tag{4.46}$$

No doubt the reader will have guessed that by using the dictionary we mentioned before, we have the formal equality,

$$\begin{aligned} a_t &= \varphi(-2i\pi)^{-n/2}(C/\sqrt{C'})^n \int_{LX} \alpha_t \wedge c, \\ b_t &= \varphi\frac{1}{\sqrt{2i\pi}}(-2i\pi)^{-n/2}(C/\sqrt{C'})^n \int_{LX} \beta_t \wedge c, \\ c_t &= -i\varphi(-2i\pi)^{-n/2}(C/\sqrt{C'})^n \int_{LX} \gamma_t \wedge c. \end{aligned} \tag{4.47}$$

Under the path integral correspondence, (4.46) is exactly the counterpart to (1.91). In this case, forms γ_t were guessed from (1.90), (1.91).

Assume now $R\pi_* E$ to be locally free. The L_2 metric induces a metric $g^{R\pi_* E}$ on $R\pi_* E$. Set

$$a_\infty = \mathrm{ch}(R^{\cdot}\pi_* E, g^{R^{\cdot}\pi_* E}). \tag{4.48}$$

A nontrivial application of (4.24) shows that as $t \to +\infty$,

$$a_t = a_\infty + \mathcal{O}(1/\sqrt{t}). \tag{4.49}$$

There is a similar result for c_t.

Definition 4.3. For $s \in \mathbf{C}$, $\Re s < 1$, set

$$R(s) = \frac{1}{\Gamma(s)} \int_0^{+\infty} t^{s-1}(c_t - c_\infty)dt. \tag{4.50}$$

Then one finds easily that $R(s)$ is a holomorphic function of s which extends holomorphically near $s = 0$. Set

$$T = \frac{\partial}{\partial s} R(s)|_{s=0}. \tag{4.51}$$

Now we state the result of [BGSo88a, Theorem 2.20] and [BK92, Theorem 3.9].

Theorem 4.4. *The form T on S is real and is a sum of forms of type (p, p). Moreover,*

$$\frac{\overline{\partial}^S \partial^S}{2i\pi} T = a_\infty - a_0, \qquad \frac{\overline{\partial}^S - \partial^S}{4i\pi} T = \widetilde{\eta}. \tag{4.52}$$

Proof. The proof follows in particular from equations (4.46).

The forms T are called holomorphic analytic torsion forms.

Let τ be the form over S in (1.94), with $\mu = c$. From (1.92), (1.95), (4.47), (4.50), and (4.51), we get the formal equality

$$T = -i\varphi(-2i\pi)^{-n/2}(C/\sqrt{C'})^n \tau. \tag{4.53}$$

which, by (1.95), can also be rewritten in the form

$$T = -i\varphi(-2i\pi)^{-n/2}(C/\sqrt{C'})^n \int_{LX} \sigma \wedge c. \tag{4.54}$$

The second equation in (4.52) is compatible with (1.97), (4.37) and (4.53).

The component of degree 0 of T, which we denote by $T^{(0)}$, is the Ray–Singer holomorphic torsion [RS73] of the fibre X, which is a spectral invariant of the fibre, used by Quillen [Q85b] to define the Quillen metric on the determinant of the cohomology. In degree 2, the first equation in (4.52) is just a consequence the curvature Theorem of [BGSo88b] for Quillen metrics.

Let us just outline a few applications of the above formal equalities. Since these equalities are formal, 'applications' should be taken here with a touch of salt.

Indeed an important question in the theory of analytic torsion forms is to understand their dependence on the given metrics. This was carried out in [BGSo88b, BK92]. The corresponding question for the currents σ was dealt with in [B90a].

We already stressed the importance of studying the compatibility of the above objects to functorial operations. For the forms T, in the case of submersions, this was carried out in [BerB94, Ma99, Ma00].

For immersions, the corresponding problem for analytic holomorphic torsion forms was studied in [BL91, B97], and for the currents σ in [B92a]. The proofs of [BL91] and [B92a] were shown in [B92b] to be parallel. Of course it is very difficult to show that the arguments one uses in the Lagrangian side can be made rigorous and imported in the rigorous Hamiltonian side.

It is now time to extend further table 3.1 on the Hamiltonian–Lagrangian correspondence, by incorporating what we learnt from the present section.

Table 4.1 The Hamiltonian–Lagrangian correspondence

Supertraces of heat operators	Integrals of differential forms on LX
$\mathrm{Tr_s}$	$\int_{LX}$
D^X	d_K^{LX}
$D^{X,2}$	L_K
$\mathrm{Tr_s}[[D^X, A]] = 0$	$\int_{LX} d_K^{LX} \mu = 0$
Local index theorem	Localization formula
Local families index	Equivariant integration along the fibre
η-invariant and $\widetilde{\eta}$-forms	Currents ε
Holomorphic torsion forms T	Bott–Chern forms τ

Note that the Ray–Singer analytic torsion in de Rham theory has a Lagrangian counterpart which has been studied in [BGo04]. The finite-dimensional theory is connected with a different kind of localization formulas.

4.5 *Lefschetz formulas*

The case of Lefschetz formulas can be also approached from the point of view of infinite-dimensional localization. In this context given a compact Lie group G acting isometrically on X, the action of G lifts to LX and commutes with the action of S^1. The relevant finite-dimensional localization formulas [B86b] now involve two commuting Killing vector fields. Their extension to infinite dimensions was established in [BGo00] and relate two versions of the holomorphic torsion forms to the finite-dimensional Bott–Chern currents of Subsection 1.7.

4.6 Towards the hypoelliptic Laplacian

We have already hinted that the Hamiltonian–Lagrangian correspondence, in spite of the fact that it is not well-defined, can be of considerable help in guessing the construction of new objects on the rigorous Hamiltonian side, and predicting their properties.

In [B05, B08], we developed a theory of the hypoelliptic Laplacian. This is a family of operators acting on the total space of the tangent or cotangent bundle of the given Riemannian manifold X, which interpolates between the classical Laplacian of X and the geodesic flow. Part of the construction was done first in the Lagrangian formulation, by adequately modifying the differential d_K^{LX}-closed forms which appear in the path integrals so as to get the desired interpolation property in the path integral formalism, and then to find the adequate object in the operator formalism which would correspond to the deformed path integrals.

References

[Al83] L. Alvarez-Gaumé. Supersymmetry and the Atiyah-Singer index theorem. *Comm. Math. Phys.*, 90(2):161–173, 1983.

[A85] M. F. Atiyah. Circular symmetry and stationary-phase approximation. *Astérisque*, (131):43–59, 1985. Colloquium in honor of Laurent Schwartz, Vol. 1 (Palaiseau, 1983).

[ABP73] M. Atiyah, R. Bott, and V. K. Patodi. On the heat equation and the index theorem. *Invent. Math.*, 19:279–330, 1973.

[APS75a] M. F. Atiyah, V. K. Patodi, and I. M. Singer. Spectral asymmetry and Riemannian geometry. I. *Math. Proc. Cambridge Philos. Soc.*, 77:43–69, 1975.

[APS75b] M. F. Atiyah, V. K. Patodi, and I. M. Singer. Spectral asymmetry and Riemannian geometry. II. *Math. Proc. Cambridge Philos. Soc.*, 78(3):405–432, 1975.

[APS76] M. F. Atiyah, V. K. Patodi, and I. M. Singer. Spectral asymmetry and Riemannian geometry. III. *Math. Proc. Cambridge Philos. Soc.*, 79(1):71–99, 1976.

[AS71] M. F. Atiyah and I. M. Singer. The index of elliptic operators. IV. *Ann. of Math. (2)*, 93:119–138, 1971.

[BeGeV92] N. Berline, E. Getzler, and M. Vergne. *Heat kernels and Dirac operators*. Grundl. Math. Wiss. Band 298. Springer-Verlag, Berlin, 1992.

[BeV83] N. Berline and M. Vergne. Zéros d'un champ de vecteurs et classes caractéristiques équivariantes. *Duke Math. J.*, 50(2):539–549, 1983.

[BerB94] A. Berthomieu and J.-M. Bismut. Quillen metrics and higher analytic torsion forms. *J. Reine Angew. Math.*, 457:85–184, 1994.

[B84a] J.-M. Bismut. The Atiyah-Singer theorems: a probabilistic approach. I. The index theorem. *J. Funct. Anal.*, 57(1):56–99, 1984.

[B84b] J.-M. Bismut. The Atiyah-Singer theorems: a probabilistic approach. II. The Lefschetz fixed point formulas. *J. Funct. Anal.*, 57(3):329–348, 1984.

[B85] J.-M. Bismut. Index theorem and equivariant cohomology on the loop space. *Comm. Math. Phys.*, 98(2):213–237, 1985.

[B86a] J.-M. Bismut. The Atiyah-Singer index theorem for families of Dirac operators: two heat equation proofs. *Invent. Math.*, 83(1):91–151, 1986.

[B86b] J.-M. Bismut. Localization formulas, superconnections, and the index theorem for families. *Comm. Math. Phys.*, 103(1):127–166, 1986.

[B90a] J.-M. Bismut. Equivariant Bott-Chern currents and the Ray-Singer analytic torsion. *Math. Ann.*, 287(3):495–507, 1990.

[B90b] J.-M. Bismut. Koszul complexes, harmonic oscillators, and the Todd class. *J. Amer. Math. Soc.*, 3(1):159–256, 1990. With an appendix by the author and C. Soulé.

[B90c] J.-M. Bismut. Superconnection currents and complex immersions. *Invent. Math.*, 99(1):59–113, 1990.

[B92a] J.-M. Bismut. Complex equivariant intersection, excess normal bundles and Bott-Chern currents. *Comm. Math. Phys.*, 148(1):1–55, 1992.

[B92b] J.-M. Bismut. On certain infinite-dimensional aspects of Arakelov intersection theory. *Comm. Math. Phys.*, 148(2):217–248, 1992.

[B97] J.-M. Bismut. Holomorphic families of immersions and higher analytic torsion forms. *Astérisque*, (244):viii+275, 1997.

[B98] J.-M. Bismut. Local index theory, eta invariants and holomorphic torsion: a survey. In *Surveys in differential geometry, Vol. III (Cambridge, MA, 1996)*, pages 1–76. Int. Press, Boston, MA, 1998.

[B05] J.-M. Bismut. The hypoelliptic Laplacian on the cotangent bundle. *J. Amer. Math. Soc.*, 18(2):379–476 (electronic), 2005.

[B08] J-.M. Bismut. The hypoelliptic Dirac operator. In *Geometry and dynamics of groups and spaces*, volume 265 of *Progr. Math.*, pages 113–246. Birkhäuser, Basel, 2008.

[BC89] J.-M. Bismut and J. Cheeger. η-invariants and their adiabatic limits. *J. Amer. Math. Soc.*, 2(1):33–70, 1989.

[BF86a] J.-M. Bismut and D.S. Freed. The analysis of elliptic families. I. Metrics and connections on determinant bundles. *Comm. Math. Phys.*, 106(1):159–176, 1986.

[BF86b] J.-M. Bismut and D.S. Freed. The analysis of elliptic families. II. Dirac operators, eta invariants, and the holonomy theorem. *Comm. Math. Phys.*, 107(1):103–163, 1986.

[BGSo88a] J.-M. Bismut, H. Gillet, and C. Soulé. Analytic torsion and holomorphic determinant bundles. II. Direct images and Bott-Chern forms. *Comm. Math. Phys.*, 115(1): 79–126, 1988.

[BGSo88b] J.-M. Bismut, H. Gillet, and C. Soulé. Analytic torsion and holomorphic determinant bundles. III. Quillen metrics on holomorphic determinants. *Comm. Math. Phys.*, 115(2):301–351, 1988.

[BGS90a] J.-M. Bismut, H. Gillet, and C. Soulé. Bott-Chern currents and complex immersions. *Duke Math. J.*, 60(1):255–284, 1990.

[BGS90b] J.-M. Bismut, H. Gillet, and C. Soulé. Complex immersions and Arakelov geometry. In *The Grothendieck Festschrift, Vol. I*, pages 249–331. Birkhäuser Boston, Boston, MA, 1990.

[BGo00] J.-M. Bismut and S. Goette. Holomorphic equivariant analytic torsions. *Geom. Funct. Anal.*, 10(6):1289–1422, 2000.

[BGo04] J.-M. Bismut and S. Goette. Equivariant de Rham torsions. *Ann. of Math.*, 159: 53–216, 2004.

[BK92] J.-M. Bismut and K. Köhler. Higher analytic torsion forms for direct images and anomaly formulas. *J. Algebraic Geom.*, 1(4):647–684, 1992.

[BL91] J.-M. Bismut and G. Lebeau. Complex immersions and Quillen metrics. *Inst. Hautes Études Sci. Publ. Math.*, (74):ii+298 pp. (1992), 1991.

[BZ93] J.-M. Bismut and W. Zhang. Real embeddings and eta invariants. *Math. Ann.*, 295(4):661–684, 1993.

[BoC65] R. Bott and S. S. Chern. Hermitian vector bundles and the equidistribution of the zeroes of their holomorphic sections. *Acta Math.*, 114:71–112, 1965.

[Ca51a] H. Cartan. La transgression dans un groupe de Lie et dans un espace fibré principal. In *Colloque de topologie (espaces fibrés), Bruxelles, 1950*, pages 57–71. Georges Thone, Liège, 1951.

[Ca51b] H. Cartan. Notions d'algèbre différentielle; application aux groupes de Lie et aux variétés où opère un groupe de Lie. In *Colloque de topologie (espaces fibrés), Bruxelles, 1950*, pages 15–27. Georges Thone, Liège, 1951.

[Ch79] J. Cheeger. Analytic torsion and the heat equation. *Ann. of Math. (2)*, 109(2): 259–322, 1979.

[ChS85] J. Cheeger and J. Simons. Differential characters and geometric invariants. In *Geometry and topology (College Park, Md., 1983/84)*, volume 1167 of *Lecture Notes in Math.*, pages 50–80. Springer, Berlin, 1985.

[DH82] J. J. Duistermaat and G. J. Heckman. On the variation in the cohomology of the symplectic form of the reduced phase space. *Invent. Math.*, 69(2):259–268, 1982.

[DH83] J. J. Duistermaat and G. J. Heckman. Addendum to: "On the variation in the cohomology of the symplectic form of the reduced phase space". *Invent. Math.*, 72(1): 153–158, 1983.

[Ge86] E. Getzler. A short proof of the local Atiyah-Singer index theorem. *Topology*, 25(1):111–117, 1986.

[Gil84] P. B. Gilkey. *Invariance theory, the heat equation, and the Atiyah-Singer index theorem*, volume 11 of *Mathematics Lecture Series*. Publish or Perish Inc., Wilmington, DE, 1984.

[GSo90] H. Gillet and C. Soulé. Arithmetic intersection theory. *Inst. Hautes Études Sci. Publ. Math.*, (72):93–174 (1991), 1990.

[GSo92] H. Gillet and C. Soulé. An arithmetic Riemann-Roch theorem. *Invent. Math.*, 110(3):473–543, 1992.

[GuSt99] V. W. Guillemin and S. Sternberg. *Supersymmetry and equivariant de Rham theory*. Mathematics Past and Present. Springer-Verlag, Berlin, 1999. With an appendix containing two reprints by Henri Cartan [MR0042426 (13,107e); MR0042427 (13,107f)].

[Hi74] N. Hitchin. Harmonic spinors. *Advances in Math.*, 14:1–55, 1974.

[Ho85] L. Hörmander. *The analysis of linear partial differential operators. III*. Grundl. Math. Wiss. Band 274. Springer-Verlag, Berlin, 1985. Pseudodifferential operators.

[L51] P. Lévy. Wiener's random function, and other Laplacian random functions. In *Proceedings of the Second Berkeley Symposium on Mathematical Statistics and Probability, 1950*, pages 171–187, Berkeley and Los Angeles, 1951. University of California Press.

[Ma99] X. Ma. Formes de torsion analytique et familles de submersions. I. *Bull. Soc. Math. France*, 127(4):541–621, 1999.

[Ma00] X. Ma. Formes de torsion analytique et familles de submersions. II. *Asian J. Math.*, 4(3):633–667, 2000.

[MS67] H. P. McKean, Jr. and I. M. Singer. Curvature and the eigenvalues of the Laplacian. *J. Differential Geometry*, 1(1):43–69, 1967.

[M78] W. Müller. Analytic torsion and R-torsion of Riemannian manifolds. *Adv. in Math.*, 28(3):233–305, 1978.

[Q85a] D. Quillen. Superconnections and the Chern character. *Topology*, 24(1):89–95, 1985.

[Q85b] D. Quillen. Determinants of Cauchy-Riemann operators on Riemann surfaces. *Functional Anal. Appl.*, 19(1):31–34, 1985.

[RS73] D. B. Ray and I. M. Singer. Analytic torsion for complex manifolds. *Ann. of Math. (2)*, 98:154–177, 1973.

[W82] E. Witten. Supersymmetry and Morse theory. *J. Differential Geom.*, 17(4):661–692 (1983), 1982.

Asymptotic equivariant index of Toeplitz operators and relative index of CR structures

Louis Boutet de Monvel, Eric Leichtnam, Xiang Tang* and Alan Weinstein†

This paper is dedicated to J.J. Duistermaat for his 65th birthday

Abstract Using equivariant Toeplitz operator calculus, we give a new proof of the Atiyah–Weinstein conjecture on the index of Fourier integral operators and the relative index of CR structures.

Key words: Index, CR manifolds, contact manifolds, Toeplitz operators

Mathematics Subject Classification (2010): Primary: 58J20; Secondary: 19L47, 32A45, 53D10, 58J40

1 Introduction

Let Ω, Ω' be two bounded Stein domains (or manifolds) with smooth strictly pseudoconvex boundaries X_0, X_0' (these are compact contact manifolds), and f_0

Louis Boutet de Monvel
UPMC-Paris 6, Institut de Mathématiques de Jussieu-UMR7586, 4, place Jussieu, F-75005, Paris, France
e-mail: boutet@math.jussieu.fr

Eric Leichtnam
CNRS, Institut de Mathématiques de Jussieu-UMR7586, 4, place Jussieu, F-75005, Paris, France
e-mail: ericleichtnam@math.jussieu.fr

Xiang Tang
Department of Mathematics, Washington University, St. Louis, MO, 63130 USA
e-mail: xtang@math.wustl.edu

Alan Weinstein
Department of Mathematics, University of California, Berkeley, CA 94720 USA
e-mail: alanw@math.berkeley.edu

* Research partially supported by NSF Grant 0703775

† Research partially supported by NSF Grant 0707137

a contact isomorphism $X_0 \to X'_0$. If $\mathbb{H}_0, \mathbb{H}'_0$ denote the spaces of CR functions (or distributions) on X_0, X'_0 (boundary values of holomorphic functions), S, S' the Szegő projectors,[3] the map $E_0 : u \mapsto S'(u \circ f_0^{-1}) : \mathbb{H}_0 \to \mathbb{H}'_0$ is Fredholm (it is an elliptic Toeplitz FIO). The index of E_0 was introduced by C. Epstein [17, 18, 19, 20], who called it the relative index of the two CR structures. A formula for the index was proposed in [27]. A special case was established in [21], and a proof of this index formula in the general case was given by C. Epstein in [19], based on an analysis of the situation using the "Heisenberg-pseudodifferential calculus." (Another proof based on deformation quantization should also be possible, using the ideas in [23] and [24].) In this paper we propose a simpler proof based on equivariant Toeplitz-operator calculus, which gives a straightforward view. Our formula is described in Section 4.4. It is essentially equivalent to the formula proposed in [27], which was stimulated by a problem in the theory of Fourier integral operators, a subject in which Hans Duistermaat was a pioneer [13].

It is awkward to keep track of the index in the setting of Toeplitz operators on X_0 and X'_0 alone, because we are dealing with several Szegő projectors, and Toeplitz-operator calculus controls the range $\mathbb{H}$ of a generalized Szegő projector at best up to a vector space of finite rank.[4]

To make up for this, we use the ball $\widetilde{\Omega} \subset \mathbb{C} \times \Omega$ defined by $t\bar{t} < \phi$, where t is the coordinate on $\mathbb{C}$, ϕ is a smooth defining function ($\phi = 0, d\phi \neq 0$ on $X_0 = \partial\Omega, \phi > 0$ inside; note that this is the opposite sign from the usual one) chosen so that $\operatorname{Log} \frac{1}{\phi}$ is strictly plurisubharmonic, so that the boundary $X = \partial\widetilde{\Omega}$ is strictly pseudoconvex; such a defining function always exists; e.g., we can choose ϕ strictly plurisubharmonic. Then X is a compact contact manifold with (positive) action of the circle group $U(1)$. We will identify X_0 with the submanifold $\{0\} \times X_0$ of X.

We perform the same construction for Ω': we will see that there exist an equivariant germ near X_0 of an equivariant contact isomorphism $f : X \to X'$ extending f_0 such that $t' \circ f$ is a positive multiple of t, and an elliptic equivariant Toeplitz FIO E extending E_0, associated[5] to the contact map f; the holomorphic spaces $\mathbb{H}, \mathbb{H}'$ split in Fourier components $\mathbb{H}_k, \mathbb{H}'_k$ on which the index is repeated infinitely many times. This construction has the advantage of taking into account the geometry of the two fillings Ω, Ω', which obviously must come into the picture.

The final result can then be expressed in terms of an asymptotic version of the relative index (G-index) of E, derived from the equivariant theory of M.F. Atiyah and I.M. Singer [4]: the asymptotic index, described in Section 4.4, ignores finite-dimensional spaces and is well defined for Toeplitz operators or Toeplitz systems; it is also preserved by suitable contact embeddings.

The asymptotic equivariant trace and index are described in Sections 2 and 3. The relative index formula is described and proved in Section 4 (Theorem 5).

[3] The definition of S requires choosing a smooth positive density on X_0; nothing of what follows depends on this choice.

[4] There is no index formula for a vector bundle elliptic Toeplitz operator, although there is one for matrix Toeplitz operators, a straightforward generalisation of the Atiyah–Singer formula; cf. [7].

[5] f is associated to E in the same manner as a canonical map is associated to an FIO.

2 Equivariant trace and index

2.1 Equivariant Toeplitz Operators

Let G be a compact Lie group with Haar measure $dg(\int dg = 1)$, $\mathfrak{g}$ its Lie algebra, and X a smooth compact cooriented contact manifold with an action of G; this means that X is equipped with a contact form λ (two forms define the same cooriented contact structure if they are positive multiples of each other); G acts smoothly on X and preserves the contact structure and coorientation, i.e., for any g the image $g_*\lambda$ is a positive multiple of λ; replacing λ by the mean $\int g_*\lambda\, dg$, we may suppose that it is invariant. The associated symplectic cone Σ is the set of positive multiples of λ in T^*X, a principal $\mathbb{R}^+$ bundle over X, a half-line bundle over X.

We also choose an invariant measure dx with smooth positive density on X, so L^2 norms are well defined. The results below will not depend on this choice.

It was shown in [10] that there always exists an invariant generalized Szegő projector S which is a self adjoint Fourier-integral projector whose microsupport is Σ, mimicking the classical Szegő projector. The projector S extends or restricts to all Sobolev spaces; for $s \in \mathbb{R}$ we will denote by $\mathbb{H}^{(s)}$ the range of S in the Sobolev space $H^s(X)$, and by $\mathbb{H}$ the union.

A Toeplitz operator of degree m on $\mathbb{H}$ is an operator of the form $f \mapsto T_Q f = SQf$, where Q is a pseudodifferential operator of degree m. Here we use pseudodifferential operators in a strict sense, i.e., in any local set of coordinates the total symbol has an asymptotic expansion $q(x,\xi) \sim \sum_{k\geq 0} q_{m-k}(x,\xi)$, where q_{m-k} is homogeneous of degree $m-k$ with respect to ξ, and the degree m and $k \geq 0$ are integers.[6] A Toeplitz operator of degree m is continuous $\mathbb{H}^{(s)} \to \mathbb{H}^{(s-m)}$ for all s. Recall that Toeplitz operators give rise to a symbolic calculus, microlocally isomorphic to the pseudodifferential calculus, that lives on Σ (cf. [10]).

In particular, the infinitesimal generators of G (vector fields determined by elements $\xi \in \mathfrak{g}$) define Toeplitz operators T_ξ of degree 1 on $\mathbb{H}$. An element P of the universal enveloping algebra $U(\mathfrak{g})$ acts as a higher-order Toeplitz operator P_X (equivariant if P is invariant), and the elements of G act as unitary Fourier integral operators, or "Toeplitz-FIO."

$\mathbb{H}$ (with its Sobolev counterparts) splits according to the irreducible representations of $G : \mathbb{H} = \widehat{\bigoplus} \mathbb{H}_\alpha$.

Below we will use the following extended notions: an *equivariant Toeplitz bundle* $\mathbb{E}$ is the range of an equivariant Toeplitz projector P of degree 0 on a direct sum $\mathbb{H}^N$. The *symbol* of $\mathbb{E}$ is the range of the principal symbol of P; it is an equivariant vector bundle on X. Any equivariant vector bundle on X is the symbol of an equivariant Toeplitz bundle (this also follows from [10]).

[6] We will occasionally use as multipliers operators of degree $m = \frac{1}{2}$ (or any other complex number), with k still an integer in the expansion.

2.2 G-trace

The G-trace and G-index (relative index in [4]) were introduced by M.F. Atiyah in his joint work with I.M. Singer [4] for equivariant pseudodifferential operators on G-manifolds. The G-trace of such an operator A is a distribution on G, describing $\operatorname{tr}(g \circ A)$. Here we adapt this to Toeplitz operators. Because the Toeplitz spaces $\mathbb{H}$ and $\mathbb{E}$ are really defined only up to a finite-dimensional space, their G-traces or indexes are ultimately defined only up to a smooth function, i.e., they are distribution singularities on G (distributions mod C^∞); they are described below, and renamed "asymptotic G-trace or index."

If $\mathbb{E}, \mathbb{F}$ are equivariant Toeplitz bundles, there is an obvious notion of Toeplitz (matrix) operator $P : \mathbb{E} \to \mathbb{F}$, and of its principal symbol $\sigma_d(P)$ (if it is of degree d), a homogeneous vector-bundle homomorphism $E \to F$ over Σ. The operator P is elliptic if its symbol is invertible; it is then a Fredholm operator $\mathbb{E}^s \to \mathbb{F}^{s-d}$ and has an index which does not depend on s.

If $\mathbb{E}$ is an equivariant Toeplitz bundle and $P : \mathbb{E} \to \mathbb{E}$ is a Toeplitz operator of trace class[7] ($\deg P < -n$), the trace function[8] $\operatorname{Tr}_P^G(g) = \operatorname{tr}(g \circ P)$ is well defined; it is a continuous function on G. It is smooth if P is of degree $-\infty$ ($P \sim 0$). If P is equivariant, its Fourier coefficient for the representation α is $\frac{1}{d_\alpha}\operatorname{tr} P|_{\mathbb{E}_\alpha}$ (with d_α the dimension of α, $\mathbb{E}_\alpha$ the α-isotypic component of $\mathbb{E}$).

Definition 1. We denote by $\operatorname{char}\mathfrak{g} \subset \Sigma$ the characteristic set of the G-action, i.e., the closed subcone where all symbols of infinitesimal operators $T_\xi, \xi \in \mathfrak{g}$, vanish (this contains the fixed-point set Σ^G). The base of $\operatorname{char}\mathfrak{g}$ is the set of points of X where all Lie generators $L_\xi, \xi \in \mathfrak{g}$, are annihilated by the contact form λ; in the sequel we will usually denote it by $Z \subset X$.

The fixed-point set X^G is the base of Σ^G because G is compact (there is an invariant section). The base Z contains the fixed-point set X^G. Note that Σ^G is always a smooth symplectic cone and its base X^G is a smooth contact manifold; $\operatorname{char}\mathfrak{g}$ and Z may be singular.

The following result is an immediate adaptation of the similar result for pseudodifferential operators in [4].

Proposition 1. *Let $P : \mathbb{E} \to \mathbb{E}$ be a Toeplitz operator, with $P \sim 0$ near $\operatorname{char}\mathfrak{g}$ (i.e., its total symbol vanishes near $\operatorname{char}\mathfrak{g}$). Then $\operatorname{Tr}_P^G = \operatorname{tr}(g \circ P)$ is well defined as a distribution on G. If P is equivariant, we have, in the sense of distributions,*

$$\operatorname{Tr}_P^G = \sum \frac{1}{d_\alpha} (\operatorname{tr} P|_{\mathbb{E}_\alpha})\, \chi_\alpha \tag{1}$$

[7] $\dim X = 2n - 1$. The Toeplitz algebra is microlocally isomorphic to the algebra of pseudodifferential operators in n real variables, and operators of degree $< -n$ are of trace class.

[8] We still denote by g the action of a group element g through a given representation, for example if we are dealing with the standard representation on functions, $gf = f \circ g^{-1}$, also denoted by $g_* f$, $g^{*-1} f$, or $g^{-1*} f$.

where α runs over the set of irreducible representations, d_α is the dimension, and χ_α the character.

We have seen above that this is true if P is of trace class. For the general case, let D_G be a bi-invariant elliptic operator of order $m > 0$ on G (e.g., the Casimir of a faithful representation, with $m = 2$). Since D_G is in the center of $U(\mathfrak{g})$, the Toeplitz operator $D_X : \mathbb{E} \to \mathbb{E}$ it defines is invariant, with characteristic set char $\mathfrak{g}$.

If $P \sim 0$ near char $\mathfrak{g}$, we can divide it repeatedly by D_X (modulo smoothing operators) and get for any N,

$$P = D_X^N Q + R \quad \text{with R} \sim 0.$$

The degree of Q is $\deg P - N \deg(D_G)$, so it is of trace class if N is large enough. We set $\mathrm{Tr}_P^G = D_G^N \mathrm{Tr}_Q^G + \mathrm{Tr}_R^G$: this is well defined as a distribution; the fact that this does not depend on the choice of D_G, N, Q, R is immediate.

Formula (1) for equivariant operators is obvious for trace class operators, and the general case follows by application of D_X^N and D_G^N. Note that the series in the formula converges in the sense of distributions, i.e., the coefficients have at most polynomial growth.

Slightly more generally, let

$$(\mathbb{E}, d) : \cdots \to \mathbb{E}_j \xrightarrow{d} \mathbb{E}_{i+1} \to \cdots$$

be an equivariant Toeplitz complex of finite length, i.e., $\mathbb{E}$ is a finite sequence $\mathbb{E}_k$ of equivariant Toeplitz bundles, $d = (d_k : \mathbb{E}_k \to \mathbb{E}_{k+1})$ a sequence of Toeplitz operators such that $d^2 = 0$. If the (degree-zero) endomorphism $P = \{P_k\}$ of the complex $\mathbb{E}$ is ~ 0 near char $\mathfrak{g}$, its supertrace $\mathrm{Tr}_P^G = \sum (-1)^k \mathrm{Tr}_{P_k}^G$ is well defined; it vanishes if $P = [P_1, P_2]$ is a supercommutator with one factor ~ 0 on char $\mathfrak{g}$.

2.3 G index

Let $\mathbb{E}_0, \mathbb{E}_1$ be two equivariant Toeplitz bundles. An equivariant Toeplitz operator $P : \mathbb{E}_0 \to \mathbb{E}_1$ is G-elliptic (relatively elliptic in [4]) if it is elliptic on char $\mathfrak{g}$, i.e., the principal symbol $\sigma(P)$, which is a homogeneous equivariant bundle homomorphism $E_0 \to E_1$, is invertible on char $\mathfrak{g}$. Then there exists an equivariant $Q : \mathbb{E}_1 \to \mathbb{E}_0$ such that $QP \sim 1_{\mathbb{E}_0}, PQ \sim 1_{\mathbb{E}_1}$ near char $\mathfrak{g}$. The G-index I_P^G is defined as the distribution $\mathrm{Tr}_{1-QP}^G - \mathrm{Tr}_{1-PQ}^G$.

More generally,[9] an equivariant complex $\mathbb{E}$ as above is G-elliptic if the principal symbol $\sigma(d)$ is exact on char $\mathfrak{g}$. Then there exists an equivariant Toeplitz operator $s = (s_k : \mathbb{E}_k \to \mathbb{E}_{k-1})$ such that $1 - [d, s] \sim 0$ near char $\mathfrak{g}([d, s] = ds + sd)$.

---

[9] This reduces to the case of a single operator where the complex is concentrated in degrees 0 and 1.

The index (Euler characteristic) is the supertrace $I^G_{(\mathbb{E},d)} = \operatorname{str}(1 - [d,s]) = \sum(-1)^j \operatorname{Tr}^G_{(1-[d,s])_j}$.

For any irreducible representation α, the restriction $P_\alpha : \mathbb{E}_{0,\alpha} \to \mathbb{E}_{1,\alpha}$ is a Fredholm operator with index I_α, (respectively the cohomology H^*_α of $d \mid_{\mathbb{E}_\alpha}$ is finite-dimensional), and we have

$$I^G_P = \sum \frac{1}{d_\alpha} I_\alpha \chi_\alpha \qquad \left(\text{respectively } I^G_{(\mathbb{E},D)} = \sum_{j,\alpha} \frac{(-1)^j}{d_\alpha} \dim H^j_\alpha \, \chi_\alpha \right).$$

The G-index I^G_A is obviously invariant under compact perturbation and deformation, so it depends only on the homotopy class of $\sigma(P)$ once $\mathbb{E}_j$ has been chosen; it does depend on a choice of $\mathbb{E}_j$ (on the projector that defines it, or on the Szegő projector), because $\mathbb{E}_j$ is determined by its symbol bundle only up to a finite-dimensional space; this inconvenience is removed with the asymptotic index below.

It is sometimes convenient to notate an index as an infinite representation (mod finite representations) $\sum n_\alpha \chi_\alpha$. For the circle group $U(1)$, all simple representations are powers of the tautological representation, denoted by J, and all representations occurring as indices have a generating series

$$\sum_{k \in \mathbb{Z}} n_k J^k \quad \text{(mod finite sums).} \tag{2}$$

In fact, the positive and negative parts of the series have a weak periodicity property: they are of the form $P_\pm(J^{\pm 1}) / \prod_i (1 - (J^{\pm 1})^{k_i})$ for a suitable polynomial $P_\pm$ and positive integers k_i.[10]

Here in our relative index problem, only very simple representations of the form $m \sum_0^\infty J^k = m(1-J)^{-1}$ (for some integer m) will occur.

3 K-theory and embedding

A crucial point in the proof of the Atiyah–Singer index theorem [2] consists in showing how one can embed an elliptic system A in a simpler manifold where the index theorem is easy to prove, preserving the index and keeping track of the K-theoretic element $[A]$. The new embedded system $F_+ A$ is analogous to a derived direct image (as in algebraic geometry), and the K-theoretic element $[F_+ A]$ is the image of $[A]$ by the Bott homomorphism constructed out of R. Bott's periodicity theorem (cf. [2]).

[10] This notation represents the series expansion in positive powers of $J^{\pm 1}$; it is obviously abusive but suggestive, especially if one thinks of the extension to a multidimensional torus; it also represents a rational function whose poles are roots of 1, and whose Taylor series has integral coefficients, of which the corresponding distribution on G is the boundary value from one or the other side of the circle in the complex plane. Something similar occurs for any compact group; cf. [4].

Here we will do the same for Toeplitz operators. The direct image F_+A is even somewhat more natural, as is its relation to the Bott homomorphism (Section 3.4). The direct image for elliptic systems does not preserve the exact index, since this is not defined (because the Toeplitz space $\mathbb{H}$ is at best defined only mod a space of finite rank); but it does preserve the asymptotic equivariant index.

3.1 A short digression on Toeplitz algebras

We use the following notation: for distributions, $f \sim g$ means that $f - g$ is C^∞; for operators, $A \sim B$ (or $A = B \bmod C^\infty$) means that $A - B$ is of degree $-\infty$, i.e., has a smooth Schwartz kernel. If M is a manifold, $T^\bullet M$ denotes the cotangent bundle deprived of its zero section; it is a symplectic cone with base $S^*M = T^\bullet M/\mathbb{R}_+$, the cotangent sphere bundle.

As mentioned above, a compact contact G-manifold always possesses an invariant generalized Szegő projector. More generally, if M is a G-manifold, $\Sigma \subset T^\bullet M$ an invariant symplectic cone, there exists an associated equivariant Szegő projector (cf. [10]). If $\Sigma \subset T^\bullet M$, $\Sigma' \subset T^\bullet M'$, and $f : \Sigma \to \Sigma'$ is an isomorphism of symplectic cones, there always exists an "adapted FIO" F which defines a Fredholm map $u \mapsto \tilde{F}u = S'(Fu) : \mathbb{H} \to \mathbb{H}'$ and an isomorphism of the corresponding Toeplitz algebras ($A \mapsto \tilde{F}A\tilde{F}^{-1}$, mod C^∞).

One can choose F equivariant if f is. Indeed, any adapted FIO can be defined using a global phase function ϕ on $T^\bullet(M \times M'^{\mathrm{op}})$ such that[11]

(1) ϕ vanishes on the graph of f, and $d\phi$ coincides with the Liouville form $\xi \cdot dx - \eta \cdot dy$ there;
(2) $\operatorname{Im}\phi \gg 0$, i.e., $\operatorname{Im}\phi > 0$ outside of the graph of f, and the transversal Hessian is $\gg 0$; replacing ϕ by its mean gives an invariant phase; we may set $Ff(x) = \int e^{i\phi} af(y)dy\,d\eta\,d\xi$, where the density $a(x,\xi,y,\eta)dy\,d\eta\,d\xi$ is a symbol, invariant and positive elliptic (F is of Sobolev degree deg $(a\,dy\,d\eta\,d\xi)$ $-\frac{3}{4}(n_x + n_y)$ (cf. L. Hörmander [22]), so a is possibly of nonintegral degree if we want F of degree 0). The transfer map from $\mathbb{H}$ to $\mathbb{H}'$ is $S'FS$.

If M is a manifold and $X = S^*M$, the cotangent sphere, X carries a canonical Toeplitz algebra, viz. the sheaf $\mathcal{E}_{S^*M}$ of pseudodifferential operators acting on the sheaf μ of microfunctions. In general, if X is a contact manifold, we will denote by $\mathcal{E}_X$ (or just $\mathcal{E}$) the algebra of Toeplitz operators on X. It is a sheaf of algebras on X acting on $\mu\mathbb{H} = \mathbb{H} \bmod C^\infty$, which is a sheaf of vector spaces on X; the pair $(\mathcal{E}_X, \mu\mathbb{H})$ is locally isomorphic to the pair of sheaves of pseudodifferential operators acting on microfunctions. If X is a G-contact manifold, we can choose the Szegő projector invariant, so G acts on $\mathcal{E}_X$ and μ_X.

For a general contact manifold, $\mathcal{E}_X$ is well defined up to isomorphism, independently of any embedding, but no better than that. The corresponding Szegő projector

[11] op in M'^{op} refers to the change of sign in the symplectic form on T^*M'.

(not mod C^∞) is defined only up to a compact operator (a little better than that; see below).

3.2 Asymptotic trace and index

The symbol bundles E_j of the Toeplitz bundles $\mathbb{E}_j$ determine these only up to a space of finite dimension (because, as mentioned above, both the projector defining them, and the Szegő projector, are not uniquely determined by their symbols). However, if $\mathbb{E}, \mathbb{E}'$ are two equivariant Toeplitz bundles with the same symbol, there exists an equivariant elliptic Toeplitz operator $U : \mathbb{E} \to \mathbb{E}'$ with quasi-inverse V (i.e., $VU \sim 1_{\mathbb{E}}, UV \sim 1_{\mathbb{E}'}$). This may be used to transport equivariant Toeplitz operators from $\mathbb{E}$ to $\mathbb{E}'$: $P \mapsto Q = UPV$. Then if $P \sim 0$ on Z, $Q = UPV$ and VUP have the same G-trace, and since $P \sim VUP$, we have $\mathrm{Tr}_P^G - \mathrm{Tr}_Q^G \in C^\infty(G)$.

Definition 2. We define the *asymptotic G-trace* of P as the singularity of Tr_P^G (i.e., $\mathrm{Tr}_P^G \bmod C^\infty(G)$).

The asymptotic trace vanishes if and only if the sequence of Fourier coefficients of Tr_P^G is of rapid decrease, i.e., $O(c_\alpha)^{-m}$ for all m, where c_α is the eigenvalue of D_G in the representation α. This is the case if P is of degree $-\infty$.

Definition 3. We will say that a system P of Toeplitz operators is *G-elliptic* (relatively elliptic in [4]) if it is elliptic on $\mathrm{char}\,\mathfrak{g}$. When this is the case, the asymptotic G-index (or $\widetilde{I}_P^G$) is defined as the singularity of I_P^G. (We will still denote it by I_P^G if there is no risk of confusion.)

We denote by $K^G(X-Z)$ the equivariant K-theory with compact support. By the excision theorem, $K^G(X-Z)$ is the same as $K_{X-Z}^G(X)$, the equivariant K-theory of X with compact support in $X-Z$, i.e., the group of stable classes of triples $d(E, F, u)$, where E, F are equivariant G-bundles on X, and u an equivariant isomorphism $E \to F$ defined near the set Z (the equivalence relation is $d(E, F, a) \sim 0$ if a is stably homotopic (near Z) to an isomorphism on the whole of X). The asymptotic index is also defined for equivariant Toeplitz complexes, exact near $\mathrm{char}\,\mathfrak{g}$.

If $u : \mathbb{E} \to \mathbb{F}$ is a G-elliptic Toeplitz system or complex, its principal symbol defines a homogeneous linear map on Σ, invertible on $\mathrm{char}\,\mathfrak{g}$. Its restriction to any equivariant section of Σ defines a K-theoretic element $[u] \in K^G(X-Z)$ (in case of a complex, u defines the same K-theoretic element as $u + u^* : \mathbb{E}^{\mathrm{even}} \to \mathbb{E}^{\mathrm{odd}}$). The asymptotic index depends only on the homotopy class of the principal symbol $\sigma(P)$, and since it is obviously additive, we get the following

Theorem 2. *The asymptotic index of u depends only on the K-theoretic element* $[u]$. *It defines an additive map from $K^G(X-Z)$ to $C^{-\infty}(G)/C^\infty(G)$, where Z is, as above, the base of $\mathrm{char}\,\mathfrak{g}$.*

Note that the sequence of Fourier coefficients $\frac{\operatorname{tr} P_\alpha}{d_\alpha}$ is in any case of polynomial growth with respect to the eigenvalues of D or D_X; if $P \sim 0$, it is of rapid decrease. The coefficients $\frac{I_\alpha}{d_\alpha}$ of the asymptotic index are integers, so they are completely determined, except for a finite number of them, by the asymptotic index.

Remark. If V is a finite-dimensional representation of G and $V \otimes P$ or $V \otimes d$ is defined in the obvious way, we have $I^G_{V \otimes P} = \chi_V I^G_P$ (i.e., Index $(V \otimes P)_\alpha =$ $(V \otimes \text{Index}\, P)_\alpha$, except at a finite number of places).

For example, let $G = \mathrm{SU}_2$ acting on the sphere X of $V = \mathbb{C}^2$ in the usual manner, and $E = S^m V$ the mth symmetric power. Then $E \times X$ is a G bundle with the action $g(v, x) = (gv, gx)$. The CR structure on the sphere gives rise to a first Szegő projector $S_1(v \cdot f) = v \cdot S(f)$, where S is the canonical Szegő projector on holomorphic functions. On the other hand, since X is a free orbit of G, the bundle $E \times X$ is isomorphic to the trivial bundle $E_0 \times X$, where E_0 is some fiber (i.e., the vector space of homogeneous polynomials of degree m, with trivial action of G). This gives rise to a second Szegő projector S_0, not equal to the first, but giving the same asymptotic index; we recover the fact that $S^m V \otimes \sum S^k V \sim (m+1) \sum S^k V$ ($=$ in degree $\geq m$).

3.3 $\mathcal{E}$-modules

For the sequel, it will be convenient to use the language of $\mathcal{E}$-modules. In the C^∞ category, $\mathcal{E}$ is not coherent; general $\mathcal{E}$-module theory is therefore not practical and not usefully related to topological K-theory. We will just stick to the two useful cases below (elliptic complexes or "good" modules).[12] Note also that the notion of ellipticity is slightly ambiguous; more precisely: a system of Toeplitz operators (or pseudodifferential operators) is obviously invertible mod C^∞ if its principal symbol is, but the converse is not true. The notion of "good" system below partly compensates for this; it is in fact indispensable for a good relationship between elliptic systems and K-theory.

If $\mathcal{M}$ is an $\mathcal{E}$-module (respectively a complex of $\mathcal{E}$ modules), it corresponds to the system of pseudodifferential (respectively Toeplitz) operators whose sheaf of solutions is Hom $(\mathcal{M}, \mu\mathbb{H})$; e.g., a locally free complex of (L, d) of $\mathcal{E}$-modules defines the Toeplitz complex $(\mathbb{E}, D) =$ Hom $(L, \mathbb{H})$.

More generally we will say that an $\mathcal{E}$-module $\mathcal{M}$ is "good" if it is finitely generated, equipped with a filtration $\mathcal{M} = \bigcup \mathcal{M}_k$ (i.e., $\mathcal{E}_p \mathcal{M}_q = \mathcal{M}_{p+q}$, $\bigcap \mathcal{M}_k = 0$) such that the symbol $\operatorname{gr} \mathcal{M}$ has a finite locally free resolution. We write $\sigma(\mathcal{M}) = \mathcal{M}_0/\mathcal{M}_{-1}$, which is a sheaf of C^∞ modules on the basis X; since there exist global elliptic sections of $\mathcal{E}$, $\operatorname{gr} \mathcal{M}$ is completely determined by the symbol, as is the resolution.

[12] Things work better in the analytic category.

A resolution of $\sigma(\mathcal{M})$ lifts to a "good resolution" of $\mathcal{M}$, i.e., a finite locally free resolution[13] of $\mathcal{M}$.

It is standard that two resolutions of $\sigma(\mathcal{M})$ are homotopic, and if $\sigma(\mathcal{M})$ has locally finite locally free resolutions it also has a global one (because we are working in the C^∞ category on a compact manifold or cone with compact support, and dispose of partitions of unity); this lifts to a global good resolution of $\mathcal{M}$.

If $\mathcal{M}$ is "good," it defines a K-theoretic element $[\mathcal{M}] \in K_Y(X)$ (where Y is the support of $\sigma(\mathcal{M})$), viz. the K-theoretic element defined by the symbol of any good resolution (this does not depend on the resolution, since any two such are homotopic).

All this works just as well in the presence of a G-action (if the filtration etc. is invariant).

As above (Section 2.2), the asymptotic G-trace Tr_A^G [using subscripts as before] is well defined if A is an endomorphism of a good locally free complex of Toeplitz modules. The same holds for a good module $\mathcal{M}$: the asymptotic trace of $A \in \mathrm{End}\,\mathcal{E}(M)$ vanishing near char $\mathfrak{g}$ is the asymptotic trace of any lifting of A to a good resolution of $\mathcal{M}$. (Such a lifting, vanishing near char $\mathfrak{g}$, exists and is unique up to homotopy, i.e., modulo supercommutators.) Likewise, the asymptotic G-index of a locally free complex exact on Z, or of a good $\mathcal{E}$- module with support outside of Z, is defined: it is the asymptotic G-trace of the identity.

Definition 3 of the asymptotic index (or Euler characteristic) extends in an obvious manner to good complexes of locally free $\mathcal{E}$-modules or to good $\mathcal{E}$-modules. The asymptotic G-index of such an object, when it is G-elliptic, depends only on the K-theoretic element which it defines on the base.

Let us note that the asymptotic trace and index are still well defined for locally free complexes or modules with a locally free resolution, not necessarily good; in that case, what no longer works is the relation to topological K-theory on the base.

3.4 Embedding

If M is a manifold, $\Sigma \subset T^\bullet M$ a symplectic subcone, the Toeplitz space $\mathbb{H}$ is the space of solutions of a pseudodifferential system mimicking $\bar{\partial}_b$. If $I \subset \mathcal{E}$ is the ideal generated by these operators (mod C^∞), and $\mathcal{M} = \mathcal{E}/I$, we have $\mu\mathbb{H} = \mathrm{Hom}_{\mathcal{E}}(\mathcal{M}, \mu)$ (as a sheaf: $f \in \mathrm{Hom}\,(\mathcal{M}, \mu) \mapsto f(1)$; here as above μ denotes the sheaf of microfunctions). For example, in the holomorphic situation, I is the ideal generated by the components of $\bar{\partial}_b$.

We have $\mathrm{End}_{\mathcal{E}}(\mathcal{M}) = [I : I]$, the set of pseudodifferential operators a such that $Ia \subset I$, acting on the right: if $a \in [I : I]$, the corresponding endomorphism of M takes f (mod I) to fa (mod I); this vanishes if and only if $a \in I$. The map which takes $a \in [I : I]$ to the endomorphism $f \mapsto af$ of $\mathbb{H}$ defines an isomorphism from

[13] The converse is not true: if d is a locally free resolution of $\mathcal{M}$, its symbol is not necessarily a resolution of the symbol of $\mathcal{M}$, if only because filtrations must be defined to define the symbol and can be modified rather arbitrarily.

$\mathrm{End}_{\mathcal{E}}(\mathcal{M})$ to the algebra of Toeplitz operators mod C^∞. Thus $\mathcal{M}$ is an $\mathcal{E}_{T^\bullet M} - \mathcal{E}_\Sigma$ bimodule (where $\mathcal{E}_\Sigma \simeq \mathrm{End}_{\mathcal{M}}$ denotes the sheaf of Toeplitz operators mod C^∞).

This extends immediately to the case in which $T^\bullet M$ is replaced by an arbitrary symplectic cone Σ'' with base X''.[14] The small Toeplitz sheaf $\mu\mathbb{H}$ can be realized as $\mathrm{Hom}_{\mathcal{E}''}(\mathcal{M}, \mu\mathbb{H}'')$, where $\mathcal{M} = \mathcal{E}''/I$ and $I \subset \mathcal{E}''$ is the annihilator of the Szegő projector S of Σ (i.e., the null-sheaf of I in $\mathrm{Hom}_{\mathcal{E}''}(\mathcal{M}, \mathbb{H}'') = \mu\mathbb{H}$). If $\mathcal{P}$ is a (good) $\mathcal{E}$-module, the transferred module is $\mathcal{M} \otimes_{\mathcal{E}} \mathcal{P}$, which has the same solution sheaf ($\mathrm{Hom}_{\mathcal{E}''}(\mathcal{M} \otimes \mathcal{P}, \mathbb{H}'') = \mathrm{Hom}_{\mathcal{E}''}(\mathcal{P}, \mathrm{Hom}_{\mathcal{E}}(\mathcal{M}, \mathbb{H}''))$ and $\mathrm{Hom}_{\mathcal{E}''}(\mathcal{M}, \mathbb{H}'') = \mathbb{H}$). Thus the transfer preserves traces and indices.

The module $\mathcal{M} = \mathcal{E}''/I$ is generated by 1 (mod I) and has a natural filtration, which is a good filtration, in the holomorphic case, the good resolution is dual to the complex $\bar\partial_b$ on $(0, *)$ forms.

In general it always has a good locally free resolution, well defined up to homotopy equivalence. In a small tubular neighborhood of Σ one can choose this so that its symbol is the Koszul complex on $\bigwedge N'$, where N' is the dual of the normal tangent bundle of Σ equipped with a positive complex structure (as in the holomorphic case). The corresponding K-theoretic element $[\mathcal{M}] \in K^G_X(X'')$ is precisely the element used to define the Bott isomorphism (with support $Y \subset \Sigma$) $K^G_Y(\Sigma) \to K^G_Y(\Sigma'')$. (Here, Y is some set containing the support of $\sigma(\mathcal{M})$ and the map is the product map: $[E] \mapsto [\mathcal{M}][E]$, where the virtual bundle $[E]$ on Σ is extended arbitrarily to some neighborhood of Σ in Σ''.)[15]

For example, if Σ'' is $\mathbb{C}^N \backslash \{0\}$, with Liouville form $\mathrm{Im}\, \bar z \cdot dz$ and base the unit sphere $X'' = \mathbb{S}^{2N-1}$, $\mathbb{H}''$ is the space of boundary values of holomorphic functions, $\Sigma \subset \Sigma''$ consists of the nonzero vectors in the subspace $z_1 = \cdots = z_k = 0$, and $X \subset X''$ is the corresponding subsphere, then $\mathbb{H}$ consists of the functions independent of $z_1, \ldots, z_k$, and I is the ideal spanned by the Toeplitz operators $T_{\partial_1}, \ldots T_{\partial_k}$. In this example the ideal I is generated by $\bar z_1, \ldots, \bar z_k$, or by $T_{\bar z_j}$, $j = 1, \ldots, k$ (on the sphere we have $T_{\partial_j} = (A + N) T_{\bar z_j}$ with $A = T_{\sum_1^N z_j \partial_j}$). The $\mathcal{E}$-module $\mathcal{M}$ itself has a global resolution with symbol the Koszul complex constructed on $\bar z_1, \ldots, \bar z_k$.

What precedes works exactly as well in the presence of a compact group action. If $\mathcal{P}$ is a good module with support outside of Z (or a complex with symbol exact on Z), the transferred module has the same property ($Z \subset Z''$), and it has the same G-index (the G-index of the complex $\mathrm{Hom}_{\mathcal{E}}(\mathcal{M}, \mathbb{H}) \simeq \mathrm{Hom}_{\mathcal{E}''}(\mathcal{M}'', \mathbb{H}'')$).

If X, X'' are (compact) contact G-manifolds, $f : X \to X''$ an equivariant embedding, $\mathcal{P}$ a good $(G - \mathcal{E})$-module with support outside of Z (the base of $\mathrm{char}\, \mathfrak{g}$ in Σ), or a Toeplitz complex, exact on Z, the transferred module on X is $f_+\mathcal{P} = \mathcal{M} \otimes_{f_*\mathcal{E}'} f_*\mathcal{P}'$. This is exact outside of $f(\Sigma)$ and has the same G-index as $\mathcal{P}$; its K-theoretic invariant $[\mathcal{P}]$ is the image of $[\mathcal{P}]$ by the equivariant Bott homomorphism. The K-theoretic element $[f_+\mathcal{P}] \in K^G_{X-Z}(X)$ is the image of $[\mathcal{P}]$ by

---

14 We use a double prime here because, eventually, we will be embedding *two* cones in a third one.

15 Toeplitz operators (mod C^∞) live on Σ and their principal symbols are homogeneous functions on Σ. However, the K-theoretic element $[u] \in K^G(X - Z)$ of a G-elliptic element lives on the base X, so as the support of "good" $\mathcal{E}$-modules or complexes, in contrast to what happens for pseudodifferential operators.

the Bott homomorphism (it is well defined, since $f(Z) \subset Z''$). Thus we have the following Theorem.

Theorem 3. *Let $f : X \to X''$ be an equivariant embedding. The Bott homomorphism $K^G_{X-Z}(X) \to K^G_{X''-Z''}(X'')$ commutes with the asymptotic G index.*[16]

It is always possible to embed a compact contact manifold in a canonical contact sphere with linear G-action. In fact, it is easier to work with the corresponding cones, as follows:

Proposition 4. *Let Σ be a G-cone (with compact base), λ a horizontal 1-form, homogeneous of degree 1, i.e., $\rho \lrcorner \lambda = 0$, and $L_\rho \lambda = \lambda$, where ρ is the radial vector field, generating homotheties. Then there exists a homogeneous embedding $x \mapsto z(x)$ of Σ in a unitary representation space V^c of G such that $\lambda = \operatorname{Im} \bar{z} \cdot dz$.*

In the proposition, $z(x)$ must be homogeneous of degree $\frac{1}{2}$. This applies of course if Σ is a symplectic cone, λ its Liouville form. (The symplectic form is $\omega = d\lambda$ and $\lambda = \rho \lrcorner \omega$.)

We first choose a smooth equivariant function $y = (y_j)$, homogeneous of degree $\frac{1}{2}$, realizing an equivariant embedding of Σ in $V - \{0\}$, where V is a real unitary G-vector space (this always exists if the base is compact (the coordinates z_j on V are homogeneous of degree $\frac{1}{2}$, so that the canonical form $\operatorname{Im} \bar{z} \cdot dz$ is of degree 1)). Then there exists a smooth function $x = (x_j)$ homogeneous of degree $\frac{1}{2}$ such that $\lambda = 2x \cdot dy$. We can suppose x equivariant, replacing it by its G-mean if need be. Since y is of degree $\frac{1}{2}$ we have $2\rho \lrcorner dy = y$, hence $x \cdot y = \rho \lrcorner \lambda = 0$. Finally, we get

$$\lambda = \operatorname{Im} \bar{z} \cdot dz \quad \text{with } z = x + iy.$$

4 Relative index

As indicated in the introduction, we are considering the index of the Fredholm map $E_0 : u \mapsto S'(u \circ f_0^{-1})$ from $\mathbb{H}_0$ to $\mathbb{H}'_0$, where X_0, X'_0 are the boundaries of two smooth strictly pseudoconvex Stein manifolds Ω, Ω', $\mathbb{H}, \mathbb{H}'$ the spaces of CR distributions ($\ker \bar{\partial}_b$, equal to space of boundary values of holomorphic functions), S, S' the Szegő projectors, and f_0 a contact isomorphism $X_0 \to X'_0$.

As announced, we modify the problem and move to the larger boundaries X, X' of "balls" $|t|^2 < \phi$, $|t'|^2 < \phi'$ in $\mathbb{C} \times \Omega$, $\mathbb{C} \times \Omega'$, on which the circle group acts ($t \mapsto e^{i\lambda} t$) (Section 4.1). We will see (Section 4.2) that the Toeplitz FIO E_0 defines almost canonically an equivariant extension F which is $U(1)$-elliptic, and Index $(F|_{\mathbb{H}_k}) =$ Index (E_0) for all k ($\mathbb{H}_k \subset \mathbb{H}(X)$ is the subspace of functions $f = t^k g(x)$), so that

16 As mentioned above, the interplay between the Bott isomorphism and embeddings of systems of differential or pseudodifferential operators lies at the root of Atiyah–Singer's proof of the index theorem; it is described in M.F. Atiyah's works [1, 2, 3, 4]; cf. also [11] in the context of holomorphic D-modules, close to the Toeplitz context.

our relative index Index (E_0) appears as an asymptotic equivariant index, easier to handle in the framework of Toeplitz operators.

In Section 4.3 we will show that the whole situation can be embedded in a large sphere, with action of $U(1)$ as in the examples above. In the final result (Section 4.4) the relative index appears as the asymptotic index of an equivariant $U(1)$-elliptic Toeplitz complex on this large sphere. In general, the equivariant index (asymptotic or not) is rather complicated to compute, but in our case the $U(1)$-action is quite simple.[17] It reduces naturally to the standard Atiyah–Singer K-theoretic index formula on a symplectic ball. The result is better stated in terms of K-theory anyway, but it can be translated via the Chern character in terms of cohomology or integrals. We give here a (rather clumsy) cohomological-integral translation, essentially equivalent to the result conjectured in [27].

We will also see below (Section 4.2) that f_0 has an almost canonical extension f near the boundary, well defined up to isotopy, not holomorphic but symplectic. We can then define a space Y by gluing together Y_+, Y_- by means of f. The space Y is not a Hausdorff manifold, but it is symplectic and both Y_+, Y_- carry orientations which agree on their intersection (as do the symplectic structures). We can further choose differential forms $\nu_\pm$ representatives of the Todd classes of $Y_\pm$ so that they are equal near the boundary X_0 (the symplectic structures agree, not the complex structures, but they define the same Todd classes).

Theorem 5. *The relative index (index of E_0) is the integral $\int_Y(\nu_+ - \nu_-)$, where $\nu_\pm$ are representatives of* Todd($Y_\pm$) *as above, so that the difference has compact support in $Y - X_0$.*

This will be explained in more detail below (Section 4.4). This formula is related to the Atiyah–Singer index formula on the glued space Y, but is not quite the same, since Y is not a symplectic manifold.

To prove the index theorem we will give an equivalent equivariant description of the situation, where the index of E_0 is repeated infinitely many times, and embed everything in a large sphere where the index is given by the K-theoretic index character (Section 4.4).

4.1 Holomorphic setting

Let Ω be a strictly pseudoconvex domain (or Stein manifold), with smooth boundary X_0; $\bar\Omega = \Omega \cup X_0$ is assumed to be compact. We write $\widetilde\Omega \subset \mathbb{C} \times \bar\Omega$ for the ball $|t|^2 < \phi$, where ϕ is a defining function ($\phi = 0, d\phi \neq 0$ on X_0, $\phi > 0$ inside). ϕ is chosen so that the boundary $X = \partial\widetilde\Omega$ is strictly pseudoconvex, i.e., Log $\frac{1}{\phi}$ is strictly plurisubharmonic (i.e., Im $\bar\partial\partial\frac{1}{\phi} \gg 0$).

The circle group $U(1)$ acts on X by $(t, x) \mapsto (e^{i\lambda}t, x)$. We choose as volume element on X the density $d\theta\, dv$, where dv is a smooth positive density on $\overline\Omega$

[17] It is free on the support of the K-theoretic symbol of our complex.

($t = e^{i\theta}|t|$). This is a smooth positive density on X; it is invariant by the action of $U(1)$ as are the Szegő projector S and its range $\mathbb{H}$, the space of boundary values of holomorphic functions.

The infinitesimal generator of the action of $U(1)$ is ∂_θ, and we denote by D the restriction to $\mathbb{H}$ of $\frac{1}{i}\partial_\theta$, which is a self-adjoint nonnegative Toeplitz operator; D is the restriction of $T_t T_{\partial_t}$.

The model case is the sphere $\mathbb{S}^{2N+1} \subset \mathbb{C}^{N+1}$ with the action

$$(t = z_0, z = (z_1, \ldots, z_N)) \mapsto (e^{i\theta}t, z).$$

The Fourier decomposition of $\mathbb{H}$,

$$\mathbb{H} = \widehat{\bigoplus}_{k\geq 0}\, \mathbb{H}_k \qquad (\mathbb{H}_k = \ker(D - k)),$$

corresponds to the Taylor expansion of holomorphic functions: the kth component of $f = \sum f_k(x)t^k \in \mathbb{H}$ is $f_k t^k$.

$\mathbb{H}_0$ identifies with the set of holomorphic functions on X_0 (it is the set of boundary values of holomorphic functions on Ω with moderate growth at the boundary, i.e., $|f| \leq c\ d(\cdot, X_0)^{-N}$ where c and N are constants, and $d(\cdot, X_0)$ is the distance to the boundary).

Remark. If $f = t^k g(x)$ with g continuous, in particular if $f \in \mathbb{H}_k$, its $L^2(X)$ norm is

$$\|f\|_{L^2(X)} = \frac{\pi}{k+1}\int_\Omega \phi^{k+1}|g(x)|^2 dv,$$

where as above, dv is the chosen as the smooth volume element on Ω. The restriction of the Szegő projector to functions of the form $t^k g(x)$ is thus identified with the orthogonal projector on holomorphic functions in $L^2(\Omega, \phi^{k+1}dv)$. Such sequences of projectors were considered by F.A. Berezin [5] and further exploited by M. Englis [14, 15, 16], whose presentation is closely related to the one used here.

For the sequel, it will be convenient to modify the factorisation $D = t\partial_t$. We begin with the easy following result.

Lemma 6. *Let $D = PQ$ be any factorisation where P, Q are Toeplitz operators and $[D, P] = P$. Then there exists a (unique) invariant invertible Toeplitz operator U such that $P = tU$, $Q = U^{-1}\partial_t$.*

Indeed it is immediate that any homogeneous function a on σ such that $\frac{1}{i}\partial_\theta a = \pm a$ is a multiple mt of t (respectively of $\bar{t}$), with m invariant. For the same reason (or by successive approximations) a Toeplitz operator A such that $[D, A] = \pm A$ is a multiple of $T_t M$ (or $M' T_t$) T_t with M or M' invariant (respectively of T_{∂_t}, on the right or on the left). Thus in the lemma above we have $P = T_t U$, $Q = U' T_{\partial_t}$, where U, U' are Toeplitz operators which necessarily commute with D, and are elliptic and inverses of each other.

Note that $D = PQ, [D, P] = P$ is equivalent to $D = PQ, [Q, P] = 1$.

In particular, since $D = D^* = T^*_{\partial_t} T^*_t$, there exists a Toeplitz operator A such that $T_{\partial_t} = A T^*_t$; A is elliptic of degree 1 (in fact invertible), positive since $D = T_t A T^*_t$ is self-adjoint ≥ 0; it is also invariant: $[D, A] = 0$.

Definition 4. We will set $\mathcal{T} = T_t A^{\frac{1}{2}}$; its symbol is denoted by $\sigma(\mathcal{T}) = \tau$.

Note that τ is homogeneous of degree $\frac{1}{2}$, and $\mathcal{T}$ is of degree $\frac{1}{2}$, so it is not a Toeplitz operator in our strict sense, but for multiplications and automorphisms $P \mapsto UPU^{-1}$ it is just as good. We have

$$\mathcal{T}^* = A^{\frac{1}{2}} T^*_t, \quad [D, \mathcal{T}] = \mathcal{T} \cdot D = \mathcal{T}\mathcal{T}^* \tag{3}$$

(for any other such factorisation $D = BB^*$ with $[D, B] = B$, B is of degree $\frac{1}{2}$, and we have $B = \mathcal{T} U$ with U invariant and unitary; $\mathcal{T}$ is the unique Toeplitz operator giving such a factorisation and such that $\mathcal{T} = T_t A'$ with A' a Toeplitz operator of degree $\frac{1}{2}$, $A' \gg 0$).

In what precedes, all = signs can be replaced by $\sim$ (= mod C^∞); we then get local statements.

The symbol $\tau = \sigma(\mathcal{T})$ is the unique homogeneous function of degree $\frac{1}{2}$ such that $\sigma(D) = |\tau|^2$, $\partial_\theta \tau = i\tau$, $\frac{\tau}{t} > 0$.

We also have the following (easy) local result:

Lemma 7. *Given any Toeplitz operator $\mathcal{K}$ (mod C^∞) on $\mathbb{H}$ such that $D \sim \mathcal{K}\mathcal{K}^*$, $[D, \mathcal{K}] = \mathcal{K}$ near the boundary, there exists a unique unitary equivariant Toeplitz FIO F such that $F|_{\mathbb{H}_0} \sim \mathrm{Id}$, $F\mathcal{T} \sim \mathcal{K}F$.*

The geometric counterpart is this: given any function k on Σ homogeneous of degree $\frac{1}{2}$ such that $\sigma(D) = k\bar{k}$ there exists a unique germ of homogeneous symplectic isomorphism f such that $f|_{\Sigma_0} = \mathrm{Id}$, $k \circ f = \tau$. This is immediate because the two Hamiltonian pairs $H_\tau, H_{\bar{\tau}}, H_k, H_{\bar{k}}$ define real 2-dimensional foliations, and an isomorphism $\Sigma \sim \Sigma_0 \times \mathbb{C}$ near Σ_0. Note that this would not work if we replaced $k, \bar{k}$ by two functions a, b such that $\sigma(D) = ab$, $\partial_\theta a = ia$ but not $b = \bar{a}$, because then the "foliation" defined by the Hamiltonian vector fields H_a, H_b, although it is formally integrable, is not real.

The operator statement follows, e.g., by successive approximations. Note that F is completely determined by its restriction F_0 if it commutes with $\mathcal{T}$. In fact in $\mathcal{E}_\Sigma$, the commutator sheaf of $\mathcal{T}$ and $\mathcal{T}^*$ identifies with the inverse image of $\mathcal{E}_{\Sigma_0}$, at least as far as the leaves of the Hamiltonian fields $H_\mathcal{T}, H_{\mathcal{T}^*}$ define a fibration over Σ_0: $\mathcal{E}_\Sigma$ is the (completed) tensor product of the Toeplitz algebra $\mathrm{Toepl}(\mathcal{T}, \mathcal{T}^*)$ generated by $\mathcal{T}$ and $\mathcal{T}^*$ and this commutator: $\mathcal{E}_\Sigma \sim \mathcal{E}_{\Sigma_0} \otimes \mathrm{Toepl}(\mathcal{T}, \mathcal{T}^*)$ (in a neighborhood of Σ_0). In this statement, $(\mathcal{T}, \mathcal{T}^*)$ cannot be replaced by (T_t, T_{∂_t}), whose commutator sheaf is defined only in the algebra of jets of infinite order along Σ_0, because the Hamiltonian leaves are complex, no longer real.

Note that in our case, the base of char $\mathfrak{g}$ is the boundary X_0 (the set of fixed points), outside of which D is elliptic.

4.2 Collar isomorphisms

Let now Ω' be another strictly pseudoconvex domain (or Stein manifold) with smooth boundary X'. We do the similar constructions $\tilde{\Omega}', \mathbb{H}'$, and D', ... as in the previous subsection. Let $f_0 : X_0 \to X'_0$ be a contact isomorphism.

We define the Fourier Toeplitz operator $E_0 : u \mapsto S'(u \circ f_0^{-1}) : \mathbb{H} \to \mathbb{H}'$, which is a Fredholm operator. It will be convenient to replace E_0 by $F_0 = (E_0 E_0^*)^{-\frac{1}{2}} E_0$, which has the same index and is unitary mod C^∞ ($E_0 E_0^*$ is an elliptic ≥ 0 Toeplitz operator on $\mathbb{H}'$); $(E_0 E_0^*)^{-\frac{1}{2}}$ is defined to be 0 on $\ker E_0^*$ (mod C^∞ would be quite enough). As for $\tilde{\Omega}$, we construct a Toeplitz operator $\mathcal{T}'$ such that $D' = \mathcal{T}'\mathcal{T}'^*, [D'\mathcal{T}'] = \mathcal{T}', T_t'^{-1}\mathcal{T}' \gg 0$.

Exactly as in Lemma 4.2, there exists a unique (unitary) Toeplitz FIO F, defined near the boundary X_0 and mod C^∞, elliptic, such that $F|_{\mathbb{H}_0} = F_0$, and $F\mathcal{T} \sim \mathcal{T}'F$ near the boundary (mod C^∞).

The geometric counterpart is this: there exists a unique equivariant germ of contact isomorphism $f : X \to X'$ (defined and invertible near the boundary) such that $f|_{X_0} = f_0,\ \tau' = \tau \circ f$.

We may extend F, using an invariant cut-off Toeplitz operator, so that it vanishes (mod C^∞) away from the boundary. There is an invariant FIO parametrix F', i.e., $F'F \sim 1_{\mathbb{H}}, FF' \sim 1_{\mathbb{H}'}$, near the boundary.

Proposition 8. *For any k, $F_k = F|_{\mathbb{H}_k}$ has an index, equal to* Index F_0.

Proof. Both $F'F$ and FF' are invertible on the boundary, so have a G-index; the index of $F_k = F|_{\mathbb{H}_k}$ is $\operatorname{tr}(1 - F'F)_k - \operatorname{tr}(1 - FF')_k$. Now $\mathcal{T}$, respectively $\mathcal{T}'$, is an isomorphism $\mathbb{H}_k \to \mathbb{H}_{k+1}$, respectively $\mathbb{H}'_k \to \mathbb{H}'_{k+1}$, and we have $\operatorname{Index}(F_{k+1}A) = \operatorname{Index}(A'F_k)$, so $\operatorname{Index} F_{k+1} = \operatorname{Index} F_k$, i.e., the index does not depend on k and is equal to Index E_0.[18]

The asymptotic index is stable by embedding; here the index is constant, and the asymptotic index of F (which is essentially a Toeplitz invariant) gives the index of F_0 itself.

4.3 Embedding

Theorem 9. *Let $f : X \to X'$ be a collar isomorphism defined in some invariant neighborhood of X_0 in X. Then for large N there exist equivariant contact embeddings $U : X \to \mathbb{S}^{2N+1}$, $U' : X' \to \mathbb{S}^{2N+1}$ such that $U = U' \circ f$ near the boundary, and $t_X, t'_{X'}$ map to positive multiples of $t_{\mathbb{S}^{2N+1}}$ (as above, the contact sphere $\mathbb{S}^{2N+1}$ is equipped with the $U(1)$-action $(t, z) \mapsto (e^{i\theta} t, z)$).*

[18] For a more general situation in which P is a Toeplitz operator elliptic on X_0, or in which the canonical Szegő projector is replaced by some other general equivariant one, we would get only that the index Index (P_k) is constant for $k \gg 0$. Here the fact that Index P_k = Index P_0 is obvious but important.

As usual, it will be more comfortable to work with the symplectic cones. The symplectic cone of X is $\Sigma = \mathbb{R}_+ \times X$, where we choose the radial coordinate invariant.

The symbol of D is $\bar{\tau}\,\tau$ with $\tau/t > 0$ as in Definition 4. The Liouville form is $\mathrm{Im}\,(\bar{\tau}d\tau) + \lambda_0$, where λ_0 is a horizontal form, i.e., the pull-back of a form on $\Sigma_b = U(1)\backslash\Sigma \simeq \mathbb{R}_+ \times \bar{\Omega}$ (equivalently: $\partial_\theta \lrcorner \lambda_0 = L_{\partial_\theta}\lambda_0 = 0$).

Lemma 4 provides an embedding $x \mapsto z_b(x)$ of Σ_b in $\mathbb{C}^{N'} - \{0\}$ (with the trivial action of $U(1)$). We now choose real functions ψ_1, ψ_2 invariant, homogeneous of degree 0, such that $\psi_1^2 + \psi_2^2 = 1$, with supp ψ_1 contained in the domain of definition of f and ψ_2 vanishing near the boundary, and we construct a new embedding z in three pieces: $z = (z_1, z_2, z_3)$ with $z_1 = \psi_1 z_0, z_2 = \psi_2 z_0, z_3 = 0$ in $\mathbb{C}^{N''}$, N'' to be defined below.

Since $\mathrm{Im}\,\bar{z}_j z_j \psi_j d\psi_j = 0$ ($\bar{z}_j z_j \psi_j d\psi_j$ is real), we still have $\mathrm{Im}\,(\bar{z}_1 \cdot dz_1 + \bar{z}_2 \cdot dz_2) = (\psi_1^2 + \psi_2^2)\mathrm{Im}\,\bar{z_0} \cdot dz_0 = \mathrm{Im}\,\bar{z_0} \cdot dz_0$ inducing λ_0. The first embedding is $U = (\tau, v) : \Sigma \to \mathbb{C}^{1+N}$ $(N = 2N' + N'')$.

Similarly there exists an embedding $x' \mapsto z_0'(x')$ of Σ_b' in $\mathbb{C}^{N''} - \{0\}$ (with the trivial action of $U(1)$).

We replace this by $z' = (z_1', z_2', z_3')$ with $z_1' = \psi_1' z_1 \circ f^{-1}, z_2' = 0, z_3' = \psi_3' z_0'$, where ψ_1', ψ_3' again are invariant, homogeneous of degree 0, $\psi'^2_1 + \psi'^2_3 = 1$, and supp ψ_1' is contained in the domain of definition of f^{-1}, ψ_3' vanishes near the boundary. This also defines an embedding $U' = (a', z') : \Sigma' \to \mathbb{C}^{N+1}$; we have $U = U' \circ f$ near the boundary, since ψ_2, ψ_3' vanish there.

4.4 Index

We are now reduced to the case in which both $U(1)$-manifolds X, X' sit in a large sphere $\mathbb{S} = \mathbb{S}^{2N+1}$ and coincide near the set of fixed points $\mathbb{S}_0$.

As in the preceding section, we can embed the $U(1)$ sheaves $\mu\mathbb{H}_X, \mu\mathbb{H}_{X'}$ as sheaves of solutions of two good equivariant $\mathcal{E}_{\mathbb{S}}$-modules $\mathcal{M}_X, \mathcal{M}_{X'}$, and the identification F gives an equivariant Toeplitz isomorphism $\widetilde{F}$ near X_0 (we can make the construction so that $\mathcal{M}_X = \mathcal{M}_{X'}$, $\widetilde{F} = \mathrm{Id}$ near X_0).

The asymptotic index then depends only on the difference element

$$d([\mathcal{M}_X], [\mathcal{M}_{X'}], \sigma(\widetilde{F})) \in K^{U(1)}(\mathbb{S} - \mathbb{S}_0).$$

Now $U(1)$ acts freely on $\mathbb{S} - \mathbb{S}_0$, with quotient space $U(1)\backslash(\mathbb{S} - \mathbb{S}_0)$ the open unit ball $\mathbb{B}_{2N} \subset \mathbb{C}^N$. We have the following result.

Proposition 10. *The pull-back map is an isomorphism $K(\mathbb{B}) \to K^{U(1)}(\mathbb{S} - \mathbb{S}_0)$.*

We have $K(\mathbb{B}) \sim \mathbb{Z}$, with generator the symbol of the Koszul complex k_x at the origin (or any point of the interior), whose index is 1.

Its pull-back is the generator of $K^{U(1)}_{\mathbb{S}-\mathbb{S}_0}(\mathbb{S})$: the symbol is the same, but now acting on $\mathbb{H}(\mathbb{S})$. Its index is $\sum_0^\infty J^k$, where (as in (2)) J is the tautological character of $U(1)$: $J(e^{i\lambda}) = e^{i\lambda}$.

The first assertion is immediate (cf. [4]): if G is a compact group acting freely on a space Y, the pull-back defines an equivalence from the category of vector bundles on $G\backslash Y$ to that of G-vector bundles on Y (an inverse equivalence is given by $E \mapsto G\backslash E$), and this induces a bijection on K-theory (with supports).

The fact that k_x defines the generator of $K(\mathbb{B})(= K_0(\mathbb{B}))$ is just a restatement of Bott's periodicity theorem. Its pull-back is then the generator of $K^{U(1)}(\mathbb{S} - \mathbb{S}_0)$: the corresponding complex of Toeplitz operators is then the standard Koszul complex, acting on holomorphic functions, whose index is the space of holomorphic functions of $z_0 = t$ alone.

Thus if $[u] \in K^{U(1)}(\mathbb{S} - \mathbb{S}_0)$, its asymptotic index is $m\sum_{k=0}^\infty J^k$, where the integer m is the value of the K-theoretic character $K(\mathbb{B})$ on the element $[u_\mathbb{B}]$ whose pull-back is $[u]$.

Let us now return to our index problem: we have constructed the difference bundle $d([\mathcal{M}_X], [\mathcal{M}_{X'}], \sigma(\widetilde{F}))$. We may replace $\mathcal{M}_X, \mathcal{M}_{X'}$ by good resolutions in small equivariant tubular neighborhoods of X, respectively X', whose K-theoretic symbol is the Bott element: the Koszul complex for a positive complex structure on the normal symplectic bundle of X, respectively X'. $\widetilde{F}$ lifts to the resolutions (uniquely up to homotopy), and the symbol of the lifting u is an isomorphism near X_0 (we can make the construction so that $u = \mathrm{Id}$ near X_0), so our K-theoretic element is $[u] = d(\beta_X, \beta_{X'}, u)$ (the equivariant K-theoretic element attached to the double complex defined by u).

Theorem 11. *Let m be the index of E_0 we are investigating. Then, notation and embeddings being as above,*

(1) the asymptotic index of our equivariant extension $\widetilde{F}$ is the asymptotic index of the difference element $[u] = d(\beta_X, \beta'_X, u) \in K^{U(1)}(\mathbb{S} - \mathbb{S}_0)$, where u is the symbol of $\widetilde{F}$ (i.e., the identity map near $\mathbb{S}_0$, where X and X' coincide);
(2) the index m itself is the value of the index character of $K(\mathbb{B})$ on the element $[u_\mathbb{B}] = d(\beta_\Omega, \beta_{\Omega'}, \bar{u})$.

The first part has just been proved. The asymptotic index is $\sim m(1 - J)^{-1}$ for some integer m.

To prove the second we go down to $\mathbb{B}_{2N}$. The bases of X, X' are the embeddings Y_+, Y_- of Ω, Ω' in $\mathbb{B}$, which coincide near the boundary, and as above, the pull-back is an isomorphism $K_{Y_\pm}(\mathbb{B}) \to K^G_{X_\pm}(\mathbb{S} - \mathbb{S}_0)$. The Bott complexes $\beta_{X_\pm}$ descend as Bott elements $\beta_{Y_\pm}$ on $\mathbb{B}$, realized as Koszul complexes of positive complex structure of the normal symplectic bundle;[19] u descends as an isomorphism near the boundary.

[19] Note that $Y_\pm$ are symplectic submanifolds, not complex; but all positive complex structures are homotopic.

The index m we are looking for is the K-theoretic index character of the difference element $d(\beta_{Y_+}, \beta_{Y_-}, u)$. This can as usual be translated in terms of cohomology, or as an integral:

$$m = \int_{\mathbb{B}} \omega,$$

where ω is a differential form with compact support, representative of the Chern character of our difference element $d(\beta_{Y_+}, \beta_{Y_-}, u)$.

We can push this down further. The construction can be made so that $u = \mathrm{Id}$ near the boundary; choose differential forms $\omega_\pm$ with support in small tubular neighborhoods of $Y_\pm$ so that they coincide near the boundary (as do the tubular neighborhoods), so that ω is the difference $\omega_+ - \omega_-$.

The integral $\nu_\pm$ of $\omega_\pm$ over the fibers of the respective tubular neighborhoods is then a representative of the Todd class of $Y_\pm$; ν_+ and ν_- coincide near the boundary, so that the difference $\nu_+ - \nu_-$ has compact support in $Y = Y_+ \cup Y_-$.

Finally, the index m is the integral $\int_Y (\nu_+ - \nu_-)$ as announced in Theorem 5.

The integral can also be thought of as the constant limit $\int_{Y_{+,\epsilon}} \nu_+ - \int_{Y_{-,\epsilon}} \nu_-$, where the subscript ϵ means that we have deleted the neighborhood $\phi < \epsilon$ in Y_+ and the corresponding image in Y_-.

4.5 Appendix

In this section we show how various symplectic extensions of f_0 are related. It is a little intriguing that although in our proof, the extension f must be chosen rather carefully so that the asymptotic index of the corresponding Toeplitz FIO E is (asymptotically) the index of E_0, the final result, expressed as an integral on the bases glued together by means of f near their boundaries, depends only on the isotopy class of f, which is unique.

4.5.1 Contact isomorphisms and base symplectomorphisms

Let X be as above, with X_0 the fixed-point set of codimension 2. Near the boundary, X is identified with $X = X_0 \times \mathbb{C}$, and the base $U(1)\backslash X \sim \Omega$ identifies with $X_0 \times \mathbb{R}_+$; we have $\phi = t\bar{t}$ and the $\mathbb{C}$-coordinate is $t = \sqrt{\phi}\, e^{i\theta}$ (it is smooth on X). The contact form is $\lambda_X = \mathrm{Im}\,(\bar{t}\, dt - \partial\phi) = \phi\, d\theta + \lambda_\Omega$, where $\lambda_\Omega = -\mathrm{Im}\,\partial\phi$ is a smooth basic form. The connection form is $\gamma = d\theta - \frac{\lambda_\Omega}{\phi}$, and the base $\Omega = X_0 \times \mathbb{R}_+$ is equipped with the (basic) symplectic curvature form

$$\mu = d\gamma \qquad \left(\text{with } \gamma = \frac{\lambda_\Omega}{\phi}, \quad \lambda_\Omega = -\mathrm{Im}\,\partial\phi\right).$$

We will still use the symplectic cone of X; this is $\Sigma = \mathrm{char}\, \mathfrak{g} \simeq \mathbb{R}_+ \times X$, with Liouville form $a\lambda_X$ and symplectic form its derivative, with the $\mathbb{R}_+$ coordinate a

defined below. With the notation of Lemma 4, we have $a = \sigma(A)$, i.e., $\sigma(D) = a\phi = \tau\bar{\tau}, \tau = t\sqrt{a}$ (as above, $D = \frac{1}{i}T_{\partial_\theta}$ denotes the infinitesimal generator of rotations). We will also write in polar coordinates $\tau = \rho\, e^{i\theta}$ ($\rho = \sqrt{\phi\, a}$).

Let F be a homogeneous equivariant symplectic transformation of Σ. Then F preserves $\sigma(D) = \tau\bar{\tau}$, so we have necessarily $F_*\tau = u\ \tau$, with u invariant, $|u| = 1$. Then F is completely determined by its restriction to the boundary, since it commutes with the two real commuting Hamiltonian vector fields Re H_τ, Im H_τ, which are linearly independent and transversal to Σ_0.

Thus there is a one-to-one correspondence between unitary functions on the base Ω and germs near $\Sigma_0 = \mathrm{char}\,\mathfrak{g}$ of equivariant symplectomorphisms inducing Id on $\mathrm{char}\,\mathfrak{g}$, or equivalently of contact automorphisms of X inducing Id on X_0.

If F is such a contact automorphism, the base map F_Ω is obviously a diffeomorphism of Ω which induces Id on the boundary X_0 and preserves the symplectic form μ.

The converse is not true. If F_Ω is a smooth symplectomorphism of Ω inducing the identity on X_0, we have $F_\Omega^*(\frac{\lambda_\Omega}{\phi}) = \frac{\lambda_\Omega}{\phi} + \alpha$ with α a closed form. It is elementary that $\alpha = c\frac{d\phi}{\phi} + \beta$, where c is a constant and β is smooth on the boundary. Locally on X_0, F_Ω lifts to X or Σ: the lifting is $F : (x, \tau) \mapsto (x', \tau' = \tau e^{i\psi})$ $(\theta' = \theta + \psi)$, where ψ is a primitive of α (this is not smooth at the boundary, only continuous). It is immediate that conversely, any α of the form above gives rise to such a contact isomorphism with smooth base map. (On Σ the horizontal (invariant) coordinates satisfy $H_{\tau e^{i\psi}} f = 0$; the horizontal part of the Hamiltonian $H_{\tau e^{i\psi}}$ is $-i\tau e^{i\psi}(\partial_\rho - H_\psi^0)$ (with $H_\psi^0 = \psi_{\xi_j}\partial_{x_j} - \psi_{x_j}\partial_{\xi_j}$); finally, $\partial_\rho - H_\psi^0$ is smooth, so the horizontal coordinates (x', ξ) are determined by smooth differential equations.) Summing up:

Theorem 12. *The map which to a germ of contact isomorphism F (near X_0) assigns the invariant unitary smooth function u such that $F^*\tau = \tau u$ is one-to-one (and continuous). In particular, the homotopy class of F is determined by that of u (an element of $H^1(X, \mathbb{Z})$).*

The map which to a smooth germ of symplectomorphism F_Ω (near X_0) assigns the closed one-form $\alpha = c\,\frac{d\phi}{\phi}$ + smooth *is one-to-one; the group of such symplectomorphisms is contractible. The contact lifting (which exists locally, and globally if α is exact) is smooth on X_0 if and only if $c = 0$.*

The fact that this group is contractible (connected) simplifies the final result. Namely, in the proof of Theorem 11 it was essential that the base map F_Ω have a smooth symplectic extension preserving $\tau > 0$; for Theorem 5, however, any symplectic F_Ω will do, since these are all isotopic.

4.5.2 Example

(A smooth symplectic automorphism of the base does not lift to a smooth equivariant contact automorphism of the sphere.)

Let $\mathbb{S}$ be the unit sphere in $\mathbb{C}^{N+1}$, with coordinates $x_0 = t, x_1, \ldots, x_N$.

$U(1)$ acts by $t \mapsto e^{i\theta}t$. The base is $\mathbb{B} = \mathbb{S}/U(1)$, the unit ball of $\mathbb{C}^N$.

The contact form is $\operatorname{Im} \bar{t}\, dt + \lambda = \phi\, d\theta + \lambda$ with $\lambda = \sum \bar{x}_j dx_j, \phi = \bar{t}t = 1 - \bar{x}x$.

The connection form is $\gamma = d\theta + \frac{\lambda}{\phi}$; its curvature is the symplectic form $\mu = d\frac{\lambda}{\phi}$ (on the interior of $\mathbb{B}$).

Let F_B be the diffeomorphism of $\mathbb{B}$ defined by $x \mapsto x' = F_B(x) = e^{ci\phi}x$, c a constant; this preserves ϕ, and the inverse is $x = e^{-ci\phi}x'$. We have

$$F_B^*\lambda = \operatorname{Im}(\bar{x}(dx + cix\, d\phi)) = \lambda + c(1-\phi)d\phi.$$

Since $d(1-\phi)\frac{d\phi}{\phi} = 0$, F_B is symplectic ($F_B^*\mu = \mu$).

But F_B does not lift to a smooth equivariant contact automorphism of $\mathbb{S}$: such a lifting F must preserve the connection form, so it is of the form

$$t \mapsto e^{-i\alpha}t \quad (\theta \mapsto \theta - \alpha) \quad \text{with } \alpha = c\operatorname{Log}\phi - \phi + \text{const}$$

($d\alpha = c(1-\phi)\frac{d\phi}{\phi}$), and this is not smooth at the boundary $t = 0$ if $c \neq 0$.

Of course the reverse works: if F is a smooth equivariant contact automorphism of the sphere $\mathbb{S}$ (or a germ of such near the fixed diameter $\mathbb{S}_0$), the base map F_B is a smooth symplectomorphism of the ball $\mathbb{B}$ (up to the boundary).

4.6 Final remarks

(1) The preceding construction applies in particular to the following situation: let V, W be two compact manifolds, and f_0 a contact isomorphism $S^*V \to S^*W$.

We may suppose V real analytic; then S^*V is contact isomorphic to the boundary of small tubular neighborhoods of V in its complexification. For example, if V is equipped with an analytic Riemannian metric, and $(x, v) \mapsto e_x(v)$ denotes the geodesic exponential map, the map $(x, v) \mapsto e_x(iv)$ is well defined for small v, and for small ϵ it realizes a contact isomorphism of the tangent (or cotangent) sphere of radius ϵ to the boundary of the complex tubular neighborhood of radius ϵ (cf. [6]).

The corresponding FIOs can be described as follows: as above, there exists a complex phase (as in [25, 26]) function ϕ on $T^*W \times T^*V^0$ such that (1) ϕ vanishes on the graph of f_0 and $d\phi = \xi.dx - \eta.dy$ there; (2) $\operatorname{Im}\phi \gg 0$, i.e., it is positive outside of the graph and the transversal Hessian is $\gg 0$. Then ϕ is a global phase function for FIO associated to f_0 (ϕ is not unique, but obviously the set of such functions is convex, hence contractible).

The elliptic FIOs we are interested in are those that can be defined by a positive symbol (or a symbol isotopic to 1):

$$f \mapsto g(x) = \int e^{i\phi} a(x, \xi, y, \eta) f(y)\, dy\, d\eta\, d\xi \quad \text{with } a > 0 \text{ on the graph.}$$

The degree of such operators depends on the degree of a, but they all have the same index, given by the formula above.

(2) The formula above extends also to vector bundle cases: if E, E' are holomorphic vector bundles (or complexes of such) on Ω, Ω', f_0 a contact isomorphism ($\partial\Omega \to \partial\Omega'$) as above, and A a smooth (not holomorphic) isomorphism $f_{0*}E \to E'$ on the boundaries, the Toeplitz operator $a \mapsto S'(Af_{0*}a)$ is Fredholm and its index is given by the same construction as above. For this construction f_0 needs to be defined only where the complexes are not exact.

In particular let Ω, Ω' have singularities (isolated singularities, since we still want smooth boundaries): we can embed Ω, Ω' in smooth strictly pseudoconvex domains $\widetilde{\Omega}, \widetilde{\Omega}'$ of the same (higher) dimension; the contact isomorphism extends at least in a small neighborhood of $\partial\Omega$ in $\partial\widetilde{\Omega}$. The coherent sheaves $\mathcal{O}_\Omega, \mathcal{O}_{\Omega'}$ have global locally free holomorphic resolutions on $\widetilde{\Omega}, \widetilde{\Omega}'$; near the boundary these are homotopy equivalent to a Koszul complex, hence equivalent.

The theorem above shows that the relative index is the K-theoretic character of the difference virtual bundle $d([\mathcal{O}_\Omega], [\mathcal{O}_{\Omega'}])$ (vanishing near the boundary). Note, however, that the virtual bundles $[\mathcal{O}_\Omega], [\mathcal{O}_{\Omega'}]$ lie in the K-theory of $\widetilde{\Omega}$ with support in Ω. This can be readily described in terms of cohomology classes on $\widetilde{\Omega}$, etc., with support in Ω, not on Ω itself (the relation between coherent holomorphic modules and topological K-theory, or K-theory and cohomology, is not good enough when there are singularities).

References

1. Atiyah, M. F. *K-theory*. W. A. Benjamin, Inc., New York-Amsterdam, 1967.
2. Atiyah, M. F.; Singer; I. M. *The index of elliptic operators. I, II, III*. Ann. of Math. (2) 87 (1968) 484–530, 531–545, 546–604.
3. Atiyah, M. F. *Bott periodicity and the index of elliptic operators*. Quart. J. Math. Oxford Ser. (2) 19 (1968), 113–140.
4. Atiyah, M. F. *Elliptic operators and compact groups*. Lecture Notes in Mathematics, Vol. 401. Springer-Verlag, Berlin-New York, 1974.
5. Berezin, F. A. *General concept of quantization*. Comm. Math. Phys. 40 (1975), 153–174.
6. Boutet de Monvel, L. *Convergence dans le domaine complexe des séries de fonctions propres*. C.R.A.S. 287 (1978), 855–856.
7. Boutet de Monvel, L. *On the index of Toeplitz operators of several complex variables*. Inventiones Math. 50 (1979), 249–272.
8. Boutet de Monvel, L. *Symplectic cones and Toeplitz operators* (actes du congrès en l'honneur de Trèves, São Carlos). Contemporary Math. vol. 205 (1997), 15–24.
9. Boutet de Monvel, L. *Vanishing of the logarithmic trace of generalized Szegő projectors*. arXiv:math.AP/0604166 v1, Proceedings of the Conference "Algebraic Analysis of Differential Equations" in honor of Prof. T. Kawai, Springer-Verlag, 2007.
10. Boutet de Monvel, L.; Guillemin, V. *The Spectral Theory of Toeplitz Operators*. Ann. of Math Studies no. 99, Princeton University Press, 1981.
11. Boutet de Monvel, L.; Malgrange, B. *Le théorème de l'indice relatif*. Ann. Sci. ENS. 23 (1990), 151–192.
12. Boutet de Monvel, L.; Sjöstrand, J. *Sur la singularité des noyaux de Bergman et de Szegő*. Astérisque 34–35 (1976), 123–164.
13. Duistermaat, J.J.; Hörmander, L. *Fourier integral operators II*. Acta Math. 128 (1972), 183–269.

14. Engliš, M. *Berezin quantization and reproducing kernels on complex domains.* Trans. Amer. Math. Soc. 349, 411–479, 1996.
15. Engliš, M. *Weighted Bergman kernels and quantization.* Comm. Math. Phys. 227 (2002), 211–241.
16. Engliš, M. *Toeplitz operators and weighted Bergman kernels.* J. Funct. Anal. 255 (2008), 1419–1457.
17. Epstein, C. *Subelliptic $Spin_c$ Dirac operators, I.* Ann. of Math. (2) 166 (2007), no. 1, 183–214.
18. Epstein, C. *Subelliptic $Spin_c$ Dirac operators, II.* Ann. of Math. (2) 166 (2007), no. 3, 723–777.
19. Epstein, C. *Subelliptic $Spin_c$ Dirac operators, III, the Atiyah–Weinstein conjecture.* Ann. of Math. 168 (2008), 299–365.
20. Epstein, C. *Cobordism, relative indices and Stein fillings.* J. Geom. Anal. 18 (2008), no. 2, 341–368.
21. Epstein, C.; Melrose, R. *Contact degree and the index of Fourier integral operators.* Math. Res. Lett. 5 (1998), no. 3, 363–381.
22. Hörmander, L. *Fourier integral operators I.* Acta Math. 127 (1971), 79–183.
23. Leichtnam, E.; Nest, R.; Tsygan, B. *Local formula for the index of a Fourier integral operator.* J. Differential Geom. 59 (2001), no. 2, 269–300.
24. Leichtnam, E.; Tang, X.; Weinstein, A. *Poisson geometry and deformation quantization near a strictly pseudoconvex boundary.* J. Eur. Math. Soc. 9 (2007), no. 4, 681–704.
25. Melin, A.; Sjöstrand, J. *Fourier Integral operators with complex valued phase functions.* Lecture Notes 459 (1975), 120–223.
26. Melin, A.; Sjöstrand, J. *Fourier integral operators with complex phase functions and parametrix for an interior boundary value problem.* Comm. P.D.E. 1:4 (1976), 313–400.
27. Weinstein, A. *Some questions about the index of quantized contact transformations.* RIMS Kokyuroku No. 1014, 1–14, 1997.

A semi-classical inverse problem I: Taylor expansions

Yves Colin de Verdière and Victor Guillemin

In memory of Hans Duistermaat

Abstract In dimension 1, we show that the Taylor expansion of a "generic" potential near a nondegenerate critical point can be recovered from the knowledge of the semi-classical spectrum of the associated Schrödinger operator near the corresponding critical value. Contrary to the work of previous authors, *we do not assume that the potential is even*. The classical Birkhoff normal form does not contain enough information to determine the potential, but the quantum Birkhoff normal form does.[1] In a companion paper [5], the first author shows how the potential itself is, without any analyticity assumption and under some mild genericity hypotheses, determined by the semi-classical spectrum.

Key words: Schrödinger operator, semi-classics, inverse spectral problem

Mathematics Subject Classification (2010): 34E20, 81Q10, 81Q20

1 Introduction

In this paper,[2] we will only consider a configuration space of dimension 1.

Yves Colin de Verdière
Institut Fourier, Unité mixte de recherche CNRS-UJF 5582, BP 74, 38402-Saint Martin d'Hères Cedex, France
e-mail: yves.colin-de-verdiere@ujf-grenoble.fr

Victor Guillemin
Math. Dept. MIT
e-mail: vwg@math.mit.edu

[1] This work started from discussions we had during the Hans conference in Utrecht (August 2007). The proofs were completed independently by both authors two months later. We decided then to write a joint paper.

[2] Many thanks to Frédéric Faure for discussions and his computations.

Let us consider a (classical) Hamiltonian

$$H(x,\xi)=\frac{1}{2}\xi^2+V(x)$$

with $V(0)=E_0$, $V'(0)=0$, $V''(0)=\pm 1$[3]. We have

$$H(x,\xi)\equiv E_0+\Omega_\pm+\sum_{j=3}^{\infty}a_jx^j$$

with $\Omega_\pm=\frac{1}{2}(\xi^2\pm x^2)$. The Hamiltonian H can be quantized as a Schrödinger operator $\hat{H}=-\frac{1}{2}\hbar^2\frac{d^2}{dx^2}+V(x)$ where the Taylor expansion of V at $x=0$ is $E_0+\sum_{j=2}^{\infty}a_jx^j$ with $a_2=\pm\frac{1}{2}$. This operator admits a semi-classical Birkhoff normal form [8] (denoted the QBNF) at the origin of which the Weyl symbol is a formal power series of the form

$$B\equiv\Omega_\pm+\sum_{2j+k\geq 2}b_{j,k}\hbar^{2j}\Omega_\pm^k. \tag{1}$$

In this paper, we are interested in the following "inverse spectral problem":

Does the QBNF, given in (1), of the Schrödinger operator determine the Taylor series of V?

We cannot hope for a positive answer, because $V(x)$ and $V(-x)$ give the same QBNF. Moreover

Remark 1.1. **The classical BNF does not suffice to determine the Taylor expansion of V at $x=0$.**

Let $y=f(x)=x+O(x^2)$ be an analytic function whose local inverse near 0 is of the form $x=y+g(y)$ with g an even function. Then the Hamiltonian $H_f=\frac{1}{2}(\xi^2+f(x)^2)$ is classically conjugate to Ω_+ near the origin, in particular all its trajectories are of period 2π: it is enough to show that the action integrals $I(E)=\int_{\xi^2+f(x)^2\leq 2E}dxd\xi$ are the same; using the change of variable $x=y+g(y)$, we get $I(E)=\int_{\xi^2+y^2\leq 2E}(1+g'(y))dyd\xi$ and using the fact that g' is odd we get the result. A simple example is $V(x)=\frac{1}{2}(\sqrt{1+2x}-1)^2$. This result is reminiscent of the well-known result for Zoll surfaces in Riemannian geometry [2].

However, an even potential can be determined by the classical BNF, as a consequence of a result of N. Abel [1][4].

Our main result is:

Theorem 1.2. *The coefficients $\pm a_3$ and a_4 are determined from $b_{0,2}$ and $b_{1,0}$ by the formulas*

[3] Assuming that $V''(0)=\pm 1$ does not affect the results below, because $a_2=V''(0)/2$ is known from the first eigenvalue if $a_2>0$ and from the density of states if $a_2<0$.

[4] We are grateful to Hans for pointing this out to us.

$$a_3 = \pm\sqrt{b_{1,0}},\ a_4 = \frac{2}{3}b_{0,2} + \frac{5}{2}b_{1,0}.$$

If a_3 does not vanish, all a_j's are determined from the $b_{0,k}$'s and the $b_{1,k}$'s once we have chosen the sign of a_3.

This result is reminiscent of the much more sophisticated results by Zelditch on the Kac problem [9]. If we use a (trivial) particular case of the result of [6], we get

Corollary 1.1. *If we know the asymptotic expansions of the eigenvalues $\lambda_n(\hbar)$ for all n's of a Schrödinger operator near the minimum $x = 0$ of the potential and $V''(0) > 0$, we know the value of $V'''(0)$ and, if $V'''(0) \neq 0$, the Taylor expansion of the potential at that point.*

In fact, we have the more precise result:

Corollary 1.2. *From the knowledge of the N first eigenvalues of $\hat{H}$ modulo $O(\hbar^{2N})$, one can recover the Taylor expansion of V to order $2N$.*

A similar result holds for a local nondegenerate maximum of V using the "density of states" techniques. This is the content of Section 10:

Corollary 1.3. *If E_0 is a nondegenerate local maximum of V and 0 is the only critical point of V on the set $V = E_0$, the knowledge of the semi-classical spectrum of $\hat{H}$ in some intervall $]E_0, E_1[$ (or $]E_1, E_0[$) determines $V'''(0)$ and, provided that $V'''(0) \neq 0$, the Taylor expansion of V at $x = 0$.*

and also in the case of a local minimum (Section 10.4):

Corollary 1.4. *If E_0 is a nondegenerate local minimum of V and 0 is the only critical point of V on the set $V = E_0$, the knowledge of the semi-classical spectrum of $\hat{H}$ in some interval $]E_1, E_2[$, with $E_1 < E_0 < E_2$, determines $V'''(0)$ and, provided that $V'''(0) \neq 0$, the Taylor expansion of V at $x = 0$.*

Knowing the semi-classical spectrum as a function of $\hbar$ seems to be a huge amount of information. As was shown in [3], this is however the case for the effective Hamiltonians driving the propagation of waves inside a stratified medium.

2 A counterexample for a general Hamiltonian

The QBNF of a general Hamiltonian, *independent of $\hbar$*, $H(x,\xi) = \Omega_\pm + O(3)$ is not enough to know the Taylor expansion of H at the singular point. It is enough to consider $H = \frac{1}{2}((\xi - 3x^2)^2 + x^2)$ which is gauge equivalent to Ω_+ by the gauge transform $u \to ue^{ix^3}$.

3 Review of the Moyal product

The Moyal product is the product rule of symbols of Weyl quantized ΨDO's; it is given by

$$a \star b \equiv \sum_{j=0}^{\infty} \frac{1}{j!} \left(\frac{\hbar}{2i}\right)^j \{a, b\}_j$$

with

$$\{a, b\}_j := \sum_{p=0}^{j} \binom{p}{j} (-1)^p \partial_x^p \partial_\xi^{j-p} a \partial_x^{j-p} \partial_\xi^p b.$$

We will also use the *Moyal bracket*,

$$[a, b]^\star = a \star b - b \star a.$$

We have

$$\frac{i}{\hbar}[a, b]^\star \equiv \sum_{j=0}^{\infty} \frac{1}{2j+1!} \left(\frac{\hbar}{2i}\right)^{2j} \{a, b\}_{2j+1}.$$

In particular, $\{a, b\}_1 = a_\xi b_x - a_x b_\xi$ is the Poisson bracket and

$$\{a, b\}_3 = a_{\xi\xi\xi} b_{xxx} - 3a_{\xi\xi x} b_{xx\xi} + 3a_{\xi xx} b_{x\xi\xi} - a_{xxx} b_{\xi\xi\xi}.$$

We have

$$\frac{i}{\hbar}[a, b]^\star \equiv \{a, b\}_1 - \frac{\hbar^2}{24}\{a, b\}_3 + \frac{\hbar^4}{1920}\{a, b\}_5 + \cdots .$$

4 The Weyl algebra

The "Weyl algebra" which consists of formal power series in $\hbar$ and (x, ξ),

$$W = \sum_{j=2}^{\infty} W_j,$$

where W_j is the space of polynomials in (x, ξ) and $\hbar$ of total degree j where the degree of $x^l \xi^m \hbar^n$ is $l + m + 2n$. W is a graded algebra for the Moyal product: we have $W_j \star W_k \subset W_{j+k}$ and hence $\frac{i}{\hbar}[W_j, W_k]^\star \subset W_{j+k-2}$. Moreover, if we define W_j^+ as the subspace of W_j which is generated by monomials of even degree in $\hbar$, we have

$$\frac{i}{\hbar}[W_j^+, W_k^+]^\star \subset W_{j+k-2}^+;$$

we will define $W^+ = \sum_{j=3}^{\infty} W_j^+$ which is a Lie algebra for the bracket $\frac{\hbar}{i}[\cdot, \cdot]^\star$. W^+ is the (formal) Lie algebra of FIO's that are tangent to the identity at the the origin. The grading is obtained by looking at the action on the (graded) vector space of *symplectic spinors*: if $F \equiv \sum_{j=0}^{\infty} \hbar^j F_j(X)$ with $F \in \mathcal{S}(\mathbb{R})$, we define $f_\hbar(x) = \hbar^{-\frac{1}{2}} F(x/\hbar)$ whose microsupport is the origin. W^+ acts on this space of functions in a graded way as differential operators of infinite degree: if $w \in W$, $w \cdot f = \mathrm{OP}_\hbar(w)(f)$.

5 Moyal versus functional QBNF

There are two different QBNF:

- The first is a *Weyl symbol* $B \equiv \sum b_{j,k} \hbar^{2j} \Omega^k$ as before, and
- the second is an *operator* which is a formal power series of the harmonic oscillator $\hat{\Omega}$ of the form $\hat{B} \equiv \sum \widehat{b}_{j,k} \hbar^{2j} \hat{\Omega}^k$.

The second is the Weyl quantization of the first. So they are equivalent. The equivalence can be made explicit in both directions by computing $\mathrm{Op}_{\mathrm{Weyl}}(\Omega^k)$ or the Weyl symbol of $\hat{\Omega}^k$. The functional form is useful in order to compute successive approximations of the eigenvalues, while the Weyl form is easier to compute using the Moyal product.

6 Useful lemmas

The following result is classical:

Lemma 6.1. *The equation $\{\Omega_\pm, P\}_1 = Q$ where Q is a given homogeneous polynomial of degree N has a solution P, a homogeneous polynomial of degree N,*

- *if N is odd,*
- *if $N = 2N'$ is even and $c_\pm(Q) = 0$, where $c_\pm$ is a linear form on the space of homogeneous polynomials of degree N that satisfies $c_\pm(\Omega_\pm^{N'}) = 1$. In particular, given Q, the equation $\{\Omega_\pm, P\}_1 = Q - c_\pm(Q)\Omega_\pm^{N'}$ has a solution.*

Remark 6.2. In the case Ω_+, $c_+(Q)\Omega_+^{N'}$ is the average of Q under the natural action of S^1 on homogeneous polynomials of degree $2N'$.

Definition 6.3. We will denote by $\Sigma_{2N-1}^\pm$ the homogeneous polynomial of degree $2N-1$ that satisfies

$$\{\Omega_\pm, \Sigma_{2N-1}^\pm\}_1 = x^{2N-1}.$$

Lemma 6.4. *We have*

$$\Sigma_{2N-1}^\pm = -\left(\pm x^{2N-2}\xi + \frac{2N-2}{3} x^{2N-4}\xi^3 + \cdots\right).$$

We can also check:

Lemma 6.5. *The polynomials* $x^{2N'}$ *are not Poisson brackets of the form* $x^{2N'} = \{\Omega_\pm, P\}_1$, *i.e.,* $c_\pm(x^{2N'}) \neq 0$.

7 The QBNF

In order to reduce to the QBNF, we will use automorphisms of W^+ of the form

$$H \to H_S = \exp(iS/\hbar) \star H \star \exp(-iS/\hbar) = \exp\left(\frac{i}{\hbar} ad(S)^\star\right) H$$

with $S = S_3 + S_4 + \cdots \in W^+$. We get

$$H_S = H + \frac{i}{\hbar}[S, H]^\star + \cdots + \frac{1}{k!}\left(\frac{i}{\hbar}\right)^k \overbrace{[S, [S, \cdots, [S, H]^\star]^\star \cdots]^\star}^{k \text{ brackets}} + \cdots,$$

which is a convergent formal power series whose k-th term is of degree $\geq k+2$. The brackets will be calculated using the Moyal bracket. We remark that the terms of degree 0 in $\hbar$ give the calculation of the classical BNF (denoted CBNF) where the brackets are now just Poisson brackets.

8 The first terms

Let us consider $V(x) = \frac{1}{2}x^2 + ax^3 + bx^4 + \cdots$ whose QBNF is $\Omega + A\Omega^2 + B\hbar^2 + O(6)$ where $O(6)$ means terms of degree ≥ 6 in the Weyl algebra. Here we assume $\Omega = \frac{1}{2}(\xi^2 + x^2)$. Our first result is:

Theorem 8.1.

$$A = -\frac{15}{4}a^2 + \frac{3}{2}b, \quad B = a^2.$$

The calculation: we start with $S = S_3 + S_4$, where $S_3(x,\xi)$ (resp. $S_4(x,\xi)$) is a homogeneous polynomial of degree 3 (resp. 4). There is no need to put terms in $\hbar^2$ in S_4 because they would be of the form $c\hbar^2$ which is in the center. We then have

$$\exp\left(\frac{i}{\hbar}[S, \cdot]^\star\right) H = \Omega + \frac{i}{\hbar}[S, H]^\star + \frac{1}{2}\left(\frac{i}{\hbar}\right)^2 [S, [S, H]^\star]^\star + 0(6).$$

By identification of terms of degree 3 and 4 and using the expression of the Moyal bracket $[\cdot, \cdot]^\star$:

$$\frac{i}{\hbar}[f, g]^\star = \{f, g\}_1 - \frac{1}{24}\hbar^2\{f, g\}_3 + \cdots,$$

we get the system of equations:

$$\begin{cases} (3)\ ax^3 + \{S_3, \Omega\}_1 = 0 \\ (4)\ bx^4 + \{S_3, ax^3\}_1 + \{S_4, \Omega\}_1 + \frac{1}{2}\{S_3, \{S_3, \Omega\}_1\}_1 - \frac{1}{24}\hbar^2\{S_3, ax^3\}_3 \\ \qquad = A\Omega^2 + B\hbar^2 \end{cases}$$

Using Equation (3), Equation (4) splits into 2 equations:

$$\begin{cases} (4') \qquad\qquad -\frac{1}{24}\{S_3, ax^3\}_3 = B \\ (4'')\ bx^4 + \frac{1}{2}\{S_3, ax^3\}_1 + \{S_4, \Omega\}_1 = A\Omega^2. \end{cases}$$

From Equation (3) and the formula for Σ_{2N-1} given in Lemma 6.3, we get

$$S_3 = -a\left(x^2\xi + \frac{2}{3}\xi^3\right). \tag{2}$$

From Equation (4′), we get $B = a^2$. From Equation (4″), we get the value of A.

9 The induction

We carry out the proof in the case of Ω_+ and $E_0 = 0$. The minus case is similar.

Let us start with

$$H' = \Omega_+ + a_3x^3 + \cdots + a_{2N-2}x^{2N-2}$$

and $S' = S_3 + S_4 + \cdots + S_{2N-2}$ with $S_j \in W_j$, so that

$$\exp\left(\frac{i}{\hbar}[S', \cdot]^\star\right) H' = \Omega_+ + B_4 + \cdots + B_{2N-2} + R_{2N-1} + R_{2N} + \cdots (:= B'),$$

with

- $B_{2j} \in W^+_{2j}$ a polynomial in $\hbar^2$ and Ω_+
- For $n = 2N-1$ and $n = 2N$, $R_n \in W_n$.

In other words S' generates the transformation that converts H' into its QBNF mod $O(2N-1)$. The polynomials H' and S' and the partial QBNF B' are known by the induction hypothesis. We are now trying to get $S'' = S_{2N-1} + S_{2N}$ so that $S = S' + S''$ converts $H = H' + ax^{2N-1} + bx^{2N}$ into the QBNF mod $O(2N+1)$. We will only consider the terms of degree 0 and 2 in $\hbar$. So we can split every polynomial P_j in W^+_j into $P_j = P^0_j + \hbar^2 P^2_j + \cdots$ with P^2_j of degree $j-4$ in (x, ξ).

The equation to solve is

$$\exp\left(\frac{i}{\hbar}[S'+S'',\cdot]^\star\right)(H'+ax^{2N-1}+bx^{2N})$$
$$=\Omega_+ + B_4 + \cdots + B_{2N-2} + B_{2N} + O(2N+1) \tag{3}$$

with $B_{2N}=b_{2N}^0\Omega_+^N + b_{2N}^2\hbar^2\Omega_+^{N-2}+\cdots$. We hope to recover a and b from b_{2N}^0 and b_{2N}^2 using what we know already at this step. The left-hand side of Equation (3) splits into

$$\exp\left(\frac{i}{\hbar}[S'+S'',\cdot]^\star\right)(ax^{2N-1}+bx^{2N}) = ax^{2N-1}+bx^{2N}$$
$$+\frac{i}{\hbar}[S_3, ax^{2N-1}]^\star + O(2N+1),$$

and

$$\exp\left(\frac{i}{\hbar}[S'+S'',\cdot]^\star\right)H' = B' + \frac{i}{\hbar}[S'', \Omega_+ + a_3x^3]^\star$$
$$+\frac{1}{2}\left(\frac{i}{\hbar}\right)^2\left([S_{2N-1},[S_3,\Omega_+]^\star]^\star + [S_3,[S_{2N-1},\Omega_+]^\star]^\star\right) + 0(2N+1),$$

so that, we get

- **In degree** $2N-1$**:**

$$ax^{2N-1} + \{S_{2N-1}^0, \Omega_+\}_1 + R_{2N-1}^0 = 0$$
$$\{S_{2N-1}^2, \Omega_+\}_1 + R_{2N-1}^2 = 0.$$

  We see that S_{2N-1}^2 is known at this step, while S_{2N-1}^0 is modulo known terms a solution of
$$\{\Omega_+, S_{2N-1}^0\}_1 = ax^{2N-1}.$$
  This equation gives, always mod known terms:
$$S_{2N-1}^0 = a\Sigma_{2N-1},$$
  with Σ_{2N-1} given by Definition 6.3.
- **In degree** $2N$**:**

$$bx^{2N} + \frac{i}{\hbar}([S_3, ax^{2N-1}]^\star + [S_{2N},\Omega_+]^\star + [S_{2N-1}, a_3x^3]^\star)$$
$$+\frac{1}{2}\left(\frac{i}{\hbar}\right)^2([S_{2N-1},[S_3,\Omega_+]^\star]^\star + [S_3,[S_{2N-1},\Omega_+]^\star]^\star) + R_{2N}$$
$$= B_{2N} + O(2N+1).$$

The previous equation gives one equation in $\hbar^0$ and one in $\hbar^2$:

– **degree** $2N$**, hbar**^{0}

$$bx^{2N} + \{S_3, ax^{2N-1}\}_1 + \{S^0_{2N}, \Omega_+\}_1 + \{S^0_{2N-1}, a_3x^3\}_1$$
$$+ \frac{1}{2}\{S^0_{2N-1}, \{S_3, \Omega_+\}_1\}_1 + \frac{1}{2}\{S_3, \{S^0_{2N-1}, \Omega_+\}_1\}_1 + R^0_{2N} = b^0_{2N}\Omega_+^N.$$

This can be simplified as

$$\{\Omega_+, S^0_{2N}\}_1 = -b^0_{2N}\Omega_+^N + bx^{2N} + \frac{a}{2}\{S_3, x^{2N-1}\}_1 + \frac{a_3}{2}\{S^0_{2N-1}, x^3\}_1$$
$$+ R^0_{2N} - \frac{1}{2}\{S_3, R^0_{2N-1}\}_1 \tag{4}$$

This gives $b^0_{2N} = \beta_N b + \gamma_N a_3 a$ modulo known terms. Moreover, Lemma 6.5 implies $\beta_N \neq 0$.

– **degree** $2N$**, hbar**^{2}

$$-\frac{1}{24}(\{S_3, ax^{2N-1}\}_3 + \{S^0_{2N-1}, a_3x^3\}_3) + \{S^2_{2N}, \Omega_+\}_1$$
$$+ \frac{1}{2}(\{S^2_{2N-1}, \{S_3, \Omega_+\}_1\}_1 + \{S_3, \{S^2_{2N-1}, \Omega_+\}_1\}_1)$$
$$- \frac{1}{48}(\{S^0_{2N-1}, \{S_3, \Omega_+\}_1\}_3 + \{S_3, \{S^0_{2N-1}, \Omega_+\}_1\}_3) + R^2_{2N} = b^2_{2N}\Omega_+^{N-2}$$

which can be simplified as

$$\{\Omega_+, S^2_{2N}\}_1 = -b^2_{2N}\Omega_+^{N-2} - \frac{1}{48}(a\{S_3, x^{2N-1}\}_3 + a_3\{S^0_{2N-1}, x^3\}_3) + R^2_{2N}, \tag{5}$$

modulo known terms. This gives, using Lemma 6.1, $b^2_{2N} = \delta_N a a_3$ modulo known terms.

From Equation (5) and the expressions for S_3 (Equation (2)) and Σ_{2N-1} (Lemma 6.4), we get

$$\{\Omega_+, S^2_{2N}\}_1 = aa_3\frac{(N-1)(2N^2-4N+3)}{3}x^{2N-4} - b^2_{2N}\Omega_+^{N-2} \text{ mod known terms} \tag{6}$$

Because x^{2N-4} is not a Poisson bracket with Ω_+ by Lemma 6.5, we get $\delta_N \neq 0$.

From the fact that β_N and δ_N do not vanish, this concludes the induction $N-1 \to N$.

10 Getting the QBNF from the density of states in case of a local extremum of the potential

10.1 $\hbar$-dependent distributions

Let $T_\hbar$ be an $\hbar$-dependent Schwartz distribution on an open interval J.

Definition 10.1. The family $T_\hbar$ is

- regular at the point $E_0 \in J$ if there exists a sequence of functions T_j which are smooth in some neighbourhood K of E_0 with $j = j_0, \ j_0 + 1, \dots$ ($j_0 \in \mathbb{Z}$), so that, for any $f \in C_o^\infty(K)$, we have the asymptotic expansion $T_\hbar(f) \equiv \sum_{j=j_0}^{+\infty} \hbar^j \int_J T_j(x) f(x) dx$.
- right regular (resp. left regular) at the point $E_0 \in J$ if there exists $E_1 > E_0$ (resp. $E_1 < E_0$) and a sequence of functions T_j which are smooth in some neighbourhood of E_0 with $j = j_0, \ j_0+1, \dots$ ($j_0 \in \mathbb{Z}$), so that, for any $f \in C_o^\infty(]E_0, E_1[)$, we have the asymptotic expansion $T_\hbar(f) \equiv \sum_{j=j_0}^{+\infty} \hbar^j \int_J T_j(x) f(x) dx$.

We will use the following notations:

Definition 10.2. If $T_\hbar$ is a family of distributions on J and $E_0 \in J$, $T_\hbar^+$ (resp. $T_\hbar^-$), the right (resp left) singular part of $T_\hbar$ is the equivalence class of $T_\hbar$ modulo families of distributions which are right-(resp. left-)regular at the point E_0.

10.2 Density of states

Consider a smooth potential $V : I \to \mathbb{R}$ where I is an open interval with $0 \in I$ and $\liminf_{x \to \partial I} V(x) = E_\infty > -\infty$ and let $\hat{H}$ be the Schrödinger operator with potential V.

Definition 10.3. The density of states is the $\hbar$-dependent Schwartz distribution $T_\hbar$ on $]-\infty, E_\infty[$ defined by

$$D_\hbar(f) := \operatorname{Trace} f(\hat{H}).$$

Lemma 10.4. *If J is an open subset of $]-\infty, E_\infty[$ that contains no critical values of V, the density of states is regular at every point of J.*

Proof. Let us denote by $H = \frac{1}{2}\xi^2 + V(x)$ the symbol of the Schrödinger operator $\hat{H}$. The operator $f(\hat{H})$ is a pseudodifferential operator whose symbol $f^\star(H)$ is given (see [4]) by

$$f^\star(H) = f(H) + \sum_{j \geq 1, \ l \geq 1} \hbar^{2j} P_{j,l}(x,\xi) f^{(l)}(H),$$

where the $P_{j,l}$'s are smooth functions locally computable from the symbol H. It is now enough to check that $f \to (2\pi\hbar)^{-1}\int\int P_{j,l}(x,\xi)f^{(l)}(H)dxd\xi$ is regular at each point of J using the fact that H has no critical value in J.

10.3 Singularity of the density of states near a local maximum of the potential

Let us assume that $V(0) = E_0 < E_\infty$, $V'(0) = 0$ and $V''(0) < 0$. Assume also that 0 is the unique critical point of V whose critical value is E_0.

We have:

Theorem 10.5. *If the QBNF of $\hat{H}$ is*

$$B \equiv \Omega_- + \sum_{2j+k\geq 2} b_{j,k}\hbar^{2j}\Omega_-^k,$$

the density of states is right and left singular at the point E_0 and one can recover the full QBNF (the coefficients $b_{j,k}$) from the right (resp. left) singular part $D_\hbar^+$ (resp. $D_\hbar^-$) of the density of states $D_\hbar$ at E_0.

In what follows, it is more convenient to use $\Omega_- = x\xi$.

10.3.1 The singularity of the density of states and the QBNF

Lemma 10.6. *If B is the QBNF of $\hat{H}$, the singular part of the density of states is the same as that of the family of distributions*

$$G_\hbar : f \to \frac{1}{2\pi\hbar}\int\int_D f^\star(B)dxd\xi,$$

where $f^\star(B)$ is the Weyl symbol of $f(\hat{B})$ and D is the square $\max(|x|,|\xi|) \leq 1$.

Proof. Let $\Pi = \mathrm{Op}_{\mathrm{Weyl}}(\omega)$ be a compactly supported ΨDO whose Weyl symbol ω is $\equiv 1$ near $(0,0)$. We have

$$D_\hbar(f) = (2\pi\hbar)^{-1}\left(\int\int \omega \star f^\star(H)dxd\xi + \int\int (1-\omega)\star f^\star(H)dxd\xi\right).$$

Using (the proof of) Lemma 10.4, the second term is a regular distribution. The first term can be transformed using the QBNF: there exists an FIO U, microlocally unitary, which transforms $\hat{H}$ into its QBNF and hence for every function f, we have

$$U^\star f(\hat{B})U = f(\hat{H})$$

microlocally near the origin. In this way, we get

$$\mathrm{Trace}(\Pi \circ f(\hat{H})) \equiv \mathrm{Trace}(\Pi U^\star f(\hat{B})U).$$

Introducing $\Pi_1 := U\Pi U^\star$ (a ΨDO whose Weyl symbol is $\equiv 1$ near the origin) and using the commutativity of the trace, we have

$$\mathrm{Trace}(\Pi \circ f(\hat{H})) \equiv \mathrm{Trace}(\Pi_1 \circ f(\hat{B}))).$$

It remains to check that, if $\Pi_1 = \mathrm{Op}_{\mathrm{Weyl}}(\omega_1)$, $f \to \int\int_{(I\times\mathbb{R})\setminus D} \omega_1 \star f^\star(B)dxd\xi$ is regular.

10.3.2 Computing some singularities

Lemma 10.7. *Let us consider the family of distributions*

$$K_\hbar(f) = \int\int_D f\left(\sum_{j=0}^{\infty} \hbar^{2j} b_j(x\xi)\right) dxd\xi$$

on $]E_0, E_1[$ *(we consider only the case* $E_1 > E_0$*, the other case is similar), where the* b_j*'s are smooth on* $[E_0, E_1]$ *and* $b_0(u) \equiv E_0 + \sum_{j=1}^{\infty} \beta_j u^j$ *with* $\beta_1 > 0$*. Then* $K_\hbar(f)$ *admits an asymptotic expansion in powers of* $\hbar$*:*

$$K_\hbar(f) \equiv \sum_{j=0}^{\infty} K_j(f)\hbar^{2j}$$

and the right singularities of $K_0, \dots, K_N$ *at the point* E_0 *determine the Taylor expansions of* $b_0, \dots, b_N$ *at the origin.*

Proof. Let us Taylor expand the integrand as

$$\begin{aligned} f\left(\sum_{j=0}^{\infty} \hbar^{2j} b_j(x\xi)\right) &\equiv f(b_0(x\xi)) + f'(b_0(x\xi))\left(\sum_{j=1}^{\infty} \hbar^{2j} b_j(x\xi)\right) \\ &\quad + \sum_{k=2}^{\infty} \frac{1}{k!} f^{(k)}(b_0(x\xi))\left(\sum_{j=1}^{\infty} \hbar^{2j} b_j(x\xi)\right)^k, \\ &\equiv f(b_0(x\xi)) + \sum_{j=1}^{\infty} \hbar^{2j}\Bigg(f'(b_0(x\xi))b_j(x\xi) \\ &\quad + \sum_l f^{(l)}(b_0(x\xi))R_{j,l}(x\xi)\Bigg), \end{aligned}$$

where the functions $R_{j,l}$ depend only on $b_1, \dots, b_{j-1}$.

We have to prove the following 2 facts:

1. The right singularity of
$$\int\int_D f(b_0(x\xi))dxd\xi$$
determines the Taylor expansion of b_0 at the origin.
2. The right singularity of
$$\int\int_D f'(b_0(x\xi))b_j(x\xi)dxd\xi$$
determines the Taylor expansion of b_j at the origin, assuming the Taylor expansion of b_0 is known.

Both are easy consequences of the following elementary calculus result:

$$\int\int_D f'(b_0(x\xi))b_j(x\xi)dxd\xi \equiv \int_{E_0}^{E_1} f'(t)b_j(c_0(t))c_0'(t)|\log(t-E_0)|dt$$

(modulo smooth distributions) where c_0 is the inverse function of b_0.

10.3.3 End of the proof of Theorem 10.5

We have

$$f^\star(B)(z_0) = \sum_{j=0}^{\infty} \frac{1}{2j!}(f^{(2j)}(B(z_0))(B-B(z_0))^{\star 2j})(z_0).$$

It is enough to check:

Lemma 10.8. *If*

$$f^\star(B) \equiv f(B) + \sum_{j=1}^{\infty} \hbar^{2j} \sum_l f^{(2l)}(B)R_{j,l},$$

the $R_{j,l}$'s depend only on $b_0, \dots, b_{j-1}$.

Proof. The $\star$-powers of $B - B(z_0)$ evaluated at z_0 start with terms in $\hbar^2$ and the b_l's, for $l \geq j$ have already an $\hbar^{2j}$ in front of them!

So everything works as if $f^\star(B) = f(B)$ and we are reduced to Lemma 10.7.

10.4 The case of a local minimum

The same strategy applies, but now the density of states is right AND left regular, with a jump singularity at the point E_0.

We get:

Theorem 10.9. *If the QBNF of $\hat{H}$ is*

$$B \equiv \Omega_+ + \sum_{2j+k\geq 2} b_{j,k}\hbar^{2j}\Omega_+^k,$$

the density of states is singular at the point E_0, and one can recover the full QBNF (the coefficients $b_{j,k}$) from the singular part of the density of states $D_\hbar$ at E_0.

The proof is very similar to the case of a local maximum. We have now a "Heaviside singularity", meaning that the density of states is right AND left regular, but the functions T_j defined by

$$D_\hbar(f) = \sum_{j=-1}^{\infty} \int f T_j \hbar^j$$

and their derivatives have jumps at the point E_0. We have only to look at the singularities of $T : f \to \int f(\Omega_+)dxd\xi$. We have $T(f) = 2\pi \int_0^{+\infty} f(u)du$, so $T = 2\pi Y$ where Y is the Heaviside function.

11 Open problems

- **Is the result still true if $a_3 = 0$?** This is the case modulo some global assumption on V (see [5]). In fact in [5], it is shown that, modulo some genericity assumptions, the potential itself is determined from its semi-classical spectrum.
- **Is the result still valid in any dimension?** We think that the answer is no; at least it does not work with the same arguments; let us assume that the quadratic part of the Hamiltonian is $H_2 = \omega_1\Omega_1 + \omega_2\Omega_2$ with Ω_1 (resp. Ω_2) harmonic oscillators in x_1 (resp. x_2).
  - Nonresonant case: ω_1 and ω_2 are independent over $\mathbb{Z}$. In degree 4, the QBNF has 4 unknown coefficients, an homogeneous polynomial of degree 2 in (Ω_1, Ω_2) and the coefficient of $\hbar^2$. On the other hand, $V_3(x_1, x_2) + V_4(x_1, x_2)$ has $9 (> 4)$ coefficients. However, it is possible that higher terms in the QBNF give other information ...
  - Resonant case: $\omega_1 = \omega_2$. In degree 4, the classical BNF has already 9 coefficients (it is a polynomial of degree 4 on $\mathbb{R}^4$ invariant by the circle action generated by the flow of $\Omega_1 + \Omega_2$), this seems promising. However, we have to take into account an $O(2)$ action by isometries in $\mathbb{R}^2$: on the one hand, we

can only expect to determine the potential up to this action; on the other hand, the QBNF is determined only up to action by $SU(2)$.

12 Homogeneity properties of the QBNF

We have the following:

Theorem 12.1. *The $b_{j,k}$'s (coefficients of $\hbar^{2j}\Omega^k$ in the QBNF) satisfy the following homogeneity properties:*

$$b_{j,k}(ta_3, t^2a_4, \ldots, t^n a_{n+2}, \ldots) = t^{2(2j+k)-2} b_{j,k}(a_3, a_4, \ldots).$$

Proof. Let us consider

$$\hat{H}_t = \frac{1}{2}\left(-\hbar^2 \frac{d^2}{dx^2} + x^2\right) + ta_3x^3 + \cdots + t^{n-2}a_n x^n + \cdots ,$$

and make the change of variable $tx = y$, $\hbar_1 = t^2\hbar$. We get a new operator

$$t^{-2}\left[\frac{1}{2}\left(-\hbar_1^2 \frac{d^2}{dy^2} + y^2\right) + a_3y^3 + \cdots + a_n y^n + \cdots\right].$$

The spectrum of the second one is then t^{-2} times that of the first one. This implies the property.

References

1. N. Abel. Auflösung einer mechanichen Aufgabe. *Journal de Crelle 1:153–157 (1826).*
2. Arthur L. Besse. Manifolds all of whose Geodesics are closed. *Springer. Ergebnisse no 93 (1978).*
3. Yves Colin de Verdière. Semi-classical analysis and passive imaging. *Nonlinearity 22:45–75 (2009).*
4. Yves Colin de Verdière. Bohr-Sommerfeld rules to all orders. *Ann. Henri Poincaré 6:925–936 (2005).*
5. Yves Colin de Verdière. A semi-classical inverse problem II. Reconstruction of the potential. *This Volume.*
6. Victor Guillemin, Thierry Paul and Alexandro Uribe. "Bottom of the well" semi-classical wave trace invariants. *ArXiv:math-SP/0608617* and *Math. Res. Lett. 14:711–719 (2007).*
7. Victor Guillemin and Alexandro Uribe. Some inverse spectral results for semi-classical Schrödinger operators. *ArXiv:math-SP/0509290* and *Math. Res. Lett. 14:623–632 (2007).*
8. San Vũ Ngọc and Laurent Charles. Spectral asymptotics via the Birkhoff normal form. *Duke Math. J. 143:463–511 (2008).*
9. Steve Zelditch. The inverse spectral problem. *Surveys in Differential Geometry IX, 401–467 (2004).*

A semi-classical inverse problem II: reconstruction of the potential

Yves Colin de Verdière

To the memory of our friend, the inspiring mathematician,
Hans Duistermaat

Abstract This paper is the continuation of our work with Victor Guillemin (previous paper in this volume); Victor and I proved that the Taylor expansion of the potential at a generic non-degenerate critical point is determined by the semi-classical spectrum of the associated Schrödinger operator near the corresponding critical value. Here I show that, under some genericity assumptions, the potential of the 1D Schroedinger operator is determined by its semi-classical spectrum. Moreover, there is an explicit reconstruction. This paper is strongly related to a paper of David Gurarie (J. Math. Phys. 36:1934–1944 (1995)).

Key words: Schrödinger operator, semi-classics, inverse spectral problem, trace formulae

Mathematics Subject Classification (2010): 34E20, 81Q10, 81Q20

1 Introduction

This paper is the continuation of [6], where Victor Guillemin and I proved the following result: the Taylor expansion of a potential $V(x)$ $(x \in \mathbb{R})$ at a non-degenerate critical point x_0 of V, satisfying $V'''(x_0) \neq 0$, is determined by the semi-classical spectrum of the associated Schrödinger operator near the corresponding critical value $V(x_0)$. Here, I prove results which are stronger in some aspects: the potential itself, without any analyticity assumption, but with some genericity

Université Grenoble I, Institut Fourier, Unité mixte de recherche CNRS-UJF 5582, BP 74, 38402-Saint Martin d'Hères Cedex (France)
e-mail: yves.colin-de-verdiere@ujf-grenoble.fr

conditions, is determined from the semi-classical spectrum. Moreover, my method gives an explicit way to reconstruct the potential.

Inverse spectral results for Sturm–Liouville operators are due to Borg, Gelfand, Levitan, Marchenko and others (see for example [12]). They need the spectra of the differential operator with two different boundary conditions in order to recover the potential. My results are different in several aspects:

- They are *local* using only the part of the spectrum included in some interval $]-\infty, E[$ in order to get V in the inverse image by V of this interval.
- They need only *approximate* spectra.
- They still apply if the operator is *essentially self-adjoint.*

After having completed the present work, I found that similar methods were already used by David Gurarie [9] in order to recover a surface of revolution from the joint spectrum of the Laplace operator and the momentum operator. In the present paper, the genericity assumptions are weaker and more explicit:

- David Gurarie assumes that the potential is a Morse function with pairwise different critical values, while I assume only a weaker nondegeneracy condition (see Section 10.1.1).
- His argument for the separation of spectra associated to the different wells is less explicit than mine which uses the semi-classical trace formula (see Section 12.3).
- He does not say a word about the problem of a nongeneric symmetry defect and explicit nonisomorphic potentials with the same semi-classical spectra (Section 7 and assumption 3 in theorem 5.1).

Semi classics has been used in inverse spectral problems since the seventies; for a recent review, the reader could look at [14].

2 Motivation I: surfaces of revolution

Let us consider on a 2−sphere the metric of revolution

$$ds^2 = dx^2 + a^4(x)dy^2$$

with $x \in [0, L]$ and $y \in \mathbb{R}/2\pi\mathbb{Z}$. We assume that $a(0) = a(L) = 0$, $a(x) > 0$ for $0 < x < L$ and a is smooth. The volume element is given by $dv = a^2(x)|dxdy|$ and the Laplace operator by

$$\Delta = -\frac{\partial^2}{\partial x^2} - \frac{2a'}{a}\frac{\partial}{\partial x} - \frac{1}{a^4}\frac{\partial^2}{\partial y^2}.$$

Using the change of function $f = aF$, we get the operator $P = a\Delta a^{-1}$ which is formally symmetric with respect to $|dxdy|$:

$$P = -\frac{\partial^2}{\partial x^2} + \frac{a''}{a} - \frac{1}{a^4}\frac{\partial^2}{\partial y^2}.$$

If $F(x, y) = \varphi(x)\exp(ily)$ with $l \in \mathbb{Z}$, we define Q_l as follows:

$$PF = l^2(Q_l\varphi)e^{ily},$$

and putting $\hbar = l^{-1}$, we get

$$Q_\hbar\varphi = -\hbar^2\varphi'' + (a^{-4} + \hbar^2 W)\varphi$$

with $W = \frac{a''}{a}$. It implies that the knowledge of the joint spectrum of Δ and ∂_y is closely related to the spectra of $Q_\hbar$ for $\hbar = 1/l$ with $l \in \mathbb{Z}\backslash 0$. This relates our paper to Gurarie's result [9].

3 Motivation II: effective surface waves Hamiltonian

In our paper [3] Section 7, we started with the following acoustic wave equation[1]

$$\begin{cases} u_{tt} - \operatorname{div}(n\, \operatorname{grad} u) = 0 \\ u(\mathbf{x}, 0, t) = 0 \end{cases} \tag{1}$$

in the halfspace $X = \mathbb{R}^{d-1}_{\mathbf{x}} \times]-\infty, 0]_z$ where $n(z) : \mathbb{R}_- \to \mathbb{R}_+$ is a nonnegative function that satisfies

$$0 < n_0 := \inf n(z) < n_\infty := \liminf_{z\to-\infty} n(z).$$

This equation describes the propagation of acoustic waves in a medium which is stratified: the variations of the density are on much smaller scales vertically than horizontally.[2] This equation admits solutions of the form $\exp(i(\omega t - \mathbf{x}\xi))v(z)$ provided that v is an eigenfunction of the operator L_ξ on the half line $z \leq 0$ defined as follows:

$$L_\xi v := -\frac{d}{dz}\left(n(z)\frac{dv}{dz}\right) + n(z)|\xi|^2 v \tag{2}$$

with Dirichlet boundary conditions and eigenvalue ω^2. These solutions are exponentially localized near the boundary provided ω^2 is in the discrete spectrum of L_ξ contained in $J :=]n_0|\xi|^2, n_\infty|\xi|^2[$.

Let us denote by $\lambda_1(\xi) < \lambda_2(\xi) < \cdots < \lambda_j(\xi) < \cdots$ the spectrum of L_ξ in the interval J and $v_j(\xi, z)$ the associated normalized eigenfunctions. The unitary maps

[1] $u = u(\mathbf{x}, z, t)$ is the pressure, $n = K/\rho$ with ρ the density and $K > 0$ the in-compressibility assumed to be a constant. The acoustic wave equation is a simplification of the elastic wave equation which holds if the medium is fluid.

[2] In [3], we took a more complicated function $n(\mathbf{x}, z) = N(\mathbf{x}, z/\varepsilon, z)$ with N smooth and ε small.

from $L^2(\partial X)$ into $L^2(X)$ defined by

$$T_j(a) := (2\pi)^{-(d-1)} \int_{\mathbb{R}^{d-1}} \hat{a}(\xi) v_j(\xi, z) e^{i\mathbf{x}\xi} d\xi,$$

with $\hat{a}(\xi) := \int_{\mathbb{R}^{d-1}} a(\mathbf{x}) e^{-i\mathbf{x}\xi} d\mathbf{x}$, satisfies

$$PT_j = T_j \mathrm{Op}(\lambda_j),$$

where $P = -\mathrm{div}(n \ \mathrm{grad} u)$ with Dirichlet boundary conditions and $\mathrm{Op}(\lambda_j)$ is an elliptic pseudodifferential operator of degree 2 and of symbol λ_j. So that, for each $j = 1, \ldots$, we get an effective surface wave Hamiltonian with the Hamiltonian λ_j. The map $T : \oplus_{j=1}^{\infty} L^2(\partial X) \to L^2(X)$ given by $T = \oplus_{j=1}^{\infty} T_j$ is an injective isometry.

We see that the high frequency surface waves are associated to the semi-classical spectrum of a Schrödinger type operator

$$\mathcal{L}_\hbar = -\hbar^2 \frac{d}{dz}\left(n(z)\frac{d}{dz}\right) + n(z),$$

with $\hbar = \|\xi\|^{-1}$.

One can try to recover $n(z)$ from the propagation of surface waves: this is equivalent to get the operator $\mathcal{L}_\hbar$ from its semi-classical spectrum.

4 Schrödinger operators and spectra

The following notation will be used everywhere in this paper.

The *interval* I is defined by $I =]a, b[$ with $-\infty \leq a < b \leq +\infty$. The *potential* $V : I \to \mathbb{R}$ is a smooth function with $-\infty < E_0 := \inf V < E_\infty = \liminf_{x \to \partial I} V(x)$.

The *Schrödinger operator* $\hat{H}$ is any self-adjoint extension of the operator $-\hbar^2 \frac{d^2}{dx^2} + V(x)$ defined on $C_o^\infty(I)$.

The *discrete spectrum* of $\hat{H}$ will be denoted by

$$\lambda_1(\hbar) < \lambda_2(\hbar) < \cdots < \lambda_l(\hbar) < \cdots .$$

Lemma 4.1. *The discrete spectra below E_∞ are, modulo $O(\hbar^\infty)$, independent of the boundary conditions.*

This comes from the fact that the eigenfunctions are $O(\hbar^\infty)$ outside the wells; this is proved, using *semi-classical ellipticity,* in [13], Section 2.9.

Using the previous lemma, we can assume that we work always with the Friedrichs extension with initial domain (the closure of) $C_o^\infty(I)$.

The *semi-classical limit* is associated to the classical Hamiltonian $H = \xi^2 + V(x)$ whose dynamics is given by the vector field

$$X_H = 2\xi \frac{\partial}{\partial x} - V'(x)\frac{\partial}{\partial \xi}$$

in the phase space $T^\star I$, the cotangent space of I with the canonical coordinates (x, ξ).

Definition 4.1. Let us give E with $E_0 < E \le E_\infty$ and a positive real number N. We say that a sequence $\mu_l(\hbar), l = 1, \dots$ is a **semi-classical spectrum of $\hat{H}$ mod $o(\hbar^N)$ in** $]-\infty, E[$ if, we have for the l's so that $\lambda_l(\hbar) < E$, **uniformly** on every compact $K \subset]-\infty, E[$,

$$\lambda_l(\hbar) = \mu_l(\hbar) + o(\hbar^N).$$

In the paper [6], it was enough to know the asymptotic expansions of the λ_l's for all l, but not uniformly in l in order to recover the Taylor expansion of V at the point x_0 where V reaches is minimum.

5 A theorem for one-well potentials

In what follows, E is given with $E_0 < E \le E_\infty$.

Theorem 5.1. *Let us assume that the potential $V : I \to \mathbb{R}$ satisfies:*

1. **A single well below** E: *for any $y < E$, the sets $I_y := \{x | V(x) \le y\}$ are compact intervals. There exists an unique x_0 so that $V(x_0) = E_0$ ($= \inf_{x \in I} V(x)$). For any y with $E_0 \le y < E$, if we define the functions $f_\pm : [E_0, E[\to \mathbb{R}$ so that*

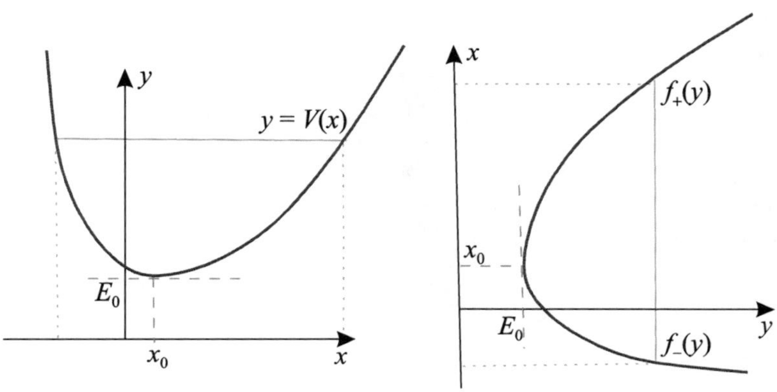

Fig. 1 The potential V and the functions f_+ and f_-.

the intervals I_y are defined by $I_y = [f_-(y), f_+(y)]$, we have $V'(x) < 0$ for $f_-(E_-) < x < x_0$ and $V'(x) > 0$ for $x_0 < x < f_+(E_-)$.

2. **A genericity hypothesis at the minimum:** *there exists $N \geq 2$ so that the N-th derivative $V^{(N)}(x_0)$ does not vanish.*
3. **A generic symmetry defect:** *if there exists $x_\pm$, satisfying $f_-(E_-) < x_- < x_0 < x_+ < f_+(E_+)$ and $\forall n \in \mathbb{N}$, $V^{(n)}(x_-) = (-1)^n V^{(n)}(x_+)$, then V is globally even with respect to $x_0 = (x_- + x_+)/2$ in the interval I_E. This is true for example if V is real analytic.*

Then the spectra modulo $o(\hbar^2)$ *in the interval* $]-\infty, E[$ *of the Schrödinger operators $\hat{H}_\hbar$, for a sequence $\hbar_j \to 0^+$, determine V in the interval I_E up to a symmetry-translation $V(x) \to V(c \pm x)$.*

6 One-well potentials: Bohr–Sommerfeld rules and a pseudodifferential trace formula

It is a classical fact (see [4]) that the semi-classical spectrum (i.e., the spectrum up to $O(\hbar^\infty)$) of $\hat{H}_\hbar$ in the interval $]-\infty, E[$ is given by the "Bohr–Sommerfeld rules":

$$\Sigma(\hbar) = \{\mu_l(\hbar) | E_0 < \mu_l(\hbar) < E \text{ and } S(\mu_l(\hbar)) = 2\pi\hbar l\},$$

where, for $E_0 < y < E$, the function $S = S_\hbar(y) :]E_0, E[\to \mathbb{R}$ admits the formal series expansion

$$S(y) \equiv S_0(y) + \hbar\pi + \hbar^2 S_2(y) + \hbar^4 S_4(y) + \cdots \tag{3}$$

(the formal series S will be called the *semi-classical action* and the remainder term in the expansion is uniform in every compact sub-interval of $]E_0, E[$). We have

- $S_0(y) = \int_{\gamma_y} \xi dx = \int_{H(x,\xi)\leq y} |dx d\xi|$ with $\gamma_y = \{(x,\xi) | H(x,\xi) = y\}$ oriented according to the classical dynamics and

$$\frac{dS_0}{dy}(y) = \int_{f_-(y)}^{f_+(y)} \frac{dx}{\sqrt{y - V(x)}}$$

  is the *period* $T(y)$ of the trajectory of energy y for the classical Hamiltonian H,
- If t is the time parametrization of γ_y (outside the caustic set $\{V(x) = y,\ \xi = 0\}$, we have $dt = dx/2\xi$),

$$S_2(y) = -\frac{1}{12}\frac{d}{dy}\int_{\gamma_y} V''(x)|dt|,$$

which can be rewritten as

$$S_2(y) = -\frac{1}{12}\frac{d}{dy}\left(\int_{f_-(y)}^{f_+(y)} \frac{V''(x)dx}{\sqrt{y - V(x)}}\right).$$

- For $j \geq 1$, $S_{2j}(y)$ is a linear combination of expressions of the form

$$\left(\frac{d}{dy}\right)^n \int_{\gamma_y} P(V', V'', \ldots)|dt|,$$

where dt is the differential of the time on γ_y.

In what follows, we will use only S_0 and S_2. It will be convenient to relate the semi-classical action to the spectra by using the following trace formula:

Theorem 6.1 (ΨDO trace formula). *Let $f \in C_o^\infty(]E_0, E[)$ and $F(y) := -\int_y^\infty f(u)du$, we have, with $Z = T^\star I$:*

$$\mathrm{Trace} F(\hat{H}) = \frac{1}{2\pi\hbar}\left(\int_Z F(H)|dxd\xi| - \hbar^2 \int_{E_0}^E f(y)(S_2(y) + \hbar^2 S_4(y) + \cdots)dy\right) + O(\hbar^\infty). \tag{4}$$

Corollary 6.1. *The functions S_0, $S_2 : [E_0, E[\to \mathbb{R}$ are determined by the semi-classical spectrum mod $o(\hbar^2)$ in $]-\infty, E[$.*

In fact, S_0 is already given from the Weyl asymptotics:

$$\#\{\lambda_l(\hbar) \leq y\} \sim_{\hbar \to 0} \frac{S_0(y)}{2\pi\hbar}.$$

The Weyl asymptotic formula can easily be deduced from the trace formula (4).

Remark 6.1. The previous trace formula can be seen as an extension to the semi-classical setting of the famous "heat trace" method introduced by Mark Kac in [11] and strongly developed by geometers as a tool in the inverse spectral problem for the Laplace–Beltrami operator (see [2]): putting $t = \hbar^2$ in the heat semi-group $\exp(-t\Delta)$, we get $\exp(-t\Delta) = F(\hbar^2\Delta)$ with F the exponential function. This way, we get an identification of the previous expansion in powers of $\hbar^2$ with the heat trace expansion in powers of t.

We give now a proof of theorem 6.1.

Proof.

1. *The case where F is compactly supported in J:*
   Defining $F^\star(H)$ by $F(\hat{H}) = \mathrm{Op}_{\mathrm{Weyl}}(F^\star(H))$, we know (see [8] lemma 4.2) that with $z_0 = (x_0, \xi_0)$ and $H_0 = H(z_0)$,

$$F^\star(H)(z_0) = F(H_0) + \frac{1}{2}F''(H_0)(H-H_0)^{\star 2}(z_0) + \frac{1}{6}F'''(H_0)(H-H_0)^{\star 3}(z_0) + O(\hbar^4).$$

Computing the Moyal powers of $H - H_0$ at the point z_0 mod $O(\hbar^4)$, we get

$$F^\star(H) = F(H) - \hbar^2\left(\frac{1}{8}f'(H)\det(H'') + \frac{1}{24}f''(H)H''(X_H, X_H)\right) + O(\hbar^4).$$

Computing the trace of $F(\hat{H})$ as $1/2\pi\hbar$ the phase space integral of the symbol $F^\star(H)$, we get:

$$\begin{aligned}&\mathrm{Trace}(F(\hat{H}))\\ &= \frac{1}{2\pi\hbar}\left[\int_Z F(H)|dxd\xi| - \hbar^2\left(\int_J f'(y)\left(\int_{\gamma_y}\det(H'')|dt|\right)|dy|\cdots\right.\right.\\ &\left.\left.+\frac{1}{24}\int_J f''(y)\left(\int_{\gamma_y} H''(X_H, X_H)|dt|\right)|dy|\right)\right] + O(\hbar^3).\end{aligned}$$

Using Stokes formula, we have

$$\int_{\gamma_y} H''(X_H, X_H)|dt| = -2\int_{H\le y}\det(H'')|dxd\xi|,$$

and the final result for $S_2(y)$ using an integration by part:

$$S_2(y) = -\frac{1}{24}\frac{d}{dy}\int_{\gamma_y}\det(H'')|dt|.$$

2. *The harmonic oscillator case* $\Omega = -\frac{1}{2}\left(\frac{d^2}{dx^2} + x^2\right)$*:*

$$\mathrm{Trace} F(\Omega) = \frac{1}{2}\sum_{n\in\mathbb{Z}}\tilde{F}\left(\left(n+\frac{1}{2}\right)\hbar\right)$$

with $\tilde{F}$ even and coinciding with F on the positive axis. We get with the Poisson summation formula:

$$\mathrm{Trace} F(\Omega) = \frac{1}{2\pi\hbar}\int\!\!\int F\left(\frac{x^2+\xi^2}{2}\right)|dxd\xi| + O(\hbar^\infty).$$

Moreover, $\int_{\gamma_y} V''|dt|$ is independent of y because the classical period of the harmonic oscillator is constant.

A more general argument holds if $H = \Omega$ in support of $F \circ H$. It is based on the identity

$$F^\star(\Omega) = F(\Omega) - \frac{\hbar^2}{8} F''(\Omega) + O(\hbar^4).$$

3. *A deformation argument:* Let us consider a new Hamiltonian K which coincides with H in $H^{-1}(J')$ where $J' \subset J$ and J' contains the support of f. Then we have $\text{Trace}(F(\hat{H})) \equiv \text{Trace}(F(\hat{K}))$. This is because the symbols $F^\star(H)$ and $F^\star(K)$ coincide with $F(E_0)$ in the domain bounded by $H^{-1}(J')$. Because the right-hand sides of Equation (4) also coincide, we can choose a suitable K in order to prove it. We will choose K, which has globally a single well and which coincides with an harmonic oscillator near its minimum.
4. *The final step:* We can assume that K is as before with $E_0 = \inf K$, K is harmonic in the energy interval $[E_0, E_0 + \alpha]$.
   We split $F = F_0 + F_1$ where F_1 is compactly supported in $]E_0, +\infty[$ and F_1 supported in $]-\infty, E_0 + \alpha[$. Equation (4) is valid for F_0 (case studied in 1.) and for F_1 (case of the harmonic oscillator).

Theorem 6.1 is closely related to (but a bit stronger) than what is proved in my paper [4]. The trace formula contains implicitly the Maslov index; it is no more valid if we replace $\hbar\pi$ by another value in the expansion of the semi-classical action given in Equation (3).

7 Two potentials with the same semi-classical spectra

We introduced a genericity Assumption 3 on symmetry defects in Theorem 5.1. Figure 2 shows two one-well potentials with the same semi-classical spectra mod $O(\hbar^\infty)$. The fact that they have the same semi-classical spectra comes from the description of Bohr–Sommerfeld rules in Section 6.

It would be nice to prove that they do NOT have the same spectra!

8 One-well potentials: the proof of Theorem 5.1

8.1 Some useful lemmas

Lemma 8.1. *If V satisfies Assumption 2 in Theorem 5.1, we have*

$$\lim_{y \to (E_0)+} \int_{\gamma_y} V''(x)|dt| = \pi\sqrt{2V''(x_0)}.$$

This holds even if the minimum is degenerate.[3]

[3] I do not know if this is still true without the genericity Assumption 2 in Theorem 5.1; it is the only place where I use it.

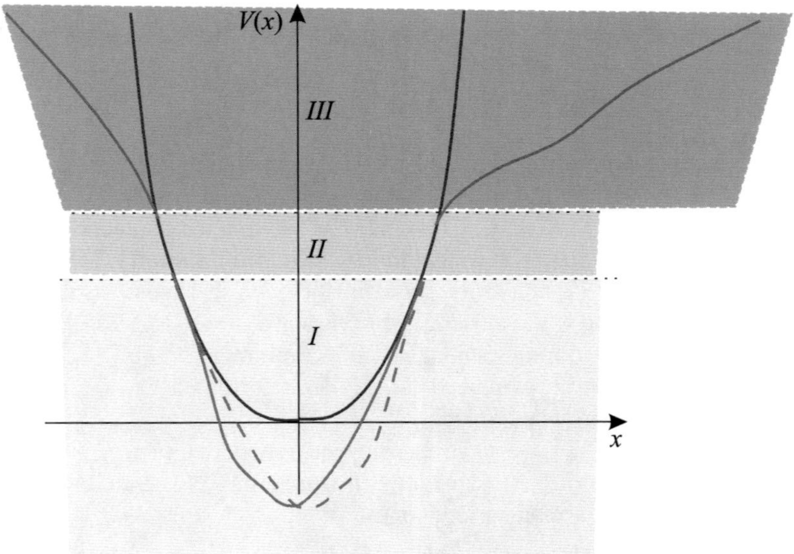

Fig. 2 The (graphs of the) two potentials are the same in the sets II and III; they are mirror images of each other in I, the potential is even in the set II.

The lemma is clear if $V''(x_0) > 0$: the limit is then $V''(x_0)$ times the period of small oscillations of a pendulum which is $\pi\sqrt{2/V''(x_0)}$.

Let us consider the case of an isolated degenerate minimum with $V(x) = E_0 + a(x-x_0)^N(1+o(1))$ $(a > 0,\ N > 2)$; we can check that the integral to be evaluated is $O((y - E_0)^{\frac{3}{2}-\frac{3}{N}}) = o(1)$.

Lemma 8.2. *We have*

$$\lim_{y \to E_0} \left(\frac{1}{f'_+(y)} - \frac{1}{f'_-(y)} \right) = 0.$$

Lemma 8.3. *If x_0 is the unique point where $V(x_0) = \inf V = E_0$, the first eigenvalue of $\hat{H}_\hbar$ satisfies $\lambda_1(\hbar) = E_0 + \hbar\sqrt{V''(x_0)/2} + o(\hbar)$.*

This is well known if $V''(x_0) > 0$ and is still true otherwise by comparison: if $E_0 \leq V(x) \leq A(x - x_0)^2$ with $A > 0$, near x_0, then $E_0 < \lambda_1(\hbar) \leq 2\pi\hbar\sqrt{A}$.

8.2 Rewriting V using F and G

We will denote by $F = \frac{1}{2}(f_+ + f_-)$ and $G = \frac{1}{2}(f_+ - f_-)$.

- The function F is smooth on $]E_0, E[$, continuous on $[E_0, E[$ (smooth in the non-degenerate case $V''(x_0) > 0$ as a consequence of the Morse Lemma), with

$F(E_0) = x_0$, and is constant if and only if V is even with respect to x_0. More generally, if F is constant on some interval, V is even on the inverse image of that interval. We call F the *parity defect*.

Lemma 8.4. *Under the Assumption 3 in Theorem 5.1, the function F' is determined up to $\pm$ by its square.*

- The function G is smooth on $]E_0, E[$, continuous at $y = E_0$. We have $G(E_0) = 0$. It is clear that, from F and G, we can recover the restriction of V to I_E.

8.3 How to get V from S_0 and S_2

Let us consider, for $E_0 < y < E$,

$$I(y) := \int_{f_-(y)}^{f_+(y)} \frac{dx}{\sqrt{y - V(x)}}$$

and

$$J(y) = \int_{f_-(y)}^{f_+(y)} \frac{V''(x)dx}{\sqrt{y - V(x)}}.$$

We have $I(y) = dS_0(y)/dy$ and $S_2(y) = -(1/12)dJ(y)/dy$. This implies that S_0, S_2, and the limit $J(E_0)$ determine I and J. The limit $J(E_0)$ is determined by $V''(x_0)$ (Lemma 8.1) which is determined by the first semi-classical eigenvalue (Lemma 8.3). We can express I and J using F and G. Using the change of variables $x = f_+(u)$ for $x > x_0$ and $x = f_-(u)$ for $x < x_0$, we get

$$I(y) = 2\int_{E_0}^{y} \frac{G'(u)du}{\sqrt{y-u}}$$

$$J(y) = \int_{E_0}^{y} \frac{d}{du}\left(\frac{1}{f_+'(u)} - \frac{1}{f_-'(u)}\right) \frac{du}{\sqrt{y-u}}.$$

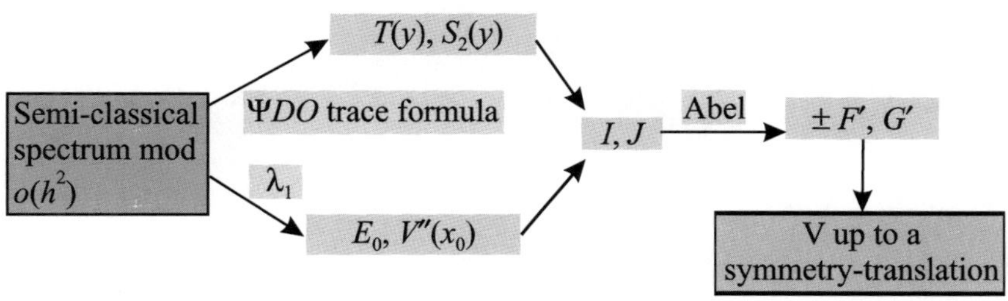

Fig. 3 The scheme of the proof.

Using Abel's result [1] (and the Appendix), we can recover G' and

$$\frac{d}{dy}\left(\frac{1}{f'_+(y)} - \frac{1}{f'_-(y)}\right) = \frac{d}{dy}\left(\frac{2G'}{G'^2 - F'^2}\right).$$

Using Lemma 8.2, we recover F'^2. The Assumption 3 implies that there exists an unique square root to F'^2 up to signs. From that we recover G' and $\pm F'$ and hence $\pm F$ and G modulo constants. This gives V up to change of x into $c \pm x$.

9 Taylor expansions

From the previous section, we see that the semi-classical spectra determine F'^2 and G even without assuming the hypothesis 3 of Theorem 5.1 on symmetry defect. It is not difficult to see that, if V satisfies the hypothesis 2 of Theorem 5.1, the parity defect F is a smooth function of $y^{2/N}$. We have the following:

Lemma 9.1. *Let us give two formal powers series* $a = \sum_{j=0}^{\infty} a_j t^j$ *and* $b = \sum_{j=0}^{\infty} b_j t^j$ *which satisfy* $a^2 = b$. *The equation* $f^2 = b$ *has exactly two solutions as formal powers series:* $f = \pm a$.

From this lemma, we deduce

Theorem 9.1. *Under the Assumptions 1 and 2 of Theorem 5.1, but without Assumption 3, the Taylor expansion of* V *at a local minimum* x_0 *is determined (up to mirror symmetry) by the semi-classical spectrum modulo* $o(\hbar^2)$ *in a fixed neighborhood of* E_0.

In some aspects, this result is stronger than the one obtained in [6], but it requires the knowledge of the semi-classical spectrum in a fixed neighborhood of E_0, while, in [6], we need only N semi-classical eigenvalues in order to get $2N$ terms in the Taylor expansion.

10 A theorem for a potential with several wells

We will extend our main result to cases including that of Figure 4: a two-wells potential with three critical values, $E_0 = 0$, E_1, and E_2. We can take any boundary condition at $x = 0$.

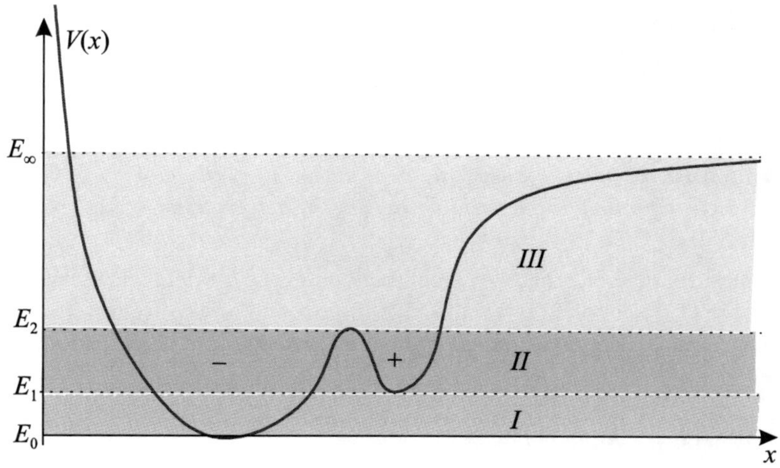

Fig. 4 A 2 wells potential V.

10.1 The genericity assumptions

In what follows, we choose E so that $E_0 < E \leq E_\infty$ and define $I_E = \{x | V(x) < E\}$. The goal is to determine the restriction of V to I_E from the semi-classical spectrum in $]-\infty, E[$.

We need the following assumptions which are generically satisfied. We introduce:

Definition 10.1. Two smooth functions $f, g : J \to \mathbb{R}$ are weakly transverse if, for every x_0 so that $f(x_0) = g(x_0)$, there exists an integer N such that the N-th derivative $(f-g)^{(N)}(x_0)$ does not vanish.

10.1.1 Assumption on critical points

- For any point x_0 so that $V'(x_0) = 0$ and $V(x_0) < E$, there exists $N \geq 2$ so that the N-th derivative $V^{(N)}(x_0)$ does not vanish.
- The *critical values* associated to different critical points are *distinct.*

The wells: Let us label the critical values of V below E_∞ as $E_0 < E_1 < \cdots < E_k < \cdots < E_\infty$ and the corresponding critical points by $x_0, x_1, \ldots$. The critical values can only accumulate at E_∞ because the critical points are isolated.

Let us denote, for $k = 1, 2, \ldots$ by $J_k =]E_{k-1}, E_k[$.

Definition 10.2. A well of order k is a connected component of $\{x \in I | V(x) < E_k\}$.

Let us denote by $N_k (\leq k)$ the number of wells of order k.

For any k, $H^{-1}(J_k)$ is an union of N_k topological annuli A_j^k and the map $H : A_j^k \to J_k$ is a fibration whose fibers $H^{-1}(y) \cap A_j^k$ are topological circles $\gamma_j^k(y)$ that are periodic trajectories of the classical dynamics: if $y \in J_k$, $H^{-1}(y) = \cup_{j=1}^{N_k} \gamma_j^k(y)$. We will denote by $T_j^k(y) = \int_{\gamma_j^k} |dt|$, the corresponding classical periods. We will often remove the index k in what follows.

We have the well-known:

Proposition 10.1. *The semi-classical spectrum in J_k is the union of N_k spectra, which are given by Bohr–Sommerfeld rules associated to actions $S_j^k(y)$ given as in Section 6.*

This comes from the fact that the eigenfunctions are $O(\hbar^\infty)$ outside the wells. This is proved in [13] Section 2.9; see also [10] for much more precise results including estimates of the exponentially small "tunneling" effects.

10.1.2 A generic symmetry defect

If there exists $x_- < x_+$, satisfying $V(x_-) = V(x_+) < E$ and, $\forall n \in \mathbb{N},\ V^{(n)}(x_-) = (-1)^n V^{(n)}(x_+)$, then V is globally *even* on I_E.

10.1.3 Separation of the wells

For any $k = 1, 2, \ldots$ and any j with $1 \leq j < l \leq N_k$, the *classical periods $T_j(y)$ and $T_l(y)$ are weakly transverse* in J_k.

10.2 Quartic potentials

If V is a polynomial of degree four with two wells like $V(x) = x^4 + ax^3 + bx^2$ with $b < 0$, the periods of the two wells (between E_1 and $E_2(= 0)$) are identical. This is because, on the complex projective compactification X_E (with $E < 0$) of $\xi^2 + V(x) = E$, the differential dx/ξ is holomorphic and the real part of X consists of two homotopic curves in X_E. One can check directly that all other actions S_{2j}, $j \geq 1$ coincide; this is proved in [7] p. 191.

10.3 Statement of the result

Our result is:

Theorem 10.1. *Under the three assumptions in Sections 10.1.1, 10.1.2, and 10.1.3, V is determined in the domain $I_E := \{x | V(x) < E\}$ by the semi-classical spectrum in $]-\infty, E[$ modulo $o(\hbar^{5/2})$ up to the following moves: I_E is an union of disjoint open intervals $I_{E,\alpha}$, each interval $I_{E,\alpha}$ is defined up to translation and the restriction of V to each $I_{E,\alpha}$ is defined up to $V(x) \to V(c - x)$.*

Remark 10.1. We need $o(\hbar^{5/2})$ in Theorem 10.1, while we needed only $o(\hbar^2)$ for the one-well case in Theorem 5.1. This is due to the way we are able to separate the spectra associated to the different wells.

11 The semi-classical trace formula

The semi-classical trace formula, also known as the "Gutzwiller trace formula", is valid for a Schrödinger operator in any dimension (see [5] for a recent review). In the one-dimensional case (and more generally in the "integrable" case), this formula can be derived from the Bohr–Sommerfeld rules, via the Poisson summation formula.

In this section, we will derive the semi-classical trace formula in dimension 1 from the Bohr–Sommerfeld rules.

Let us start with the following application of the Poisson summation formula:

Lemma 11.1. *Let $S : J \to \mathbb{R}$ be a smooth function with $S' > 0$, then we have the following identity as Schwartz distributions in J; i.e., equality holds when applying both sides to a test function $\phi \in C_o^\infty(J)$,*

$$\sum_{l\in\mathbb{Z}} \delta(y - S^{-1}(2\pi\hbar l)) = \frac{1}{2\pi\hbar}\sum_{m\in\mathbb{Z}} e^{imS(y)/\hbar} S'(y). \tag{5}$$

Let us insist that the identity (5) is valid for any fixed value of $\hbar$.

We will now develop semi-classical approximations of the identity (5). Let us start with:

Definition 11.1. Let $D_\hbar$ be an $\hbar$–dependent distribution on the interval J. We will write $D_\hbar = o(\hbar^N)$ if for any $\hbar$–pseudo-differential operator $P = \mathrm{Op}_\hbar(p)$ with $p \in C_o^\infty(T^\star J)$, we h5ave

$$\|P D_\hbar\|_{L^2(J)} = o(\hbar^N).$$

With the previous definition, we get:

Lemma 11.2. *Let us give two sequences $\lambda_l(\hbar)$ and $\mu_l(\hbar)$ in J so that*

- *$\mu_l(\hbar) = \lambda_l(\hbar) + o(\hbar^N)$ uniformly on every compact of J*
- *$\#\{\lambda_l(\hbar) \in K\} = O(1/\hbar)$ for any K compact subset of J,*

then

$$\sum_{l\in\mathbb{Z}} \delta(y - \mu_l(\hbar)) - \delta(y - \lambda_l(\hbar)) = o(\hbar^{N-\frac{5}{2}}).$$

Proof. Let us consider first the operator Q with symbol $a(\eta)\chi(y)$ with $a \in C_o^\infty(\mathbb{R})$ and $\chi \in C_o^\infty(J)$. The L^2 norm of Qu is equal to the L^2 norm of $a(\eta)$ times the $\hbar$–Fourier transform of χu. In our case, this is the L^2 norm of

$$a(\eta)\frac{1}{\sqrt{2\pi\hbar}}\sum_{l\in\mathbb{Z}}(\chi(\lambda_l(\hbar))e^{-i\eta\lambda_l(\hbar)/\hbar}-\chi(\mu_l(\hbar))e^{-i\eta\mu_l(\hbar)/\hbar}),$$

which is $o(\hbar^{N-\frac{5}{2}})$. Any other P is of the form $P = PQ$ for some suitable Q. The conclusion follows by the $\hbar$−uniform L^2 continuity of P.

Let us give another:

Definition 11.2. The L^2-Microsupport of a family of distributions $T_\hbar$ in the interval I is the closed subset of $T^\star I$ denoted $\mathrm{MS}(T_\hbar)$ given by

$$((x,\xi)\notin \mathrm{MS}(T_\hbar)) \text{ if and only if}$$
$$(\exists p\in C_o^\infty(T^\star I),\ p(x,\xi)\neq 0 \text{ and } \mathrm{Op}_\hbar(p)T_\hbar = o(1)).$$

We get the following statement of the semi-classical trace formula for the general statement:

Theorem 11.1. *As distributions on J_k, we have, if $\mu_l(\hbar)$ is a semi-classical spectrum modulo $o(\hbar^{5/2})$,*

$$\sum_{l\in\mathbb{Z}}\delta(y-\mu_l(\hbar))) = \frac{1}{2\pi\hbar}\sum_{j=1}^{N_k}\sum_{m\in\mathbb{Z}}(-1)^m e^{imS_0^j(y)/\hbar}T_j(y)(1+im\hbar S_2^j(y))+o(1). \tag{6}$$

This means that $\sum_{l\in\mathbb{Z}}\delta(y-\mu_l(\hbar))$ is mod($o(1)$) a (locally finite in the cotangent space) sum of the WKB functions

$$Z_{m,j} = \frac{1}{2\pi\hbar}(-1)^m e^{imS_0^j(y)/\hbar}T_j(y)(1+im\hbar S_2^j(y))$$

associated to the Lagrangian manifolds (the micro-support of $Z_{m,j}$)

$$L_{m,j} := \{(y, mT_j(y))|y\in I\}.$$

Proof. The trace formula is a consequence of Equation (5) applied to the spectra given by the Bohr–Sommerfeld rules (see Section 6) and Lemma 11.2.

12 The case of several wells: the proof of Theorem 10.1

12.1 What can be read from Weyl's asymptotics?

Lemma 12.1. *Under Assumption 10.1.1, the singular (nonsmooth) points of the function $y\to A(y)=\int_{H(x,\xi)\leq y}|dxd\xi|$ in $]-\infty,E[$ are exactly the critical values $E_0<E_1<\ldots(<E)$ of V. Moreover,*

- *the function $A(y)$ in smooth on $]E_k - c, E_k]$, with $c > 0$, if and only if x_k is a local minimum of V,*
- *from the singularity of $A(y)$ at E_k, on can read the value of $V''(x_k)$.*

The function $A(y)$ is determined by the semi-classical spectrum; this is a consequence of the Weyl asymptotics:

$$\#\{\lambda_l(\hbar) \leq y\} \sim_{\hbar \to 0} \frac{A(y)}{2\pi\hbar}.$$

12.2 The scheme of the reconstruction

The proof is by "induction" on E.

We start by constructing the piece of V where $V(x) < E_1$ using Theorem 5.1.

We then want to construct V where $E_1 \leq V(x) < E_2$.

There are two cases:

1. *x_1 is not an extremum:* We know then V in the interval $\{V(x) \leq E_1\}$ by continuity. We can then extend the proof of Theorem 5.1 using the fact that we know, using Section 12.4, the limits of $\int_{\gamma_y} V''(x)|dt|$ and $f'_\pm(y)$ as $y \to E_1^+$. We can reduce to an Abel transform starting from E_1 using, for $E_1 < y < E_2$,

$$\int_{V(x)\leq y} = \int_{V(x)\leq E_1} + \int_{E_1\leq V(x)\leq y}$$

   where the first part is known from the knowledge of $V(x)$ in $\{x|V(x) \leq E_1\}$.
2. *x_1 is a local minimum:* Using the separation of spectra (Section 12.3) and Theorem 5.1, we can construct the 2 wells of order 2 if we know $V''(x_1)$ (lemma 12.1).

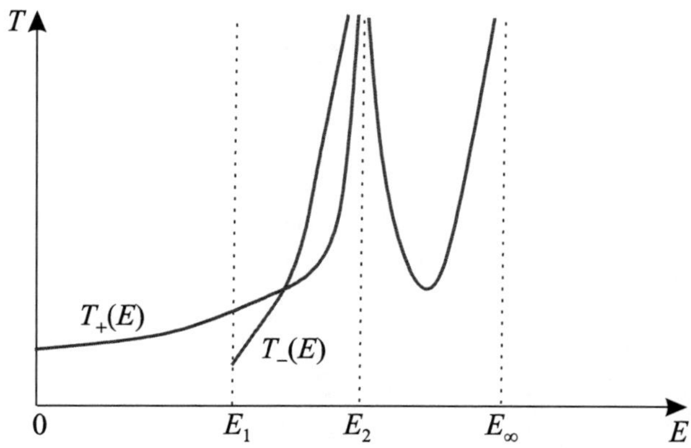

Fig. 5 The primitive periods as functions of y for the Example of Figure 4.

We then proceed to the interval $[E_2, E_3]$. A new case arises when x_2 *is a local maximum.* Then we need to glue together the wells of order 2. This case works then as before.

12.3 Separation of spectra

The main input of the proof of theorem 10.1 is the fact that the assumption 10.1.3 allows to split the semi-classical trace formulas in the interval J_k into the contributions of the N_k wells: from the spectra mod $o(\hbar^{5/2})$ in J_k, we will recover the WKB functions $Z_{1,j}$ for $j = 1, \ldots, N_k$.

Let the distributions $D_\hbar = \sum_l \delta(y - \mu_l(\hbar))$ be given modulo $o(1)$ in the interval $J = J_k$ by Equation (6). The distributions $D_\hbar$ are determined mod $o(1)$ by the semi-classical spectra mod $o(\hbar^{5/2})$. Let us denote by B the set defined by

$$B := \{y \in J_k | \exists j \neq l,\ T_j(y) = T_l(y)\};$$

using assumption 10.1.3, we see that B is a discrete subset of J_k. Let us denote by $Z_\hbar$ the finite sum defined by the r.h.s of Equation (6) restricted to $m = 1$, i.e.,

$$Z_\hbar = \sum_{j=1}^{N_k} Z_{1,j}.$$

We have:

Lemma 12.2. *Using the weak transversality assumption of Section 10.1.3, the set B and the distributions $Z_\hbar$ mod $o(1)$ are determined by the distributions $D_\hbar$ mod $o(1)$.*

Proof. The difficulty is that there are possible cancellations in the trace formula: we do not assume the weak transversality of the nonprimitive periods mT_j.

Let us denote by $\tau_1(y) = \inf_j T_j(y)$; the function τ_1 is piecewise smooth. The nonsmooth points belong to B. Let us take a maximal (open) interval K where τ_1 is smooth. On K, $\tau_1 = T_{j_0}$ for a unique j_0, and $D_\hbar = Z_{1,j_0} + o(1)$ near the graph L_1 of τ_1, meaning that

$$\mathrm{MS}(D_\hbar - Z_{1,j_0}) \cap L_1 = \emptyset.$$

This is clear because the Lagrangian curves $L_{m,j}$ for $j \neq j_0$ and for $j = j_0,\ m \neq 1$ are disjoint from L_1. So, we can recover L_1 as

$$L_1 = \{(y, \eta) | \eta = \inf\{\eta' > 0, (y, \eta') \in \mathrm{MS}(D_\hbar)\}\}.$$

From Z_{1,j_0}, we recover, for any $m \in \mathbb{Z}$, the Z_{m,j_0}'s, and we introduce a new distribution $D_\hbar^1$ in K defined by

$$D_\hbar^1 = D_\hbar - \sum_{m\in\mathbb{Z}} Z_{m,j_0}.$$

We do the same constructions with $D_\hbar^1$ in K: this allows us to split again K into sub-intervals separated by points of B where $T_j(y) = T_l(y)$ for some $j \neq l,\ j \neq j_0,\ l \neq j_0$ and to get a function τ_2 and distributions $D_\hbar^2$. After a finite number of steps, the new distributions $D_\hbar^N$ is $o(1)$. We have $N_k = N$, and B is the union of all points of nonsmoothness of the τ_j's.

We will need:

Lemma 12.3. *There is an unique splitting of $Z_\hbar$ as a sum*

$$Z_\hbar(y) = \frac{1}{2\pi\hbar}\sum_{j=1}^{N_k}(a_j(y) + \hbar b_j(y))e^{iS_j(y)/\hbar} + o(1),$$

where the S_j's are smooth and the a_j's do not vanish.

Proof. The L^2-Microsupport of $Z_\hbar$ is the union of the graphs of the S_j': this decomposition is unique due to Assumption 10.1.3. Hence the decomposition of $Z_\hbar$ as a finite sum of smooth WKB functions is unique.

From the two previous lemmas, it follows that, with Assumption 10.1.3, the spectrum in J_k modulo $o(\hbar^{5/2})$ determines the actions S_j and $S_{j,2}(y)$.

12.4 Limit values of some integrals

Using the trick of Section 8.3, we can use Abel's result (Section 13.4) once we know the following limits (or asymptotic behaviors) as $y \to E_j^+ (j = 0, 1, \ldots)$:

- $f_\pm^j(y)$
- $\int_{H^{-1}(y)} V''(x)|dt|$ where $H = \xi^2 + V(x)$ is the classical Hamiltonian.
- $f_\pm^{\prime j}(y)$

All of them are determined by the knowledge of V in the set $\{x|V(x) \leq E_j\}$.

It is clear, except for the second one, we have:

Lemma 12.4. *Let us assume that V satisfies assumption 1 of Section 10.1. If E_j is a critical value of V which is not a local minimum and $\tau(z) := \int_{H^{-1}(E_j+z)} V''(x)|dt| - \int_{H^{-1}(E_j-z)} V''(x)|dt|$, then $\lim_{z\to 0^+}\tau(z) = 0$.*

Proof. We cut the integrals into pieces. One piece near each critical point, and another piece far from them. Far from the critical points, the convergence is clear.

- *Local maximum:* let us take a critical point where $V(x) = E_j - A(x-x_0)^{2N}(1+o(1))$ with $N \geq 1$ and $A > 0$. We use a smooth change of variable $x = \psi(y)$ with $\psi(0) = x_0$ so that $V(\psi(y)) = E_j - y^{2N}$. We are reduced to check that

$$\lim_{\varepsilon \to 0^+} \left(\int_0^1 \frac{W(y)dy}{\sqrt{\varepsilon + y^{2N}}} - \int_{\varepsilon^{1/2N}}^1 \frac{W(y)dy}{\sqrt{y^{2N} - \varepsilon}} \right) = 0,$$

  assuming that $W(y) = O(y^{2N-2})$.
- *Other critical points:* let us take a critical point where $V(x) = E_j + A(x-x_0)^{2N+1}(1+o(1))$ with $N \geq 1$ and $A > 0$. We use the same method.

13 Extensions to other operators

13.1 The statement

Let us indicate in this section how to extend the previous results to the operator

$$\mathcal{L}_\hbar = -\hbar^2 \frac{d}{dx}\left(n(x) \frac{d}{dx} \right) + n(x)$$

which was introduced in Section 3. We want to recover the function $n(x)$. Let us sketch the one-well case for which we will get:

Theorem 13.1. *Assuming that*

- *the function* $n(x)$ *admits a* non-degenerate minimum $n(x_0) = E_0 > 0$,
- *the function* $n(x)$ *has* no critical values in $]E_0, E_1[$ *with* $E_1 \leq \liminf_{x \to \partial I} n(x)$,
- *the function* $n(x)$ *has* a generic symmetry defect *as in Theorem 5.1,*

then the function n is determined in $\{x | n(x) \leq E_1\}$ *by the semi-classical spectrum of* $\mathcal{L}_\hbar$ *modulo* $o(\hbar^2)$.

The proof works along the same lines as that of Theorem 5.1 except that we get an integral transform which is not exactly Abel's transform.

13.2 The Weyl symbol and the actions

The Weyl symbol l of $\mathcal{L}$ can be computed, using the Moyal product, as $l = \xi \star n \star \xi + n$. We get

$$l(x,\xi) = n(x)(1+\xi^2) + \frac{\hbar^2}{4} n''(x).$$

The action S_0 satisfies

$$\frac{dS_0}{dy}(y) = T(y) = \int_{n(x)\leq y} \frac{dx}{\sqrt{n(x)(y-n(x))}}.$$

The action S_2 is given from [4] by

$$S_2(y) = -\frac{1}{12}\frac{d}{dy}\int_{\gamma_y}\left(yn'' - 2\left(\frac{y}{n}-1\right)n'^2\right)|dt| - \frac{1}{4}\int_{\gamma_y} n''|dt|,$$

which we rewrite as

$$S_2(y) = -\frac{1}{12}\frac{d}{dy}J(y) - \frac{1}{4}K(y).$$

As in Sections 5 and 8.2, using n instead of V, we introduce the functions $f_\pm$, F and G.

13.3 Recovering G

We have

$$T(y) = 2\int_{E_0}^{y} \frac{G'(z)}{\sqrt{z}}\frac{dz}{\sqrt{y-z}}.$$

The function T is the Abel transform, starting from E_0, of the continuous function $G'(z)/\sqrt{z}$ (E_0 is > 0). Using the inversion of Abel transform, we get G.

13.4 Recovering ±F

- **The integral J:**

$$J(y) = \int_{x_-(y)}^{x_+(y)}\left(yn'' - 2\left(\frac{y}{n}-1\right)n'^2\right)\frac{dx}{\sqrt{n(y-n)}}$$

Using $x = f_\pm(y)$ as in Section 5 and

$$\Phi(y) = \frac{1}{f'_+(y)} - \frac{1}{f'_-(y)},$$

we get $J(y) = (\mathcal{J}\Phi)(y)$, with

$$(\mathcal{J}\Phi)(y) = \int_{E_0}^{y}\left(y\Phi'(u) - 2\left(\frac{y}{u}-1\right)\Phi(u)\right)\frac{du}{\sqrt{u(y-u)}}.$$

• **The integral K:**

$$K(y) = \int_{E_0}^{y} \Phi'(u) \frac{du}{\sqrt{u(y-u)}}$$

and

$$K(y) = 2\frac{d}{dy}\int_{E_0}^{y} \Phi'(u) \frac{\sqrt{y-u}\,du}{\sqrt{u}}$$

which is rewritten as

$$K(y) = 2\frac{d}{dy}(\mathcal{K}\Phi)(y).$$

• **An integral transform:**

Lemma 13.1. *If* $0 < E_0 < E_1$, *the kernel of* $A := \mathcal{J} + 6\mathcal{K}$ *on the space of continuous function on* $[E_0, E_1]$ *at most two-dimensional, and all functions in this kernel are smooth.*

Proof. We have

$$A\Phi(y) = \int_{E_0}^{y} \left((7y - 6u)\Phi'(u) - 2\left(\frac{y}{u} - 1\right)\Phi(u)\right) \frac{du}{\sqrt{u(y-u)}}. \tag{7}$$

We compute $T \circ A$ with the operator T defined by $T\psi(y) = \int_{E_0}^{y} \frac{\psi(u)du}{\sqrt{y-u}}$. We will need the easy:

Lemma 13.2. *We have*

$$\int_{E_0}^{y} \frac{udu}{\sqrt{y-u}} \int_{E_0}^{u} f(t)\frac{dt}{\sqrt{u-t}} = \frac{\pi}{2}\int_{E_0}^{y} (t+y) f(t)dt,$$

and

$$\int_{E_0}^{y} \frac{du}{\sqrt{y-u}} \int_{E_0}^{u} f(t)\frac{dt}{\sqrt{u-t}} = \pi \int_{E_0}^{y} f(t)dt.$$

Applying the previous formulas, we get

$$T \circ A\ (\Phi)(y) = \frac{\pi}{2}\int_{E_0}^{y} \left[(t+y)\left(7\Phi'(t) - 2\frac{\Phi(t)}{t}\right) + 2(-6t\Phi'(t) + 2\Phi(t))\right]\frac{dt}{\sqrt{t}}.$$

Taking two derivatives:

$$\frac{\pi}{y^{3/2}}\frac{d^2}{dy^2}((T \circ A)\ \Phi)(y) = y^2\Phi''(y) + 4y\Phi'(y) - \Phi(y).$$

From S_2 and $A\Phi(E_0)$, we get $A\Phi$, then we get $P(\Phi)$ where $P\phi = y^2\phi'' + 4y\phi' - \phi$ is a nonsingular linear differential equation (recall that $E_0 > 0$). So, if we know also $\Phi(E_0)$ and the asymptotic behavior of $\Phi'(E_0)$, we can get Φ. Let us assume $n''(x_0) = a > 0$. Then we have:

- $A\Phi(E_0) = 2\pi\sqrt{aE_0}$
- $\Phi(E_0) = 0$
- $\Phi'(y) \sim 4\sqrt{a}/\sqrt{y - E_0}$.

Appendix: Abel's result

Let us consider the linear operator T that acts on continuous functions on $[E_0, E[$ defined by

$$Tf(x) = \int_{E_0}^{x} \frac{f(y)dy}{\sqrt{x-y}}.$$

Then $T^2 f(x) = \pi \int_{E_0}^{x} f(y)dy$. This implies that T is injective! This is the content of [1].

References

1. Niels Abel. Auflösung einer mechanichen Aufgabe. *Journal de Crelle* **1***:153–157 (1826).*
2. Marcel Berger, Paul Gauduchon and Edmond Mazet. Le spectre d'une variété riemannienne comapcte. *Springer Lecture Notes in Mathematics, Vol.* **194** *(1971).*
3. Yves Colin de Verdière. Semi-classical analysis and passive imaging. *Nonlinearity* **22***:45–75 (2009).*
4. Yves Colin de Verdière. Bohr–Sommerfeld rules to all orders. *Ann. Henri Poincaré* **6***:925–936 (2005).*
5. Yves Colin de Verdière. Spectrum of the Laplace operators and periodic geodesics: thirty years after. *Ann. Inst. Fourier* **57***:2429–2463 (2007).*
6. Yves Colin de Verdière and Victor Guillemin. A semi-classical inverse problem I: Taylor expansions. Geometric Aspects of Analysis and Mechanics: In Honor of the 65th Birthday of Hans Duistermaat, (Eds) J. A. C. Kolk and Erik P. van den Ban, *Birkhäuser, Boston, MA, 81–95 (2011).*
7. Eric Delabaere and Frédéric Pham. Unfolding the quartic oscillator. *Annals of Physics* **261***:180–218 (1997).*
8. Alfonso Gracia-Saz. The symbol of a function of a pseudo-differential operator. *Ann. Inst. Fourier* **55***:2257–2284 (2005).*
9. David Gurarie. Semi-classical eigenvalues and shape problems on surface of revolution. *J. Math. Phys.* **36***:1934–1944 (1995).*
10. Bernard Helffer and Johannes Sjöstrand. Multiple wells in the semi-classical limit I, *Commun. in PDE,* **9***(4):337–408 (1984).*
11. Mark Kac. Can one hear the shape of a drum? *Amer. Math. Monthly,* **73***(4):1–23 (1966).*
12. B. M. Levitan and M. G. Gasymov. Determination of a differential equation by two of its spectra. *Russian Math. Surveys* **19***(2):1–63 (1964).*
13. André Martinez. An Introduction to semi-classical Theory. *Springer (2002).*
14. Steve Zelditch. The inverse spectral problem. *Surveys in Differential Geometry* **IX***:401–467 (2004).*

On the solvability of systems of pseudodifferential operators

Nils Dencker

In Honor of the Memory of Hans Duistermaat

Abstract This paper studies the solvability for square systems of classical pseudodifferential operators. We assume that the system is of principal type, i.e., the principal symbol vanishes of first order on the kernel. We shall also assume that the eigenvalues of the principal symbol close to zero have constant multiplicity. We prove that local solvability for the system is equivalent to condition (Ψ) on the eigenvalues of the principal symbol. This condition rules out any sign changes from $-$ to $+$ of the imaginary part of the eigenvalue when going in the positive direction on the bicharacteristics of the real part. Thus we need no conditions on the lower order terms. We obtain local solvability by proving a localizable a priori estimate for the adjoint operator with a loss of 3/2 derivatives (compared with the elliptic case).

Key words: Solvability, pseudodifferential operator, principal type, systems of differential equations

Mathematics Subject Classification (2010): Primary: 35A01; Secondary: 35S05, 47G30, 58J40

1 Introduction

In this paper we shall study the question of local solvability for square systems of classical pseudodifferential operators $P \in \Psi^m_{cl}(M)$ on a C^∞ manifold M. These are the pseudodifferential operators which have an asymptotic expansion in homogeneous terms, where the highest order term is called the principal symbol. We shall only consider operators acting on distributions $\mathcal{D}'(M, \mathbf{C}^N)$ with values in $\mathbf{C}^N$, but since the results are local and invariant under base changes, they

Centre for Mathematical Sciences, University of Lund, Box 118, SE-221 00 Lund, Sweden
e-mail: dencker@maths.lth.se

immediately carry over to operators on sections of vector bundles. We shall assume that P is of principal type, so that the principal symbol vanishes of first order on the kernel, see Definition 2.1.

Local solvability for a $N \times N$ system P at a compact set $K \subseteq M$ means that the equation

$$Pu = v \tag{1.1}$$

has a local weak solution $u \in \mathcal{D}'(M, \mathbf{C}^N)$ in a neighborhood of K for all $v \in C^\infty(M, \mathbf{C}^N)$ in a subset of finite codimension. We say that P is microlocally solvable at a compactly based cone $K \subset T^*M$ if there exists an integer N, such that for every $f \in H^{loc}_{(N)}(M, \mathbf{C}^N)$, there exists $u \in \mathcal{D}'(M, \mathbf{C}^N)$ so that $K \bigcap \mathrm{WF}(Pu - f) = \emptyset$, see [14, Definition 26.4.3]. Here $H_{(s)}$ is the usual L^2 Sobolev space and $H^{loc}_{(s)}$ is the localized Sobolev space, i.e., those $f \in \mathcal{D}'$ such that $\phi f \in H_{(s)}$ for any $\phi \in C_0^\infty$.

Hans Lewy's famous counterexample [25] from 1957 showed that not all smooth linear differential operators are solvable. It was conjectured by Nirenberg and Treves [28] in 1970 that local solvability for principal type *scalar* pseudodifferential operators is equivalent to condition (Ψ) on the principal symbol p, which means that

$$\begin{gathered}\operatorname{Im}(ap) \text{ does not change sign from } - \text{ to } + \\ \text{along the oriented bicharacteristics of } \operatorname{Re}(ap)\end{gathered} \tag{1.2}$$

for any $0 \neq a \in C^\infty(T^*M)$. Recall that the operator is of principal type if $dp \neq 0$ when $p = 0$ with non-radial Hamilton vector field and the oriented bicharacteristics are the positive flow-outs of the Hamilton vector field

$$H_{\operatorname{Re}(ap)} = \sum_j \partial_{\xi_j} \operatorname{Re}(ap)\partial_{x_j} - \partial_{x_j} \operatorname{Re}(ap)\partial_{\xi_j}$$

on $\operatorname{Re}(ap) = 0$ (also called the semi-bicharacteristics of p). Condition (1.2) is obviously invariant under symplectic changes of coordinates and multiplication with nonvanishing factors. Thus the condition is invariant under conjugation of P with elliptic Fourier integral operators. It actually suffices to check the condition with some $0 \neq a \in C^\infty$ such that $d(\operatorname{Re} ap) \neq 0$, see [14, Lemma 26.4.10]. Recall that p satisfies condition $(\overline{\Psi})$ if $\overline{p}$ satisfies condition (Ψ), and that p satisfies condition (P) if there are no sign changes on the semi-bicharacteristics, that is, p satisfies both condition (Ψ) and $(\overline{\Psi})$.

The necessity of (Ψ) for local solvability of scalar pseudodifferential operators was proved by Moyer [27] in 1978 for the two-dimensional case, and by Hörmander [13] in 1981 for the general case. The sufficiency of condition (Ψ) for solvability of scalar pseudodifferential operators in two dimensions was proved by Lerner [18] in 1988. The Nirenberg–Treves conjecture was finally proved by the author [8], giving solvability with a loss of two derivatives (compared with the elliptic case). This has been improved to a loss of arbitrarily more than $3/2$ derivatives by the author [9], and to a loss of exactly $3/2$ by Lerner [24]. Observe that there only exist counterexamples showing a loss of $1+\varepsilon$ derivatives for arbitrarily small $\varepsilon > 0$, see Lerner [19].

For partial differential operators, condition (Ψ) is equivalent to condition (P). The sufficiency of (P) for local solvability of scalar partial differential operators with a loss of one derivative was proved in 1973 by Beals and Fefferman [1], introducing the Beals–Fefferman calculus. In the case of operators that are *not* of principal type, conditions corresponding to (Ψ) are neither necessary nor sufficient for local solvability, see [3].

For systems there is no corresponding conjecture for solvability. By looking at diagonal systems, one finds that condition (Ψ) for the eigenvalues of the principal symbol is necessary for solvability. But when the principal symbol is not diagonalizable, condition (Ψ) is not sufficient, see Example 2.14 below. It is not even known if condition (Ψ) is sufficient in the case when the principal symbol is C^∞-diagonalizable. We shall consider the case when the principal symbol has constant characteristics; then the eigenvalue close to the origin has constant multiplicity, see Definition 2.5. In that case, the eigenvalue is a C^∞-function and condition (Ψ) on the eigenvalues is well defined. The main result of the paper is that classical square systems of pseudodifferential operators of principal type having constant characteristics are solvable (with a loss of $3/2$ derivatives) if and only if the eigenvalues of the principal symbol has non-radial Hamilton vector field and satisfies condition (Ψ), see Theorem 2.7.

2 Statement of results

We say that the system $P \in \Psi^m_{cl}$ is *classical* if the symbol of P is an asymptotic sum $P_m + P_{m-1} + \cdots \in S^m_{cl}$. Here $P_j(x,\xi) \in S^j$ is homogeneous of degree j in ξ; and P_m is called the *principal symbol* of P. Recall that the eigenvalues of the principal symbol are the solutions to the characteristic equation

$$|P_m(x,\xi) - \lambda \operatorname{Id}_N| = 0,$$

where $|A|$ is the determinant of the matrix. In the following, we shall denote by $\operatorname{Ker} A$ the kernel and $\operatorname{Ran} A$ the range of the matrix A, and let $w = (x,\xi)$. The definition of principal type for systems is similar to the one for scalar operators.

Definition 2.1. We say that the $N \times N$ system $P(w) \in C^1$ is of *principal type* at w_0 if

$$\partial_\nu P(w_0) : \operatorname{Ker} P(w_0) \mapsto \operatorname{Coker} P(w_0) = \mathbf{C}^N / \operatorname{Ran} P(w_0) \qquad (2.1)$$

is bijective for some ν; here $\partial_\nu P = \langle \nu, dP \rangle$ and the mapping (2.1) is given by $u \mapsto \partial_\nu P(w_0)u$ modulo $\operatorname{Ran} P(w_0)$. We say that $P \in \Psi^m_{cl}$ is of principal type at w_0 if the principal symbol $P_m(x,\xi)$ is of principal type at w_0.

Remark 2.2. If $P(w) \in C^1$ is of principal type and $A(w)$, $B(w) \in C^1$ are invertible, then APB is of principal type. We also have that P is of principal type if and only if the adjoint P^* is of principal type.

In fact, Leibniz's rule gives

$$d(APB) = (dA)PB + A(dP)B + APdB \tag{2.2}$$

and $\operatorname{Ran}(APB) = A(\operatorname{Ran} P)$ and $\operatorname{Ker}(APB) = B^{-1}(\operatorname{Ker} P)$ when A and B are invertible, which gives the invariance under left and right multiplication. Since $\operatorname{Ker} P^*(w_0) = \operatorname{Ran} P(w_0)^\perp$ we find that P satisfies (2.1) if and only if

$$\operatorname{Ker} P(w_0) \times \operatorname{Ker} P^*(w_0) \ni (u, v) \mapsto \langle \partial_\nu P(w_0)u, v\rangle \tag{2.3}$$

is a nondegenerate bilinear form. Since $\langle \partial_\nu P^* u, v\rangle = \overline{\langle \partial_\nu P v, u\rangle}$ we then obtain that P^* is of principal type.

Observe that only square systems can be of principal type since

$$\operatorname{Dim} \operatorname{Ker} P = \operatorname{Dim} \operatorname{Coker} P + M - N$$

if P is an $N \times M$ system. In general, if the system is of principal type and has constant multiplicity of the eigenvalues, then there are no nontrivial Jordan boxes, see Definition 2.3 and Proposition 2.10. Then we also have that the eigenvalues λ are of principal type: $d\lambda \neq 0$ when $\lambda = 0$. When the multiplicity is equal to one, this condition is sufficient. In fact, by using the spectral projection one can find invertible systems A and B so that

$$APB = \begin{pmatrix} \lambda & 0 \\ 0 & E \end{pmatrix}$$

with E invertible $(N-1) \times (N-1)$ system, and this system is obviously of principal type.

Definition 2.3. Let A be an $N \times N$ matrix and λ an eigenvalue of A. The multiplicity of λ as a root of the characteristic equation $|A - \lambda \operatorname{Id}_N| = 0$ is called the *algebraic* multiplicity of the eigenvalue, and the dimension of $\operatorname{Ker}(A - \lambda \operatorname{Id}_N)$ is called the *geometric* multiplicity.

Observe that if the matrix $P(w)$ depend continuously on a parameter w, then the eigenvalues $\lambda(w)$ also depend continuously on w. We will call such a continuous function $\lambda(w)$ of eigenvalues a *section of eigenvalues of* $P(w)$.

Remark 2.4. If the section of eigenvalues $\lambda(w)$ of the $N \times N$ system $P(w) \in C^\infty$ has constant *algebraic* multiplicity, then $\lambda(w) \in C^\infty$. In fact, if k is the multiplicity, then $\lambda = \lambda(w)$ solves $\partial_\lambda^{k-1}|P(w) - \lambda \operatorname{Id}_N| = 0$ so $\lambda(w) \in C^\infty$ by the Implicit Function Theorem.

This is *not* true for constant geometric multiplicity, for example $P(t) = \left(\begin{smallmatrix} 0 & 1 \\ t & 0 \end{smallmatrix}\right)$, $t \in \mathbf{R}$, has geometric multiplicity equal to one for the eigenvalues $\pm\sqrt{t}$. Observe the geometric multiplicity is lower or equal to the algebraic, and for symmetric systems they are equal. We shall assume that the eigenvalues close to zero have constant algebraic and geometric multiplicities by the following definition.

Definition 2.5. The $N \times N$ system $P(w) \in C^\infty$ has *constant characteristics* near w_0 if there exists an $\varepsilon > 0$, so that any section of eigenvalues $\lambda(w)$ of $P(w)$ has both constant algebraic and geometric multiplicity when $|\lambda(w)| < \varepsilon$ and $|w - w_0| < \varepsilon$.

Definition 2.5 is invariant under changes of bases: $P \mapsto E^{-1}PE$ where E is an invertible system, since this preserves the multiplicities of the eigenvalues of the system. It is also invariant under taking adjoints, since $|P^*(w) - \lambda^*(w)\,\mathrm{Id}\,| = \overline{|P(w) - \lambda(w)\,\mathrm{Id}\,|}$ and $\operatorname{Dim}\operatorname{Ker}(P^*(w) - \lambda^*(w)\,\mathrm{Id}) = \operatorname{Dim}\operatorname{Ker}(P(w) - \lambda(w)\,\mathrm{Id})$. The definition is *not* invariant under multiplication of the system with invertible systems, even in the case when $P(w) = \lambda(w)\,\mathrm{Id}$ since $A(w)P(w) = \lambda(w)A(w)$ need not have constant characteristics.

Observe that generically the eigenvalues of a system have constant multiplicity, but not necessarily when equal to zero. For example, the system

$$P(w) = \begin{pmatrix} w_1 & w_2 \\ w_2 & -w_1 \end{pmatrix}$$

is symmetric and of principal type with eigenvalues $\pm\sqrt{w_1^2 + w_2^2}$, which have constant multiplicity except when equal to 0.

Definition 2.6. Let the $N \times N$ system $P \in \Psi_{cl}^m$ be of principal type and constant characteristics. We say that P satisfies condition (Ψ) or (P) if the eigenvalues of the principal symbol satisfies condition (Ψ) or (P).

Observe that the eigenvalue close to the origin is a uniquely defined C^∞-function of principal type by Definition 2.3 and Proposition 2.10. Thus, the semi-bicharacteristics of the eigenvalues are well defined near the characteristic set $\{\, w : |P(w)| = 0 \,\}$, so the conditions (Ψ) and (P) on the eigenvalues are well defined. Also well defined is the condition that the Hamilton vector field of an eigenvalue λ does not have the radial direction when $\lambda = 0$.

To get local solvability at a point $x_0 \in M$ we shall also assume a strong form of the non-trapping condition at x_0 for the eigenvalues λ of P:

$$\lambda = 0 \implies \partial_\xi \lambda \neq 0. \tag{2.4}$$

This means that all nontrivial semi-bicharacteristics of λ are transversal to the fiber $T_{x_0}^* M$, which originally was the condition for principal type of Nirenberg and Treves [28]. Microlocally, in a conical neighborhood of $(x, \xi) \in T^*M$, we can always obtain (2.4) after a canonical transformation if the Hamilton vector field is not radial. In the following, we shall use the usual L^2 Sobolev norm $\|u\|_{(s)}$ and the L^2 norm $\|u\| = \|u\|_{(0)}$.

Theorem 2.7. *Let $P \in \Psi^m_{cl}(M)$ be an $N \times N$ system of principal type and constant characteristics near $(x_0, \xi_0) \in T^*M$, such that the Hamilton vector field of an eigenvalue λ does not have the radial direction when $\lambda = 0$. Then P is microlocally solvable near (x_0, ξ_0) if and only if condition (Ψ) is satisfied near (x_0, ξ_0), and then*

$$\|u\| \leq C(\|P^*u\|_{(3/2-m)} + \|Ru\| + \|u\|_{(-1)}) \qquad u \in C_0^\infty(M, \mathbf{C}^N). \tag{2.5}$$

Here $R \in \Psi^{1/2}(M)$ is a $K \times N$ system such that $(x_0, \xi_0) \notin \mathrm{WF}(R)$, which gives microlocal solvability of P at (x_0, ξ_0) with a loss of at most $3/2$ derivatives. If P is of principal type and constant characteristics with eigenvalues satisfying condition (Ψ) and (2.4) near $x_0 \in M$, then we obtain (2.5) with $x \neq x_0$ in $\mathrm{WF}(R)$, which gives local solvability of P at x_0 with a loss of at most $3/2$ derivatives.

As usual, $\mathrm{WF}(R)$ is the smallest smallest conical set in $T^*M\setminus 0$ such that $R \in \Psi^{-\infty}$ in the complement. The conditions in Theorem 2.7 are invariant under conjugation with scalar Fourier integral operators since they only depend on the principal symbol of the system. They are also invariant under the base change: $P \mapsto E^{-1}PE$ with invertible system E, since this preserves the multiplicities of the eigenvalues of the principal symbol. The conditions of Theorem 2.7 are more or less necessary; of course condition (Ψ) is necessary even in the scalar case. Examples 2.14 and 2.15 show that we need both the condition of principal type and of constant multiplicity in order to get this result.

We shall postpone the proof of Theorem 2.7 to Section 4. The proof of the necessity is essentially the classical Moyer–Hörmander proof for the scalar case. The proof of the sufficiency will be an adaption of the proof for the scalar case in [8], using some of the ideas of Lerner [24]. In fact, since the normal form of the operator will have a scalar principal symbol, the multiplier will essentially be the same as in [8]. But since we lose more than one derivative in the estimate we also have to consider the lower-order matrix-valued terms in the expansion of the operator. This is done in Section 7 and is the main new part of the paper. In Section 3 we review the Weyl calculus and state the estimates we will use in the proof of Theorem 2.7. But we shall postpone the proof of the semiclassical estimate of Proposition 3.6 until Section 7. In Section 4 we prove Theorem 2.7 by a microlocal reduction to a normal form using the estimates in Section 3. In Section 5 we define the symbol classes and weights we are going to use. In Section 6 we review the Wick quantization, introduce the function spaces and the multiplier estimate that we will use for the proof of Proposition 3.6. Finally, in Section 7 we prove Proposition 3.6 by estimating the contributions of the lower order terms. The proof of Theorem 2.7 in Section 4 also gives the following results.

Remark 2.8. If P is of principal type with constant characteristics satisfying condition (P), then we get the estimate (2.5) with $3/2$ replaced by 1. If P satisfies condition $(\overline{\Psi})$ and some repeated Poisson bracket of the real and imaginary parts of the eigenvalue close to the origin is nonvanishing, then we obtain a subelliptic estimate for P with $3/2$ replaced by $k/k+1$ in (2.5) for some $k \in \mathbf{Z}_+$, see [14, Chapter 27].

The Poisson bracket of f and g is defined by $\{f, g\} = H_f g$. Theorem 2.7 has applications to scalar non-principal type pseudodifferential operators by the following result.

Theorem 2.9. *Let $Q \in \Psi^1_{cl}(M)$ be a scalar operator of principal type near $(x_0, \xi_0) \in T^*M$ and let $A_j \in \Psi^0_{cl}(M)$, $j = 1, \dots, N$ be scalar. Then the equation*

$$Pu = Q^N u + \sum_{j=0}^{N-1} A_j Q^j u = f \tag{2.6}$$

is locally solvable near (x_0, ξ_0) if and only if $\sigma(Q)$ satisfies condition (Ψ) near (x_0, ξ_0).

Proof. This is a standard reduction to a first-order system. For scalar $u \in \mathcal{D}'$ we let $u_{j+1} = Q^j u$ for $0 \le j < N$. Then (2.6) holds if and only if $U = {}^t(u_1, \dots, u_N)$ solves

$$\mathbb{P} U = F, \tag{2.7}$$

where

$$\mathbb{P} = \begin{pmatrix} Q & -1 & 0 & 0 & & \dots \\ 0 & Q & -1 & 0 & & \dots \\ 0 & 0 & Q & -1 & & \dots \\ \dots & & & & & \\ A_0 & A_1 & A_2 & \dots & & Q + A_{N-1} \end{pmatrix}$$

and $F = {}^t(0, 0, \dots, f)$. Now the equation (2.6) is locally solvable if and only if the system (2.7) is locally solvable. In fact, to solve (2.7) we first put $u_1 = 0$, $u_2 = -f_1$ and recursively $u_{j+1} = Qu_j - f_j$ for $1 \le j < N$. Then we only have to solve (2.6) for $u = v_1$ with f depending on f_j, and add $Q^{j-1} v_1$ to u_j. Now $\sigma(\mathbb{P}) = \sigma(Q)\,\mathrm{Id}_N$ which is of principal type with constant characteristics, so it is locally solvable if and only if Q satisfies condition (Ψ) according to Theorem 2.7.

We shall conclude the section with some examples. But first we prove a result about the characterization of systems of principal type.

Proposition 2.10. *Assume that $P(w) \in C^\infty$ is an $N \times N$ system such that $|P(w_0)| = 0$ and there exists an $\varepsilon > 0$ such that the eigenvalue λ of $P(w)$ with $|\lambda| < \varepsilon$ has constant algebraic multiplicity in a neighborhood of w_0. Let $\lambda(w) \in C^\infty$ be the unique eigenvalue for $P(w)$ near w_0 satisfying $\lambda(w_0) = 0$ by Remark 2.4. Then $P(w)$ is of principal type at w_0 if and only if $d\lambda(w_0) \neq 0$ and the geometric multiplicity of the eigenvalue λ is equal to the algebraic multiplicity at w_0.*

Thus, if $P(w)$ is of principal type having constant characteristics, then all sections of eigenvalues $\lambda(w)$ are of principal type, and we have no nontrivial Jordan boxes in the normal form. This means that for symmetric systems having constant characteristics it suffices that the eigenvalues are of principal type. If $P(w)$ does not have constant characteristics, then this is no longer true; in fact the eigenvalues need not even be differentiable, see Example 2.15.

Observe that if $P(w)$ is of principal type and has constant characteristics, then $P(w) - z \operatorname{Id}_N$ is of principal type for $|z| \ll 1$. In fact, the algebraic and geometric multiplicities are constant for the eigenvalue λ and $d\lambda \neq 0$ when $\lambda = 0$.

Now the eigenvalue $\lambda(w)$ in Proposition 2.10 is the unique C^∞ solution to $\partial_\lambda^{k-1}|P(w) - \lambda \operatorname{Id}_N| = 0$ according to Remark 2.4, where k is the algebraic multiplicity. Thus we find that $d\lambda(w) \neq 0$ if and only if

$$\partial_w \partial_\lambda^{k-1}|P(w) - \lambda \operatorname{Id}_N| \neq 0 \qquad \text{when } \lambda = \lambda(w).$$

We only need this condition for a symmetric system with constant multiplicity to be of principal type.

Example 2.11. Let

$$P(w) = \begin{pmatrix} w_1 + iw_2^2 & w_2 \\ 0 & w_1 + iw_2^2 \end{pmatrix} \qquad w = (w_1, w_2) \in \mathbf{R}^2,$$

then P is of principal type, has constant algebraic multiplicity of the eigenvalue $w_1 + iw_2^2$ but not constant geometric multiplicity. In fact, $\partial_{w_1} P = \operatorname{Id}_2$, $P(w)$ has nontrivial kernel only when $w_2 = 0$, but the geometric multiplicity of the eigenvalue is equal to one when $w_2 \neq 0$.

Example 2.12. Let

$$P = p(x, D_x) \operatorname{Id}_N + B(x, D_x) + P_0(x, D_x),$$

where $p \in S^1$ is a scalar homogeneous symbol of principal type, $B \in \Psi^1_{cl}$ with nilpotent homogeneous principal symbol $\sigma(B)$ and $P_0 \in \Psi^0_{cl}$. Then p is the only eigenvalue to $\sigma(P)$ and P is of principal type if and only if $\sigma(B) = 0$ when $p = 0$ by Proposition 2.10.

Remark 2.13. Observe that the conclusion of Proposition 2.10 does not hold if the algebraic multiplicity is not constant. For example,

$$P(w) = \begin{pmatrix} w_1 & 1 \\ w_2 & w_1 \end{pmatrix} \qquad w = (w_1, w_2) \in \mathbf{R}^2$$

has determinant equal to $w_1^2 - w_2$ and eigenvalues $w_1 \pm \sqrt{w_2}$, so the geometric but not the algebraic multiplicity is constant near $w_2 = 0$. Since

$$\begin{pmatrix} 0 & 1 \\ 1 & 0 \end{pmatrix} P(w) = \begin{pmatrix} w_2 & w_1 \\ w_1 & 1 \end{pmatrix}$$

we find that $P(w)$ is of principal type at $(0, 0)$ by invariance.

Proof of Proposition 2.10. First we note that $P(w)$ is of principal typ at w_0 if and only if

$$\partial_\nu^k |P(w_0)| \neq 0 \qquad k = \operatorname{Dim} \operatorname{Ker} P(w_0) \tag{2.8}$$

for some $\nu \in T(T^*\mathbf{R}^n)$. Observe that $\partial^j |P(w_0)| = 0$ for $j < \operatorname{Dim} \operatorname{Ker} P(w_0)$. In fact, by choosing bases for $\operatorname{Ker} P(w_0)$ and $\operatorname{Im} P(w_0)$ respectively, and extending to bases of $\mathbf{C}^N$, we obtain matrices A and B so that

$$AP(w)B = \begin{pmatrix} P_{11}(w) & P_{12}(w) \\ P_{21}(w) & P_{22}(w) \end{pmatrix},$$

where $|P_{22}(w_0)| \neq 0$ and P_{11}, P_{12} and P_{21} all vanish at w_0. By the invariance, P is of principal type if and only if $\partial_\nu P_{11}$ is invertible for some ν, so by expanding the determinant we obtain (2.8).

Now since the eigenvalue $\lambda(w)$ has constant algebraic multiplicity near w_0, we find that

$$|P(w) - \lambda \operatorname{Id}_N| = (\lambda(w) - \lambda)^m e(w, \lambda)$$

near w_0, where $\lambda(w_0) = 0$, $e(w, \lambda) \neq 0$ and $m \geq \operatorname{Dim} \operatorname{Ker} P(w_0)$ is the algebraic multiplicity. By putting $\lambda = 0$ we obtain that $\partial_\nu^j |P(w_0)| = 0$ if $j < m$ and $\partial_\nu^m |P(w_0)| = (\partial_\nu \lambda(w_0))^m e(w_0, 0)$ which proves Proposition 2.10.

The following example is an unsolvable system of principal type with real eigenvalues, but it does not have constant characteristics.

Example 2.14. Let

$$P = \begin{pmatrix} D_{x_1} & B(x, D_x) \\ -1 & D_{x_1} + R(D_x) \end{pmatrix}, \tag{2.9}$$

where $R(\xi) = \xi_2^2/|\xi|$ and $\sigma(B)(x, \xi) = \xi_2 B_0(x, \xi)$ with $B_0 \in S^0$ homogeneous in ξ. The eigenvalues of the principal symbol $\sigma(P)$ are ξ_1 and $\xi_1 + R(\xi)$ that are real and coincide when $\xi_2 = 0$. Since $\partial_{\xi_1} \sigma(P) = \operatorname{Id}_2$ and $\sigma(B)$ vanish when $\xi_2 = 0$, we find that P is of principal type. If $t \mapsto \operatorname{Im} B_0(t, x', 0, 0, \xi'')$ changes sign at $t = x_1$, then P is not microlocally solvable at $(x, 0, 0, \xi'')$; here $x = (x_1, x') = (x_1, x_2, x'')$ and $\xi'' \neq 0$. In fact, the system $PU = F$ with $U = {}^t(u_1, u_2)$ and $F = {}^t(f_1, f_2)$ is equivalent to the equation

$$Qu_2 = (D_{x_1}(D_{x_1} + R(D_x)) + B(x, D_x))u_2 = f_1 + D_{x_1} f_2 \tag{2.10}$$

if we put $u_1 = (D_{x_1} + R(D_x))u_2 - f_2$. Thus the system P is solvable if and only if Q is solvable. That Q is not solvable follows from using the construction of approximate solutions to the adjoint in [26], replacing D_{x_2} with $R(D_x)$.

We can also generalize this to the case where

$$R(\xi) = \xi_2^k |\xi|^{1-k}$$

$\sigma(B)(x, \xi) = \xi_2^j B_j(x, \xi)$ with $j < k$, $B_j \in S^{1-j}$ homogeneous of degree $1 - j$ in ξ, and satisfying the same conditions as B_0. On the other hand, if $\sigma(B)(x, \xi) =$

$\xi_2^j B_j(x,\xi)$ with $j \geq k$, then we can write

$$B(x, D_x) \cong A(x, D_x)R(D_x) \qquad \text{modulo } \Psi^0$$

for some $A \in \Psi^0$, and then

$$\begin{pmatrix} 1 & -A \\ 0 & 1 \end{pmatrix} P \begin{pmatrix} 1 & A \\ 0 & 1 \end{pmatrix} \cong \begin{pmatrix} D_{x_1} & 0 \\ 0 & D_{x_1} + R(D_x) \end{pmatrix} \qquad \text{modulo } \Psi^0$$

which is solvable. In fact, the principal symbol is on diagonal form with real diagonal elements of principal type, giving L^2 estimates of the adjoint that can be perturbed by lower-order terms.

Finally, we have an example of an unsolvable operator which is diagonalizable and self-adjoint, but not of principal type.

Example 2.15. Take real $b(t) \in C^\infty(\mathbf{R})$, and define the symmetric system

$$P = \begin{pmatrix} D_t + b(t)D_x & (t - ib(t))D_x \\ (t + ib(t))D_x & -D_t + b(t)D_x \end{pmatrix} = P^* \qquad (t, x) \in \mathbf{R}^2.$$

Eigenvalues of $\sigma(P)$ are $b(t)\xi \pm \sqrt{\tau^2 + (t^2 + b^2(t))\xi^2}$ which are zero for $(\tau, \xi) \neq 0$ only if $t = \tau = 0$. The eigenvalues coincide for $(\tau, \xi) \neq 0$ if and only if $b(t) = t = \tau = 0$. We have that

$$Q = \frac{1}{2}\begin{pmatrix} 1 & -i \\ 1 & i \end{pmatrix} P \begin{pmatrix} 1 & 1 \\ -i & i \end{pmatrix} = \begin{pmatrix} D_t - itD_x & 2b(t)D_x \\ 0 & D_t + itD_x \end{pmatrix}$$

which is not locally solvable at $t = 0$ for any choice of $b(t)$. In fact, $D_t + itD_x$ is not locally solvable since condition (Ψ) is not satisfied when $\xi > 0$. The eigenvalues of the principal symbol $\sigma(Q)$ are $\tau \pm it\xi$. By the invariance, P is of principal type if and only if $b(0) = 0$. When $b(t) \neq 0$ we find that $\sigma(P)$ is diagonalizable and self-adjoint, but not of principal type. When $b \equiv 0$ the system is symmetric of principal type, but does not have constant characteristics.

3 The multiplier estimates

In this section we shall prove multiplier estimates for microlocal normal forms of the adjoint operator, which we shall use in the proof of Theorem 2.7. We shall consider the model operators

$$P_0 = (D_t + iF(t, x, D_x))\,\mathrm{Id}_N + F_0(t, x, D_x), \tag{3.1}$$

where $F \in C^\infty(\mathbf{R}, \Psi^1_{cl}(\mathbf{R}^n))$ is scalar with the real homogeneous principal symbol $\sigma(F) = f$, and $F_0 \in C^\infty(\mathbf{R}, \Psi^0_{cl})$ is an $N \times N$ system. In the following, we shall assume that P_0 satisfies condition $(\overline{\Psi})$:

$$f(t,x,\xi) > 0 \quad \text{and } s > t \implies f(s,x,\xi) \ge 0 \tag{3.2}$$

for any t, $s \in \mathbf{R}$ and $(x,\xi) \in T^*\mathbf{R}^n$. This means that the adjoint P_0^* satisfies condition (Ψ) for the eigenvalue $\tau - if(t,x,\xi)$. Observe that if $\chi \ge 0$, then χf also satisfies (3.2); thus the condition can be localized.

Remark 3.1. We may also consider symbols $f \in L^\infty(\mathbf{R}, S^1(\mathbf{R}^n))$, that is, $f(t,x,\xi)$ is measurable and bounded in $S^1(\mathbf{R}^n)$ for almost all t. Then we say that P_0 satisfies condition $(\overline{\Psi})$ if for every (x,ξ) condition (3.2) holds for almost all s, $t \in \mathbf{R}$.

Observe that, since $(x,\xi) \mapsto f(t,x,\xi)$ is continuous for almost all t, it suffices to check (3.2) for (x,ξ) in a countable dense subset of $T^*\mathbf{R}^n$. Then we find that f has a representative satisfying (3.2) for any t, s and (x,ξ) after putting $f(t,x,\xi) \equiv 0$ for t in a null set.

In order to prove Theorem 2.7 we shall make a second microlocalization using the specialized symbol classes of the Weyl calculus, and the Weyl quantization of symbols $a \in \mathcal{S}'(T^*\mathbf{R}^n)$ defined by

$$(a^w u, v) = (2\pi)^{-n} \iint \exp\left(i\langle x-y,\xi\rangle\right) a\left(\frac{x+y}{2},\xi\right) u(y)\overline{v}(x)\, dx dy d\xi$$

$$u, v \in \mathcal{S}(\mathbf{R}^n)$$

Observe that $\operatorname{Re} a^w = (\operatorname{Re} a)^w$ is the symmetric part and $i \operatorname{Im} a^w = (i \operatorname{Im} a)^w$ the antisymmetric part of the operator a^w. Also, if $a \in S^m_{cl}(\mathbf{R}^n)$, then $a^w(x,D_x) = a(x,D_x)$ modulo $\Psi^{m-1}_{cl}(\mathbf{R}^n)$ by [14, Theorem 18.5.10]. The same holds for $N \times N$ systems of operators.

We recall the definitions of the Weyl calculus: let g_w be a Riemannean metric on $T^*\mathbf{R}^n$, $w = (x,\xi)$; then we say that g is slowly varying if there exists $c > 0$ so that $g_{w_0}(w - w_0) < c$ implies

$$1/C \le g_w/g_{w_0} \le C,$$

that is, $g_w \cong g_{w_0}$. Let σ be the standard symplectic form on $T^*\mathbf{R}^n$, $g^\sigma(w)$ the dual metric of $w \mapsto g(\sigma(w))$, and assume that $g^\sigma(w) \ge g(w)$. We say that g is σ temperate if it is slowly varying and

$$g_w \le C g_{w_0}(1 + g^\sigma_w(w - w_0))^N \qquad w, w_0 \in T^*\mathbf{R}^n.$$

A positive real-valued function $m(w)$ on $T^*\mathbf{R}^n$ is g-continuous if there exists $c > 0$ so that $g_{w_0}(w - w_0) < c$ implies $m(w) \cong m(w_0)$. We say that m is σ, g-temperate if it is g-continuous and

$$m(w) \le C m(w_0)(1 + g^\sigma_w(w - w_0))^N \qquad w, w_0 \in T^*\mathbf{R}^n.$$

If m is σ, g temperate, then m is a weight for g and we can define the symbol classes: $a \in S(m, g)$ if $a \in C^\infty(T^*\mathbf{R}^n)$ and

$$|a|_j^g(w) = \sup_{T_i \neq 0} \frac{|a^{(j)}(w, T_1, \dots, T_j)|}{\prod_1^j g_w(T_i)^{1/2}} \leq C_j m(w) \qquad w \in T^*\mathbf{R}^n \qquad j \geq 0, \tag{3.3}$$

which defines the seminorms of $S(m, g)$. Of course, these symbol classes can also be defined locally. For matrix-valued symbols, we use the matrix norms. If $a \in S(m, g)$, then we say that the corresponding Weyl operator $a^w \in \operatorname{Op} S(m, g)$. For more results on the Weyl calculus, see [14, Section 18.5].

Definition 3.2. Let m be a weight for the metric g. We say that $a \in S^+(m, g)$ if $a \in C^\infty(T^*\mathbf{R}^n)$ and $|a|_j^g \leq C_j m$ for $j \geq 1$.

Observe that if $a \in S^+(m, g)$, then a is a symbol. In fact, since $g \leq g^\sigma$ we find by integration that

$$\begin{aligned} |a(w) - a(w_0)| &\leq C_1 \sup_{\theta \in [0,1]} m(w_\theta) g_{w_\theta}(w - w_0)^{1/2} \\ &\leq C_N m(w_0)(1 + g_{w_0}^\sigma(w - w_0))^{N_0}, \end{aligned}$$

where $w_\theta = \theta w + (1 - \theta) w_0$, which implies that $m + |a|$ is a weight for g. Clearly, $a \in S(m + |a|, g)$, so the operator a^w is well defined.

Lemma 3.3. *Assume that m_j is a weight for for the σ-temperate conformal metrics $g_j = h_j g^\sharp \leq g^\sharp = (g^\sharp)^\sigma$ and $a_j \in S^+(m_j, g_j)$, $j = 1, 2$. Let $g = (g_1 + g_2)/2$ and $h^2 = \sup g_1/g_2^\sigma = \sup g_2/g_1^\sigma$. Then we find that $h^2 = h_1 h_2$ and*

$$a_1^w a_2^w - (a_1 a_2)^w \in \operatorname{Op} S(m_1 m_2 h, g). \tag{3.4}$$

We also obtain the usual expansion of (3.4) *with terms in* $S(m_1 m_2 h^k, g)$, $k \geq 1$.

Observe that by Proposition 18.5.7 and (18.5.14) in [14] we find that g is σ-temperate and $g/g^\sigma \leq (h_1 + h_2)^2/4 \leq 1$.

Proof. As shown after Definition 3.2 we have that $m_j + |a_j|$ is a weight for g_j and $a_j \in S(m_j + |a_j|, g_j)$, $j = 1, 2$. Thus

$$a_1^w a_2^w \in \operatorname{Op} S((m_1 + |a_1|)(m_2 + |a_2|), g)$$

is given by Proposition 18.5.5 in [14]. We find that $a_1^w a_2^w - (a_1 a_2)^w = a^w$ with

$$a(w) = E\left(\frac{i}{2}\sigma(D_{w_1}, D_{w_2})\right) \frac{i}{2}\sigma(D_{w_1}, D_{w_2}) a_1(w_1) a_2(w_2)\big|_{w_1 = w_2 = w},$$

where $E(z) = (e^z - 1)/z = \int_0^1 e^{\theta z}\, d\theta$. We have that $\sigma(D_{w_1}, D_{w_2}) a_1(w_1) a_2(w_2) \in S(M, G)$ where

$$M(w_1, w_2) = m_1(w_1)m_2(w_2)h_1^{1/2}(w_1)h_2^{1/2}(w_2)$$

and $G_{w_1,w_2}(z_1, z_2) = g_{1,w_1}(z_1) + g_{2,w_2}(z_2)$. Now the proof of Theorem 18.5.5 in [14] works also when $\sigma(D_{w_1}, D_{w_2})$ is replaced by $\theta\sigma(D_{w_1}, D_{w_2})$, uniformly in $0 \le \theta \le 1$. By using Proposition 18.5.7 in [14] and integrating over $\theta \in [0, 1]$ we obtain that $a(w)$ has an asymptotic expansion in $S(m_1 m_2 h^k, g)$, which proves the lemma.

Remark 3.4. The conclusions of Lemma 3.3 also hold if a_1 has values in $\mathcal{L}(B_1, B_2)$ and a_2 has values in B_1 where B_1 and B_2 are Banach spaces (see Section 18.6 in [14]).

For example, if $\{a_j\}_j \in S(m_1, g_1)$ with values in ℓ^2, and $b_j \in S(m_2, g_2)$ uniformly in j, then $\{a_j^w b_j^w\}_j \in \mathrm{Op}(m_1 m_2, g)$ with values in ℓ^2. Observe that if $\{\phi_j\}_j \in S(1, g)$ is a partition of unity so that $\sum_j \phi_j^2 = 1$ and $a \in S(m, g)$, then $\{\phi_j a\}_j \in S(m, g)$ has values in ℓ^2.

Example 3.5. The standard symbol class $S^\mu_{\varrho,\delta}$ defined by

$$|\partial_x^\alpha \partial_\xi^\beta a(x, \xi)| \le C_{\alpha\beta} \langle\xi\rangle^{\mu+\delta|\alpha|-\varrho|\beta|}$$

has σ-temperate metric if $0 \le \delta \le \varrho \le 1$ and $\delta < 1$.

In the proof of Theorem 2.7 we shall microlocalize near (x_0, ξ_0) and put $h^{-1} = \langle\xi_0\rangle = 1 + |\xi_0|$. Then after doing a symplectic dilation: $(x, \xi) \mapsto (h^{-1/2}x, h^{1/2}\xi)$, we find that $S^k_{1,0} = S(h^{-k}, hg^\sharp)$ and $S^k_{1/2,1/2} = S(h^{-k}, g^\sharp)$, $k \in \mathbf{R}$, where $g^\sharp = (g^\sharp)^\sigma$ is the Euclidean metric. We shall prove a semiclassical estimate for a microlocal normal form of the operator.

Let $\|u\|$ be the L^2 norm on $\mathbf{R}^{n+1}$, and (u, v) the corresponding sesquilinear inner product. As before, we say that $f \in L^\infty(\mathbf{R}, S(m, g))$ if $f(t, x, \xi)$ is measurable and bounded in $S(m, g)$ for almost all t. The following is the main estimate that we shall prove.

Proposition 3.6. *Assume that*

$$P_0 = (D_t + if^w(t, x, D_x))\,\mathrm{Id}_N + F_0^w(t, x, D_x),$$

where $f \in L^\infty(\mathbf{R}, S(h^{-1}, hg^\sharp))$ is real satisfying condition $(\overline{\Psi})$ given by (3.2), and $F_0 \in L^\infty(\mathbf{R}, S(1, hg^\sharp))$ is an $N \times N$ system; here $0 < h \le 1$ and $g^\sharp = (g^\sharp)^\sigma$ are constant. Then there exists $T_0 > 0$ and $N \times N$ symbols $b_T(t, x, \xi) \in L^\infty(\mathbf{R}, S(h^{-1/2}, g^\sharp) \bigcap S^+(1, g^\sharp))$ such that $\mathrm{Im}\, b_T \in L^\infty(\mathbf{R}, S(h^{1/2}, g^\sharp))$ uniformly for $0 < T \le T_0$, and

$$h^{1/2}(\|b_T^w u\|^2 + \|u\|^2) \le C_0 T\, \mathrm{Im}(P_0 u, b_T^w u) \tag{3.5}$$

for $u(t, x) \in \mathcal{S}(\mathbf{R} \times \mathbf{R}^n, \mathbf{C}^N)$ having support where $|t| \le T \le T_0$. The constants C_0, T_0 and the seminorms of b_T only depend on the seminorms of f and F_0.

Remark 3.7. It follows from the proof that $b_T = \widetilde{b}_T E^* E$, modulo $S(h^{1/2}, g^\sharp)$, where $E \in S(1, hg^\sharp)$ is an invertible $N \times N$ system, $\widetilde{b}_T$ is scalar and $|\widetilde{b}_T| \le CH^{-1/2}$. Here H is a weight for $g^\sharp$ such that $h \le H \le 1$, and $G = Hg^\sharp$ is σ-temperate (see Claim 3.9, Definition 5.3 and Proposition 6.3).

Observe that it follows from (3.5) and the Cauchy–Schwarz inequality that

$$\|u\| \le CTh^{-1/2}\|P_0 u\|,$$

which will give a loss of 3/2 derivatives after microlocalization. Proposition 3.6 will be proved in Section 7.

There are two difficulties present in estimates of the type (3.5). The first is that b_T is not C^∞ in the t variables. Therefore one has to be careful not to involve b_T^w in the calculus with symbols in all the variables. We shall avoid this problem by using tensor products of operators and the Cauchy–Schwarz inequality. The second difficulty lies in the fact that we could have $|b_T| \gg h^{1/2}$, so it is not obvious that cut-off errors can be controlled.

Lemma 3.8. *The estimate* (3.5) *can be perturbed with terms in* $L^\infty(\mathbf{R}, S(h^{1/2}, hg^\sharp))$ *in the expansion of* P_0 *for small* T. *Also, it can be microlocalized: if* $\phi(w) \in S(1, hg^\sharp)$ *is real valued and independent of* t, *then we have*

$$\operatorname{Im}(P_0\phi^w u, b_T^w \phi^w u) \le \operatorname{Im}(P_0 u, \phi^w b_T^w \phi^w u) + Ch^{1/2}\|u\|^2 \qquad u(t,x) \in \mathcal{S}(\mathbf{R}^{n+1}, \mathbf{C}^N) \tag{3.6}$$

where $\phi^w b_T^w \phi^w$ *satisfies the same conditions as* b_T^w.

Proof. In the following, we shall say that a system is *real* if it is a real multiple of the identity matrix. It is clear that we may perturb (3.5) with terms in $L^\infty(\mathbf{R}, S(h^{1/2}, g^\sharp))$ in the expansion of P_0 for small enough T. Now, we can also perturb with real terms $r^w \in L^\infty(\mathbf{R}, \operatorname{Op} S(1, hg^\sharp))$. In fact, if $r \in S(1, hg^\sharp)$ is real and $B \in S^+(1, g^\sharp)$ is symmetric modulo $S(h^{1/2}, g^\sharp)$, then

$$|\operatorname{Im}(r^w u, B^w u)| \le |([(\operatorname{Re} B)^w, r^w]u, u)|/2 + |(r^w u, (\operatorname{Im} B)^w u)| \le Ch^{1/2}\|u\|^2. \tag{3.7}$$

In fact, we have $[(\operatorname{Re} B)^w, r^w] \in \operatorname{Op} S(h^{1/2}, g^\sharp)$ by Lemma 3.3.

If $\phi(w) \in S(1, hg^\sharp)$, then

$$[P_0, \phi^w \operatorname{Id}_N] = \{f, \phi\}^w \operatorname{Id}_N \text{ modulo } L^\infty(\mathbf{R}, \operatorname{Op} S(h, hg^\sharp)),$$

where $\{f, \phi\} \in L^\infty(\mathbf{R}, S(1, hg^\sharp))$ is real valued. By using (3.7) with $r^w = \{f, \phi\}^w \operatorname{Id}_N$ and $B^w = b_T^w \phi^w$, we obtain (3.6) since $b_T^w \phi^w \in \operatorname{Op} S^+(1, g^\sharp)$ is symmetric modulo $\operatorname{Op} S(h^{1/2}, g^\sharp)$ for almost all t by Lemma 3.3. Since Lemma 3.3 also gives that

$$\phi^w b_T^w \phi^w = \phi^w (b_T \phi)^w = (b_T \phi^2)^w$$

modulo $L^\infty(\mathbf{R}, \operatorname{Op} S(h^{1/2}, g^\sharp))$ we find that $\phi^w b_T^w \phi^w$ satisfies the same conditions as b_T^w.

Claim 3.9. When proving the estimate (3.5) we may assume that

$$F_0 = \langle d_w f, R_0 \rangle = \sum_j \partial_{w_j} f R_{0,j} \qquad \text{modulo } L^\infty(\mathbf{R}, S(h, hg^\sharp)), \tag{3.8}$$

where $R_{0,j} \in L^\infty(\mathbf{R}, S(h^{1/2}, hg^\sharp))$ are $N \times N$ systems, for all j.

Proof. By conjugation with $(E^{\pm 1})^w \in \operatorname{Op} S(1, hg^\sharp)$ we find that

$$(E^{-1})^w P E^w = (E^{-1})^w E^w (D_t + i f^w) \operatorname{Id}_N + (E^{-1}(D_t E + H_f E + F_0 E))^w = \widetilde{P}$$

modulo $L^\infty(\mathbf{R}, S(h, hg^\sharp))$. By solving

$$\begin{cases} D_t E + F_0 E = 0 \\ E\big|_{t=0} = \operatorname{Id}_N \end{cases}$$

we obtain (3.8) for $\widetilde{P}$ with $\langle d_w f, R_0 \rangle = E^{-1} H_f E$. From the calculus we obtain that

$$E^w (E^{-1})^w = 1 = (E^{-1})^w E^w \qquad \text{modulo } \operatorname{Op} S(Th, hg^\sharp)$$

uniformly when $|t| \le T$. Thus, for small enough T we obtain that $(E^{\pm 1})^w$ is invertible in L^2. Since the metric $hg^\sharp$ is trivially strongly σ-temperate in the sense of [2, Definition 7.1], we find from [2, Corollary 7.7] that there exists $A \in L^\infty(\mathbf{R}, S(1, hg^\sharp))$ such that $E^w A^w = 1$. Thus, if we prove the estimate (3.5) for $\widetilde{P}$ and substitute $u = A^w v$, we obtain the estimate for P with b_T replaced by $((E^{-1})^w)^* b_T^w A^w$. Since $A = E^{-1}$ modulo $S(h, hg^\sharp)$ we find from Lemma 3.3 as before that the symbol of this multiplier is in $S(h^{-1/2}, g^\sharp) \bigcap S^+(1, g^\sharp)$ and that it is symmetric modulo $S(h^{1/2}, g^\sharp)$.

We shall see from the proof that if F_0 is on the form (3.8), then $b_T = b_T \operatorname{Id}_N$ is real. Thus, in general the symbol of the multiplier will be on the form $b_T (E^{-1})^* E^{-1}$ modulo $S(h^{1/2}, g^\sharp)$ with invertible E and a real scalar b_T. In the following, we shall use the partial Sobolev norms:

$$\|u\|_s = \|\langle D_x \rangle^s u\|. \tag{3.9}$$

We shall now prove the estimate that is going to be used in the proof of Theorem 2.7.

Proposition 3.10. *Assume that*

$$P_0 = (D_t + i F^w(t, x, D_x)) \operatorname{Id}_N + F_0^w(t, x, D_x)$$

with $F^w \in L^\infty(\mathbf{R}, \Psi^1_{cl}(\mathbf{R}^n))$ *having the real principal symbol* f *satisfying condition* $(\overline{\Psi})$ *given by* (3.2) *and* $F_0 \in L^\infty(\mathbf{R}, \Psi^0_{cl}(\mathbf{R}^n))$ *is an* $N \times N$ *system. Then there exists* $T_0 > 0$ *and* $N \times N$ *symbols* $B_T(t, x, \xi) \in L^\infty(\mathbf{R}, S^1_{1/2,1/2}(\mathbf{R}^n))$ *with*

$$\nabla B_T = (\partial_x B_T, |\xi| \partial_\xi B_T) \in L^\infty(\mathbf{R}, S^1_{1/2,1/2}(\mathbf{R}^n))$$

and $\operatorname{Im} B_T(t,x,\xi) \in L^\infty(\mathbf{R}, S^0_{1/2,1/2}(\mathbf{R}^n))$ *uniformly for* $0 < T \le T_0$*, such that*

$$\|B^w_T u\|^2_{-1/2} + \|u\|^2 \le C_0(T \operatorname{Im}(P_0 u, B^w_T u) + \|u\|^2_{-1}) \tag{3.10}$$

for $u \in \mathcal{S}(\mathbf{R}^{n+1}, \mathbf{C}^N)$ *having support where* $|t| \le T \le T_0$*. The constants* T_0*,* C_0 *and the seminorms of* B_T *only depend on the seminorms of* F *and* F_0*.*

Since $\nabla B_T \in L^\infty(\mathbf{R}, S^1_{1/2,1/2})$ we find that the commutators of B^w_T with scalar operators in $L^\infty(\mathbf{R}, \Psi^0_{1,0})$ are in $L^\infty(\mathbf{R}, \Psi^0_{1/2,1/2})$. This will make it possible to localize the estimate. The idea to include the first term in (3.10) is due to Lerner [24].

Proof that Proposition 3.6 gives Proposition 3.10. Choose real symbols $\{\phi_j(x,\xi)\}_j$ and $\{\psi_j(x,\xi)\}_j \in S^0_{1,0}(\mathbf{R}^n)$ having values in ℓ^2, such that $\sum_j \phi_j^2 = 1$, $\psi_j\phi_j = \phi_j$ and $\psi_j \ge 0$. We may assume that the supports are small enough so that $\langle\xi\rangle \cong \langle\xi_j\rangle$ in $\operatorname{supp}\psi_j$ for some ξ_j, and that there is a fixed bound on number of overlapping supports. Then, after doing a symplectic dilation

$$(y,\eta) = (x\langle\xi_j\rangle^{1/2}, \xi/\langle\xi_j\rangle^{1/2})$$

we obtain that $S^m_{1,0}(\mathbf{R}^n) = S(h_j^{-m}, h_j g^\sharp)$ and $S^m_{1/2,1/2}(\mathbf{R}^n) = S(h_j^{-m}, g^\sharp)$ in $\operatorname{supp}\psi_j$, $m \in \mathbf{R}$, where $h_j = \langle\xi_j\rangle^{-1} \le 1$ and $g^\sharp(dy,d\eta) = |dy|^2 + |d\eta|^2$ is constant.

By using the calculus in the y variables, we find $\phi^w_j P_0 = \phi^w_j P_{0j}$ modulo $\operatorname{Op} S(h_j, h_j g^\sharp)$, where

$$\begin{aligned} P_{0j} &= (D_t + i(\psi_j F)^w(t,y,D_y)) \operatorname{Id}_N + (\psi_j F_0)^w(t,y,D_y) \\ &= (D_t + i f^w_j(t,y,D_y)) \operatorname{Id}_N + F^w_j(t,y,D_y) \end{aligned} \tag{3.11}$$

with $f_j = \psi_j f \in L^\infty(\mathbf{R}, S(h_j^{-1}, h_j g^\sharp))$ satisfying (3.2), and $F_j \in L^\infty(\mathbf{R}, S(1, h_j g^\sharp))$ uniformly in j. Then, by using Proposition 3.6 and Lemma 3.8 for P_{0j} we obtain symbols $b_{j,T}(t,y,\eta) \in L^\infty(\mathbf{R}, S(h_j^{-1/2}, g^\sharp) \bigcap S^+(1, g^\sharp))$ such that $\operatorname{Im} b_{j,T} \in S(h_j^{1/2}, g^\sharp)$ uniformly for $0 < T \ll 1$, and

$$\|b^w_{j,T}\phi^w_j u\|^2 + \|\phi^w_j u\|^2 \le C_0 T(h_j^{-1/2} \operatorname{Im}(P_0 u, \phi^w_j b^w_{j,T}\phi^w_j u) + \|u\|^2) \qquad \forall j \tag{3.12}$$

for $u(t,y) \in \mathcal{S}(\mathbf{R} \times \mathbf{R}^n, \mathbf{C}^N)$ having support where $|t| \le T$. Here and in the following, the constants are independent of T.

By substituting $\psi^w_j u$ in (3.12) and summing up, we obtain

$$\|B^w_T u\|^2_{-1/2} + \|u\|^2 \le C_0 T(\operatorname{Im}(P_0 u, B^w_T u) + \|u\|^2) + C_1\|u\|^2_{-1} \tag{3.13}$$

for $u(t,x) \in \mathcal{S}(\mathbf{R} \times \mathbf{R}^n, \mathbf{C}^N)$ having support where $|t| \le T$. Here

$$B_T^w = \sum_j h_j^{-1/2} \psi_j^w \phi_j^w b_{j,T}^w \phi_j^w \psi_j^w = \sum_j B_{j,T}^w \in L^\infty(\mathbf{R}, \Psi^1_{1/2,1/2})$$

so $\operatorname{Im} B_T \in L^\infty(\mathbf{R}, \Psi^0_{1/2,1/2})$. In fact, since $d\psi_j = 0$ on $\operatorname{supp} \phi_j$, we have

$$\{\phi_j^w [P_{0j}^w, \psi_j^w]\}_j \in \Psi^{-1}_{1,0}(\mathbf{R}^n)$$

with values in ℓ^2 for almost all t. Also, $\sum_j \phi_j^2 = 1$ so $\sum_j \phi_j^w \phi_j^w = 1$ modulo $\Psi^{-1}(\mathbf{R}^n)$, and by the finite overlap of supports we find that

$$\begin{aligned}(\langle D_x\rangle^{-1/2} B_T^w)^* \langle D_x\rangle^{-1/2} B_T^w &= (B_T^w)^* \langle D_x\rangle^{-1} B_T^w \\ &= \sum_{|j-k|\le K} (B_{j,T}^w)^* \langle D_x\rangle^{-1} B_{k,T}^w \qquad \text{modulo } \Psi^{-2}\end{aligned}$$

for some K, which implies that

$$\|B_T^w u\|^2_{-1/2} \le C_K \left(\sum_k \|B_{k,T}^w u\|^2_{-1/2} + \|u\|^2_{(-1)} \right).$$

We also have that $\langle D_x\rangle^{-1/2} h_j^{-1/2} \psi_j^w \phi_j^w \in \Psi^0(\mathbf{R}^n)$ uniformly, which gives

$$\|B_{k,T}^w u\|_{-1/2} \le C \|b_{k,T}^w \phi_k^w \psi_k^w u\| \qquad \forall k.$$

We find that $\nabla B_T \in S^1_{1/2,1/2}$ since Lemma 3.3 gives

$$B_T = \sum_j h_j^{-1/2} b_{j,T} \phi_j^2 \in S^1_{1/2,1/2} \qquad \text{modulo } S^0_{1/2,1/2},$$

where $\phi_j \in S(1, h_j g^\sharp)$ and $b_{j,T} \in S^+(1, g^\sharp)$ for almost all t. For small enough T we obtain (3.10) and the corollary.

4 Proof of Theorem 2.7

In order to prove the theorem, we first need a preparation result so that we can get the system on a normal form.

Proposition 4.1. *Assume that $P \in S^m_{cl}(M)$ is an $N \times N$ system of principal type having constant characteristics near $(x_0, \xi_0) \in T^*M$. Then there exist elliptic $N \times N$ systems A and $B \in S^0_{cl}(M)$ such that*

$$A^w P^w B^w = Q^w = \begin{pmatrix} Q_{11}^w & 0 \\ 0 & Q_{22}^w \end{pmatrix} \in \Psi_{cl}^m$$

microlocally near (x_0, ξ_0). We have that $\sigma(Q_{11}) = \lambda \operatorname{Id}_K$ where the section of eigenvalues $\lambda(w) \in C^\infty$ of $P(w)$ is of principal type, and Q_{22}^w is elliptic.

Thus we obtain the system on a block form. Observe that if $K = 0$, then P is elliptic at (x_0, ξ_0). Since P is of principal type we find by the invariance given by (2.2) that Q is of principal type, so λ vanishes of first order on its zeros.

Proof. Since P_m has constant characteristics by the assumptions, we find that the characteristic equation

$$|P_m(w) - \lambda \operatorname{Id}_N| = 0$$

has a unique local solution $\lambda(w) \in C^\infty$ of multiplicity $K > 0$. Since $P_m(w)$ is of principal type, Proposition 2.10 gives that $d\lambda(w_0) \neq 0$ and the geometric multiplicity $\operatorname{Dim}\operatorname{Ker}(P_m(w) - \lambda(w)\operatorname{Id}_N) \equiv K$ in a neighborhood of $w_0 = (x_0, \xi_0)$. Since the dimension is constant, we may choose a C^∞-base for $\operatorname{Ker}(P_m(w) - \lambda(w)\operatorname{Id}_N)$ in a neighborhood of w_0. By orthogonalizing it, extending to a orthonormal C^∞-base for $\mathbf{C}^N$ and using homogeneity, we obtain orthogonal homogeneous E such that

$$E^* P_m E = \begin{pmatrix} \lambda(w)\operatorname{Id}_K & P_{12} \\ 0 & P_{22} \end{pmatrix} = \widetilde{P}_m = \sigma((E^w)^* P^w E^w).$$

Clearly $\operatorname{Ker} \widetilde{P}_m = \{ (z_1, \dots, z_N) : z_j = 0 \text{ for } j > K \}$ when $\lambda = 0$ and $d\widetilde{P}_m$ is equal to multiplication with $d\lambda$ on $\operatorname{Ker} \widetilde{P}_m$. Since $\widetilde{P}_m$ is of principal type when $\lambda = 0$, we find that $\operatorname{Im} \widetilde{P}_m \bigcap \operatorname{Ker} \widetilde{P}_m = \{0\}$ at w_0, which implies that P_{22} is invertible. In fact, if it were not invertible, there would exist $0 \neq z'' \in \mathbf{C}^{N-K}$ so that $P_{22} z'' = 0$; then

$$0 \neq \widetilde{P}_m{}^t(0, z'') = {}^t(P_{12} z'', 0) \in \operatorname{Im} \widetilde{P}_m \bigcap \operatorname{Ker} \widetilde{P}_m,$$

giving a contradiction. By multiplying $\widetilde{P}_m$ from the left with

$$\begin{pmatrix} \operatorname{Id}_K & -P_{12} P_{22}^{-1} \\ 0 & \operatorname{Id}_{N-K} \end{pmatrix}$$

we obtain $P_{12} \equiv 0$. Thus, we find that

$$A^w P^w B^w = \begin{pmatrix} Q_{11}^w & Q_{12}^w \\ Q_{21}^w & Q_{22}^w \end{pmatrix} \in \Psi_{cl}^m$$

where $\sigma(Q_{11}) = \lambda \operatorname{Id}_K$, $|\sigma(Q_{22})| \neq 0$ and $Q_{12}, Q_{21} \in \Psi_{cl}^{m-1}$. Choose a microlocal parametrix $B_{22}^w \in \Psi_{cl}^{-m}$ to Q_{22}^w so that $B_{22}^w Q_{22}^w = Q_{22}^w B_{22}^w = \operatorname{Id}_{N-K}$ modulo C^∞ near w_0. By multiplying from the left with

$$\begin{pmatrix} \mathrm{Id}_K & -Q_{12}^w B_{22}^w \\ 0 & \mathrm{Id}_{N-K} \end{pmatrix} \in \Psi_{cl}^0,$$

we obtain that $Q_{12} \in S^{-\infty}$. By multiplying from the right with

$$\begin{pmatrix} \mathrm{Id}_K & 0 \\ -B_{22}^w Q_{21}^w & \mathrm{Id}_{N-K} \end{pmatrix} \in \Psi_{cl}^0,$$

we obtain $Q_{21} \in S^{-\infty}$. Note that these multiplications do not change the principal symbols of Q_{jj} for $j = 1, 2$, which finishes the proof.

Proof of Theorem 2.7. Observe that since P satisfies condition (Ψ) we find that the adjoint P^* satisfies condition $(\overline{\Psi})$. By multiplying with an elliptic pseudodifferential operator, we may assume that $m = 1$. Let P^* have the expansion $P_1 + P_0 + \cdots$ where $P_1 = \sigma(P^*) \in S^1$; then it is clear that it suffices to consider $w_0 = (x_0, \xi_0) \in |P_1|^{-1}(0)$; otherwise $P^* \in \Psi_{cl}^1(M)$ is elliptic near w_0 so (2.5) holds and P is microlocally solvable. Now P^* is of principal type having constant characteristics, so we find by using Proposition 4.1 that

$$P^* = \begin{pmatrix} Q_{11}^w & 0 \\ 0 & Q_{22}^w \end{pmatrix} \in \Psi_{cl}^1$$

microlocally near w_0, where $\sigma(Q_{11}) = \lambda \,\mathrm{Id}_K$ with $\lambda \in C^\infty$ an eigenvalue of $\sigma(P^*)$ of principal type and Q_{22}^w is elliptic. Since Q_{22}^w is elliptic, it is trivially solvable, so we only have to investigate the solvability of Q_{11}^w. Now λ is of principal type by the invariance, so if it does not satisfy condition $(\overline{\Psi})$, then the proof of [14, Theorem 26.4.7] can easily be adapted to this case, since the principal part of the operator is a scalar symbol times the identity matrix.

To prove solvability when condition $(\overline{\Psi})$ is satisfied, we shall prove that there exists ϕ and $\psi \in S_{1,0}^0(T^*M)$ such that $\phi = 1$ in a conical neighborhood of (x_0, ξ_0), and for any $T > 0$, there exists a $K \times N$ system $R_T \in S_{1,0}^{1/2}(M)$ with the property that $\mathrm{WF}(R_T^w) \bigcap T_{x_0}^* M = \emptyset$ and

$$\|\phi^w u\| \le C_1 \left(\|\psi^w P^* u\|_{(1/2)} + T\|u\| \right) + \|R_T^w u\| + C_0 \|u\|_{(-1)}$$
$$u \in C_0^\infty(M, \mathbf{C}^N). \tag{4.1}$$

Here $\|u\|_{(s)}$ is the L^2-Sobolev norm and the constants are independent of T. Then for small enough T we obtain (2.5) and microlocal solvability, since $(x_0, \xi_0) \notin \mathrm{WF}(1-\phi)^w$. In the case where the eigenvalue satisfies condition (Ψ) and (2.4) near x_0, we may choose finitely many $\phi_j \in S_{1,0}^0(M)$ such that $\sum \phi_j \ge 1$ near x_0 and $\|\phi_j^w u\|$ can be estimated by the right-hand side of (4.1) for some suitable ψ and R_T. By elliptic regularity of $\{\phi_j\}$ near x_0, we then obtain the estimate (2.5) for small enough T with $x \ne x_0$ in $\mathrm{WF}(R)$.

Observe that in the case when λ satisfies condition (P), we obtain the estimate (4.1) for $P^* = \lambda(x, D_x)\,\mathrm{Id}_N$ with $3/2$ replaced with 1 and $C_1 = \mathcal{O}(T)$ from the Beals–Fefferman estimate, see [1]. Since this estimate can be perturbed with terms in Ψ^0_{cl} for small enough T we get the estimate and solvability in this case. A similar argument gives subelliptic estimates if λ satisfies condition $(\overline{\Psi})$ and the bracket condition, see [14, Chapter 27]. This gives Remark 2.8.

It remains to consider the case $P_1 = \lambda \operatorname{Id}_N$, where λ satisfies condition $(\overline{\Psi})$. It is clear that by multiplying with an elliptic factor we may assume that $\partial_\xi \operatorname{Re} \lambda(w_0) \neq 0$, in the microlocal case after a conical transformation. Then, we may use Darboux's theorem and the Malgrange preparation theorem to obtain microlocal coordinates $(t, y; \tau, \eta) \in T^*\mathbf{R}^{n+1}$ so that $w_0 = (0, 0; 0, \eta_0)$, $t = 0$ on $T^*_{x_0}M$ and $\lambda = q(\tau + if)$ in a conical neighborhood of w_0, where $f \in C^\infty(\mathbf{R}, S^1_{1,0})$ is real and homogeneous satisfying condition (3.2), and $0 \neq q \in S^0_{1,0}$, see Theorem 21.3.6 in [14]. By using the Malgrange preparation theorem and homogeneity we find that

$$P_0(t, x; \tau, \xi) = Q_{-1}(t, x; \tau, \xi)(\tau + if(t, x, \xi)) \operatorname{Id}_N + F_0(t, x, \xi),$$

where Q_{-1} is homogeneous of degree -1 and F_0 is homogeneous of degree 0 in the ξ variables. By conjugation with elliptic Fourier integral operators and using the Malgrange preparation theorem successively on lower order terms, we obtain that

$$P^* = Q^w(D_t \operatorname{Id}_N + i\,(\chi F)^w) + R^w \tag{4.2}$$

microlocally in a conical neighborhood Γ of w_0 as in the proof of Theorem 26.4.7′ in [14]. Here we find that $F \in C^\infty(\mathbf{R}, S^1_{1,0}(\mathbf{R}^n))$ has the real principal symbol $f \operatorname{Id}_N$ satisfying (3.2), $Q \in S^0_{1,0}(\mathbf{R}^{n+1})$ has the principal symbol $q \operatorname{Id}_N \neq 0$ in Γ and $R \in S^1_{1,0}(\mathbf{R}^{n+1})$ satisfies $\Gamma \bigcap \operatorname{WF}(R^w) = \emptyset$. Also, $\chi(\tau, \eta) \in S^0_{1,0}(\mathbf{R}^{n+1})$ is equal to 1 in Γ and $|\tau| \leq C|\eta|$ in $\operatorname{supp} \chi(\tau, \eta)$. By cutting off in the t variable we may assume that $F \in L^\infty(\mathbf{R}, S^1_{1,0}(\mathbf{R}^n))$. Now, we can follow the proof of Theorem 1.4 in [10]. As before, we shall choose ϕ and ψ so that $\phi = 1$ conical neighborhood of w_0, $\psi = 1$ on $\operatorname{supp} \phi$ and $\operatorname{supp} \psi \subset \Gamma$. Also, we shall choose

$$\phi(t, y; \tau, \eta) = \chi_0(t, \tau, \eta)\phi_0(y, \eta)$$

where $\chi_0(t, \tau, \eta) \in S^0_{1,0}(\mathbf{R}^{n+1})$, $\phi_0(y, \eta) \in S^0_{1,0}(\mathbf{R}^n)$, $t \neq 0$ in $\operatorname{supp} \partial_t \chi_0$, $|\tau| \leq C|\eta|$ in $\operatorname{supp} \chi_0$ and $|\tau| \cong |\eta|$ in $\operatorname{supp} \partial_{\tau,\eta} \chi_0$.

Since $|\sigma(Q)| \neq 0$ and $R = 0$ on $\operatorname{supp} \psi$ it is no restriction to assume that $Q \equiv \operatorname{Id}_N$ and $R \equiv 0$ when proving the estimate (4.1). Now, by Theorem 18.1.35 in [14] we may compose $C^\infty(\mathbf{R}, \Psi^m_{1,0}(\mathbf{R}^n))$ with operators in $\Psi^k_{1,0}(\mathbf{R}^{n+1})$ having symbols vanishing when $|\tau| \geq c(1 + |\eta|)$; and we obtain the usual asymptotic expansion in $\Psi^{m+k-j}_{1,0}(\mathbf{R}^{n+1})$ for $j \geq 0$. Since $|\tau| \leq C|\eta|$ in $\operatorname{supp} \chi$ and $\chi = 1$ on $\operatorname{supp} \psi$, it suffices to prove (4.1) for $P^* = D_t + iF^w$.

By using Proposition 3.10 on $\phi^w u$, we obtain that

$$\begin{aligned} &\|B^w_T \phi^w u\|^2_{-1/2} + \|\phi^w u\|^2 \\ &\quad \leq C_0 T (\operatorname{Im}(\phi^w P^* u, B^w_T \phi^w u) + \operatorname{Im}([P^*, \phi^w \operatorname{Id}_N]u, B^w_T \phi^w u)) + C_1 \|\phi^w u\|^2_{-1}, \end{aligned} \tag{4.3}$$

where $B_T^w \in L^\infty(\mathbf{R}, \Psi^1_{1/2,1/2}(\mathbf{R}^n))$ is an $N \times N$ system with $\nabla B_T \in L^\infty(\mathbf{R}, S^1_{1/2,1/2}(\mathbf{R}^n))$, and $\|u\|_s = \|\langle D_y\rangle^s u\|$ is the partial Sobolev norm in the y variables. Since $|\tau| \le C|\eta|$ in $\operatorname{supp}\phi$ we find that $\|\phi^w u\|_{-1} \le C\|u\|_{(-1)}$ For any $u,\ v \in \mathcal{S}(\mathbf{R}^n, \mathbf{C}^N)$ we have that

$$|(v, B_T^w u)| = |(\langle D_y\rangle^{1/2} v, \langle D_y\rangle^{-1/2} B_T^w u)| \le C(\|v\|^2_{1/2} + \|B_T^w u\|^2_{-1/2}), \qquad (4.4)$$

where $\langle D_y\rangle = 1 + |D_y|$. Now $\phi^w = \phi^w \psi^w$ modulo $\Psi^{-2}_{1,0}(\mathbf{R}^{n+1})$, and thus we find from (4.4) that

$$|(\phi^w P^* u, B_T^w \phi^w u)| \le C(\|\psi^w P^* u\|^2_{1/2} + \|u\|^2 + \|B_T^w \phi^w u\|^2_{-1/2}) \qquad (4.5)$$

where the last term can be cancelled for small enough T in (4.3). We also have to estimate the commutator term $\operatorname{Im}\big([P^*, \phi^w \operatorname{Id}_N]u, B_T^w \phi^w u\big)$ in (4.3). We find

$$[P^*, \phi^w \operatorname{Id}_N] = -(i\partial_t \phi^w - \{ f, \phi \}^w) \operatorname{Id}_N \in \Psi^0_{1,0}(\mathbf{R}^{n+1}),$$

modulo $\Psi^{-1}_{1,0}(\mathbf{R}^{n+1})$ by the expansion, where the error term can be estimated by (4.4). Since $\phi = \chi_0\phi_0$ we find that $\{ f, \phi \} = \phi_0\{ f, \chi_0 \} + \chi_0\{ f, \phi_0 \}$, where $\phi_0\{ f, \chi_0 \} = R_0 \in S^0_{1,0}(\mathbf{R}^{n+1})$ is supported when $|\tau| \cong |\eta|$ and $\psi = 1$. Now $(\tau + if)^{-1} \in S^{-1}_{1,0}(\mathbf{R}^{n+1})$ when $|\tau| \cong |\eta|$; thus by [14, Theorem 18.1.35] we find that $R_0^w = A_1^w \psi^w P^*$ modulo $\Psi^{-1}_{1,0}(\mathbf{R}^{n+1})$ where $A_1 = R_0(\tau + if)^{-1} \in S^{-1}_{1,0}(\mathbf{R}^{n+1})$. As before, we find from (4.4) that

$$\begin{aligned} |(R_0^w u, B_T^w \phi^w u)| &\le C(\|R_0^w u\|^2_{1/2} + \|B_T^w \phi^w u\|^2_{-1/2}) \\ &\le C_0(\|\psi^w P^* u\|^2_{-1/2} + \|B_T^w \phi^w u\|^2_{-1/2} + \|u\|^2_{-1/2}), \end{aligned} \qquad (4.6)$$

and also

$$|(\partial_t \phi^w u, B_T^w \phi^w u)| \le \|R_1^w u\|^2 + \|B_T^w \phi^w u\|^2_{-1/2},$$

where $R_1^w = \langle D_y\rangle^{1/2} \partial_t \phi^w \in \Psi^{1/2}_{1,0}(\mathbf{R}^{n+1})$; thus $t \ne 0$ in $\mathrm{WF}(R_1^w)$.

It only remains to estimate the term $\operatorname{Im}((\{ f, \phi_0 \}\chi_0)^w u, B_T^w \phi^w u)$. Here $(\{ f, \phi_0 \}\chi_0)^w = \{ f, \phi_0 \}^w \chi_0^w$ and $\phi^w = \phi_0^w \chi_0^w$ modulo $\Psi^{-1}_{1,0}(\mathbf{R}^{n+1})$. As in (4.4) we find that

$$|(R^w u, B_T^w v)| = |(\langle D_y\rangle R^w u, \langle D_y\rangle^{-1} B_T^w v)| \le C(\|u\|^2 + \|v\|^2)$$

for $R \in S^{-1}_{1,0}(\mathbf{R}^{n+1})$; thus we find that

$$|\operatorname{Im}((\{ f, \phi_0 \}\chi_0)^w u, B_T^w \phi^w u)| \le |\operatorname{Im}(\{ f, \phi_0 \}^w \chi_0^w u, B_T^w \phi_0^w \chi_0^w u)| + C\|u\|^2.$$

The calculus gives $B_T^w \phi_0^w = (B_T\phi_0)^w$ and

$$2i \operatorname{Im}\big((B_T\phi_0)^w \{ f, \phi_0 \}^w\big) = \{ B_T\phi_0, \{ f, \phi_0 \} \}^w = 0,$$

modulo $L^\infty(\mathbf{R}, \Psi^0_{1/2,1/2}(\mathbf{R}^n))$ since $\nabla(B_T\phi_0) \in L^\infty(\mathbf{R}, S^1_{1/2,1/2}(\mathbf{R}^n))$ and $\{f, \phi_0\}$ is real. Thus, we obtain that

$$|\operatorname{Im}(\{f, \phi_0\}^w \chi_0^w u, B_T^w \phi_0^w \chi_0^w u)| \le C\|\chi_0^w u\|^2 \le C'\|u\|^2, \tag{4.7}$$

and the estimate (4.1) for small enough T, which completes the proof of Theorem 2.7.

5 The symbol classes and weights

In this section we shall define the symbol classes to be used. Assume that $f \in L^\infty(\mathbf{R}, S(h^{-1}, hg^\sharp))$ is scalar and satisfies (3.2). Here $0 < h \le 1$ and $g^\sharp = (g^\sharp)^\sigma$ are constant. It is no restriction to change h so that $|f|_1^{g^\sharp} \le h^{-1/2}$, which we assume in what follows. The results shall be uniform in the usual sense; i.e., they will only depend on the seminorms of f in $L^\infty(\mathbf{R}, S(h^{-1}, hg^\sharp))$. Let

$$X_+(t) = \{\, w \in T^*\mathbf{R}^n : \exists s \le t,\ f(s, w) > 0\,\} \tag{5.1}$$

$$X_-(t) = \{\, w \in T^*\mathbf{R}^n : \exists s \ge t,\ f(s, w) < 0\,\}. \tag{5.2}$$

Clearly, $X_\pm(t)$ are open in $T^*\mathbf{R}^n$, $X_+(s) \subseteq X_+(t)$ and $X_-(s) \supseteq X_-(t)$ when $s \le t$. By condition $(\overline{\Psi})$ we obtain that $X_-(t) \bigcap X_+(t) = \emptyset$ and $\pm f(t, w) \ge 0$ when $w \in X_\pm(t)$, $\forall t$. Let $X_0(t) = T^*\mathbf{R}^n \setminus (X_+(t) \bigcup X_-(t))$ which is closed in $T^*\mathbf{R}^n$. By the definition of $X_\pm(t)$ we have $f(t, w) = 0$ when $w \in X_0(t)$. Let

$$d_0(t_0, w_0) = \inf\{\, g^\sharp(w_0 - z)^{1/2} :\ z \in X_0(t_0)\,\} \tag{5.3}$$

be the $g^\sharp$ distance in $T^*\mathbf{R}^n$ to $X_0(t_0)$ for fixed t_0; it is equal to $+\infty$ in the case that $X_0(t_0) = \emptyset$. By taking the infimum over z we find that $w \mapsto d_0(t, w)$ is Lipschitz continuous with respect to $g^\sharp$ for fixed t when $d_0 < \infty$, i.e.,

$$\sup_{w \ne z \in T^*\mathbf{R}^n} |\delta_0(t, w) - \delta_0(t, z)|/g^\sharp(w - z)^{1/2} \le 1.$$

Definition 5.1. We define the signed distance function $\delta_0(t, w)$ by

$$\delta_0 = \operatorname{sgn}(f)\min(d_0, h^{-1/2}), \tag{5.4}$$

where d_0 is given by (5.3) and

$$\operatorname{sgn}(f)(t, w) = \begin{cases} \pm 1, & w \in X_\pm(t) \\ 0, & w \in X_0(t) \end{cases} \tag{5.5}$$

so that $\operatorname{sgn}(f) f \ge 0$.

Remark 5.2. The signed distance function $w \mapsto \delta_0(t, w)$ given by Definition 5.1 is Lipschitz-continuous with respect to the metric $g^\sharp$ with Lipschitz constant equal to 1, for all t. We also find that $t \mapsto \delta_0(t, w)$ is nondecreasing, $\delta_0 f \geq 0$, $|\delta_0| \leq h^{-1/2}$ and when $|\delta_0| < h^{-1/2}$ we find that $|\delta_0| = d_0$ is given by (5.3).

In fact, it suffices to show the Lipschitz-continuity of $w \mapsto \delta_0(t, w)$ on $\complement X_0(t)$, and then it follows from the Lipschitz continuity of $w \mapsto d_0(t, w)$ when $d_0 < \infty$. Clearly $\delta_0 f \geq 0$, and since $X_+(t)$ is nondecreasing and $X_-(t)$ is nonincreasing when t increases, we find that $t \mapsto \delta_0(t, w)$ is nondecreasing.

In the following, we shall treat t as a parameter which we shall suppress, and we shall denote $f' = \partial_w f$ and $f'' = \partial_w^2 f$. We shall also in the following assume that we have chosen $g^\sharp$ orthonormal coordinates so that $g^\sharp(w) = |w|^2$ and $|f'| \leq h^{-1/2}$.

Definition 5.3. Let

$$H^{-1/2} = 1 + |\delta_0| + \frac{|f'|}{|f''| + h^{1/4}|f'|^{1/2} + h^{1/2}} \tag{5.6}$$

and $G = Hg^\sharp$.

Observe that $\langle \delta_0 \rangle = 1 + |\delta_0| \leq H^{-1/2}$ and

$$1 \leq H^{-1/2} \leq 1 + |\delta_0| + h^{-1/4}|f'|^{1/2} \leq 3h^{-1/2} \tag{5.7}$$

since $|f'| \leq h^{-1/2}$ and $|\delta_0| \leq h^{-1/2}$. This gives that $hg^\sharp \leq 9G$.

Definition 5.4. Let

$$M = |f| + |f'|H^{-1/2} + |f''|H^{-1} + h^{1/2}H^{-3/2}. \tag{5.8}$$

Then we have that $h^{1/2} \leq M \leq C_3 h^{-1}$.

The metric G and weight M have the following properties according to Propositions 3.7 and 3.8 in [8] and Proposition 3.5 in [10].

Proposition 5.5. *We find that $H^{-1/2}$ is Lipschitz-continuous, G is σ-temperate such that $G = H^2 G^\sigma$ and*

$$H(w) \leq C_0 H(w_0)(1 + G_w(w - w_0)), \tag{5.9}$$

We have that M is a weight for G such that $M \leq CH^{-1}$, $f \in S(M, G)$, and

$$M(w) \leq C_1 M(w_0)(1 + G_{w_0}(w - w_0))^{3/2}. \tag{5.10}$$

Since $G \leq g^\sharp \leq G^\sigma$ we find that the conditions (5.9) and (5.10) are stronger than the property of being σ-temperate (in fact, strongly σ-temperate in the sense of [2, Definition 7.1]). Note that $f \in S(M, Hg^\sharp)$ for any choice of $H \geq h$ in Definition 5.4. The following property of G is the most important for the proof.

Proposition 5.6. *Let $H^{-1/2}$ be given by Definition 5.3 for $f \in S(h^{-1}, hg^\sharp)$. There exists $\kappa_1 > 0$ so that if $\langle\delta_0\rangle = 1 + |\delta_0| \le \kappa_1 H^{-1/2}$, then*

$$f = \alpha_0 \delta_0, \tag{5.11}$$

where $\kappa_1 M H^{1/2} \le \alpha_0 \in S(MH^{1/2}, G)$, which implies that $\delta_0 = f/\alpha_0 \in S(H^{-1/2}, G)$.

This follows directly from Proposition 3.9 in [8]. Next, we shall define the weight m to be used.

Definition 5.7. For $(t, w) \in \mathbf{R} \times T^*\mathbf{R}^n$ we let

$$\begin{aligned} m(t,w) = \inf_{t_1 \le t \le t_2} \big\{ &|\delta_0(t_1, w) - \delta_0(t_2, w)| \\ &+ \max\big(H^{1/2}(t_1, w)\langle\delta_0(t_1, w)\rangle^2, H^{1/2}(t_2, w)\langle\delta_0(t_2, w)\rangle^2\big)/2 \big\} \end{aligned} \tag{5.12}$$

where $\langle\delta_0\rangle = 1 + |\delta_0|$.

This weight essentially measures how much $t \mapsto \delta_0(t, w)$ changes between the minima of $t \mapsto H^{1/2}(t, w)\langle\delta_0(t, w)\rangle^2$, which will give restrictions on the sign changes of the symbol. When $t \mapsto \delta_0(t, w)$ is constant for fixed w, we find that $t \mapsto m(t, w)$ is equal to the largest quasiconvex minorant of $t \mapsto H^{1/2}(t, w)\langle\delta_0(t, w)\rangle^2/2$, i.e., $\sup_I m = \sup_{\partial I} m$ for compact intervals $I \subset \mathbf{R}$, see [15, Definition 1.6.3].

The main difference between this weight and the weight in [8] is the use of $H^{1/2}\langle\delta_0\rangle^2$ in the definition of m instead of $H^{1/2}\langle\delta_0\rangle$, and this is due to Lerner [24]. The weight has the following properties according to Propositions 4.3 and 4.4 in [10].

Proposition 5.8. *We have that $m \in L^\infty(\mathbf{R} \times T^*\mathbf{R}^n)$, $w \mapsto m(t, w)$ is uniformly Lipschitz-continuous, for all t, and*

$$h^{1/2}\langle\delta_0\rangle^2/6 \le m \le H^{1/2}\langle\delta_0\rangle^2/2 \le \langle\delta_0\rangle/2. \tag{5.13}$$

There exists $C > 0$ so that

$$m(t_0, w) \le Cm(t_0, w_0)(1 + |w - w_0|/\langle\delta_0(t_0, w_0)\rangle)^3; \tag{5.14}$$

thus m is a weight for $g^\sharp$.

The following result will be essential for the proof of Proposition 3.6 in Section 7, and it follows from Proposition 4.5 in [10].

Proposition 5.9. *Let the weight M be given by Definition 5.4 and m by Definition 5.7. Then there exists $C_0 > 0$ such that*

$$MH^{3/2}\langle\delta_0\rangle^2 \le C_0 m. \tag{5.15}$$

We have the following convexity property of $t \mapsto m(t, w)$, which will be important for the construction of the multiplier.

Proposition 5.10. *Let m be given by Definition 5.7. Then*

$$\sup_{t_1 \le t \le t_2} m(t, w) \le \delta_0(t_2, w) - \delta_0(t_1, w) + m(t_1, w) + m(t_2, w) \qquad \forall w \qquad (5.16)$$

Proof. Since $t \mapsto \delta_0(t, w)$ is monotone, we find that

$$\inf_{\pm(t-t_0)\ge 0} (|\delta_0(t, w) - \delta_0(t_0, w)| + H^{1/2}(t, w)\langle\delta_0(t, w)\rangle^2/2) \le m(t_0, w). \quad (5.17)$$

Let $t \in [t_1, t_2]$, then by using (5.17) for $t_0 = t_1$, t_2, and taking the infima, we obtain that

$$\begin{aligned} m(t, w) \le &\inf_{r \le t_1 < t_2 \le s} \delta_0(s, w) - \delta_0(r, w) + H^{1/2}(s, w)\langle\delta_0(s, w)\rangle^2/2 \\ &+ H^{1/2}(r, w)\langle\delta_0(r, w)\rangle^2/2 \\ \le &\ \delta_0(t_2, w) - \delta_0(t_1, w) + m(t_1, w) + m(t_2, w) \end{aligned}$$

which gives (5.16) after taking the supremum.

Next, we shall construct the pseudo-sign $B = \delta_0 + \varrho_0$, which we shall use in Proposition 6.3 to construct the multiplier of Proposition 3.6.

Proposition 5.11. *Assume that δ_0 is given by Definition 5.1 and m is given by Definition 5.7. Then for $T > 0$ there exists real-valued $\varrho_T(t, w) \in L^\infty(\mathbf{R} \times T^*\mathbf{R}^n)$ with the property that $w \mapsto \varrho_T(t, w)$ is uniformly Lipschitz-continuous, and*

$$|\varrho_T| \le m \qquad (5.18)$$

$$T\partial_t(\delta_0 + \varrho_T) \ge m/2 \qquad \text{in } \mathcal{D}'(\mathbf{R}) \qquad (5.19)$$

when $|t| < T$.

Proof. (We owe this argument to Lars Hörmander [17].) Let

$$\varrho_T(t, w) = \sup_{-T \le s \le t} \left(\delta_0(s, w) - \delta_0(t, w) + \frac{1}{2T}\int_s^t m(r, w)\, dr - m(s, w)\right) \qquad (5.20)$$

for $|t| \le T$. Then

$$\begin{aligned} \delta_0(t, w) + \varrho_T(t, w) = &\sup_{-T \le s \le t} \left(\delta_0(s, w) - \frac{1}{2T}\int_0^s m(r, w)\, dr - m(s, w)\right) \\ &+ \frac{1}{2T}\int_0^t m(r, w)\, dr, \end{aligned}$$

which immediately gives (5.19) since the supremum is nondecreasing. Since $w \mapsto \delta_0(t, w)$ and $w \mapsto m(t, w)$ are uniformly Lipschitz-continuous by Proposition 5.8, we find by taking the supremum that $w \mapsto \varrho_T(t, w)$ is uniformly Lipschitz-continuous. We find from Proposition 5.10 that

$$\delta_0(s, w) - \delta_0(t, w) + \frac{1}{2T} \int_s^t m(r, w)\, dr - m(s, w) \leq m(t, w) \quad -T \leq s \leq t \leq T.$$

By taking the supremum, we obtain that $-m(t, w) \leq \varrho_T(t, w) \leq m(t, w)$ when $|t| \leq T$, which proves the result.

6 The Wick quantization

In order to define the multiplier we shall use the Wick quantization, and we shall also define the function spaces that we shall use. As before, we shall assume that $g^\sharp = (g^\sharp)^\sigma$ and the coordinates are chosen so that $g^\sharp(w) = |w|^2$. For $a \in L^\infty(T^*\mathbf{R}^n)$ we define the Wick quantization:

$$a^{\text{Wick}}(x, D_x)u(x) = \int_{T^*\mathbf{R}^n} a(y, \eta) \Sigma^w_{y,\eta}(x, D_x)u(x)\, dy d\eta \qquad u \in \mathcal{S}(\mathbf{R}^n), \quad (6.1)$$

using the projections $\Sigma^w_{y,\eta}(x, D_x)$ with Weyl symbol

$$\Sigma_{y,\eta}(x, \xi) = \pi^{-n} \exp(-g^\sharp(x - y, \xi - \eta))$$

(see [7, Appendix B] or [20, Section 4]). We find that $a^{\text{Wick}}\colon \mathcal{S}(\mathbf{R}^n) \mapsto \mathcal{S}'(\mathbf{R}^n)$ so that $(a^{\text{Wick}})^* = (\overline{a})^{\text{Wick}}$,

$$a \geq 0 \implies (a^{\text{Wick}}(x, D_x)u, u) \geq 0 \qquad u \in \mathcal{S}(\mathbf{R}^n) \quad (6.2)$$

and $\|a^{\text{Wick}}(x, D_x)\|_{\mathcal{L}(L^2(\mathbf{R}^n))} \leq \|a\|_{L^\infty(T^*\mathbf{R}^n)}$, which is the main advantage with the Wick quantization (see [20, Proposition 4.2]). Now if $a_t(x, \xi) \in L^\infty(\mathbf{R} \times T^*\mathbf{R}^n)$ depends on a parameter t, then we find that

$$\int_{\mathbf{R}} (a_t^{\text{Wick}} u, u)\phi(t)\, dt = (A_\phi^{\text{Wick}} u, u) \qquad u \in \mathcal{S}(\mathbf{R}^n), \quad (6.3)$$

where $A_\phi(x, \xi) = \int_{\mathbf{R}} a_t(x, \xi)\phi(t)\, dt$. In fact, if $a \in L^1$, then this follows from the Fubini theorem; in general we obtain this by cutting off a_t on large sets in $T^*\mathbf{R}^n$ and using dominated convergence. We obtain from the definition that $a^{\text{Wick}} = a_0^w$ where

$$a_0(w) = \pi^{-n} \int_{T^*\mathbf{R}^n} a(z) \exp(-|w - z|^2)\, dz \quad (6.4)$$

is the Gaussian regularization; thus Wick operators with symmetric symbols have symmetric Weyl symbols.

We also have the following result about the composition of Wick operators according to Proposition 5.2 in [10].

Remark 6.1. Let $a(w)$, $b(w) \in L^\infty$, and let m_1, m_2 be bounded weights for $g^\sharp$. If $|a| \le m_1$ and $|b'| = |\partial b| \le m_2$, then

$$a^{\mathrm{Wick}} b^{\mathrm{Wick}} = (ab)^{\mathrm{Wick}} + r^w \tag{6.5}$$

with $r \in S(m_1 m_2, g^\sharp)$. In the case when a, b are real valued, $|a| \le m_1$ and $|b''| \le m_2$, we obtain that

$$\mathrm{Re}(a^{\mathrm{Wick}} b^{\mathrm{Wick}}) = \left(ab - \frac{1}{2} a' \cdot b'\right)^{\mathrm{Wick}} + r^w \tag{6.6}$$

with $r \in S(m_1 m_2, g^\sharp)$. Here a' is the distributional derivative of $a \in L^\infty$ and b' is Lipschitz continuous, so the product is well defined in L^∞.

If $A \in L^\infty(T^*\mathbf{R}^n)$ is an $M \times N$ system, then we can define A^{Wick} by (6.1) on $u \in \mathcal{S}(\mathbf{R}^n, \mathbf{C}^N)$. These operators have the same properties as the scalar operators, but of course we need that $M = N$ in order for (6.2) to hold.

In the following, we shall assume that $G = Hg^\sharp \le g^\sharp$ is a slowly varying metric satisfying

$$H(w) \le C_0 H(w_0)(1 + |w - w_0|)^{N_0} \tag{6.7}$$

and that m is a weight for G satisfying (6.7) with H replaced by m. This means that G and m are strongly σ-temperate in the sense of [2, Definition 7.1]. Recall the symbol class $S^+(1, g^\sharp)$ defined by Definition 3.2.

Proposition 6.2. *Assume that $a \in L^\infty(T^*\mathbf{R}^n)$ is an $N \times N$ system such that $|a| \le Cm$. Then $a^{\mathrm{Wick}} = a_0^w$ where $a_0 \in S(m, g^\sharp)$ is given by (6.4). If $a \in S(m, G)$ for $G = Hg^\sharp$, then $a_0 = a$ modulo symbols in $S(mH, G)$. If $|a| \le Cm$ and $a = 0$ in a fixed G ball with center w, then $a_0 \in S(mH^N, G)$ near w for any N. If a is Lipschitz-continuous, then we have $a_0 \in S^+(1, g^\sharp)$. If $a(t, w)$ and $g(t, w) \in L^\infty(\mathbf{R} \times T^*\mathbf{R}^n)$ are $N \times N$ systems and $\partial_t a(t, w) \ge g(t, w)$ in $\mathcal{D}'(\mathbf{R})$ for almost all $w \in T^*\mathbf{R}^n$, then we find $(\partial_t(a^{\mathrm{Wick}})u, u) \ge (g^{\mathrm{Wick}}u, u)$ in $\mathcal{D}'(\mathbf{R})$ for $u \in \mathcal{S}(\mathbf{R}^n, \mathbf{C}^N)$.*

Observe that the results are uniform in the metrics and weights. By localization we find, for example, that if $|a| \le Cm$ and $a \in S(m, G)$ in a G neighborhood of w_0, then $a_0 = a$ modulo $S(mH, G)$ in a smaller G neighborhood of w_0. These results are well known, but for convenience we give a short proof.

Proof. Since a is measurable satisfying $|a| \le Cm$, where $m(z) \le C_0 m(w)(1 + |z - w|)^{N_0}$ by (6.7), we find that $a^{\mathrm{Wick}} = a_0^w$ where $a_0 = \mathcal{O}(m)$ is given by (6.4). By differentiating on the exponential factor, we find $a_0 \in S(m, g^\sharp)$.

If $a = 0$ in a G ball of radius $\varepsilon > 0$ and center at w, then we can write

$$\pi^n a_0(w) = \int_{|z-w| \ge \varepsilon H^{-1/2}(w)} a(z) \exp(-|w - z|^2)\, dz = \mathcal{O}(m(w) H^N(w))$$

for any N even after repeated differentiation. If $a \in S(m, G)$, then Taylor's formula gives

$$a_0(w) = a(w) + \pi^{-n} \int_0^1 \int_{T^*\mathbf{R}^n} (1-\theta)\langle a''(w+\theta z)z, z\rangle e^{-|z|^2}\, dz d\theta,$$

where $a'' \in S(mH, G)$ because $G = Hg^\sharp$. Since $m(w+\theta z) \le C_0 m(w)(1+|z|)^{N_0}$ and $H(w+\theta z) \le C_0 H(w)(1+|z|)^{N_0}$ when $|\theta| \le 1$, we find that $a_0(w) = a(w)$ modulo $S(mH, G)$. Now, the Lipschitz continuity of a means that $\partial a \in L^\infty(T^*\mathbf{R}^n)$. Since $\partial a_0(w) = \pi^{-n} \int_{T^*\mathbf{R}^n} \partial a(z) \exp(-|w-z|^2)\, dz$, we obtain that $a_0 \in S^+(1, g^\sharp)$.

For the final claim, we note that $-\int a(t,w)\phi'(t)\, dt \ge \int g(t,w)\phi(t)\, dt$ for all $0 \le \phi \in C_0^\infty(\mathbf{R})$ and almost all $w \in T^*\mathbf{R}^n$, which by (6.2) and (6.3) gives

$$-\int (a^{\text{Wick}}(t,x,D_x)u, u)\phi'(t)\, dt \ge \int (g^{\text{Wick}}(t,x,D_x)u, u)\phi(t)\, dt \qquad 0 \le \phi \in C_0^\infty(\mathbf{R})$$

for $u \in \mathcal{S}(\mathbf{R}^n, \mathbf{C}^N)$.

We shall compute the Weyl symbol for the Wick operator $(\delta_0 + \varrho_T)^{\text{Wick}}$, where ϱ_T is given by Proposition 5.11. In the following we shall suppress the t variable.

Proposition 6.3. *Let $B = \delta_0 + \varrho_0$, where δ_0 is given by Definition 5.1 and ϱ_0 is real valued and Lipschitz-continuous, satisfying $|\varrho_0| \le m$, where $m \le H^{1/2}\langle\delta_0\rangle^2/2 \le \langle\delta_0\rangle/2$ is a weight for $g^\sharp$. Then we find*

$$B^{\text{Wick}} = b^w,$$

where $b = \delta_1 + \varrho_1 \in S(\langle\delta_0\rangle, g^\sharp) \bigcap S^+(1, g^\sharp)$ is real, $\delta_1 \in S(H^{-1/2}, g^\sharp) \bigcap S^+(1, g^\sharp)$, and $\varrho_1 \in S(m, g^\sharp) \bigcap S^+(1, g^\sharp)$. Also, there exists $\kappa_2 > 0$ so that $\delta_1 = \delta_0$ modulo $S(H^{1/2}, G)$ when $\langle\delta_0\rangle \le \kappa_2 H^{-1/2}$, which gives $b = \delta_0$ modulo $S(H^{1/2}\langle\delta_0\rangle^2, g^\sharp)$. For any $\lambda > 0$ we find that $|\delta_0| \ge \lambda H^{-1/2}$ and $H^{1/2} \le \lambda/3$ imply that $\operatorname{sgn}(B) = \operatorname{sgn}(\delta_0)$ *and* $|B| \ge \lambda H^{-1/2}/3$.

Proof. Let $\delta_0^{\text{Wick}} = \delta_1^w$ and $\varrho_0^{\text{Wick}} = \varrho_1^w$. Since $|\delta_0| \le \langle\delta_0\rangle \le H^{-1/2}$, $|\varrho_0| \le m \le \langle\delta_0\rangle/2$ and the symbols are real-valued, we obtain from Proposition 6.2 that $b \in S(\langle\delta_0\rangle, g^\sharp)$, $\delta_1 \in S(H^{-1/2}, g^\sharp)$ and $\varrho_1 \in S(m, g^\sharp)$ are real-valued. Since δ_0 and ϱ_0 are uniformly Lipschitz-continuous, we find that δ_1 and $\varrho_1 \in S^+(1, g^\sharp)$ by Proposition 6.2.

If $\langle\delta_0\rangle \le \kappa H^{-1/2}$ at w_0 for sufficiently small $\kappa > 0$, then we find by the Lipschitz-continuity of δ_0 and the slow variation of G that $\langle\delta_0\rangle \le C_0\kappa H^{-1/2}$ in a fixed G neighborhood ω_κ of w_0 (depending on κ). For $\kappa \ll 1$ we find $\delta_0 \in S(H^{-1/2}, G)$ in ω_κ by Proposition 5.6; thus $\delta_1 = \delta_0$ modulo $S(H^{1/2}, G)$ near w_0 by Proposition 6.2 after localization.

When $|\delta_0| \ge \lambda H^{-1/2} > 0$ at w_0, we find that

$$|\varrho_0| \le m \le \langle\delta_0\rangle/2 \le (1 + H^{1/2}/\lambda)|\delta_0|/2.$$

We obtain that $|\varrho_0| \le 2|\delta_0|/3$ so $\operatorname{sgn}(B) = \operatorname{sgn}(\delta_0)$ and $|B| \ge |\delta_0|/3 \ge \lambda H^{-1/2}/3$ when $H^{1/2} \le \lambda/3$, which completes the proof.

Let m be given by Definition 5.7. Then m is a weight for $g^\sharp$ according to Proposition 5.8. We are going to use the symbol classes $S(m^k, g^\sharp)$, $k \in \mathbf{R}$.

Definition 6.4. Let $H(m^k, g^\sharp)$, be the Hilbert space given by [2, Definition 4.1] so that

$$u \in H(m^k, g^\sharp) \iff a^w u \in L^2 \qquad \forall a \in S(m^k, g^\sharp) \qquad k \in \mathbf{R}. \tag{6.8}$$

We let $\|u\|_k$ be the norm of $H(m^k, g^\sharp)$.

This Hilbert space has the following properties: $\mathcal{S}$ is dense in $H(m^k, g^\sharp)$, the dual of $H(m^k, g^\sharp)$ is naturally identified with $H(m^{-k}, g^\sharp)$, and if $u \in H(m^k, g^\sharp)$, then $u = a_0^w v$ for some $v \in L^2(\mathbf{R}^n)$ and $a_0 \in S(m^{-k}, g^\sharp)$ (see [2, Corollary 6.7]). It follows that $a^w \in \operatorname{Op} S(m^k, g^\sharp)$ is bounded:

$$u \in H(m^j, g^\sharp) \mapsto a^w u \in H(m^{j-k}, g^\sharp) \tag{6.9}$$

with bound only depending on the seminorms of a.

We recall Proposition 5.5 in [10], which shows that the topology in $H(m^{1/2}, g^\sharp)$ can be defined by the operator m^{Wick}. Recall that $m \ge h^{1/2}\langle\delta_0\rangle^2/6$.

Proposition 6.5. *Let $B = \delta_0 + \varrho_0$, where δ_0 is given by Definition 5.1 and $|\varrho_0| \le m$. Then there exist positive constants c_1, c_2, and C_0 such that*

$$c_1 h^{1/2}(\|B^{\mathrm{Wick}} u\|^2 + \|u\|^2) \le c_2\|u\|_{1/2}^2 \le (m^{\mathrm{Wick}} u, u) \le C_0\|u\|_{1/2}^2$$
$$u \in \mathcal{S}(\mathbf{R}^n) \tag{6.10}$$

The constants only depend on the seminorms of f in $L^\infty(\mathbf{R}, S(h^{-1}, hg^\sharp))$.

In the following, we let $\|u(t)\|$ be the L^2-norm of $x \mapsto u(t,x) \in \mathbf{C}^N$ in $\mathbf{R}^n$ for fixed t, and $(u(t), v(t))$ the corresponding sesquilinear inner product. Let $\mathcal{B} = \mathcal{B}(L^2(\mathbf{R}^n), \mathbf{C}^N)$ be the set of bounded operators $L^2(\mathbf{R}^n, \mathbf{C}^N) \mapsto L^2(\mathbf{R}^n, \mathbf{C}^N)$. We shall use operators that depend measurably on t in the following sense.

Definition 6.6. We say that $t \mapsto A(t)$ is weakly measurable if $A(t) \in \mathcal{B}$ for all t and $t \mapsto A(t)u$ is weakly measurable for every $u \in L^2(\mathbf{R}^n, \mathbf{C}^N)$, i.e., $t \mapsto (A(t)u, v)$ is measurable for any $u, v \in L^2(\mathbf{R}^n, \mathbf{C}^N)$. We say that $A(t) \in L^\infty_{loc}(\mathbf{R}, \mathcal{B})$ if $t \mapsto A(t)$ is weakly measurable and locally bounded in $\mathcal{B}$.

If $A(t) \in L^\infty_{loc}(\mathbf{R}, \mathcal{B})$, then we find that the function $t \mapsto (A(t)u, v) \in L^\infty_{loc}(\mathbf{R})$ has weak derivative $\frac{d}{dt}(Au, v) \in \mathcal{D}'(\mathbf{R})$ for any $u, v \in L^2(\mathbf{R}^n, \mathbf{C}^N)$, given by

$$\frac{d}{dt}(Au, v)(\phi) = -\int (A(t)u, v)\phi'(t)\,dt \qquad \phi(t) \in C_0^\infty(\mathbf{R}).$$

If $u(t), v(t) \in L^\infty_{loc}(\mathbf{R}, L^2(\mathbf{R}^n, \mathbf{C}^N))$ and $A(t) \in L^\infty_{loc}(\mathbf{R}, \mathcal{B})$, then

$$t \mapsto (A(t)u(t), v(t)) \in L^\infty_{loc}(\mathbf{R})$$

is measurable. We shall use the following multiplier estimate from [8].

Proposition 6.7. *Let $P = D_t + iF(t)$ with $F(t) \in L^\infty_{loc}(\mathbf{R}, \mathcal{B})$. Assume that $B(t) = B^*(t) \in L^\infty_{loc}(\mathbf{R}, \mathcal{B})$, such that*

$$\frac{d}{dt}(Bu, u) + 2\,\mathrm{Re}(Bu, Fu) \geq (Mu, u) \ \ \textit{in } \mathcal{D}'(I) \qquad \forall\, u \in \mathcal{S}(\mathbf{R}^n, \mathbf{C}^N), \tag{6.11}$$

where $M(t) = M^*(t) \in L^\infty_{loc}(\mathbf{R}, \mathcal{B})$ and $I \subseteq \mathbf{R}$ is open. Then we have

$$\int (Mu, u)\, dt \leq 2 \int \mathrm{Im}(Pu, Bu)\, dt \tag{6.12}$$

for $u \in C_0^1(I, \mathcal{S}(\mathbf{R}^n, \mathbf{C}^N))$.

Proof. Since $B(t) \in L^\infty_{loc}(\mathbf{R}, \mathcal{B})$, we may for $u, v \in \mathcal{S}(\mathbf{R}^n, \mathbf{C}^N)$ define the regularization

$$(B_\varepsilon(t)u, v) = \varepsilon^{-1} \int (B(s)u, v)\phi((t-s)/\varepsilon)\, ds = (Bu, v)(\phi_{\varepsilon,t}) \qquad \varepsilon > 0,$$

where $\phi_{\varepsilon,t}(s) = \varepsilon^{-1}\phi((t-s)/\varepsilon)$ with $0 \leq \phi \in C_0^\infty(\mathbf{R})$ satisfying $\int \phi(t)\, dt = 1$. Then $t \mapsto (B_\varepsilon(t)u, v)$ is in $C^\infty(\mathbf{R})$ with derivative equal to $\frac{d}{dt}(Bu, v)(\phi_{\varepsilon,t}) = -(Bu, v)(\phi'_{\varepsilon,t})$. Let I_0 be an open interval such that $I_0 \Subset I$. Then for small enough $\varepsilon > 0$ and $t \in I_0$ we find from condition (6.11) that

$$\frac{d}{dt}(B_\varepsilon(t)u, u) + 2\,\mathrm{Re}(Bu, Fu)(\phi_{\varepsilon,t}) \geq (Mu, u)(\phi_{\varepsilon,t}) \qquad u \in \mathcal{S}(\mathbf{R}^n, \mathbf{C}^N). \tag{6.13}$$

In fact, $\phi_{\varepsilon,t} \geq 0$ and $\operatorname{supp}\phi_{\varepsilon,t} \in C_0^\infty(I)$ for small enough ε when $t \in I_0$.

Now for $u(t) \in C_0^1(I_0, \mathcal{S}(\mathbf{R}^n, \mathbf{C}^N))$ and $\varepsilon > 0$ we define

$$B_{\varepsilon,u}(t) = (B_\varepsilon(t)u(t), u(t)) = \varepsilon^{-1} \int (B(s)u(t), u(t))\phi((t-s)/\varepsilon)\, ds. \tag{6.14}$$

For small enough ε we obtain $B_{\varepsilon,u}(t) \in C_0^1(I_0)$, with derivative

$$\frac{d}{dt} B_{\varepsilon,u} = \left(\left(\frac{d}{dt} B_\varepsilon \right) u, u \right) + 2\,\mathrm{Re}\,(B_\varepsilon u, \partial_t u)$$

since $B(t) \in L^\infty_{loc}(\mathbf{R}, \mathcal{B})$. By integrating with respect to t, we obtain the vanishing average

$$0 = \int \frac{d}{dt} B_{\varepsilon,u}(t)\, dt = \int \left(\left(\frac{d}{dt} B_\varepsilon \right) u, u \right) dt + \int 2\,\mathrm{Re}(B_\varepsilon u, \partial_t u)\, dt \tag{6.15}$$

when $u \in C_0^1(I_0, \mathcal{S}(\mathbf{R}^n, \mathbf{C}^N))$. We obtain from (6.13) and (6.15) that

$$0 \geq \iint ((M(s)u(t), u(t)) \\ + 2\operatorname{Re}(B(s)u(t), \partial_t u(t) - F(s)u(t)))\phi((t-s)/\varepsilon)\, ds dt/\varepsilon$$

By letting $\varepsilon \to 0$, we find by dominated convergence that

$$0 \geq \int (M(t)u(t), u(t)) + 2\operatorname{Re}(B(t)u(t), \partial_t u(t) - F(t)u(t))\, dt$$

since $u \in C_0^1(I_0, \mathcal{S}(\mathbf{R}^n, \mathbf{C}^N))$ and $M(t)$, $B(t)$, $F(t) \in L^\infty_{loc}(\mathbf{R}, \mathcal{B})$. Here $\partial_t u - Fu = iPu$ and $2\operatorname{Re}(Bu, iPu) = -2\operatorname{Im}(Pu, Bu)$, thus we obtain (6.12) for $u \in C_0^1(I_0, \mathcal{S}(\mathbf{R}^n, \mathbf{C}^N))$. Since I_0 is an arbitrary open subinterval with compact closure in I, this completes the proof of the proposition.

7 The lower bounds

In this section shall prove Proposition 3.6, which means obtaining lower bounds on

$$2\operatorname{Im}(P_0 u, b_T^w u) = (\partial_t b_T^w u, u) + 2\operatorname{Re}(F^w u, b_T^w u),$$

where $P_0 = D_t \operatorname{Id}_N + iF^w(t, x, D_x)$ with

$$F(t, w) = f(t, w)\operatorname{Id}_N + F_0(t, w). \tag{7.1}$$

Here $f \in L^\infty(\mathbf{R}, S(h^{-1}, hg^\sharp))$ is real-valued satisfying condition $(\overline{\Psi})$ given by (3.2), $F_0 \in C^\infty(\mathbf{R}, S(1, hg^\sharp))$ and $b_T^w = B_T^{\mathrm{Wick}}$ is the symmetric scalar operator given by Propositions 5.11 and 6.3 for this f. Since Proposition 5.11 and Proposition 6.2 give lower bounds on the first term:

$$\partial_t b_T^w = \partial_t B_T^{\mathrm{Wick}} \geq m^{\mathrm{Wick}}/2T \qquad \text{in } L^2 \quad |t| \leq T$$

it only remains to obtain comparable lower bounds on $\operatorname{Re} b_T^w F^w$ by Proposition 6.5.

By Claim 3.9 we may also assume that

$$F_0 = \langle d_w f, R\rangle = \sum_j \partial_{w_j} f R_j \qquad \text{modulo } S(h, hg^\sharp) \qquad \forall t, \tag{7.2}$$

where $R_j \in S(h^{1/2}, hg^\sharp)$ are $N \times N$ systems, $\forall j$. Observe that since $d_w f \in S(MH^{1/2}, G)$, $hg^\sharp \leq 9G$ and $h \leq MH^{1/2}h^{1/2}$ by (5.8) we find that $F_0 \in S(MH^{1/2}h^{1/2}, G) \subseteq S(1, G)$ and thus $F \in S(M, G)$, $G = Hg^\sharp$.

In the following, the results will hold for almost all $|t| \leq T$ and will only depend on the seminorms of f and F_0. We shall suppress the t-variable and assume the coordinates chosen so that $g^\sharp(w) = |w|^2$. In order to prove Proposition 3.6 we need to prove the following result.

Proposition 7.1. *Assume that F is given by (7.1)–(7.2) and $B = \delta_0 + \varrho_0$. Here δ_0 is given by Definition 5.1, ϱ_0 is real-valued and Lipschitz-continuous satisfying $|\varrho_0| \le m$, where $m \le \langle\delta_0\rangle/2$ is given by Definition 5.7. Then we have*

$$\operatorname{Re}(B^{\mathrm{Wick}} F^w u, u) \ge (C^w u, u) \qquad \forall\, u \in \mathcal{S}(\mathbf{R}^n, \mathbf{C}^N) \tag{7.3}$$

for some $N \times N$ system $C \in S(m, g^\sharp)$.

Proof of Proposition 3.6. Let $B_T = \delta_0 + \varrho_T$, where $\delta_0 + \varrho_T$ is the pseudo-sign for f given by Proposition 5.11 for $0 < T \le 1$, so that $|\varrho_T| \le m$ and

$$\partial_t(\delta_0 + \varrho_T) \ge m/2T \qquad \text{in } \mathcal{D}'\big(]-T, T[\big). \tag{7.4}$$

If we put $B_T \equiv 0$ when $|t| > T$, then $B_T^{\mathrm{Wick}} = b_T^w$ where $b_T(t, w) \in L^\infty(\mathbf{R}, S(H^{-1/2}, g^\sharp) \bigcap S^+(1, g^\sharp))$ uniformly by Proposition 6.3. We find by Proposition 6.2 and (7.4) that

$$((\partial_t B_T)^{\mathrm{Wick}} u, u) \ge (m^{\mathrm{Wick}} u, u)/2T \qquad \text{in } \mathcal{D}'\big(]-T, T[\big) \tag{7.5}$$

when $u \in \mathcal{S}(\mathbf{R}^n)$. We obtain from Proposition 6.5 that there exist positive constants c_1 and c_2 so that

$$(m^{\mathrm{Wick}} u, u) \ge c_2 \|u\|_{1/2}^2 \ge c_1 h^{1/2}(\|b_T^w u\|^2 + \|u\|^2) \qquad u \in \mathcal{S}(\mathbf{R}^n). \tag{7.6}$$

Here $\|u\|_{1/2}$ is the norm of the Hilbert space $H(m^{1/2}, g^\sharp)$ given by Definition 6.4. By Proposition 7.1, we find for almost all $t \in [-T, T]$ that

$$\operatorname{Re}((B_T^{\mathrm{Wick}} F^w)\big|_t u, u) = (C^w(t)u, u) \quad u \in \mathcal{S}(\mathbf{R}^n, \mathbf{C}^N), \tag{7.7}$$

here the $N \times N$ system $C(t) \in S(m, g^\sharp)$ uniformly. We obtain from (6.9), (7.6) and duality that there exists a positive constant c_3 such that

$$|(C^w(t)u, u)| \le \|u\|_{1/2}\|C^w(t)u\|_{-1/2} \le c_3\|u\|_{1/2}^2 \le c_3(m^{\mathrm{Wick}} u, u)/c_2 \tag{7.8}$$

for $u \in \mathcal{S}(\mathbf{R}^n, \mathbf{C}^N)$ and $|t| \le T$. We find from (7.5)–(7.8) that

$$(\partial_t b_T^w u, u) + 2\operatorname{Re}(F^w u, b_T^w u) \ge (1/2T - 2c_3/c_2)(m^{\mathrm{Wick}} u, u) \ \text{ in } \mathcal{D}'\big(]-T, T[\big)$$

for $u \in \mathcal{S}(\mathbf{R}^n, \mathbf{C}^N)$. By using Proposition 6.7 with $P = D_t \operatorname{Id}_N + iF^w(t, x, D_x)$, $B = b_T^w$ and $M = m^{\mathrm{Wick}}/4T$ we obtain that

$$c_1 h^{1/2} \int \|b_t^w u\|^2 + \|u\|^2\, dt \le \int (m^{\mathrm{Wick}} u, u)\, dt \le 8T \int \operatorname{Im}(P_0 u, b_T^w u)\, dt$$

if $u \in \mathcal{S}(\mathbf{R} \times \mathbf{R}^n, \mathbf{C}^N)$ has support where $|t| < T \le c_2/8c_3$. This finishes the proof of Proposition 3.6.

Proof of Proposition 7.1. First we note that since $B^{\mathrm{Wick}} = b^w \in \operatorname{Op} S(\langle\delta_0\rangle, g^\sharp)$ by Proposition 6.3 and $h^{1/2}\langle\delta_0\rangle^2 \le 6m$ by (5.13), we find $B^{\mathrm{Wick}} R^w \in \operatorname{Op} S(m, g^\sharp)$ when $R \in S(h^{1/2}, g^\sharp)$. Since $\operatorname{Im} F = \frac{1}{2i}(F - F^*) \in S(1, hg^\sharp)$ we find that

$$2\operatorname{Re}(B^{\mathrm{Wick}} i(\operatorname{Im} F)^w) = i[b^w, (\operatorname{Im} F)^w] \in \operatorname{Op} S(h^{1/2}, g^\sharp);$$

thus it suffices to consider symmetric F satisfying (7.2).

We shall localize in $T^*\mathbf{R}^n$ with respect to the metric $G = Hg^\sharp$, and estimate the localized operators. We shall use the neighborhoods

$$\omega_{w_0}(\varepsilon) = \{\, w : |w - w_0| < \varepsilon H^{-1/2}(w_0) \,\} \qquad \text{for } w_0 \in T^*\mathbf{R}^n. \tag{7.9}$$

In the following we may assume that ε is small enough so that $w \mapsto H(w)$ and $w \mapsto M(w)$ only vary with a fixed factor in $\omega_{w_0}(\varepsilon)$. Then by the uniform Lipschitz-continuity of $w \mapsto \delta_0(w)$ we can find $\kappa_0 > 0$ with the following property: for $0 < \kappa \le \kappa_0$ there exist positive constants c_κ and ε_κ so that for any $w_0 \in T^*\mathbf{R}^n$ we have

$$|\delta_0(w)| \le \kappa H^{-1/2}(w) \qquad w \in \omega_{w_0}(\varepsilon_\kappa) \qquad \text{or} \tag{7.10}$$

$$|\delta_0(w)| \ge c_\kappa H^{-1/2}(w) \qquad w \in \omega_{w_0}(\varepsilon_\kappa). \tag{7.11}$$

In fact, we have by Lipschitz-continuity that $|\delta_0(w) - \delta_0(w_0)| \le \varepsilon_\kappa H^{-1/2}(w_0)$ when $w \in \omega_{w_0}(\varepsilon_\kappa)$. Thus, if $\varepsilon_\kappa \ll \kappa$ we obtain that (7.10) holds when $|\delta_0(w_0)| \ll \kappa H^{-1/2}(w_0)$ and (7.11) holds when $|\delta_0(w_0)| \ge c\kappa H^{-1/2}(w_0)$.

Let κ_1 be given by Proposition 5.6, κ_2 by Proposition 6.3, and let ε_κ and c_κ be given by (7.10)–(7.11) for $\kappa = \min(\kappa_0, \kappa_1, \kappa_2)/2$. Using Proposition 6.3 with $\lambda = c_\kappa$ we obtain that $\operatorname{sgn}(B) = \operatorname{sgn}(\delta_0)$ and

$$|B| \ge c_\kappa H^{-1/2}/3 \qquad \text{in } \omega_{w_0}(\varepsilon_\kappa) \tag{7.12}$$

if $H^{1/2} \le c_\kappa/3$ and (7.11) holds in $\omega_{w_0}(\varepsilon_\kappa)$.

Choose real symbols $\{\,\psi_j(w)\,\}_j$ and $\{\,\Psi_j(w)\,\}_j \in S(1, G)$ with values in ℓ^2, such that $\sum_k \psi_j^2 \equiv 1$, $\psi_j \Psi_j = \psi_j$, $\Psi_j = \phi_j^2 \ge 0$ for some $\{\,\phi_j(w)\,\}_j \in S(1, G)$ with values in ℓ^2 so that

$$\operatorname{supp} \phi_j \subseteq \omega_j = \omega_{w_j}(\varepsilon_\kappa).$$

Recall that $B^{\mathrm{Wick}} = b^w$ where $b = \delta_1 + \varrho_1$ is given by Proposition 6.3. In particular, $\delta_1 \in S(H^{-1/2}, G)$ when $H^{1/2} \le \kappa_2/2$ and (7.10) holds, since then $\langle\delta_0\rangle \le \kappa_2 H^{-1/2}$.

Lemma 7.2. *We find that* $A_j = \Psi_j b \operatorname{Re} F \in S(MH^{-1/2}, g^\sharp) \bigcap S^+(M, g^\sharp)$ *uniformly in* j, *and*

$$\mathrm{Re}(b^w F^w) = \sum_j \psi_j^w A_j^w \psi_j^w \qquad \textit{modulo } \mathrm{Op}\, S(m, g^\sharp), \tag{7.13}$$

We have $A_j^w = \mathrm{Re}\, b^w F_j^w$ *modulo* $\mathrm{Op}\, S(m, g^\sharp)$ *uniformly in* j*, where* $F_j = \Psi_j F$.

Proof. Since $b \in S(H^{-1/2}, g^\sharp) \bigcap S^+(1, g^\sharp)$, $\psi_j \in S(1, G)$ and $F_j \in S(M, G)$ we obtain that $A_j \in S(MH^{-1/2}, g^\sharp) \bigcap S^+(M, g^\sharp)$ uniformly in j. Proposition 5.9 gives that

$$MH^{3/2}\langle\delta_0\rangle^2 \le Cm; \tag{7.14}$$

thus we may ignore terms in $\mathrm{Op}\, S(MH^{3/2}\langle\delta_0\rangle^2, g^\sharp)$. Observe that since $b \in S(H^{-1/2}, g^\sharp)$, $\{\psi_k\}_k \in S(1, G)$ has values in ℓ^2 and $A_k \in S(MH^{-1/2}, g^\sharp)$ uniformly, Lemma 3.3 and Remark 3.4 gives that the symbols of $b^w F^w$, $b^w F_j^w$ and $\sum_k \psi_k^w A_k^w \psi_k^w$ have expansions in $S(MH^{j/2}, g^\sharp)$. Also observe that in the domains ω_j where $H^{1/2} \ge c > 0$, we find from Remark 3.4 that the symbols of $\sum_k \psi_k^w A_k^w \psi_k^w$, $b^w F_j^w$ and $b^w F^w$ are in $S(MH^{3/2}, g^\sharp)$ giving the result in this case. Thus, in the following, we shall assume that $H^{1/2} \ll 1$, and we shall consider the neighborhoods where (7.10) or (7.11) holds.

If (7.11) holds, then we find that $\langle\delta_0\rangle \cong H^{-1/2}$ so $S(MH^{1/2}, g^\sharp) \subseteq S(m, g^\sharp)$ in ω_j by (7.14). Since $b \in S^+(1, g^\sharp)$ and $A_j \in S^+(M, g^\sharp)$ we find from Lemma 3.3 and Remark 3.4 that the symbols of both $\mathrm{Re}\, b^w F^w$ and $\sum_k \psi_k^w A_k^w \psi_k^w$ are equal to $\sum_k \psi_k^2 A_k = \mathrm{Re}\, bF$ modulo $S(MH^{1/2}, g^\sharp)$ in ω_j. We also find that the symbol of $\mathrm{Re}\, b^w F_j^w$ is equal to A_j modulo $S(MH^{1/2}, g^\sharp)$, which proves the result in this case.

Next, we consider the case when (7.10) holds with $\kappa = \min(\kappa_0, \kappa_1, \kappa_2)/2$ and $H^{1/2} \le \kappa_2/2$ in ω_j. Then $\langle\delta_0\rangle \le \kappa_2 H^{-1/2}$ so $b = \delta_1 + \varrho_1 \in S(H^{-1/2}, G) + S(m, g^\sharp)$ in ω_j by Proposition 6.3. Now b is real and F is symmetric modulo $S(MH, G)$. Thus, by taking the symmetric part of $b^w F^w = \delta_1^w F^w + \varrho_1^w F^w$ we obtain from Lemma 3.3 that the symbol of $\mathrm{Re}(b^w F^w - (bF)^w)$ is in $S(MH^{3/2}, G) + S(MHm, g^\sharp) \subseteq S(m, g^\sharp)$ in ω_j since $M \le CH^{-1}$. Similarly, we find that $A_j^w = \mathrm{Re}\, b^w F_j^w$ modulo $S(m, g^\sharp)$. Since $A_j \in S(MH^{-1/2}, G) + S(Mm, g^\sharp)$ uniformly, we find that the symbol of $\sum_k \psi_k^w A_k^w \psi_k^w$ is equal to $\mathrm{Re}\, bF$ modulo $S(m, g^\sharp)$ in ω_j by Remark 3.4, which proves (7.13) and Lemma 7.2.

Next, we shall show that there exists $N \times N$ system $C_j \in S(m, g^\sharp)$ uniformly, such that

$$(A_j^w u, u) \ge (C_j^w u, u) \qquad u \in \mathcal{S}(\mathbf{R}^n, \mathbf{C}^N). \tag{7.15}$$

Then we obtain from (7.13) and (7.15) that

$$\mathrm{Re}(b^w F^w u, u) \ge \sum_j (\psi_j^w C_j^w \psi_j^w u, u) + (R^w u, u) \qquad u \in \mathcal{S}(\mathbf{R}^n, \mathbf{C}^N),$$

where $\sum_j \psi_j^w C_j^w \psi_j^w$ and $R^w \in \mathrm{Op}\, S(m, g^\sharp)$, which will prove Proposition 7.1.

Thus it remains to show that there exists $C_j \in S(m, g^\sharp)$ satisfying (7.15). As before we are going to consider the cases when $H^{1/2} \cong 1$ or $H^{1/2} \ll 1$, and when (7.10) or (7.11) holds in $\omega_j = \omega_{w_j}(\varepsilon_\kappa)$ for $\kappa = \min(\kappa_0, \kappa_1, \kappa_2)/2$. When $H^{1/2} \ge c > 0$ we find that $A_j \in S(MH^{3/2}, g^\sharp) \subseteq S(m, g^\sharp)$ uniformly by (7.14) which gives the lemma with $C_j = A_j$ in this case. Thus we may assume that

$$H^{1/2} \le \kappa_4 = \min(\kappa_0, \kappa_1, \kappa_2, \kappa_3)/2 \qquad \text{in } \omega_j \tag{7.16}$$

with $\kappa_3 = 2c_\kappa/3$ so that (7.12) follows from (7.11).

First, we consider the case when $H^{1/2} \le \kappa_4$ and (7.11) holds in ω_j. Since $|\delta_0(w)| \ge c_\kappa H^{-1/2}(w)$, we find $\langle\delta_0\rangle \cong H^{-1/2}$ in ω_j. As before we may ignore terms in $S(MH^{1/2}, g^\sharp) \subseteq S(m, g^\sharp)$ in ω_j by (7.14). Let $f_j = \Psi_j f$, since $\operatorname{sgn}(f) = \operatorname{sgn}(\delta_0) = \operatorname{sgn}(B)$ in ω_j by Proposition 6.3 we find that $f_j B \ge 0$. Since $f_j \in S(M, G)$, we find $f_j^w = f_j^{\text{Wick}}$ modulo $\operatorname{Op} S(MH, G)$ by Proposition 6.2; thus we may replace f_j^w with f_j^{Wick}. Since $F_{0,j} \in S(MH^{1/2}h^{1/2}, G)$ by (7.2) we find that $B^{\text{Wick}} F_{0,j}^w \in \operatorname{Op} S(MH^{1/2}, g^\sharp)$. Since $|B| \le CH^{-1/2}$ and $B \in S^+(1, g^\sharp)$, we find from (6.6) in Remark 6.1 that

$$A_j^w = \operatorname{Re} B^{\text{Wick}} f_j^{\text{Wick}} = (Bf_j)^{\text{Wick}} \ge 0 \qquad \text{in } L^2 \text{ modulo } \operatorname{Op} S(MH^{1/2}, g^\sharp)$$

which gives (7.15) in this case.

Finally, we consider the case when (7.10) holds with $\kappa = \min(\kappa_0, \kappa_1, \kappa_2)/2$ and $H^{1/2} \le \kappa_4 \le \kappa$ in ω_j. Then $\langle\delta_0\rangle \le 2\kappa H^{-1/2}$ so we obtain from Proposition 5.6 that $\delta_0 \in S(H^{-1/2}, G) \bigcap S(\langle\delta_0\rangle, g^\sharp)$ in ω_j. We have that $b^w = (\delta_0 + \varrho_0)^{\text{Wick}} = B^{\text{Wick}}$, where

$$|\varrho_0| \le m \le H^{1/2}\langle\delta_0\rangle^2/2 \le \langle\delta_0\rangle/2 \tag{7.17}$$

by Propositions 5.8 and 5.11. Also, we find from Lemma 7.2 that $A_j^w = \operatorname{Re} B^{\text{Wick}} F_j^w$ modulo $\operatorname{Op} S(m, g^\sharp)$.

Take $\chi(t) \in C^\infty(\mathbf{R})$ such that $0 \le \chi(t) \le 1$, $|t| \ge 2$ in $\operatorname{supp} \chi(t)$ and $\chi(t) = 1$ for $|t| \ge 3$. Let $\chi_0 = \chi(\delta_0)$, then $2 \le |\delta_0|$ and $\langle\delta_0\rangle/|\delta_0| \le 3/2$ in $\operatorname{supp} \chi_0$. Thus

$$1 + \chi_0\varrho_0/\delta_0 \ge 1 - \chi_0\langle\delta_0\rangle/2|\delta_0| \ge 1/4. \tag{7.18}$$

Since $|\delta_0| \le 3$ in $\operatorname{supp}(1 - \chi_0)$ we find by (7.17) that

$$B = \delta_0 + \chi_0\varrho_0 = \delta_0(1 + \chi_0\varrho_0/\delta_0)$$

modulo terms that are $\mathcal{O}(H^{1/2})$. Since $|\delta_0'| \le 1$ and

$$|\chi_0\varrho_0/\delta_0| \le \chi_0 H^{1/2}\langle\delta_0\rangle^2/2|\delta_0| \le 3H^{1/2}\langle\delta_0\rangle/4$$

we find from (6.5) in Remark 6.1 that

$$B^{\text{Wick}} = \delta_0^{\text{Wick}} B_0^{\text{Wick}} \qquad \text{modulo } \operatorname{Op} S(H^{1/2}\langle\delta_0\rangle, g^\sharp), \tag{7.19}$$

where $B_0 = 1 + \chi_0\varrho_0/\delta_0 = \mathcal{O}(1)$. Proposition 6.3 gives $(\chi_0\varrho_0/\delta_0)^{\text{Wick}} \in \operatorname{Op} S(H^{1/2}\langle\delta_0\rangle, g^\sharp)$ and $\delta_0^{\text{Wick}} = \delta_1^w$ where $\delta_1 \in S(H^{-1/2}, g^\sharp)$ and $\delta_1 = \delta_0$ modulo $\operatorname{Op} S(H^{1/2}, G)$ in ω_j. Thus Lemma 3.3 and (7.19) gives

$$B^{\mathrm{Wick}} = \delta_1^w B_0^{\mathrm{Wick}} = \delta_0^w B_0^{\mathrm{Wick}} + c^w B_0^{\mathrm{Wick}} \qquad \text{modulo } \operatorname{Op} S(H^{1/2}\langle\delta_0\rangle, g^\sharp), \tag{7.20}$$

where $c \in S(H^{-1/2}, g^\sharp)$ such that $\operatorname{supp} c \bigcap \omega_j = \emptyset$.

We find from Proposition 5.6 that $f = \alpha_0\delta_0$, where $\kappa_1 M H^{1/2} \le \alpha_0 \in S(MH^{1/2}, G)$, so Leibniz's rule gives $\alpha_0^{1/2} \in S(M^{1/2}H^{1/4}, G)$. Let $f_j = \Psi_j f$ and

$$a_j = \alpha_0^{1/2}\phi_j\delta_0 \in S(M^{1/2}H^{-1/4}, G) \bigcap S(M^{1/2}H^{1/4}\langle\delta_0\rangle, g^\sharp). \tag{7.21}$$

Since $\Psi_j = \phi_j^2$ we find $a_j^2 = f_j\delta_0$ and the calculus gives

$$a_j^w(\alpha_0^{1/2}\phi_j)^w = f_j^w \qquad \text{modulo } \operatorname{Op} S(MH, G). \tag{7.22}$$

Since $\operatorname{supp} f_j \bigcap \operatorname{supp} c = \emptyset$ we find that $f_j^w c^w \in \operatorname{Op} S(MH^{3/2}, g^\sharp)$. We also have

$$\operatorname{Re} f_j^w \delta_0^w = a_j^w a_j^w \qquad \text{modulo } \operatorname{Op} S(MH^{3/2}, G). \tag{7.23}$$

and $\operatorname{Im} f_j^w\delta_0^w \in \operatorname{Op} S(MH^{1/2}, G)$. We obtain from (7.20) and (7.22) that

$$\begin{aligned} f_j^w B^{\mathrm{Wick}} &= f_j^w(\delta_0^w B_0^{\mathrm{Wick}} + r^w) \\ &= f_j^w\delta_0^w B_0^{\mathrm{Wick}} + a_j^w r_j^w \text{ modulo } \operatorname{Op} S(m, g^\sharp), \end{aligned} \tag{7.24}$$

where $r \in S(H^{1/2}\langle\delta_0\rangle, g^\sharp)$. This gives

$$r_j^w = (\alpha_0^{1/2}\phi_j)^w r^w \in \operatorname{Op} S(M^{1/2}H^{3/4}\langle\delta_0\rangle, g^\sharp)$$

If $\operatorname{Re} A = \frac{1}{2}(A + A^*)$, $\operatorname{Im} A = \frac{1}{2i}(A - A^*)$ and $B^* = B$ then

$$\operatorname{Re}(AB) = \operatorname{Re}((\operatorname{Re} A)B) + i[\operatorname{Im} A, B]/2.$$

By taking $A = f_j^w\delta_0^w$ and $B = B_0^{\mathrm{Wick}}$, we find from (7.23) that

$$\operatorname{Re}(f_j^w\delta_0^w B_0^{\mathrm{Wick}}) = \operatorname{Re}(a_j^w a_j^w B_0^{\mathrm{Wick}}) \qquad \text{modulo } \operatorname{Op} S(m, g^\sharp). \tag{7.25}$$

In fact, $B_0 = 1 + \chi_0\varrho_0/\delta_0$ and $(\chi_0\varrho_0/\delta_0)^{\mathrm{Wick}} \in \operatorname{Op} S(H^{1/2}\langle\delta_0\rangle, g^\sharp)$ by Proposition 6.2. Thus

$$[a^w, B_0^w] = [a^w, (\chi_0\varrho_0/\delta_0)^{\mathrm{Wick}}] \in \operatorname{Op} S(MH^{3/2}\langle\delta_0\rangle, g^\sharp)$$

when $a \in S(MH^{1/2}, G)$. Similarly, we find from (7.21) that

$$a_j^w a_j^w B_0^{\mathrm{Wick}} = a_j^w(B_0^{\mathrm{Wick}} a_j^w + s_j^w) \qquad \text{modulo } \operatorname{Op} S(m, g^\sharp), \tag{7.26}$$

where $s_j = [a_j^w, B_0^{\mathrm{Wick}}] \in S(M^{1/2}H^{3/4}\langle\delta_0\rangle, g^\sharp)$. Next, we shall use an argument by Lerner [24]. Since $B_0 \geq 1/4$ by (7.18) we find from (7.24)–(7.26) that

$$\operatorname{Re} f_j^w B^{\mathrm{Wick}} \geq \frac{1}{4} a_j^w a_j^w + \operatorname{Re} a_j^w S_j^w \qquad \text{in } L^2 \text{ modulo } \operatorname{Op} S(m, g^\sharp) \tag{7.27}$$

where $S_j = r_j + s_j \in S(M^{1/2}H^{3/4}\langle\delta_0\rangle, g^\sharp)$. Then by completing the square, we find that

$$\operatorname{Re} f_j^w B^{\mathrm{Wick}} \geq \frac{1}{4}(a_j^w + 2S_j^w)^*(a_j^w + 2S_j^w) \geq 0 \quad \text{in } L^2 \text{ modulo } \operatorname{Op} S(m, g^\sharp) \tag{7.28}$$

since $(S_j^w)^* S_j^w = \overline{S}_j^w S_j^w \in \operatorname{Op} S(MH^{3/2}\langle\delta_0\rangle^2, g^\sharp)$.

But we must also consider $\operatorname{Re} F_{0,j}^w B^{\mathrm{Wick}}$, where F_0 satisfies (7.2) thus

$$F_{0,j} = \Psi_j F_0 \in S(MH^{1/2}h^{1/2}, G). \tag{7.29}$$

We shall prove that

$$\operatorname{Re} F_{0,j}^w B^{\mathrm{Wick}} = \operatorname{Re} a_j^w R_j^w \qquad \text{modulo } \operatorname{Op} S(m, g^\sharp), \tag{7.30}$$

where $R_j \in S(M^{1/2}H^{3/4}, g^\sharp)$, which can then be included in the term given by S_j in (7.27). Since $b = \delta_0 \in S(H^{-1/2}, G)$ modulo $S(H^{1/2}\langle\delta_0\rangle^2, g^\sharp)$ in ω_j by Proposition 6.3, we find that

$$\operatorname{Re} F_{0,j}^w B^{\mathrm{Wick}} = \operatorname{Re} F_{0,j}^w b^w = (\operatorname{Re} F_{0,j}\delta_0)^w$$

modulo $\operatorname{Op} S(m, g^\sharp)$. We find from (7.21) that $\operatorname{Im} a_j = 0$, so

$$\operatorname{Re} F_{0,j}\delta_0 = \operatorname{Re} \phi_j^2 F_0 \delta_0 = a_j R_j,$$

where

$$R_j = \operatorname{Re} \phi_j F_0 / \alpha_0^{1/2} \in S(M^{1/2}H^{1/4}h^{1/2}, G) \subseteq S(M^{1/2}H^{3/4}, G).$$

This gives $(\operatorname{Re} F_{0,j}\delta_0)^w = a_j^w R_j^w$ modulo $\operatorname{Op} S(MHh^{1/2}, G) \subseteq \operatorname{Op} S(m, g^\sharp)$, so we obtain (7.30). By adding R_j to S_j in (7.27) and completing the square as in (7.28), we obtain (7.15) in this case. This completes the proof of Proposition 7.1.

Remark 7.3. It follows from the proof of Proposition 7.1 that in order to obtain the estimate (7.3) it suffices that the lower-order term $F_0 \in S(MH, g^\sharp) \subseteq S(1, g^\sharp)$.

References

1. R. Beals and C. Fefferman, *On local solvability of linear partial differential equations*, Ann. of Math. **97** (1973), 482–498.

2. J.-M. Bony and J.-Y. Chemin, *Espace fonctionnels associés au calcul de Weyl-Hörmander*, Bull. Soc. Math. France **122** (1994), 77–118.
3. F. Colombini, L. Pernazza, and F. Treves, *Solvability and nonsolvability of second-order evolution equations*, Hyperbolic problems and related topics, Grad. Ser. Anal., Int. Press, Somerville, MA, 2003, 111–120.
4. N. Dencker, *On the propagation of singularities for pseudo-differential operators of principal type*, Ark. Mat. **20** (1982), 23–60.
5. ———, *Preparation theorems for matrix valued functions*, Ann. Inst. Fourier (Grenoble) **43** (1993), 865–892.
6. ———, *The solvability of non L^2 solvable operators*, Journées "Equations aux Dérivées Partielles", St. Jean de Monts, France, 1996.
7. ———, *A sufficient condition for solvability*, International Mathematics Research Notices **1999:12** (1999), 627–659.
8. ———, *The resolution of the Nirenberg–Treves conjecture*, Ann. of Math. **163** (2006), 405–444.
9. ———, *The solvability of pseudo-differential operators*, Phase space analysis of partial differential equations, Vol. I, Pubbl. Cent. Ric. Mat. Ennio Giorgi, Scuola Norm. Sup., Pisa, 2004, 175–200.
10. ———, *On the solvability of pseudo-differential operators*, Sémin. Equ. Dériv. Partielles, Ecole Polytech., Palaiseau, 2005–2006.
11. ———, *The pseudospectrum of systems of semiclassical operators*, Anal. PDE **1** (2008), 323–373.
12. L. Hörmander, *The Weyl calculus of pseudo-differential operators*, Comm. Partial Differential Equations **32** (1979), 359–443.
13. ———, *Pseudo-differential operators of principal type*, Nato Adv. Study Inst. on Sing. in Bound. Value Problems, Reidel Publ. Co., Dordrecht, 1981, 69–96.
14. ———, *The Analysis of Linear Partial Differential Operators*, vol. I–IV, Springer Verlag, Berlin, 1983–1985.
15. ———, *Notions of Convexity*, Birkhäuser, Boston, 1994.
16. ———, *On the solvability of pseudodifferential equations*, Structure of solutions of differential equations (M. Morimoto and T. Kawai, eds.), World Scientific, New Jersey, 1996, 183–213.
17. ———, *The proof of the Nirenberg–Treves conjecture according to N. Dencker and N. Lerner*, preprint, 2005.
18. N. Lerner, *Sufficiency of condition* (Ψ) *for local solvability in two dimensions*, Ann. of Math. **128** (1988), 243–258.
19. ———, *Nonsolvability in L^2 for a first order operator satisfying condition* (Ψ), Ann. of Math. **139** (1994), 363–393.
20. ———, *Energy methods via coherent states and advanced pseudo-differential calculus*, Multidimensional complex analysis and partial differential equations (P. D. Cordaro, H. Jacobowitz, and S. Gidikin, eds.), Amer. Math. Soc., Providence, R.I., USA, 1997, 177–201.
21. ———, *Perturbation and energy estimates*, Ann. Sci. Ecole Norm. Sup. **31** (1998), 843–886.
22. ———, *The Wick calculus of pseudo-differential operators and some of its applications.* Cubo Mat. Educ. **5** (2003), 213–236.
23. ———, *Factorization and solvability*, preprint, 1997.
24. ———, *Cutting the loss of derivatives for solvability under condition (Ψ)*, Bull. Soc. Math. France **134** (2006), 559–631.
25. H. Lewy, *An example of a smooth linear partial differential equation without solution*, Ann. of Math. **66** (1957), 155–158.
26. G. A. Mendoza and G. Uhlmann, *A necessary condition for local solvability for a class of operators with double characteristics*, J. Funct. Anal. **52** (1983), 252–256.
27. R. D. Moyer, *Local solvability in two dimensions: Necessary conditions for the principal-type case*, mimeographed manuscript, University of Kansas, 1978.

28. L. Nirenberg and F. Treves, *On local solvability of linear partial differential equations. Part I: Necessary conditions*, Comm. Partial Differential Equations **23** (1970), 1–38, *Part II: Sufficient conditions*, Comm. Pure Appl. Math. **23** (1970), 459–509; *Correction*, Comm. Pure Appl. Math. **24** (1971), 279–288.
29. K. Pravda-Starov, *Etude du pseudo-spectre d'opérateurs non auto-adjoints*, Ph.D. thesis, Université de Rennes I, 2006.
30. J.-M. Trépreau, *Sur la résolubilité analytique microlocale des opérateurs pseudodifférentiels de type principal*, Ph.D. thesis, Université de Reims, 1984.

The Darboux process and a noncommutative bispectral problem: some explorations and challenges

F. Alberto Grünbaum*

To Hans for the good times we had

Abstract The Darboux process, also known by many other names, played a very important role in some extremely enjoyable joint work that Hans and I did many years ago. I revisit a version of this problem in a case when scalars are replaced by matrices, i.e., elements of a non-commutative ring. Many of the issues considered here can be pushed to the case of a ring with identity, but my emphasis is on very concrete examples involving 2×2 matrices. This paper could be seen as an invitation for further work.

Key words: Darboux process, matrix-valued orthogonal polynomials

Mathematics Subject Classification (2010): 33-XX, 34-XX, 46-XX

1 The bispectral problem

Almost 30 years ago I was extremely lucky. I gave a talk at Berkeley where I mentioned the following problem:

Find all nontrivial instances where a function $\varphi(x,k)$ satisfies

$$L\left(x,\frac{d}{dx}\right)\varphi(x,k) \equiv (-D^2+V(x))\varphi(x,k) = k^2\varphi(x,k)$$

as well as

$$B\left(k,\frac{d}{dk}\right)\varphi(x,k) \equiv \left(\sum_{i=0}^{M} b_i(k)\left(\frac{d}{dk}\right)^i\right)\varphi(x,k) = \Theta(x)\varphi(x,k).$$

Department of Mathematics, University of California, Berkeley, CA, 94720
e-mail: grunbaum@math.berkeley.edu

* The author was supported in part by NSF Grant # DMS 0603901

All the functions $V(x), b_i(k), \Theta(x)$ are, in principle, arbitrary except for smoothness assumptions. Notice that here M is arbitrary (finite).

I was fortunate that Hans was in the audience, and about a week later he came up with a tool to attack this problem. In a few weeks we started seeing a rich landscape of examples and we were delighted to see the central role that the Darboux process played in throwing light on our problem. After many months of intense work, mainly by slow mail, we found ourselves with a rather nice picture.

The complete answer to the problem is given as follows:

Theorem 1 ([DG]). *If $M = 2$, then $V(x)$ is (except for translation) either c/x^2 or ax, i.e., we have a Bessel or an Airy case. If $M > 2$, there are two families of solutions.*

(a) L is obtained from $L_0 = -D^2$ by a finite number of Darboux transformations ($L = AA^* \to \tilde{L} = A^*A$). In this case V is a rational solution of the Korteweg–de Vries equation and all rational solutions of KdV decaying at infinity show up in this fashion.
(b) L is obtained from $L_0 = -D^2 + \frac{1}{4x^2}$ after a finite number of rational Darboux transformations.

It was later observed in [MZ] that in the second case one is dealing with rational solutions of the Virasoro or master symmetries of KdV.

In case (a) the space of common solutions has dimension one; in case (b) it has dimension two. One refers to these as the rank-one and rank-two situations. In [DG] one finds several other equivalent descriptions of the solution such as those in terms of the monodromy group of the equation.

Observe that the "trivial cases" in which $M = 2$ are self-dual in the sense that one can get B by a simple replacement in L.

My reasons for raising the above problem could be traced back to an effort to understand some work on "time- and band-limiting" that had led me to isolate certain properties of well-known special function. For an example relating to orthogonal polynomials see [G1]. For more up-to-date versions of this connection between the bispectral problem and the issue of time- and band-limiting, see [G2, G3, GY2].

The work with Hans gave rise to a large number of papers by other people, some of which can be found in an arXiv version of this paper, containing a longer list of references. Even that longer listing is far from complete, and I apologize for the omissions. For another guide to some of the work inspired by [DG], see the citations in MathScinet, MR 0826863.

It may be appropriate to observe that what we are calling the Darboux process has been reinvented many times, including in the work of some rather well known people; see for instance [Sc, IH]. Reference [YZ] talks about the Geronimus transformation, from 1940, and its inverse the Christoffel transform. It is clear that the first one (as noticed in [YZ]) has a lot in common with what we are calling the Darboux transformation. See also [SVZ, Z].

2 What is the main purpose of this paper?

Now that I have reviewed the basic facts of my joint work with Hans, it may be appropriate to address openly the question in the title of this section.

In the rest of the paper I will mention a few analytical results on certain issues that can be seen as an outgrowth of the work described in the previous section. However, the main intention is to take the problem into a noncommutative situation and to point out a number of avenues for further work. Nothing would make me happier than seeing that some other workers in this field find these problems challenging and worth pursuing.

In particular, and answering a pointed question from a very careful and constructive referee, the role of the Darboux process which unlocked the whole story in the scalar case (provided one starts at the correct starting potentials) is seen here to be a useful tool to obtain interesting bispectral situations. I make no claim that all instances of a bispectral situation in the noncommutative case can be organized neatly by repeated applications of this (or natural extensions of this) process.

In raising this question and in many other ways the referee has been very helpful in making this into a clearer and more readable paper.

In reference to topics that will appear in the next few sections it is important to note that in the scalar case, repeated applications of the Darboux process may take one outside the class of polynomials orthogonal with respect to a measure. One can easily obtain polynomials that have only an orthogonality functional, such as derivatives (of certain orders) of delta measures. The Darboux process when applied as a tool in the matrix-valued case leads to very similar phenomena.

3 The Bochner–Krall problem

In a series of papers with Luc Haine, [GH1, GH2, GH3] and then with Luc and Emil Horozov, see [GHH1, GHH2], we noticed that a large class of polynomials

$$p_n(k)$$

that satisfy three-term recurrence relations in the variable n, as well as differential equations in the variable k, can be obtained by an application of a similar Darboux transformation starting from the so-called classical orthogonal polynomials of Jacobi, Laguerre, and Hermite. In this case one goes from a tridiagonal matrix L_0 (or a function of it) factorized as a product of two bidiagonal matrices,

$$L_0 = AB,$$

to a new tridiagonal matrix

$$L = BA.$$

As indicated in [GH1], some form of this method is given in [MS1].

In this case one runs into the Toda flows and its master symmetries. Further work on these lines can be found in [GY1] and [GH4, GH5], and for a very nice survey of all this material see [H2].

The origins of this line of work are contained in papers such as [Bo, Koo, Kra1]. As usual, tracing the roots of a problem is not easy, and I owe to my friend Mourad Ismail a pointer to reference [R], where these issues were already looked at.

4 A matrix-valued version of the Darboux process for a difference operator

Consider the block tridiagonal matrix

$$L_0 = \begin{pmatrix} B_0 & I & & \\ A_1 & B_1 & I & \\ & \ddots & \ddots & \ddots \end{pmatrix},$$

where all the matrices A_i, B_i are of size $N \times N$ and I denotes the $N \times N$ identity matrix.

If we try to factorize this in the form

$$L_0 = \alpha\beta,$$

where

$$\alpha = \begin{pmatrix} \alpha_0 & I & & \\ 0 & \alpha_1 & I & \\ & \ddots & \ddots & \ddots \end{pmatrix}$$

and

$$\beta = \begin{pmatrix} I & & \\ \beta_1 & I & \\ & \ddots & \ddots \end{pmatrix}$$

with all the matrices α_i, β_i of size $N \times N$, and then define the matrix

$$L = \beta\alpha,$$

we have that

$$L = \begin{pmatrix} \tilde{B}_0 & I & & \\ \tilde{A}_1 & \tilde{B}_1 & I & \\ & \ddots & \ddots & \ddots \end{pmatrix},$$

where all the matrices $\tilde{A}_i, \tilde{B}_i$ are of size $N \times N$.

This gives

$$\tilde{B}_n = \alpha_n + \beta_n, \tilde{A}_n = \beta_n \alpha_{n-1}, \tilde{A}_1 = \beta_1 \alpha_0.$$

Under the appropriate invertibility conditions these relations can be rewritten in terms of α_n and β_n as follows:

$$\beta_n = B_{n-1} - \alpha_{n-1}, \alpha_n = A_n \beta_n^{-1},$$

which then gives

$$\tilde{B}_n = B_n - \beta_{n+1} + \beta_n = B_{n-1} - \alpha_{n-1} + \alpha_n$$

and

$$\tilde{A}_n = \beta_n A_{n-1} \beta_{n-1}^{-1} = \alpha_n^{-1} A_n \alpha_{n-1}.$$

These expressions are valid for $n = 2, 3, \ldots$ in the case of $\tilde{A}_n$, and for $n = 1, 2, \ldots$ in the case of $\tilde{B}_n$.

Above we take $\beta_0 = 0$, so that $\tilde{B}_0 = \alpha_0 = B_0 - \beta_1$. We also need to take, as observed above, $\tilde{A}_1 = (B_0 - \alpha_0)\alpha_0$.

A moment's thought gives that once L_0 is given, the only free parameter is the *matrix* α_0. This all important difference with the scalar case will reappear in Section 10.

Just as in [GH1], and in spite of the fact that one is dealing here with a semi-infinite block tridiagonal matrix, it is possible to see the connection between this construction and that in [MS1]. One puts

$$\beta_n = -\phi_n \phi_{n-1}^{-1}$$

and then notices that this amounts to choosing ϕ in the null-space of L. Since L is not doubly infinite, we seem to have lost some freedom in picking this subspace, but this can be remedied as in [GH1] by considering L as a limit of an appropriate doubly infinite matrix with a rich null-space.

5 Fancier versions of the Darboux process

It is well known that it is useful to extend the original method of Darboux consisting in going from

$$L_0 = AB$$

to

$$L = BA$$

in an appropriate way.

Notice that in the standard case we have

$$BL_0 = LB.$$

For those interested in the history of these developments, the celebrated book by Darboux makes it clear that the process is not due to him but to Moutard. So much for names and historical accuracy.

The new idea is to allow for an arbitrary banded matrix (or a differential operator) U and to declare L a Darboux transform of L_0 as long as we have

$$UL_0 = LU.$$

In several of the uses of Darboux's original method one needs to apply it repeatedly, and this fancier version of the method takes care of that.

One should also keep in mind the results in [GHH1, GHH2, H2], where the usual factorization followed by a reversal of the factors is applied not directly to L but to a constant-coefficient polynomial in L.

I thank Jose Liberati for pointing out to me that at the very end of [GGRW] one finds an application of the theory of quasideterminants (a notion that goes back to Cayley) to obtain expressions for matrix-valued orthogonal polynomials in terms of their matrix-valued moments. Many of these results, as well as others, have been derived independently by making good use of the notion of Schur complements in L. Miranian's Berkeley thesis. The main new results are contained in [M]. In the next section we give a very brief look into the theory of matrix-valued orthogonal polynomials and a short guide to the literature that is relevant to us.

One should remark that this same theory of quasideterminants has been studied in connection with a certain Darboux process for a matrix Schrödinger equation in [GV]. In this case one needs to consider this fancier version. For a nice use of quasideterminants in our context see [BL].

The matrix version of the Darboux process for the difference operator discussed in the previous section could be extended in this fancier fashion too.

6 Matrix-valued orthogonal polynomials

Given a self-adjoint positive definite matrix-valued weight function $W(x)$, M.G. Krein, see [K1, K2], considers the skew-symmetric bilinear form defined for any pair of matrix-valued functions $P(x)$ and $Q(x)$ by the matrix

$$\langle P, Q\rangle = \langle P, Q\rangle_W = \int_{\mathbb{R}} P(x)W(x)Q^*(x)dx,$$

where $Q^*(x)$ denotes the conjugate transpose of $Q(x)$.

Proceeding as in the case of a scalar valued inner product, Krein proves that there exists a sequence $(P_n)_n$ of matrix polynomials, orthogonal with respect to W, with P_n of degree n and monic.

Krein goes on to prove that any sequence of monic orthogonal matrix-valued polynomials $(P_n)_n$ satisfies a three-term recurrence relation

$$A_n P_{n-1}(x) + B_n P_n(x) + P_{n+1}(x) = x P_n(x), \tag{1}$$

where P_{-1} is the zero matrix and P_0 is the identity matrix. These coefficient matrices enjoy certain properties: in particular, the A_n are nonsingular.

The equation above can be rewritten as

$$\mathscr{L} P_n(x) = x P_n(x)$$

with a matrix $\mathscr{L}$ such as the one that has appeared in previous sections.

To place ourselves in the context of the *bispectral problem* we consider matrix-valued polynomials $(P_n)_n$ satisfying not only the equation above but also "right-hand side" differential equations of the form

$$P_n D = \Lambda_n P_n \qquad \text{for all} \quad n \geq 0 \tag{2}$$

with Λ_n a matrix-valued eigenvalue and D a differential operator of order s with matrix coefficients given by

$$D = \sum_{i=0}^{s} \partial^i F_i(x), \qquad \partial = \frac{d}{dx},$$

which acts on $P(x)$ by means of

$$PD = \sum_{i=0}^{s} \partial^i (P)(x) F_i(x).$$

This problem in the matrix case was raised first in [D] and then further studied in [DG1, G10, GI, GPT1, GPT2, GPT3, GT] and in a few other places.

One can see that the differential operators that correspond to a fixed family of polynomials form an associative algebra which in general is noncommutative; see [CG2, GT, T]. The problem of exhibiting elements of this algebra that have a minimal order will occupy us in a few examples in the next two sections. For early matrix valued instances of the bispectral problem and related work see [Z1, Z2, Z3, Z4].

7 A few examples

Here we consider in detail a few examples of the matrix version of the basic Darboux process described above.

For $\lambda > 3/2$ consider the monic matrix-valued polynomials which are orthogonal with respect to the weight matrix

$$W(x) = ((2-x)x)^{\lambda-3/2}\begin{pmatrix} 1 & x-1 \\ x-1 & 1 \end{pmatrix}, \quad x \in [0,2].$$

Let

$$L_0 = \begin{pmatrix} B_0 & I & & \\ A_1 & B_1 & I & \\ & \ddots & \ddots & \ddots \end{pmatrix}$$

be the corresponding block tridiagonal matrix with

$$B_n = \frac{1}{2}\frac{\lambda-1}{(n+\lambda)(n+\lambda-1)}S + I,$$

$$A_n = \frac{n(n+2\lambda-2)}{4(n+\lambda-1)^2}I.$$

Here $S = \left(\begin{smallmatrix} 0 & 1 \\ 1 & 0 \end{smallmatrix}\right)$ and $I = \left(\begin{smallmatrix} 1 & 0 \\ 0 & 1 \end{smallmatrix}\right)$.

These polynomials can be seen to be joint eigenfunctions of a first-order differential operator, an observation that was made for the special value $\lambda = 1$ in [CG1, CG2].

If α_0 is an arbitrary matrix we can consider the monic polynomials that result from one application of the Darboux process to the block tridiagonal matrix L_0 with *free matrix-valued parameter* α_0.

We can see that for an invertible symmetric α_0, the new orthogonality weight is given by

$$\widetilde{W}(x) = (2-x)^{\lambda-3/2}x^{\lambda-5/2}\begin{pmatrix} 1 & x-1 \\ x-1 & 1 \end{pmatrix}$$
$$-\frac{-2^{2\lambda}\mathrm{Be}\left(\frac{2\lambda-1}{2},\frac{2\lambda-1}{2}\right)}{4(2\lambda-3)}\left(\begin{pmatrix} 2\lambda-2 & -1 \\ -1 & 2\lambda-2 \end{pmatrix} - (2\lambda-3)\alpha_0^{-1}\right)\delta_0(x).$$

Here Be stands for the usual beta function.

We display below some examples that illustrate that for appropriate values of λ the new polynomials are joint eigenfunctions of some higher order differential operators, i.e., we get new bispectral situations. This appears to have little to do with α_0 being symmetric.

Example 1.

$$\lambda = 5/2, \quad \alpha_0 = \begin{pmatrix} 5 & 2 \\ 3 & 1 \end{pmatrix}.$$

Here we find one differential operator D satisfying

$$P_n D = \Lambda_n P_n$$

with

$$D = \sum_{r=0}^{4} \partial^r F_r$$

and

$$F_4 = (x-2)^2 x^2 \begin{pmatrix} 1 & -1 \\ -1 & 1 \end{pmatrix},$$

$$F_3 = 4(x-2)x(3x-2) \begin{pmatrix} 1 & -1 \\ -1 & 1 \end{pmatrix},$$

$$F_2 = \frac{24}{5} x(7x-9) \begin{pmatrix} 1 & -1 \\ -1 & 1 \end{pmatrix},$$

$$F_1 = \frac{8}{5} \begin{pmatrix} 5x+6 & -8x \\ -2(5x+3) & 1x3 \end{pmatrix},$$

$$F_0 = \begin{pmatrix} -\frac{8\times 11}{5} & -8 \\ -\frac{32}{5} & 0 \end{pmatrix},$$

$$\Lambda_n = \begin{pmatrix} \frac{(n+2)(5n^3+20n^2+3n-44)}{5} & -\frac{5n^4+30n^3+43n^2-14n+40}{5} \\ -\frac{5n^4+30n^3+43n^2+2n+32}{5} & \frac{n(5n^3+30n^2+43n+26}{5} \end{pmatrix}.$$

There are no operators of lower order in the algebra.

Example 2.

$$\lambda = 7/2, \quad \alpha_0 = \begin{pmatrix} 3 & -1 \\ 5 & 7 \end{pmatrix}.$$

Here the corresponding operator is given by

$$D = \sum_{r=0}^{6} \partial^r F_r$$

with

$$F_6 = \frac{(x-2)^3 x^3}{15} \begin{pmatrix} 1 & -1 \\ -1 & 1 \end{pmatrix},$$

$$F_5 = \frac{2(x-2)^2 x^2 (5x-4)}{5} \begin{pmatrix} 1 & -1 \\ -1 & 1 \end{pmatrix},$$

$$F_4 = 4(x-2)x(5x^2-8x+2)\begin{pmatrix} 1 & -1 \\ -1 & 1 \end{pmatrix},$$

$$F_3 = 16x(5x^2-12x+6)\begin{pmatrix} 1 & -1 \\ -1 & 1 \end{pmatrix},$$

$$F_2 = \frac{16x(131x-148)}{39}\begin{pmatrix} 1 & -1 \\ -1 & 1 \end{pmatrix},$$

$$F_1 = \frac{32}{19}\begin{pmatrix} -4(2x-7) & 3(x-6) \\ (9x-28) & -2(2x-9) \end{pmatrix},$$

$$F_0 = \frac{1}{19}\begin{pmatrix} -13\times 32 & -3\times 64 \\ 11\times 32 & 0 \end{pmatrix},$$

and we have

$$P_n D = \Lambda_n P_n$$

with

$$\Lambda_n = \left(\frac{19n^6+285n^5+1615n^4+4275n^3+2446n^2}{285}\right)\begin{pmatrix} 1 & -1 \\ -1 & 1 \end{pmatrix}$$
$$+\frac{1}{285}\begin{pmatrix} -12480n-6240 & 10080n-2880 \\ 12960n+5280 & -10560n \end{pmatrix}.$$

Example 3.

$$\lambda = 9/2, \quad \alpha_0 = \begin{pmatrix} 1 & 0 \\ 0 & 0 \end{pmatrix}.$$

In this case there is one operator of order eight whose corresponding Λ_n is given by

$$\Lambda_n = \begin{pmatrix} (n-3)(n+6)(n+7)(n+8)\alpha_n & -(n-2)n(n+7)(n+8)\beta_n \\ -(n+1)(n+6)\gamma_n & (n-1)n(n+1)(n+10)\delta_n \end{pmatrix}$$

with

$$\begin{aligned}
\alpha_n &= n^4+10n^3+59n^2+170n+840,\\
\beta_n &= n^4+14n^3+95n^2+322n+1080,\\
\gamma_n &= n^6+21n^5+169n^4+651n^3+1198n^2+840n-20160,\\
\delta_n &= n^4+18n^3+143n^2+558n+1512.
\end{aligned}$$

Once again, this corresponds to the lowest-order differential operator possible in the corresponding algebra.

8 A few Jacobi-type examples

A different avenue for exploring the similarities as well as the differences between the use of the Darboux process in the scalar and in the matrix-valued case is given by the examples in this section.

First recall that in the scalar case it follows from groundbreaking results in [Koo] and then further work in [KoKo1, Z, H2, GY1] that the polynomials orthogonal to the weight $\mu(x)$ consisting of a Jacobi density plus two possible delta masses of nonnegative strengths W, V at the ends of the interval, i.e.,

$$\mu(x) = (1-x)^{\alpha}(1+x)^{\beta} + W\delta_1(x) + V\delta_{-1}(x),$$

satisfy differential equations when α and β satisfy certain natural integrality conditions. We refer, along with other authors, to these polynomials as Koornwinder polynomials, not to be confused with some BC_n extensions of Macdonald polynomials, which are also due to the same author. If the weight at 1 is the only one that is present, then the order is $2\alpha + 4$. If both delta weights are thrown in, then the order is $2\alpha + 2\beta + 6$. The results can be obtained by an application of the Darboux process as shown in [H2, GY1].

We consider now a small collection of situations analogous to those above.

The weight matrices will, as before, consist of a matrix weight density plus a pair of deltas at the endpoints weighted by certain matrices W, V, i.e., we have

$$\widetilde{W}(x) = (1-x)^{\alpha}(1+x)^{\beta}\begin{pmatrix} 1 & x \\ x & 1 \end{pmatrix} + W\delta_1(x) + V\delta_{-1}(x).$$

For the first batch of examples we will assume that α, β are both 0. If V and W coincide with the matrix $\left(\begin{smallmatrix} a & 0 \\ 0 & 0 \end{smallmatrix}\right)$, then we find two linearly independent operators of order 5 and one of order 6 as well as other operators of higher order. There are no other operators of lower order.

If V is the matrix $\left(\begin{smallmatrix} a & 0 \\ 0 & 0 \end{smallmatrix}\right)$ and W is the matrix $\left(\begin{smallmatrix} 0 & 0 \\ 0 & b \end{smallmatrix}\right)$ or the matrix $\left(\begin{smallmatrix} b & 0 \\ 0 & 0 \end{smallmatrix}\right)$, then we find two linearly independent operators of order 6 as well as other operators of higher order. There are no other operators of lower order.

More generally, if V is the matrix $\begin{pmatrix} a^2 & ab \\ ab & b^2 \end{pmatrix}$ and W is the matrix $\begin{pmatrix} c^2 & cd \\ cd & d^2 \end{pmatrix}$, both rank-deficient, then we have the same situation as in the last example.

In general if V and W are of the form $\left(\begin{smallmatrix} a & b \\ b & c \end{smallmatrix}\right)$ and $\left(\begin{smallmatrix} d & e \\ e & f \end{smallmatrix}\right)$, then the lowest-order operator in the algebra is just one operator of order 8.

We come now to a different sort of example.

Assume that α and β (> -1) are arbitrary, but insist on picking W and V to be arbitrary (and not necessarily equal) nonnegative multiples of the matrix $\left(\begin{smallmatrix} 1 & 1 \\ 1 & 1 \end{smallmatrix}\right)$.

In this case there is a very nice second-order differential operator in the algebra which is *independent* of the choice of the scalar factors that appear in front of the matrix above to give W and V. There is no lower order operator in the algebra. When the deltas are both missing, then the algebra contains an operator of order 1.

The right-handed differential operator alluded to above is a scalar operator of the usual Jacobi type, with coefficients $(1-x^2)$ and $(\alpha+\beta-1)-x(\alpha+\beta+3)$ multiplied on the right by the matrix $\begin{pmatrix} 1 & -1 \\ -1 & 1 \end{pmatrix}$.

The eigenvalue is $-n(n+\alpha+\beta+2)$ multiplied by this same matrix.

9 An explicit differential operator

The paper [DG] contains a proof that in the continuous–continuous case, when all operators in question are differential operators, the so-called ad-condition

$$\mathrm{ad} L^{m+1}(\Theta) = 0$$

is necessary and sufficient to have what has been called a bispectral situation, i.e., a solution of the original problem tackled with Hans.

This condition gives a set of nonlinear equations that need to be solved in the unknowns L, Θ.

It is important to see that this condition can be easily adapted to other situations, including the present noncommutative one. This approach was taken up in [GI] and in [GT]. In the second of these papers the "ad-condition" is shown to be equivalent, once again, to bispectrality.

In general, finding the differential operators of lowest possible order that appear in a bispectral situation is not easy. By repeated applications of the Darboux process one may obtain elements of the corresponding algebra that are not necessarily of the lowest possible order. This issue has surfaced in several different papers, starting with [DG], and a nice account is given in [H2].

In [GT] one finds an explicit construction of a differential operator that results from the condition

$$\mathrm{ad} L^{m+1}(\Lambda) = 0.$$

The operator D is given by

$$D = \sum_{r=0}^{m} \partial^r(P) \frac{S_{m-r}}{r!}$$

with matrix coefficients $S_k = S_k(x)$ given by

$$S_k = ((L - xI)^{m-k} \Lambda P)_0.$$

In particular, we display here some of the coefficients. The lowest one is

$$S_0 = ((L - xI)^m \Lambda P)_0,$$

and at the other end,

$$S_m = (\Lambda P)_0 = \Lambda_0,$$

and the operator D, of course, satisfies the desired condition

$$P_n D = \Lambda_n P_n, \quad n \geq 0.$$

The subscript 0 above refers to the first entry of the corresponding "vector" with matrix-valued entries.

10 Toda flows with matrix-valued time

As seen in [GH2], and certainly in other places too, repeated application of the scalar Darboux process introduces "times and flows" that are related to the Toda flows. Since these times appear as the free parameters in each application of the process, it is only natural to raise the issue of "matrix-valued times" and the corresponding flows. This is a vast and really unexplored area.

11 Electrostatics: Heine, Stieltjes, Darboux

In a remarkable paper that follows on earlier work of Heine, Stieltjes came up with a nice electrostatic interpretation for the zeros of the Jacobi polynomials. Later work of Stieltjes as well as of I. Schur and D. Hilbert showed similar interpretations in the case of the Laguerre and Hermite polynomials.

In [G7, G8] I raise the possibility of some relation between the Darboux process, where the orthogonality functional gets more and more complicated with every application of the process, and the corresponding electrostatic interpretation of the families of polynomials that appear along the way.

It would be interesting to see what if anything of this picture can be developed in the matrix-valued case.

12 Markov chains

In [DRSZ, G4, G5, G6, G9, G11, GdI] one finds examples of interesting quasi-birth-and-death processes that can be studied by exploiting their connection with certain specific examples of matrix-valued orthogonal polynomials. In particular, in [G5, G6, G11] one finds examples in which the recurrence of the process is related to the presence of a matrix-valued delta weight at 1. Since the appearance of these delta weights is one of the main characteristics of an application of the Darboux process, one may wonder about a probabilistic interpretation of the relation that may exist between two Markov chains whose transition probability matrices are related by a Darboux transformation.

13 Things that appear before their time

One of the most surprising phenomena uncovered in [DG] has to do with what was called "the cusps," namely degenerate situations that correspond to degeneracies of "higher-order operators" yielding "lower-order" ones. To put this in the context of scalar-valued orthogonal polynomials, consider the simplest case of the Koornwinder polynomials which are orthogonal to Lebesgue measure in $[-1, 1]$ plus a pair of delta masses at the endpoints of the interval. In this case one knows that the corresponding orthogonal polynomials are the common eigenfunctions of a sixth-order differential operator.

In the degenerate case in which the strength of the two delta masses agrees one gets an operator of order four, and one can say that in a search according to the order of these operators, this example, just like "the cusps" in [DG], appears before its time.

We made a tentative exploration of the situation in the matrix-valued case, and examples of this phenomenon are seen in Section 7.

14 The multivariable case

In this section we mention that in [G9] one finds a specific random walk introduced by Hoare and Rahman, see [HR], which we show leads to a bispectral situation in terms of polynomials of two variables. For more recent work see [GR]. I also want to mention that in the multivariable case one finds a version of the Darboux process to obtain interesting deformations of the two-dimensional Chebyshev measure; see [GI1]. All of these phenomena can be studied in the case of matrix-valued orthogonal polynomials.

15 Conclusion

It is clear that very little of the development that I have tried to outline here could have happened were it not for my good fortune in teaming up with Hans at the beginning of this journey. As a small token of gratitude for his influence on my own work I offer to him, and to others, this collection of (mostly) open problems.

References

[Bo] Bochner, S., *Über Sturm–Liouvillesche Polynomsysteme*, Math. Z **29** (1929), 730–736.

[BL] Boyallian, C. and Liberati, J., *Matrix valued bispectral operators and quasideterminants*, J. Phys. A. **41** (2008), 11 pp.

[CG1] Castro, M. M. and Grünbaum, F. A., *Orthogonal matrix polynomials satisfying first order differential equations: a collection of instructive examples*, J. of Nonlinear Mathematical Physics, Supplement 2, Vol. **12** (2005), 63–76.

[CG2] Castro, M. and Grünbaum, F. A., *The algebra of differential operators associated to a family of matrix valued orthogonal polynomials: five instructive examples*, International Math. Research Notices, **7** (2006), 1–33.

[DRSZ] Dette, H., Reuther, B., Studden, W. and Zygmunt, M., *Matrix measures and random walks with a block tridiagonal transition matrix*, SIAM J. Matrix Anal. Applic. **29**, No. 1 (2006), 117–142.

[DG] Duistermaat, J. J. and Grünbaum, F. A., *Differential equations in the spectral parameter*, Comm. Math. Phys. **103** (1986), 177–240.

[D] Duran, A., *Matrix inner product having a matrix symmetric second order differential operators*, Rocky Mountain Journal of Mathematics **27**, No. 2, Spring 1997.

[DG1] Duran, A. J. and Grünbaum, F. A., *Orthogonal matrix polynomials satisfying second order differential equations*, Internat. Math. Research Notices, 2004: **10** (2004), 461–484.

[GGRW] Gelfand, I., Gelfand, S., Retakh, V., and Wilson, R., *Quasideterminants*, Advances in Mathematics **193** (2005), No. 1, 56–141.

[GI1] Geronimo, J. and Iliev, P., *Two variable deformations of the Chebyshev measure*, arXiv:math/0612489v1, Dec 2006.

[GV] Goncharenko, V. M. and Veselov, A. P., *Monodromy of the matrix Schrödinger equations and Darboux transformations*, J. Phys. A. Math. Gen. **31** (1998), 5315–5326.

[GH1] Grünbaum, F. A. and Haine, L., "Orthogonal polynomials satisfying differential equations: the role of the Darboux transformation" in *Symmetries and Integrability of Difference Equations*, D. Levi, L. Vinet, and P. Winternitz, eds., CRM Proc. Lecture Notes **9**, Amer. Math. Soc., Providence, 1996, 143–154.

[GH2] Grünbaum, F. A. and Haine, L., *A theorem of Bochner, revisited*, in *Algebraic Aspects of Integrable Systems: In memory of Irene Dorfman*, A. S. Fokas and I. M. Gelfand, eds., Progr. Nonlinear Differential Equations **26**, Birkhäuser, Boston, 1996, 143–172.

[GH3] Grünbaum, F. A. and Haine, L., *Some functions that generalize the Askey–Wilson polynomials*, Comm. Math. Phys. **184** (1997), 173–202.

[GH4] Grünbaum, F. A. and Haine, L., *Bispectral Darboux transformations: an extension of the Krall polynomials*, Internat. Math. Res. Notices (1997), No. 8, 359–392.

[GH5] Grünbaum, F. A. and Haine, L., *Associated polynomials, spectral matrices and the bispectral problem*, Methods and Applications of Analysis **6** (2) (1999), 209–224.

[GHH1] Grünbaum, F. A., Haine, L., and Horozov, E., *On the Krall–Hermite and the Krall–Bessel polynomials*, IMRN (1997), No. 19, 953–966.

[GHH2] Grünbaum, F. A., Haine, L., and Horozov, E., *Some functions that generalize the Krall–Laguerre polynomials*, J. Compl. Appl. Math. **106** (1999), 271–297.

[GdI] Grünbaum, F. A. and de la Iglesia, M. D., *Matrix valued orthogonal polynomials arising from group representation theory and a family of quasi-birth-and-death processes*, SIAM J. Matrix Anal. Applic. **30** (2) (2008), 741–763.

[GI] Grünbaum, F. A. and Iliev, P., *A noncommutative version of the bispectral problem*, J. of Computational and Appl. Math. **161** (2003), 99–118.

[GPT1] Grünbaum, F. A., Pacharoni, I., and Tirao, J. A., *Matrix valued spherical functions associated to the complex projective plane*, J. Functional Analysis **188** (2002), 350–441.

[GPT2] Grünbaum, F. A., Pacharoni, I., and Tirao, J. A., *Matrix valued orthogonal polynomials of the Jacobi type*, Indag. Mathem. **14** 3,4 (2003), 353–366.

[GPT3] Grünbaum, F. A., Pacharoni, I., and Tirao, J. A., *Matrix valued orthogonal polynomials of the Jacobi type: The role of group representation theory*, Ann. Inst. Fourier, Grenoble **55** 6 (2005), 2051–2068.

[GT] Grünbaum, F. A. and Tirao, J. A., *The algebra of differential operators associated to a weight matrix*, Integral equations and operator theory **58** (2007), 449–475.

[GY1] Grünbaum, F. A. and Yakimov, M., *Discrete bispectral Darboux transformations from Jacobi polynomials*, Pacific J. of Math., Vol. **204** (2002), No. 2, 395–431.

[GY2] Grünbaum, F. A. and Yakimov, M., *The Prolate Spheroidal phenomenon as a consequence of Bispectrality*, CRM Proc. and Lect. Notes, Vol. **37** (2004), 301–312.

[G1] Grünbaum, F. A., *A new property of reproducing kernels for classical orthogonal polynomials*, J. Math. Anal. Applic. **95** (1983), 491–500.

[G2] Grünbaum, F. A., *Time-band limiting and the bispectral problem*, Commun. on Pure and Applied Math. **47** (1994).

[G3] Grünbaum, F. A., *Band-time-band limiting integral operators and commuting differential operators*, Algebra y Analiz **8** (1996), 122–126.

[G4] Grünbaum, F. A., *Random walks and orthogonal polynomials: some challenges*, Probability, Geometry and Integrable Systems, M. Pinsky and B. Birnir, editors, MSRI Publication, volume 55 (2007), 241–260, see arXiv math.PR/0703375.

[G5] Grünbaum, F. A., *Block tridiagonal matrices and a beefed-up version of the Ehrenfest urn model*, Proceedings of the meeting in Odessa, Ukraine, 2007, in honor of M. G. Krein.

[G6] Grünbaum, F. A., *QBD processes and matrix valued orthogonal polynomials: some explicit examples* Dagstuhl Seminar Proceedings 07461 (2007), Numerical Methods in Structured Markov Chains, http://drops.dagstuhl.de/portals/07461/.

[G7] Grünbaum, F. A., *Variations on a theme of Heine and Stieltjes: an electrostatic interpretation of the zeros of certain polynomials*, JCAM **99** (1998), 189–194.

[G8] Grünbaum, F. A., *Electrostatic interpretation of the zeros of certain polynomials and the Darboux process*, JCAM **133** (2001), 397–412.

[G9] Grünbaum, F. A., *The Rahman polynomials are bispectral*, SIGMA **3** (2007), 065.

[G10] Grünbaum, F. A., *Matrix valued Jacobi polynomials*, Bull. Sci. Math. **127** (2003), 207–214.

[GPT1] Grünbaum, F. A., Pacharoni, I., and Tirao, J. A., *Matrix valued spherical functions associated to the complex projective plane*, J. Functional Analysis **188** (2002), 350–441.

[GPT2] Grünbaum, F. A., Pacharoni, I., and Tirao, J. A., *Matrix valued orthogonal polynomials of the Jacobi type*, Indag. Mathem. **14** 3,4 (2003), 353–366.

[GPT3] Grünbaum, F. A., Pacharoni, I., and Tirao, J. A., *Matrix valued orthogonal polynomials of the Jacobi type: The role of group representation theory*, Ann. Inst. Fourier, Grenoble **55** 6 (2005), 2051–2068.

[GPT4] Grünbaum, F. A., Pacharoni, I. and Tirao, J. A., *Some families of matrix valued Jacobi polynomials*, to appear in the Askey–Bateman project.

[GR] Grünbaum, F. A. and Rahman, M., On a family of 2-variable orthogonal Krawtch ouk polynomials, SIGMA 6 (2010) 090, 12 pages, DOI 10.3842/SIGMA.2010.090.

[G11] Grünbaum, F. A., A spectral weight matrix for a discrete version of Walsh's spider Operator Theory Advances and Applications, Vol. 202 (2010), 253–264.

[H2] Haine, L., *The Bochner–Krall problem: some new perspectives*, in *Special Functions 2000*, J. Bustoz et al. (editors), 141–178, Kluwer Acad. Publishers.

[HR] Hoare, M. and Rahman, M., *A probabilistic origin for a new class of bivariate polynomials*, SIGMA 4(2008) 089, 18 pages, arXiv: 0812.3879.

[IH] Infeld, L. and Hull, T., *The factorization method*, Rev. Modern Phys. **23** (1951), 21–68.

[KoKo1] Koekoek, J. and Koekoek, R., *Differential equations for generalized Jacobi polynomials*, JCAM **126** (2000), 1–31.

[Koo] Koornwinder, T. H., *Orthogonal polynomials with weight function* $(1-x)^{\alpha}(1+x)^{\beta} + M\delta(x+1) + N\delta(x-1)$, Canad. Math. Bull. **27** (1984), 205–214.

[Kra1] Krall, H. L., *Certain differential equations for Tchebycheff polynomials*, Duke Math. J. **4** (1938), 705–718.

[K1] Krein, M. G., *Fundamental aspects of the representation theory of hermitian operators with deficiency index* (m, m), AMS Translations, Series 2, Vol. 97, Providence, Rhode Island (1971), 75–143.

[K2] Krein, M. G., *Infinite* J*-matrices and a matrix moment problem*, Dokl. Akad. Nauk SSSR **69** (1949), Nr. 2, 125–128.

[MZ] Magri, F. and Zubelli, J., *Differential equations in the spectral parameter, Darboux transformations and a hierarchy of master symmetries for KdV*, Comm. Math. Physics **141** (1991), No. 2, 329–351.

[MS1] Matveev, V. B. and Salle, M. A., *Darboux transformations and Solitons*, Springer Series in Nonlinear Dynamics, 1991.

[M] Miranian, L., *Matrix valued orthogonal polynomials on the real line: some extensions of the classical theory*, J. Phys. A: Math. Gen. **38** (2005), 5731–5749.

[R] Routh, E., *On some properties of certain solutions of a differential equation of the second order*, Proc. London Math. Soc. **16** (1884), 245–261.

[Sc] Schrödinger, E., *A method of determining quantum mechanical eigenvalues and eigenfunctions*, Proc. Roy. Irish Acad. Sect. A **46** (1940), 9–16.

[SVZ] Spiridonov, V., Vinet, L., and Zhedanov, A., "Bispectrality and Darboux transformations in the theory of orthogonal polynomials," in J. Harnad and A. Kasman (eds), *The Bispectral Problem (Montréal, PQ, 1997)*, 111–122, CRM Proc. Lecture Notes 14, Amer. Math. Soc., Providence, 1998.

[T] Tirao, J. A., The algebra of differential operators associated to a weight matrix: a first example. XVIII Latin American Algebra Colloquium, Contem. Math. (CONM), Amer. Math. Soc. To appear, 2010.

[YZ] Yermolayeva, O. and Zhedanov, A., *Spectral transformations and generalzied Pollaczek polynomials*, Methods and Applications of Analysis, Vol. **6** (1999), No. 3, 1–20.

[Z] Zhedanov, A., *A method of constructing Krall's polynomials*, J. Comput. Appl. Math. **107** (1999), 1–20.

[Z1] Zubelli, J., Differential Equations in the Spectral Parameter for Matrix Differential Operators. Physica D Nonlinear Phenomena **43** 2–3 (1990), 269–287.

[Z2] Zubelli, J., Rational Solutions of Nonlinear Evolution Equations, Vertex Operators and Bispectrality. Journal of Differential Equations **97** 1 (1992), 71–98.

[Z3] Zubelli, J., On a Zero Curvature Problem Related to the ZS-AKNS Operator. Journal of Mathematical Physics **33** 11 (1992), 3666–3675.

[Z4] Zubelli, J., Differential Equations in the Spectral Parameter, Darboux Transformations and a Hierarchy of Master Symmetries for KdV. Joint with F. Magri. Communications in Mathematical Physics **141** (2) (1991), 329–351.

Conjugation spaces and edges of compatible torus actions

Jean-Claude Hausmann and Tara Holm*

In honor of the memory of Hans Duistermaat

Abstract Duistermaat introduced the notion of *real locus* of a symplectic manifold, and subsequently a variety of techniques have been generalized to these lagrangian submanifolds. Together with Puppe, the authors of this paper generalized these results to the topological category, introducing *conjugation spaces*. In this paper, we review the definition and basic properties of conjugation spaces, and then give a topological criterion for recognizing a conjugation space.

Key words: Conjugation spaces, group actions, equivariant cohomology

Mathematics Subject Classification (2010): 58D19, 55N91

1 Introduction

Duistermaat introduced the *real locus* of a Hamiltonian manifold [Du]. In this and in others' subsequent works [BGH, Go, GH, HH, Ho, OS, Sd], it has been shown that many of the techniques developed in the symplectic category can be used to study real loci, so long as the coefficient ring is restricted to the integers modulo 2. As we will see, these results seem not necessarily to depend on the ambient symplectic structure, but rather are topological in nature. This observation

Jean-Claude Hausmann
Mathématiques-Université, B.P. 240, CH-1211 Genève 24, Switzerland
e-mail: Jean-Claude.Hausmann@unige.ch

Tara Holm
Department of Mathematics, Cornell University, Ithaca, NY 14853-4201, USA
e-mail: tsh@math.cornell.edu

* The second author was supported in part by NSF grant DMS-0604807. In addition, both authors are grateful for support from the Swiss National Funds for Scientific Research.

prompts the definition of *conjugation space* in [HHP]. We now give a brief survey of the results in symplectic geometry that motivated the definition of a conjugation space.

A *symplectic manifold* is a manifold M together with a 2-form $\omega \in \Omega^2(M)$ that is closed ($d\omega = 0$) and nondegenerate (for each nonzero tangent vector $X \in T_pM$ there exists $Y \in T_pM$ such that $\omega_p(X, Y) \neq 0$). Let G be a compact Lie group acting on M preserving ω, $\mathfrak{g}$ the Lie algebra of G, $\mathfrak{g}^*$ its dual, and $\langle \cdot, \cdot \rangle : \mathfrak{g}^* \times \mathfrak{g} \to \mathbb{R}$ the natural pairing. For each $X \in \mathfrak{g}$, we let $X^\#$ denote the vector field on M generated by the one-parameter subgroup $\exp(tX)$. We say that the G-action on M is *Hamiltonian* if there is a *moment map*

$$\Phi : M \to \mathfrak{g}^*$$

that satisfies

1. $\iota_{X^\#}\omega = d\langle \Phi, X \rangle$ for all $X \in \mathfrak{g}$; and
2. Φ is equivariant with respect to the given G action on M and the coadjoint action of G on $\mathfrak{g}^*$.

The function $\Phi_X = \langle \Phi, X \rangle$ is called the *Hamiltonian function* for the vector field $X^\#$.

When $G = T$ is a torus, the second condition on Φ requires that it be a T-invariant map. In this special case, we have

Theorem 1.1 ([A], [GS]). *If M is a compact Hamiltonian T-manifold, then $\Phi(M)$ is a convex polytope. It is the convex hull of $\Phi(M^T)$, the images of the T-fixed points.*

More generally, Kirwan and many others have explored analogues for nonabelian groups.

A by-product of Atiyah's proof of Theorem 1.1 is that any of the Hamiltonian functions Φ_X is a perfect Morse function on M, in the sense of Bott, and for generic X, the critical set is M^T. More precisely,

$$H^*(M; \mathbb{R}) = \sum_{i=1}^{N} H^{*-d_i}(F_i; \mathbb{R}), \tag{1}$$

where the F_i are the connected components of M^T and d_i is the Morse–Bott index of F_i. This statement is also true over $\mathbb{Z}$, provided that the cohomology of each F_i is torsion-free, or when the stabilizers of the torus action satisfy some additional hypotheses.

Duistermaat introduced the concept of real locus to this framework [Du]. Let M be a Hamiltonian T-manifold, and $\tau : M \to M$ an antisymplectic involution that is *compatible* with the action; that is, it satisfies

$$\tau(t \cdot p) = t^{-1} \cdot \tau(p),$$

for all $t \in T$ and $p \in M$. Then if it is nonempty, the submanifold M^τ of τ-fixed points is a Lagrangian submanifold of M called the *real locus* of the involution. The primary example of such an involution is the one induced by complex conjugation on a complex projective variety defined over $\mathbb{R}$. For example, if $M = \mathbb{C}P^n$ equipped with the Fubini–Study symplectic form and the standard T^n action, then the real locus for complex conjugation consists of the real points $\mathbb{R}P^n$, whence the name real locus. The main results in [Du] generalize Theorem 1.1 and Atiyah's Morse-theoretic results.

Theorem 1.2 ([Du]). *If M is a compact Hamiltonian T-manifold, and τ a smooth compatible antisymplectic involution, then*

1. *The real locus has full moment image: $\Phi(M^\tau) = \Phi(M)$ is a convex polytope; and*
2. *When the coefficients are taken in $\mathbb{Z}_2$, components Φ_X of the moment map are perfect Morse functions on M^τ, in the sense of Bott, and for generic components the critical set is $M^\tau \cap M^T$.*

We have the following immediate corollary, a real locus version of equation (1), that generalizes classical results on real projective space and real flag varieties.

Corollary 1.3. *If M is a compact Hamiltonian T-manifold, and τ a smooth compatible antisymplectic involution, then*

$$H^*(M^\tau;\mathbb{Z}_2) = \sum_{i=1}^{N} H^{*-\frac{d_i}{2}}((F_i)^\tau;\mathbb{Z}_2). \tag{2}$$

where the F_i are the connected components of M^T and d_i is the Morse–Bott index of F_i (in M).

Duistermaat's work began a flurry of activity on properties of real loci. We provide a brief account here; a more detailed record is available in [Sj]. Davis and Januszkiewicz studied the real loci of toric and quasitoric varieties in their own right [DJ], independent of Duistermaat's work. The first author and Knutson analyze a large class of examples of real loci in their account of planar and spatial polygon spaces [HK]. O'Shea and Sjamaar generalized Kirwan's nonabelian convexity results to real flag manifolds and real loci [OS]. This has recently been extended by Goldberg [Go].

Schmid and independently Biss, Guillemin, and the second author generalized (2) to the equivariant setting: the idempotents $T_2 = \{t \in T \mid t^2 = 1\}$ act on the real locus, and many results in T-equivariant symplectic geometry may be generalized to T_2 equivariant geometry of real loci (with coefficients restricted to $\mathbb{Z}_2$) [BGH, Sd]. This work yields an explicit description of the T_2-equivariant cohomology for the fixed set of the Chevalley involution on certain coadjoint orbits, and on the real locus of a toric variety, using localization methods. These results were strengthened to include the fixed set of the Chevalley involution on all coadjoint orbits in [HHP].

Following this, Goldin and the second author [GH] proved that there is a natural involution on an abelian symplectic reduction of a symplectic manifold with involution. Moreover, the T_2 equivariant cohomology of the original real locus surjects onto the ordinary cohomology of the real locus of the symplectic reduction. This includes a comprehensive description of toric varieties and their real loci from yet a third perspective.

In all of these papers, a common theme is that there is a degree-halving isomorphism

$$H^{2*}(M;\mathbb{Z}_2) \to H^*(M^\tau;\mathbb{Z}_2).$$

As we now describe, this can be seen as part of a purely topological framework, that of a *conjugation space*, introduced in [HHP]. The remainder of this article is organized as follows. In Section 2, we review the definitions and properties of conjugation spaces. Our main theorem gives a criterion for recognizing when a topological space is a conjugation space; this is stated in Section 3, along with two noteworthy corollaries. We then prove some basic facts in Section 4, and prove the main theorems in Section 5.

Acknowledgments: The authors would like to thank Volker Puppe for many useful discussions, in particular regarding Lemmas 4.7 and 4.8. We are grateful to the referee for a very careful reading and valuable suggestions for improvements and simplifications of the first version of this paper.

Note. For the remainder of the paper, the cohomology is taken with coefficients in the field $\mathbb{Z}_2$: $H^*(X) = H^*(X;\mathbb{Z}_2)$.

2 A review of conjugation spaces

Let X be a G-space for a topological group G. The equivariant cohomology $H^*_G(X)$ is defined as the (singular) cohomology of the Borel construction:

$$H^*_G(X) = H^*(X \times_G BG).$$

Hence, $H^*_G(X)$ is a $H^*(BG)$-algebra. When $G = C$ is the group of order two, $BC = \mathbb{R}P^\infty$ and $H^*(BC) = \mathbb{Z}_2[u]$, with u a class in degree 1. Thus, $H^*_C(X)$ is a $\mathbb{Z}_2[u]$-algebra.

Let τ be a continuous involution on a space X. Let $\rho\colon H^{2*}_C(X) \to H^{2*}(X)$ and $r\colon H^*_C(X) \to H^*_C(X^\tau)$ be the restriction homomorphisms, where $C = \{\mathrm{id}, \tau\}$.

A *cohomology frame* or an *H^*-frame* for X is a pair (κ, σ), where

(a) $\kappa\colon H^{2*}(X) \to H^*(X^\tau)$ is an additive isomorphism dividing the degrees in half; and
(b) $\sigma\colon H^{2*}(X) \to H^{2*}_C(X)$ is an additive section of ρ.

Moreover, κ and σ must satisfy the *conjugation equation*

$$r \circ \sigma(a) = \kappa(a)u^m + \ell t_m \qquad (3)$$

for all $a \in H^{2m}(X)$ and all $m \in \mathbb{N}$, where ℓt_m denotes any polynomial in the variable u of degree less than m. An involution admitting a H^*-frame is called a *conjugation*. An even cohomology space (i.e. $H^{odd}(X) = 0$) together with a conjugation is called a *conjugation space*. Conjugation spaces were introduced in [HHP] and studied further in [FP] and [Ol]. The main examples of conjugations are given by the complex conjugation on flag manifolds, the Chevalley involution on coadjoint orbits of compact Lie groups, and other natural involutions, e.g., on toric manifolds or polygon spaces. We now enumerate some important properties of conjugation spaces.

(a) If (κ, σ) is an H^*-frame, then κ and σ are ring homomorphisms [HHP, Theorem 3.3]. The ring homomorphism κ also commutes with the Steenrod squares: $\kappa \circ \mathrm{Sq}^{2i} = \mathrm{Sq}^i \circ \kappa$, [FP, Theorem 1.3].
(b) H^*-frames are natural for τ-equivariant maps [HHP, Proposition 3.11]. In particular, if an involution admits an H^*-frame, it is unique [HHP, Corollary 3.12].
(c) For a conjugate-equivariant complex vector bundle η ("real bundle" in the sense of Atiyah) over a conjugation space X, the isomorphism κ sends the total Chern class of η to the total Stiefel–Whitney class of its fixed bundle.

Duistermaat's Corollary 1.3 admits the following generalization, proved in [HHP, Theorem 8.3].

Theorem 2.1. *Let M be a compact symplectic manifold equipped with a Hamiltonian action of a torus T and a compatible smooth antisymplectic involution τ. If M^T is a conjugation space, then M is a conjugation space.*

The proof of Theorem 2.1 involves properties of conjugations compatible with T-actions which are interesting in their own right. The involution $g \mapsto g^{-1}$ on the torus T induces an involution on ET. Using this involution together with τ, we get an involution on $X \times ET$ which descends to an involution, still called τ, on X_T. To a torus T is associated its 2-torus, i.e., the set of idempotent elements of T:

$$T_2 = \{g \in T \,|\, g^2 = 1\}.$$

The compatibility implies that T_2 acts on X^τ. The following lemma is proved in [HHP, Lemma 7.3].

Lemma 2.2. $(X_T)^\tau = (X^\tau)_{T_2}$.

The following theorem is proved in [HHP, Theorem 7.5]. For a partial converse of Theorem 2.3, see Proposition 5.2 in Section 5.

Theorem 2.3. *Let X be a conjugation space together with a compatible action of a torus T. Then the involution induced on X_T is a conjugation.*

Using Lemma 2.2, one gets the following corollary of Theorem 2.3.

Corollary 2.4. *Let X be a conjugation space together with a compatible T-action. Then there is a ring isomorphism*

$$\bar{\kappa}\colon H_T^{2*}(X) \xrightarrow{\approx} H_{T_2}^*(X^\tau).$$

3 Conjugation spaces and 1-skeleta

The new results in this paper consist of criteria to verify that an involution τ is a conjugation, in the case that τ is compatible with an action of a torus T. They are conditions on the equivariant 1-skeleton of the action of T on X and of the inherited action of the associated 2-torus T_2.

Let X be a topological space, together with a continuous action of a group G, where G is a torus or a 2-torus (finite elementary abelian 2-group). We define the *G-equivariant i-skeleton* $\mathrm{Sk}_i^G(X)$ *of the G-action on X* to be

$$\mathrm{Sk}_i^G(X) = \{x \in X | \mathrm{codim}\,(G_x \subset G) \le i\}, \tag{4}$$

where G_x denotes the G-isotropy group of x. In (4), the "codimension" is interpreted as the codimension of a manifold if G is a torus, and the codimension of a $\mathbb{Z}_2$-vector subspace if G is a 2-torus (and hence isomorphic to a $\mathbb{Z}_2$-vector space). In particular, $\mathrm{Sk}_0^G(X)$ is equal to the subspace X^G of fixed points. An *edge* (of the G-action) is the closure of a connected component of the set $\mathrm{Sk}_1^G(X) - \mathrm{Sk}_0^G(X)$. The word "edge" is inspired by Hamiltonian geometry: if X is a closed Hamiltonian T-manifold, the edges are critical points of the moment map whose images are 1-dimensional faces of the moment polytope (including the so-called *internal edges* of the polytope).

Let T be a torus and T_2 the subgroup of idempotents. A T-action on a space X induces a T_2-action on X that satisfies $\mathrm{Sk}_i^T(X) \subset \mathrm{Sk}_i^{T_2}(X)$. For example, $X^T \subset X^{T_2}$.

A continuous action of a topological group G on a space X is called *good* if X has the G-equivariant homotopy type of a finite G-CW-complex. For instance, a smooth action of a compact Lie group on a closed manifold is good. A continuous involution τ is called *good* if the corresponding action of the cyclic group $C = \{\mathrm{id}, \tau\}$ is good.

Let X be a topological space, and let τ be a continuous involution on X that is compatible with a continuous action of a torus T. Then the involution τ preserves the T-equivariant skeleta and sends each edge to a (possibly different) edge. Moreover, the real locus $X^\tau = X^C$ inherits an action of T_2. Our main results are the following.

Theorem 3.1 (Main theorem). *Let X be an even cohomology space, together with a good involution τ which is compatible with a good action of a torus T. Suppose that*

(a) (X^T, τ) *is a conjugation space.*
(b) each edge of the T*-action is preserved by* τ *and is a conjugation space.*
(c) $\mathrm{Sk}_i^T(X) = \mathrm{Sk}_i^{T_2}(X)$ *for* $i = 0, 1$.

Then X *is a conjugation space.*

Recall that a T-action on a space X is called a *GKM* action if each edge is a 2-sphere on which T acts by rotation around some axis, via a nontrivial character $T \to S^1$. One consequence of this assumption is that X^T is discrete.

Corollary 3.2. *Let* X *be an even cohomology space, together with a good involution* τ *which is compatible with a good GKM action of a torus* T*, satisfying* $\mathrm{Sk}_i^T(X) = \mathrm{Sk}_i^{T_2}(X)$ *for* $i = 0, 1$. *Suppose that* τ *acts trivially on* X^T *and preserves each edge. Then* X *is a conjugation space.*

Corollary 3.3. *Let* X *be an even cohomology space, together with a good involution* τ *which is compatible with a good action of a torus* T*, satisfying* $\mathrm{Sk}_i^T(X) = \mathrm{Sk}_i^{T_2}(X)$ *for* $i = 0, 1$. *Suppose that*

(a) (X^T, τ) *is a conjugation space.*
(b) each edge of the T*-action is preserved by* τ *and is a Hamiltonian* T*-manifold on which* τ *acts smoothly and is antisymplectic.*

Then X *is a conjugation space.*

See Section 6 for comments about the condition $\mathrm{Sk}_i^T(X) = \mathrm{Sk}_i^{T_2}(X)$ for $i = 0, 1$.

4 Preliminaries

This section is devoted to the technical details that we will need in the proof of Theorem 3.1.

4.1 Compatibility

Let X be a topological space endowed with a continuous involution τ which is compatible with a continuous action of a torus T. Then the involution τ induces an involution on the fixed-point set X^T. In addition, the associated 2-torus T_2 of T acts on X^τ and $X^\tau \cap X^T \subset (X^\tau)^{T_2}$. Condition (c) of Theorem 3.1 will play an important role.

Lemma 4.1. *Suppose that* $\mathrm{Sk}_i^T(X) = \mathrm{Sk}_i^{T_2}(X)$. *Then* $\mathrm{Sk}_i^T(X)^\tau = \mathrm{Sk}_i^{T_2}(X^\tau)$.

Proof. One has

$$\mathrm{Sk}_i^T(X)^\tau = \mathrm{Sk}_i^T(X) \cap X^\tau \subset \mathrm{Sk}_i^{T_2}(X^\tau) = X^\tau \cap \mathrm{Sk}_i^{T_2}(X) = X^\tau \cap \mathrm{Sk}_i^T(X), \quad (5)$$

which implies that $\mathrm{Sk}_i^T(X)^\tau = \mathrm{Sk}_i^{T_2}(X^\tau)$.

4.2 Equivariantly formal spaces

Let X be a space with a continuous action of a compact Lie group G. The terminology *equivariant formality* was first introduced in [GKM] for G a torus and complex coefficients, and was later developed for other coefficients where the concept is seen to be rather subtle; see [HHP, (2.3)] and [Fr, Section 8]. We define a G-space X to be *G-equivariantly formal (over $\mathbb{Z}_2$)* if the map $X \to EG \times_G X$ is totally nonhomologous to zero. This means that the restriction homomorphism $j^*\colon H_G^*(X) \to H^*(X)$ is surjective. A space X with an involution τ is called *τ-equivariantly formal* if it is C-equivariantly formal for the group $C = \{\mathrm{id}, \tau\}$. The following results are classical but may not perhaps be found in the literature with exactly our hypotheses. Let $R = H_G^*(pt)$; the map $X_G \to BG$ gives a ring homomorphism $p^*\colon R \to H_G^*(X)$, making $H_G^*(X)$ an R-module.

Proposition 4.2. *Let X be a G-space. The following conditions are equivalent:*

(i) X is G-equivariantly formal (over $\mathbb{Z}_2$).
(ii) The group G acts trivially on $H^*(X)$ and the Serre spectral sequence for the cohomology of the fibration $X \to EG \times_G X \to BG$ collapses at the term E_2.
(iii) There is an additive homomorphism $\sigma\colon H^*(X) \to H_G^*(X)$ that satisfies $j^*\circ\sigma = \mathrm{id}$ and that $p^* \otimes \sigma\colon R \otimes H^*(X) \to H_G^*(X)$ is an isomorphism of R-modules. In particular, $H_G^*(X)$ is a free R-module.
(iv) The ring homomorphism $H_G^*(X) \to H^*(X)$ descends to a ring isomorphism $H_G^*(X) \otimes_R \mathbb{Z}_2 \xrightarrow{\approx} H^*(X)$.

Proof. This proof is for mod-2 cohomology, but it works for the cohomology with coefficients in any field.

(i) is equivalent to (ii): The ring homomorphism $j^*\colon H_G^*(X) \to H^*(X)$ is the composition:

$$H_G^*(X) \twoheadrightarrow E_\infty^{0,*} \subset E_2^{0,*} = H^0(BG; H^*(X)) = H^*(X)^G \subset H^*(X). \quad (6)$$

If these inclusions are equalities, then j^* is onto, which shows that (ii) implies (i). Conversely, if j^* is onto, this shows that $H^*(X)^G = H^*(X)$ and $E_\infty^{0,*} = E_2^{0,*}$. Since the differentials are morphisms of R-modules, this implies that $E_\infty^{*,*} = E_2^{*,*}$ (see [McC, p. 148]). Hence (i) implies (ii).

(i) implies (iii): Since j^* is surjective, there exists a $\mathbb{Z}_2$-linear section σ of j^*. Since G is a compact Lie group, $H_G^p(pt)$ is a finite-dimensional $\mathbb{Z}_2$-vector space for all p. The Leray–Hirsch theorem [McC, Theorem 5.9] then implies that

$$p^* \otimes \sigma : R \otimes H^*(X) \to H_G^*(X)$$

is an isomorphism of R-modules. Note that this immediately implies that $H_G^*(X)$ is a free R-module.

(iii) implies (iv): The homomorphism $j^* \circ p^* : R \to H^*(X)$ coincides with the projection $R \otimes_R \mathbb{Z}_2 = \mathbb{Z}_2$. Therefore, j^* factors through a ring homomorphism $\bar{j}^* : H_G^*(X) \otimes_R \mathbb{Z}_2 \to H^*(X)$. On the other hand, $j^* \circ \sigma = \mathrm{id}$. Hence, one has a commutative diagram

$$\begin{array}{ccccc} R \otimes H^*(X) & \longrightarrow & (R \otimes H^*(X)) \otimes_R \mathbb{Z}_2 & = & H^*(X) \\ p^*{\otimes}\sigma \downarrow\approx & & \downarrow\approx & & \downarrow \mathrm{id} \\ H_G^*(X) & \longrightarrow & H_G^*(X) \otimes_R \mathbb{Z}_2 & \xrightarrow{\bar{j}^*} & H^*(X), \end{array}$$

which proves that $\bar{j}^*$ is an isomorphism.

(iv) implies (i): This implication is trivial.

Proposition 4.3. *Let X be a good G-space which is G-equivariantly formal (over $\mathbb{Z}_2$). Suppose that one of the following hypotheses holds:*

(a) G is a torus and $X^G = X^{G_2}$.
(b) G is a 2-torus.

Then the restriction homomorphism $H_G^*(X) \to H_G^*(X^G)$ is injective.

Remark 4.4. In Case (a), Proposition 4.3 is false without the assumption $X^G = X^{G_2}$. For example, consider the $G = S^1$ action on $X = S^2 \subset \mathbb{C} \times \mathbb{R}$ by $g(z,t) = (g^2 z, t)$. This has $X^G = \{(0, \pm 1)\}$. Let $U_+ = X - \{(0,-1)\}$ and $U_- = X - \{(0,-1)\}$. The intersection $U_+ \cap U_-$ is G-homotopy equivalent to the homogeneous space G/G_2, and then $H_G^*(U_+ \cap U_-) = H^*(BG_2)$. The Mayer–Vietoris sequence for (X, U_+, U_-) then gives an exact sequence

$$0 \to H^1(BG_2) \to H_G^2(X) \to H_G^2(X^G),$$

where $H^1(BG_2) = \mathbb{Z}_2$. Proposition 4.3 (a) also follows from [FP3, Theorem 2.1].

Proof of Proposition 4.3: Let $R_{(0)}$ be the field of fractions of R, that is, R localized at $S = R - \{0\}$. By our assumptions, the multiplicative set S is central in R. Let

$$X^S = \{x \in X \mid H^*(BG) \to H^*(BG_x) \text{ is injective}\},$$

where G_x is the isotropy group of x. The localization theorem ([AP, Theorem 3.1.6], [Al, Theorem 3.1.7]) asserts that the inclusion $X^S \subset X$ induces an isomorphism of $R_{(0)}$-vector spaces

$$S^{-1}H_G^*(X) \approx S^{-1}H_G^*(X^S). \tag{7}$$

In Case (b), if G_x is a proper subgroup of G, then $H^2(BG) \to H^2(BG_x)$ is not injective; hence $X^S = X^G$. For Case (a), we use that for each $x \in X$, there is an isomorphism $\psi_x : G \xrightarrow{\approx} (S^1)^m$ such that $\psi_x(G_x) = C_1 \times \cdots \times C_m$, where C_j is a subgroup of S^1. In order to have $H^2(BG) \to H^2(BG_x)$ injective, each C_j should be either S^1 or a finite cyclic group of even order. Then $X^G \subseteq X^S \subseteq X^{G_2} = X^G$. In all cases, we have that

$$S^{-1}H_G^*(X) \to S^{-1}H_G^*(X^G) \tag{8}$$

is an isomorphism. Therefore, $\ker(H_G^*(X) \to H_G^*(X^G))$ is the R-torsion of $H_G^*(X)$. But the R-torsion vanishes because $H_G^*(X)$ is a free R-module by Proposition 4.2.

Proposition 4.5. *Let X be a good G-space. Suppose that Condition (a) or (b) of Proposition 4.3 is satisfied. Then $\dim_{\mathbb{Z}_2} H^*(X) \geq \dim_{\mathbb{Z}_2} H^*(X^G)$ and X is G-equivariantly formal if and only if equality holds.*

Proof. As in the proof of Proposition 4.3, consider $S = R - \{0\}$ and $R_{(0)} = S^{-1}R$. We apply S^{-1} to the terms of the Serre spectral sequence, following [Al, proof of Corollary 3.10]. When G is a torus, it acts trivially on $H^*(X)$, which implies that $E_2^{*,*} \approx H^*(BG; H^*(X))$ as an R-module and there is an isomorphism of $R_{(0)}$-vector spaces $R_{(0)} \otimes_{\mathbb{Z}_2} H^*(X) \approx S^{-1}E_2$. Therefore, using equation (8), we get

$$\begin{aligned}
\dim_{\mathbb{Z}_2} H^*(X) &= \dim_{R_{(0)}}(S^{-1}E_2) \\
&\geq \dim_{R_{(0)}}(S^{-1}E_\infty) \\
&= \dim_{R_{(0)}}(S^{-1}H_G^*(X)) \\
&= \dim_{R_{(0)}}(S^{-1}H_G^*(X^G)) \\
&= \dim_{\mathbb{Z}_2}(H^*(X^G)).
\end{aligned} \tag{9}$$

By Proposition 4.2, the inequality in (9) is an equality if and only if X is equivariantly formal. Finally, when G is a 2-torus, Proposition 4.5 follows from [AP, Theorem 3.10.4].

Lemma 4.6. *Let X be a topological space, together with a good involution τ which is compatible with a good action of a torus T. Suppose that X is T-equivariantly formal, X^T is τ-equivariantly formal, and $X^T = X^{T_2}$. Then X is τ-equivariantly formal and X^τ is T_2-equivariantly formal.*

Proof. In what follows, dim denotes $\dim_{\mathbb{Z}_2}$. By our hypotheses, Proposition 4.5, and Lemma 4.1, we have

$$\dim H^*(X) = \dim H^*(X^T) = \dim H^*((X^T)^\tau)$$
$$= \dim H^*((X^\tau)^{T_2}) \le \dim H^*(X^\tau) \le \dim H^*(X),$$

which implies that

$$\dim H^*(X^\tau) = \dim H^*(X) \quad \text{and} \quad \dim H^*((X^\tau)^{T_2}) = \dim H^*(X^\tau). \qquad (10)$$

Lemma 4.6 thus follows from Proposition 4.5.

4.3 The image of the 1-skeleton

We shall need the following two lemmas, first proved by Chang and Skjelbred for rational cohomology and torus action [CS].

Lemma 4.7. *Let X be a topological space endowed with a good action of a 2-torus G. Suppose that X is G-equivariantly formal. Then the restriction homomorphisms on the mod-2 cohomology, $H^*_G(X) \to H^*_G(X^G)$ and $H^*_G(\mathrm{Sk}^G_1(X)) \to H^*_G(X^G)$, have same image.*

Proof. Using the equivalence (i) $\Leftrightarrow$ (iii) in Lemma 4.2, we know that $H^*_G(X)$ is a free $H^*_G(pt)$-module. By [Hs, Corollary, p. 63], the homomorphism

$$H^*_G(X, X^G) \to H^*_G(\mathrm{Sk}^G_1(X), X^G)$$

is injective. The H^*_G-sequences of the pairs (X, X^G) and $(\mathrm{Sk}^G_1(X), X^G)$ are part of a commutative diagram

$$\begin{array}{ccccccccc}
0 & \longrightarrow & H^*_G(X) & \longrightarrow & H^*_G(X^G) & \longrightarrow & H^{*+1}_G(X, X^G) & \longrightarrow & 0 \\
 & & \downarrow & & \downarrow = & & \downarrow & & \\
 & & H^*_G(\mathrm{Sk}^G_1(X)) & \longrightarrow & H^*_G(X^G) & \longrightarrow & H^{*+1}_G(\mathrm{Sk}^G_1(X), X^G) & \longrightarrow & 0.
\end{array}$$

Therefore, the injectivity of the last vertical arrow implies the lemma.

The following lemma follows from [FP3, Theorem 2.1].

Lemma 4.8. *Let X be a topological space endowed with a good action of a torus T. Suppose that X is T-equivariantly formal and that $\mathrm{Sk}^T_i(X) = \mathrm{Sk}^{T_2}_i(X)$ for $i = 0, 1$. Then the restriction homomorphisms on the mod-2 cohomology, $H^*_T(X) \to H^*_T(X^T)$ and $H^*_T(\mathrm{Sk}^T_1(X)) \to H^*_T(X^T)$, have same image.*

5 Proof of the main results

The main theorem will follow from two propositions.

Proposition 5.1. *Let X be an even cohomology space, together with a good involution τ which is compatible with a good action of a torus T. Suppose that Conditions (a), (b), and (c) of the main theorem are satisfied. Then X_T is a conjugation space.*

Proof. The proof is decomposed into three steps.

Step 1. X_T is τ-equivariantly formal. For G a topological group and $k \in \mathbb{N}$, we consider the G-principal bundle $G \to E_k G \to B_k G$ obtained as the kth step in the Milnor construction. If X is a G-space, the associated bundle with fibre X gives a bundle $X \to X_{G,k} \to B_k G$, where $X_{G,k} = E_k G \times_G X$.

For a torus T of dimension n, $B_k T \approx (\mathbb{C}P^k)^n$. The involution $\tau(g) = g^{-1}$ on T gives an involution τ on $B_k T$ which makes $B_k T$ a conjugation space with $(B_k T)^\tau \approx (\mathbb{R}P^k)^n \approx B_k T_2$.

We first prove that $X_{T,k}$ is τ-equivariantly formal. Since X and $B_k T$ are even cohomology spaces, the spectral sequence of $X \to X_{T,k} \to B_k T$ degenerates at the E^2-term and $H^*(X_{T,k}) \approx H^*(X) \otimes H^*(B_k T)$. As a consequence, $\dim H^*(X_{T,k}) = \dim H^*(X) \cdot \dim H^*(B_k T) < \infty$. We have assumed that X is τ-equivariantly formal, and the fact that X is an even cohomology space implies that X is T-equivariantly formal. So we may apply Lemma 4.6 to deduce

$$\begin{aligned} \dim H^*(X_{T,k}) &= \dim H^*(X) \cdot \dim H^*(B_k T) \\ &= \dim H^*(X^\tau) \cdot \dim H^*(B_k T_2). \end{aligned} \tag{11}$$

Lemma 4.6 also implies that X^τ is T_2-equivariantly formal, and so we have the following commutative diagram:

$$\begin{array}{ccc} H^*((X^\tau)_{T_2}) & \xrightarrow{\rho^\tau_{T_2}} & H^*(X^\tau) \\ \downarrow & & \downarrow = \\ H^*((X^\tau)_{T_2,k}) & \xrightarrow{\rho^\tau_{T_2,k}} & H^*(X^\tau) \end{array}$$

which shows that $\rho^\tau_{T_2,k}$ is surjective and thus $H^*((X^\tau)_{T_2,k}) \approx H^*(X^\tau) \otimes H^*(B_k T_2)$. As in Lemma 2.2, we may prove that $(X_{T,k})^\tau = (X^\tau)_{T_2,k}$, and so

$$\dim H^*((X_{T,k})^\tau) = \dim H^*((X^\tau)_{T_2,k}) = \dim H^*(X^\tau) \cdot \dim H^*(B_k T_2). \tag{12}$$

Combining (11) and (12) gives $\dim H^*(X_{T,k}) = \dim H^*((X_{T,k})^\tau)$, and together with Proposition 4.5, this implies that $X_{T,k}$ is equivariantly formal.

Now given $n \in \mathbb{N}$, there exists $k \in \mathbb{N}$ such that $H^n(X_T) \approx H^n(X_{T,k})$. The commutative diagram

$$\begin{array}{ccc} H^n(X_T) & \xrightarrow{\rho} & H^n(X) \\ \downarrow \approx & & \downarrow = \\ H^n(X_{T,k}) & \xrightarrow{\rho_k} & H^n(X) \end{array}$$

shows that ρ is surjective in degree n. This can be done for each n, so X_T is equivariantly formal.

Step 2. Construction of the ring isomorphism $\kappa_T\colon H^{2*}(X_T) \to H^*((X_T)^\tau)$. By Lemma 2.2, it is equivalent to construct a ring isomorphism

$$\kappa_T\colon H_T^{2*}(X) \to H_{T_2}^*(X^\tau).$$

By Corollary 2.4, such an isomorphism $\kappa_{\mathrm{fix}}\colon H_T^{2*}(X^T) \to H_{T_2}^*((X^T)^\tau)$ exists, since X^T is a conjugation space. Since $(X^T)^\tau = (X^\tau)^{T_2}$ by Lemma 4.1, we may view κ_{fix} as a map from $H_T^{2*}(X^T)$ to $H_{T_2}^*((X^\tau)^{T_2})$. Let us consider the following diagram:

$$\begin{array}{ccc} H_T^{2*}(X) & \xrightarrow{q} & H_T^{2*}(X^T) \\ & & \approx \downarrow \kappa_{\mathrm{fix}} \\ H_{T_2}^*(X^\tau) & \xrightarrow{q^\tau} & H_{T_2}^*((X^\tau)^{T_2}). \end{array} \tag{13}$$

By Proposition 4.3, the restriction homomorphisms q and q^τ are injective. Therefore, in order to construct $\kappa_T\colon H_T^{2*}(X) \to H_{T_2}^*(X^\tau)$, it is enough to show that $A^\tau = \kappa_{\mathrm{fix}}(A)$, where $A = \mathrm{image}(q)$ and $A^\tau = \mathrm{image}(q^\tau)$.

Let N be the 1-skeleton of X. Enumerate the edges $E_1, \dots, E_n$ of X and define

$$\hat{E}_i = E_i \cup X^T = E_i \sqcup (X^T - E_i^T). \tag{14}$$

For $1 \le k \le n$, define $N_k = \hat{E}_1 \cup \cdots \cup \hat{E}_k$. Thus $N_k \cap E_{k+1} = X^T$, and the (equivariant) Mayer–Vietoris sequence fits into a commutative diagram:

$$\begin{array}{ccccccccc} & & H_T^*(N_{k+1}) & \longrightarrow & H_T^*(N_k) \oplus H_T^*(\hat{E}_{k+1}) & \longrightarrow & H_T^*(X^T) & & \\ & & \downarrow & & \downarrow & & \downarrow \approx & & \\ 0 & \longrightarrow & H_T^*(X^T) & \xrightarrow{\mathrm{diag}} & H_T^*(X^T) \oplus H_T^*(X^T) & \xrightarrow{+} & H_T^*(X^T) & \longrightarrow & 0 \end{array}$$

The bottom row is clearly exact (we do not need the usual sign because of the $\mathbb{Z}_2$ coefficients). We may conclude that the image of $H_T^*(N_{k+1})$ into $H_T^*(X^T)$ is the intersection of the image of $H_T^*(N_k)$ with that of $H_T^*(\hat{E}_{k+1})$. Since $N = N_n$, this, together with Lemma 4.8, shows that

$$\mathrm{Image}(H^*_T(N) \to H^*_T(X^T)) = \bigcap_{i=1}^{n} \mathrm{Image}(H^*_T(\hat{E}_i) \to H^*_T(X^T)). \qquad (15)$$

In the same way, using Lemma 4.7, we have that

$$\mathrm{Image}(H^*_{T_2}(N^\tau) \to H^*_{T_2}((X^\tau)^{T_2})) = \bigcap_{i=1}^{n} \mathrm{Image}(H^*_{T_2}(\hat{E}_i^\tau) \to H^*_{T_2}((X^\tau)^{T_2})). \quad (16)$$

By naturality, an H^*-frame preserves the connected components. Since X^T is a conjugation space, so is any union of its connected components. The disjoint union decomposition of (14) implies that $\hat{E}_i$ is a conjugation space for each i. Hence,

$$\kappa_{\mathrm{fix}}(\mathrm{Image}(H^*_T(\hat{E}_i) \to H^*_T(X^T))) = \mathrm{Image}(H^*_{T_2}(\hat{E}_i^\tau) \to H^*_{T_2}((X^\tau)^{T_2})).$$

Using (15) and (16), this implies that $A^\tau = \kappa_{\mathrm{fix}}(A)$.

Step 3. Construction of a section $\sigma_T \colon H^*(X_T) \to H^*_C(X_T)$ so that (κ_T, σ_T) is an H^*-frame for (X_T, τ). Let $(\kappa_{\mathrm{fix}}, \sigma_{\mathrm{fix}})$ be the H^*-frame for X^T. The desired section σ_T will fit in the commutative diagram

$$\begin{array}{ccccccc}
H^*(X_T) & \underset{\sigma_T}{\overset{\rho_T}{\leftrightarrows}} & H^*_C(X_T) & \overset{r_T}{\rightarrowtail} & H^*_C((X_T)^\tau) & \overset{\approx}{\longleftarrow} & H^*((X_T)^\tau)[u] \\
\downarrow q & & \downarrow q_C & & \downarrow q_C^\tau & & \downarrow q^\tau[u] \\
H^*((X^T)_T) & \underset{\sigma_{\mathrm{fix}}}{\overset{\rho_{\mathrm{fix}}}{\leftrightarrows}} & H^*_C((X^T)_T) & \overset{r_{\mathrm{fix}}}{\rightarrowtail} & H^*_C(((X^T)_T)^\tau) & \overset{\approx}{\longleftarrow} & H^*(((X^T)_T)^\tau)[u]
\end{array}$$

where the vertical arrows are induced by the inclusion $X^T \hookrightarrow X$ (the notation coincides with that of diagram (13)). We must justify that the last two vertical arrows are injective. But using the identifications

$$H^*((X_T)^\tau) = H^*((X^\tau)_{T_2}) = H^*_{T_2}(X^\tau)$$

and

$$H^*(((X^T)_T)^\tau) = H^*(((X^T)^\tau)_{T_2}) = H^*_{T_2}((X^T)^\tau) = H^*_{T_2}((X^\tau)^{T_2}),$$

the map $q^\tau[u]$ coincides with the homomorphism $H^*_{T_2}(X^\tau) \to H^*_{T_2}((X^\tau)^{T_2})$ induced by the inclusion $(X^\tau)^{T_2} \hookrightarrow X^\tau$.

Note that we just need to construct a section $\sigma_T \colon H^*(X_T) \to H^*_C(X_T)$ such that $q_C \circ \sigma_T = \sigma_{\mathrm{fix}} \circ q$. Indeed, if $a \in H^{2m}(X_T)$, the conjugation equation for $(\kappa_{\mathrm{fix}}, \sigma_{\mathrm{fix}})$ implies

$$q_C^\tau \circ r_T \circ \sigma_T(a) = r_{\mathrm{fix}} \circ \sigma_{\mathrm{fix}} \circ q(a) = \kappa_{\mathrm{fix}} \circ q(a) u^m + \ell t_m.$$

Since q_C^τ is injective, this implies that

$$r_T \circ \sigma_T(a) = \tilde{a} u^m + \ell t_m,$$

with $\tilde{a} \in H^m((X_T)^\tau)$ satisfying $q^\tau(\tilde{a}) = \kappa_{\text{fix}}(a)$. By construction of κ_T, one has $q^\tau \circ \kappa_T(a) = \kappa_{\text{fix}} \circ q(a)$. Since q^τ is injective, this implies that $\tilde{a} = \kappa_T(a)$. Hence the pair (κ_T, σ_T) automatically satisfies the conjugation equation and is therefore an H^*-frame.

We construct an additive section σ_T inductively. Consider the statement $\mathcal{H}_m$: *for $k \le m$, there exists a section $\sigma_T : H^{2k}(X_T) \to H_C^{2k}(X_T)$ of ρ_T such that $q_C \circ \sigma_T = \sigma_{\text{fix}} \circ q$*. Statement $\mathcal{H}_0$ is clearly satisfied: we may assume without loss of generality that X is arc-connected; and we may then define $\sigma_T(1) = 1$, where $1 \in H^0(-)$ is the unit of $H^*(-)$. Now assume by induction that $\mathcal{H}_{m-1}$ holds. The space X_T is τ-equivariantly formal by Step 1, so there exists a section $\sigma_0 : H^{2m}(X_T) \to H_C^{2m}(X_T)$ of ρ_T. We have $\rho_{\text{fix}} \circ q_C \circ \sigma_0 = q$. Therefore, for any $a \in H^{2m}(X_T)$, we know that $q_C \circ \sigma_0(a) \equiv \sigma_{\text{fix}} \circ q(a)$ modulo $\ker \rho_{\text{fix}}$. This kernel is the ideal generated by u. Since $H_C^{\text{odd}}((X^T)_T) = 0$, only even powers of u occur and moreover

$$q_C \circ \sigma_0(a) = \sigma_{\text{fix}} \circ q(a) + \sum_{i=0}^{m} \sigma_{\text{fix}}(b_{2m-2i}) u^{2i}, \tag{17}$$

where b_{2j} are classes in $H^{2j}((X^T)_T)$ depending on the choice of σ_0. We will modify σ_0 by successive steps until $b_{2j} = 0$ for all $j = m, m-1, \ldots, 0$.

The conjugation equation for $(\kappa_{\text{fix}}, \sigma_{\text{fix}})$ implies

$$r_{\text{fix}} \circ q_C \circ \sigma_0(a) = \kappa_{\text{fix}} \circ q(a) u^m + \ell t_m(a) + \sum_{i=0}^{m} (\kappa_{\text{fix}}(b_{2m-2i}) u^{m+i} + \ell t_{m-i}(b_{2m-2i})). \tag{18}$$

Since q_C^τ is injective, this implies that

$$r_T \circ \sigma_0(a) = c_0 u^{2m} + \ell t_m, \tag{19}$$

with $c_0 \in H^0((X_T)^\tau)$ satisfying $q^\tau(c_0) = \kappa_{\text{fix}}(b_0)$. Since κ_T is an isomorphism, there exists $\tilde{c}_0 \in H^0(X_T)$ with $\kappa_T(\tilde{c}_0) = c_0$. Define a new section

$$\sigma_1 : H^{2m}(X_T) \to H_C^{2m}(X_T)$$

of ρ_T by $\sigma_1(a) = \sigma_0(a) + \sigma_T(\tilde{c}_0) u^{2m}$. By the induction hypothesis, $q_C \circ \sigma_T(\tilde{c}_0) = \sigma_{\text{fix}} \circ q(\tilde{c}_0)$. By construction of κ_T, one has $q^\tau \circ \kappa_T = \kappa_{\text{fix}} \circ q$. Therefore,

$$\begin{aligned} r_{\text{fix}} \circ q_C \circ \sigma_1(a) &= r_{\text{fix}} \circ q_C \circ \sigma_0(a) + r_{\text{fix}}(q_C \circ \sigma_T(\tilde{c}_0)) u^{2m} \\ &= r_{\text{fix}} \circ q_C \circ \sigma_0(a) + r_{\text{fix}}(\sigma_{\text{fix}} \circ q(\tilde{c}_0)) u^{2m} \\ &= r_{\text{fix}} \circ q_C \circ \sigma_0(a) + \kappa_{\text{fix}} \circ q(\tilde{c}_0)) u^{2m} \\ &= r_{\text{fix}} \circ q_C \circ \sigma_0(a) + q^\tau \circ \kappa_T(\tilde{c}_0)) u^{2m} \\ &= r_{\text{fix}} \circ q_C \circ \sigma_0(a) + q^\tau(c_0) u^{2m} \end{aligned}$$

$$
\begin{aligned}
&= r_{\mathrm{fix}} \circ q_C \circ \sigma_0(a) + \kappa_{\mathrm{fix}}(b_0) u^{2m} \\
&= \kappa_{\mathrm{fix}} \circ q(a) u^m + \ell t_m(a) \\
&\quad + \sum_{i=0}^{m-1} (\kappa_{\mathrm{fix}}(b_{2m-2i}) u^{m+i} + \ell t_{m-i}(b_{2m-2i})).
\end{aligned}
$$

The injectivity of r_{fix} implies that equation (17) is replaced by

$$q_C \circ \sigma_1(a) = \sigma_{\mathrm{fix}} \circ q(a) + \sum_{i=0}^{m-1} \sigma_{\mathrm{fix}}(b_{2m-2i}) u^{2i}. \tag{20}$$

We thus have modified σ_0 so that $b_0 = 0$. Now, using as above the injectivity of q_C^τ, this permits us to transform (19) into

$$r_T \circ \sigma_1(a) = c_1 u^{2m-1} + \ell t_{m-1}, \tag{21}$$

with $c_1 \in H^1((X_T)^\tau)$ satisfying $q^\tau(c_1) = \kappa_{\mathrm{fix}}(b_0)$. Again, write $c_1 = \kappa_T(\tilde{c}_1)$ with $\tilde{c}_1 \in H^2(X_T)$ and define a new section $\sigma_2 \colon H^{2m}(X_T) \to H_C^{2m}(X_T)$ of ρ_T by $\sigma_2(a) = \sigma_1(a) + \sigma_T(\tilde{c}_1) u^{2m-2}$. Proceeding as above, we prove that if we replace σ_1 by σ_2 in (20), the summation index runs only until $m-2$, i.e., $b_0 = b_2 = 0$. If we continue as long as possible, we finally get $\sigma_m \colon H^{2m}(X_T) \to H_C^{2m}(X_T)$ with $b_0 = b_2 = \cdots = b_{2m} = 0$. Extending σ_T in degree $2m$ by σ_m proves that the statement $\mathcal{H}_m$ holds. So by induction, we have constructed the desired section σ_T.

The second proposition is a partial converse of Theorem 2.3.

Proposition 5.2. *Let X be an even cohomology space, together with a good involution τ which is compatible with a good action of a torus T. Suppose that X^T is τ-equivariantly formal and that $X^T = X^{T_2}$. Then X is a conjugation space if and only if X_T is a conjugation space.*

Proof. The "only if" part is the content of Theorem 2.3. For the converse, let us assume that X_T is a conjugation space.

We first construct the ring isomorphism $\kappa \colon H^{2*}(X) \to H^*(X^\tau)$. We know that X is T-equivariantly formal because it is an even cohomology space. By part (iv) of Proposition 4.2, the ring epimorphism $\psi \colon H_T^{2*}(X) \twoheadrightarrow H^{2*}(X)$ descends to an isomorphism

$$H_T^{2*}(X) \otimes_{H_T^{2*}(pt)} \mathbb{Z}_2 \xrightarrow{\approx} H^{2*}(X). \tag{22}$$

By Lemma 4.6, X^τ is T_2-equivariantly formal, and so Proposition 4.2 again tells us that the ring epimorphism $\psi^\tau \colon H_{T_2}^*(X^\tau) \twoheadrightarrow H^*(X^\tau)$ descends to a graded ring isomorphism

$$H_{T_2}^*(X^\tau) \otimes_{H_{T_2}^*(pt)} \mathbb{Z}_2 \xrightarrow{\approx} H^*(X^\tau). \tag{23}$$

By Corollary 2.4, the ring isomorphism $\kappa_T \colon H_T^{2*}(X) \xrightarrow{\approx} H_{T_2}^*(X^\tau)$ is an isomorphism of modules over the ring isomorphism $H_T^{2*}(pt) \to H_{T_2}^*(pt)$. Therefore, it descends to a graded ring isomorphism $\kappa\colon H^{2*}(X) \xrightarrow{\approx} H^*(X^\tau)$. With this definition, the equation

$$\psi^\tau \circ \kappa_T = \kappa \circ \psi \tag{24}$$

is satisfied.

We now construct the section $\sigma: H^*(X) \to H^*_C(X)$ so that (κ, σ) is an H^*-frame. Consider the commutative diagram

$$\begin{array}{ccccc}
H^*(X_T) & \underset{\sigma_T}{\overset{\rho_T}{\rightleftarrows}} & H^*_C(X_T) & \xrightarrow{r_T} & H^*_C((X_T)^\tau) \\
s \uparrow \downarrow \psi & & \downarrow \psi_C & & \downarrow \psi^\tau_C \\
H^*(X) & \underset{\sigma}{\overset{\rho}{\rightleftarrows}} & H^*_C(X) & \xrightarrow{r} & H^*_C(X^\tau).
\end{array}$$

Since X is T-equivariantly formal, we may choose an additive section $s\colon H^*(X) \to H^*(X_T)$ of ψ and define $\sigma: H^*(X) \to H^*_C(X)$ by $\sigma = \psi_C \circ \sigma_T \circ s$. The linear map σ is an additive section of ρ, and for $a \in H^{2m}(X)$, we have

$$\begin{aligned}
r \circ \sigma(a) &= r \circ \psi_C \circ \sigma_T \circ s(a) \\
&= \psi^\tau_C \circ r_T \circ \sigma_T \circ s(a) \\
&= \psi^\tau_C(\kappa_T \circ s(a) u^m + \ell t_m) \\
&= (\psi^\tau_C \circ \kappa_T) \circ s(a)\, u^m + \ell t_m \\
&= (\kappa \circ \psi) \circ s(a)\, u^m + \ell t_m \\
&= \kappa(a) u^m + \ell t_m.
\end{aligned}$$

From the fourth to the fifth line, we have used the fact that $\psi^\tau_C \circ \kappa_T = \kappa \circ \psi$. This comes from equation (24), using that under the identifications $H^*_C((X_T)^\tau) = H^*((X_T)^\tau)[u]$ and $H^*_C(X^\tau) = H^*(X^\tau)[u]$, ψ^τ_C is the obvious polynomial extension of ψ^τ. Therefore, the conjugation equation is satisfied and (κ, σ) is an H^*-frame for X.

Proof of the main theorem: Theorem 3.1 follows directly from Propositions 5.1 and 5.2.

Proof of Corollary 3.2: The hypotheses imply that the restriction of τ to an edge E, which is a 2-sphere, is conjugate to a reflection (through an equatorial plane). This follows from the classical result that a continuous involution on S^2 is topologically conjugate to a linear one; see, e.g., [CK, Theorem 4.1]. Therefore, each edge is a conjugation 2-sphere in the sense of [HHP, Example 3.6]. Hence, each edge is a conjugation space, and the hypotheses of Theorem 3.1 are satisfied.

Proof of Corollary 3.3: By [HHP, Remark 3.1], τ preserves each arc-connected component of X^T. In consequence, for each edge E of X, hypothesis (a) of

Corollary 3.3 implies that E^T is a conjugation space. By Theorem 2.1, each edge is then a conjugation space. The hypotheses of Theorem 3.1 are therefore satisfied.

6 Remarks

6.1 A case when the skeleta differ

The following example shows that the condition $\mathrm{Sk}_0^T(X) = \mathrm{Sk}_0^{T_2}(X)$ does not imply that $\mathrm{Sk}_1^T(X) = \mathrm{Sk}_1^{T_2}(X)$, even for spaces like those occurring in Corollary 3.2 or 3.3. We consider the Hamiltonian action of S^1 on $S^2 \subset \mathbb{C}\times\mathbb{R}$ given by $g\cdot(z,t) = (gz,t)$, compatible with the involution $(z,t)^\tau = (\bar{z},t)$. Points of S^2 will be denoted by x, y, etc. Let $p_\pm = (0,\pm 1)$ be the north and south poles. Let $T = S^1 \times S^1$ act on $X = S^2 \times S^2$ by

$$(g,h)\cdot(x,y) = (gh\cdot x, gh^{-1}\cdot y).$$

The fixed-point sets for T and T_2 are equal, $\mathrm{Sk}_0^T(X)^\tau = \mathrm{Sk}_0^{T_2}(X^\tau)$, each consisting of the four points $\{p_\pm\}\times\{p_\pm\}$. By Proposition 4.5, X is T-equivariantly formal and X^τ is T_2-equivariantly formal.

The T-equivariant 1-skeleton is a graph of four 2-spheres

$$\mathrm{Sk}_1^T(X) = \{(x,y) | x = p_\pm \text{ or } y = p_\pm\}.$$

Therefore, X is a GKM space. On the other hand, $\mathrm{Sk}_1^T(X) \neq \mathrm{Sk}_1^{T_2}(X)$, since $\mathrm{Sk}_1^{T_2}(X) = X$. Moreover, $\mathrm{Sk}_1^T(X)^\tau \neq \mathrm{Sk}_1^{T_2}(X^\tau)$ because $\mathrm{Sk}_1^{T_2}(X^\tau) = X^\tau$. It would be interesting to know whether the conclusion of Lemma 4.8 holds for the T-space X (a question raised by the referee).

6.2 The relationship to previous work

The condition $\mathrm{Sk}_i^T(X) = \mathrm{Sk}_i^{T_2}(X)$ for $i = 0,1$ of our main theorems is already implicitly present in earlier papers [Sd, BGH] that deal with GKM Hamiltonian manifolds. In [Sd], one requires that for each point of $x \in X^T$, the characters involved in the 2-spheres adjacent to x are pairwise independent over $\mathbb{Z}_2$. In [BGH, p. 373], the authors require that $X^T = X^{T_2}$ and that "the real locus of the one-skeleton is the same as the one-skeleton of the real locus." In general, these conditions are weaker than $\mathrm{Sk}_i^T(X) = \mathrm{Sk}_i^{T_2}(X)$ for $i = 0,1$ (see Lemma 4.1), but they are equivalent hypotheses for Hamiltonian GKM manifolds. To see this, we may work with the local normal coordinates about a T-fixed point. In this model the T-action and the involution are linear.

References

[Al] Allday C. Notes on the localization theorem with applications to symplectic torus actions, Proceedings of the Winter School on Transformation Groups, Indian Statistical Institute, Calcutta 1998, to appear.

[AP] Allday C. and Puppe V. Cohomological methods in transformation groups. *Cambridge University Press*, 1993.

[A] M. Atiyah, "Convexity and commuting Hamiltonians." *Bull. London Math. Soc.* **14** (1982) no. 1, 1–15.

[BGH] Biss D., Guillemin V., and Holm T. The mod 2 cohomology of fixed point sets of anti-symplectic involutions. *Advances in Mathematics* **185** (2004) 370–399.

[CS] Chang T. and Skjelbred T. The topological Schur lemma and related results. *Ann. of Math.* 100 (1974) 307–321.

[CK] Constantin A. and Kolev B. The theorem of Kerékjártó on periodic homeomorphisms of the disc and the sphere. *Enseign. Math.* 40 (1994) 193–204.

[DJ] Davis M. and Januszkiewicz T. Convex polytopes, Coxeter orbifolds and torus actions. *Duke Math. J.* **62** (1991) 417–451.

[Du] Duistermaat J.J. "Convexity and tightness for restrictions of Hamiltonian functions to fixed point sets of an antisymplectic involution." *Trans. Amer. Math. Soc.* **275** (1983) 417–429.

[Fr] Franz M. Koszul duality and equivariant cohomology. *Documenta Math.* **11** (2006) 243–269.

[FP] Franz M. and Puppe V. Steenrod squares on conjugation spaces. *C. R. Math. Acad. Sci. Paris* **342** (2006) 187–190.

[FP2] Franz M. and Puppe V. Exact cohomology sequences with integral coefficients for torus actions. *Transformation Groups* **12** (2007) 65–76.

[FP3] Franz M. and Puppe V. Exact sequences for equivariantly formal spaces. Preprint (2003) arXiv:math/0307112.

[Go] Goldberg T. "A convexity theorem for the involution fixed set of a Borel invariant sub-variety." *Proc. Amer. Math. Soc.* **137** (2009), no. 4, 1447–1458.

[GH] Goldin R. and Holm T. Real loci of symplectic reductions. *Trans. AMS* **356** (2004) 2105–2124.

[GKM] Goresky M., Kottwitz R., and MacPherson R. Equivariant cohomology, Koszul duality, and the localization theorem. *Invent. Math.* **131** (1998) 25–83.

[GS] Guillemin V. and Sternberg S. Convexity properties of the moment mapping. *Invent. Math.* **67** (1982) no. 3, 491–513.

[HH] Harada M. and Holm T. The equivariant cohomology of hypertoric varieties and their real loci. *Comm. Anal. Geom.* **13** (2005), no. 3, 527–559.

[HHP] Hausmann J-C., Holm T., and Puppe V. Conjugation spaces. *Algebr. Geom. Topol.* **5** (2005) 923–964

[HK] Hausmann J-C. and Knutson A. The cohomology ring of polygon spaces. *Annales de l'Institut Fourier* (1998) 281–321.

[Ho] Ho N.-K. The real locus of an involution map on the moduli space of flat connections on a Riemann surface. *Int. Math. Res. Not.* (2004) no. 61, 3263–3285.

[Hs] Hsiang Wu Yi. Cohomology theory of topological transformation groups. *Springer-Verlag* (1975).

[McC] McCleary J. A user's guide to spectral sequences. *Cambridge University Press, 2nd edition* (2001).

[Ol] Olbermann M. Conjugations in 6-manifolds. *Thesis, University of Heidelberg* (2007).

[OS] O'Shea L. and Sjamaar R. Moment maps and Riemannian symmetric pairs. *Math. Ann.* **317** (2000) 415–457.

[Sd] Schmid C. Cohomologie équivariante de certaines variétés hamiltoniennes et de leur partie réelle. *Thesis, University of Geneva (2001). http://www.unige.ch/math/biblio/these/tschmid.pdf.*

[Sj] R. Sjamaar, *Real symplectic geometry,* Lecture notes from the Séminaire Itinérant de Géometrie et Physique Mathématique V. Available at the author's homepage.

Nonabelian localization for U(1) Chern–Simons theory

Lisa Jeffrey and Brendan McLellan

Dedicated to Hans Duistermaat on the occasion of his 65th birthday

Abstract This article studies an analogue of the work of Beasley and Witten [BW] for the Chern–Simons partition function on a Seifert Manifold for $U(1)$ gauge group. A key point is that our gauge group is not simply connected, whereas this is an essential assumption in Beasley and Witten's [BW] work. We are still able to use Beasley and Witten's results, however, to derive a definition of a $U(1)$ Chern–Simons partition function. We then compare this result to a definition of the $U(1)$ Chern–Simons partition function given by Mihaela Manoliu [M], and find that the two definitions agree up to some undetermined multiplicative constant. These results lead to a natural interpretation of the Ray-Singer analytic (Reidemeister) torsion as a symplectic volume form on the moduli space of flat $U(1)$ connections over a Seifert three-manifold.

Key words: Chern–Simons gauge theory, Reidemeister-Ray-Singer torsion, symplectic volume form, localization

Mathematics Subject Classification (2010): Primary: 58J28; Secondary: 53D30, 81T20

Lisa Jeffrey
Department of Mathematics, University of Toronto, Canada
e-mail: jeffrey@math.toronto.edu

Brendan McLellan
Department of Mathematics, University of Toronto, Canada
e-mail: mclellan@math.toronto.edu

1 Introduction

This article studies the nonabelian localization results of Beasley and Witten in [BW], and considers the analogue of these results when the gauge group G is the abelian group $G = \mathrm{U}(1)$. In the finite dimensional case, Duistermaat and Heckman studied the stationary phase approximation for integrals of the form

$$\int_M e^{it\mu^Y(x)}\beta(x),$$

where $\beta(x) = \frac{\omega^l}{l!(2\pi)^l}$, (M, ω, G, μ) is a Hamiltonian G-space, and Y is an element of the Lie algebra of G and μ^Y the component of the moment map in the direction of Y. They proved that the stationary phase approximation in this case is exact, and in the special case of isolated fixed points they obtained the (abelian) localization formula

$$\int_M e^{it\mu^Y(x)}\beta(x) = \left(\frac{i}{t}\right)^l \sum_{p\in M^Y} \frac{e^{it\mu^Y(p)}}{\sqrt{\det \mathcal{L}_p(Y)}},$$

where $M^Y = \{p \in M | Y_M(p) = 0\}$ is the fixed-point set of the vector field generated by Y.

The case of U(1) Chern–Simons theory is another situation in which the stationary phase approximation is exact. In [BW], Beasley and Witten study Chern–Simons gauge theory on a Seifert manifold X, with a gauge group G that is nonabelian, compact, connected, simply connected, and simple. These assumptions imply that the principal G-bundle over X

$$\begin{array}{ccc} G & \hookrightarrow & P \\ & & \downarrow \\ & & X \end{array}$$

is trivial. This is not the case for $G = \mathrm{U}(1)$, for which some of these assumptions are not valid.

The authors of [BW] then apply the technique of nonabelian localization to the Chern–Simons path integral

$$Z_X(k) = \int_{\mathcal{M}} \mathcal{D}A\ \exp\left[i\frac{k}{4\pi}\int_X \mathrm{Tr}\left(A\wedge dA + \frac{2}{3}A\wedge A\wedge A\right)\right]. \quad (1)$$

Here $\mathcal{M}$ is the moduli space of *all* connections on X; through localization, the authors of [BW] reformulate the partition function as an integral over the space of gauge equivalence classes of flat connections. They are able to compute this partition function in terms of topological data on the moduli space of flat connections, $\mathcal{M}_0$, in several cases, specifically related to $SU(2)$.

Some results for the U(1) Chern–Simons theory are already known; see [M] and [MPR]. We study the results of Manoliu in [M], where the partition function for U(1) Chern–Simons theory has been calculated independently, and compare her results to those of [BW]. Manoliu studies U(1) Chern–Simons theory for arbitrary 3-manifolds; on the other hand, Beasley and Witten study Chern–Simons theory for simple simply connected gauge groups G for Seifert 3-manifolds (manifolds which are the total space of a circle bundle over a 2-manifold or a 2-dimensional orbifold). As noted above, one of the main differences of the U(1) theory from the setting of [BW] is the fact that there exist nontrivial principal U(1)-bundles over X. This difference occurs explicitly in the expression for the partition function in the U(1) case. Manoliu gives the following formula for the Chern–Simons partition function in [M]:

$$Z_X = \frac{k^{m_X}}{|\operatorname{Tors} H^2(X,\mathbb{Z})|} \sum_{p \in \operatorname{Tors} H^2(X,\mathbb{Z})} \sigma_{X,p} \int_{\mathcal{M}_0} (T_X)^{1/2}, \tag{2}$$

where $\operatorname{Tors} H^2(X,\mathbb{Z})$ is the torsion subgroup of $H^2(X,\mathbb{Z})$ and is isomorphic to $H^1(X,\mathbb{R}/\mathbb{Z})$. Note that a principal U(1)-bundle $P \to X$ has flat connections if and only if the first Chern class $c_1(P)$—if nonzero—is a torsion class in $H^2(X,\mathbb{Z})$. The sum over p passes over different topological types for the bundle over the 3-manifold X. We have calculated the dependence of the results of both [BW] and [M] on the Chern–Simons parameter k. In our comparison we are looking only at the component of the partition function that comes from the contribution of the trivial bundle $P = \mathrm{U}(1) \times X$ in the partition function of [M], since the results of [BW] apply only to this case. For groups G which are simply connected, all G-bundles over three-manifolds are trivial, whereas this is not true for U(1) bundles.

We compare the expressions for the Chern–Simons partition functions found by Beasley–Witten and by Manoliu in their respective situations. Both partition functions are expressed as integrals of top-degree differential forms over the moduli space $\mathcal{M}$ of gauge equivalence classes of flat connections over the 3-manifold. We restrict to $G = \mathrm{U}(1)$, and compare

- the power of k appearing in the integrand;
- the integrand (in both cases it is the symplectic volume form).

Proposition 1.1. *The* U(1) *Chern–Simons partition function from (5.17) of [M] and the specialization of Beasley and Witten's Chern–Simons partition function (5.172) of [BW] to* $G = \mathrm{U}(1)$ *are both proportional to*

$$k^{(2g-1)/2} \int_{\mathrm{U}(1)^{2g}} \omega^g$$

(for the symplectic form ω *on* $\mathrm{U}(1)^{2g}$*).*

An important ingredient in our study of the partition functions is the appearance of the Reidemeister torsion (R-torsion) in [M]. We provide a section devoted specifically to the R-torsion, and also study the relationship of the R-torsion to the symplectic volume of $\mathcal{M}_0$.

2 *k*-dependence

In this section we compare the results of [BW] and [M] for the k-dependence of their partition functions. In particular, we look at eq. (5.172) of [BW], and eq. (5.17) of [M], and derive their respective dependences on the Chern–Simons coupling constant k.

Let us begin with [M]. Equation (5.17) of [M] reads

$$Z_X = \frac{k^{m_X}}{|\operatorname{Tors} H^2(M,\mathbb{Z})|} \sum_{p \in \operatorname{Tors} H^2(M,\mathbb{Z})} \sigma_{X,p} \int_{\mathcal{M}_0} (T_X)^{1/2} \tag{3}$$

$$= k^{m_X} \sum_{p \in \operatorname{Tors} H^2(M,\mathbb{Z})} \int_{\mathcal{M}_{0,p}} \sigma_{X,p} (T_X)^{1/2} \tag{4}$$

$$= k^{m_X} \int_{\mathcal{M}_0} \sigma_X (T_X)^{1/2}, \tag{5}$$

where

$$m_X = \frac{1}{2}(\dim H^1(X,\mathbb{R}) - \dim H^0(X,\mathbb{R})). \tag{6}$$

Here if p corresponds to a trivial bundle $P = \mathrm{U}(1) \times X$, then $\sigma_{X,p}(A) = e^{ikCS(A)}$ is the Chern–Simons function of the connection A, raised to the power k. We note that if P is a trivial bundle and A is a critical point for the Chern–Simons functional, then $dA = 0$, so $\sigma_{X,p}(A) = 1$.

The k dependence comes only from the factor k^{m_X}. The value of m_X is as follows. Since X is connected, $\dim H^0(X,\mathbb{R}) = 1$. The values of $\dim H^1(X,\mathbb{R})$ are stated in [FS]; for completeness we provide a short proof.

Proposition 2.1 ([FS]).

$$\dim H^1(X,\mathbb{R}) = \begin{cases} 2g, & n \geq 1, \\ 2g+1, & n = 0, \end{cases}$$

where n is the degree of the $\mathrm{U}(1)$*-bundle X, and g is the genus of the base space Σ:*

$$\begin{array}{ccc} U(1) & \hookrightarrow & X \\ & & \downarrow \\ & & \Sigma \end{array}$$

Proof. By the universal coefficient theorem (UCT),

$$H^1(X,\mathbb{R}) \simeq \operatorname{Hom}(H_1(X,\mathbb{Z}),\mathbb{R}), \tag{7}$$

i.e., the UCT implies that

$$0 \to \mathrm{Ext}(H_0(X,\mathbb{Z}),\mathbb{R}) \to H^1(X,\mathbb{R}) \to \mathrm{Hom}(H_1(X,\mathbb{Z}),\mathbb{R}) \to 0$$

is exact. Also

$$\mathrm{Ext}(H_0(X,\mathbb{Z}),\mathbb{R}) \simeq \mathrm{Ext}(\mathbb{Z},\mathbb{R}) \simeq 0,$$

since $\mathbb{Z}$ is free.

Thus we compute $\mathrm{Hom}(H_1(X,\mathbb{Z}),\mathbb{R})$. By Hurewicz,

$$H_1(X,\mathbb{Z}) \simeq \frac{\pi_1(X)}{[\pi_1(X),\pi_1(X)]}.$$

We assume that X is a Seifert fibered manifold which fibers over a 2-manifold rather than over an orbifold, since this is the setting of (5.172) of [BW]. Hence we have the following presentation of $\pi_1(X)$([O1]):

$$\pi_1(X) \simeq \left\langle a_p, b_p, h \,|\, [a_p,h] = [b_p,h] = 1, \prod_{p=1}^{g}[a_p,b_p] = h^n \right\rangle,$$

where g is the genus of the base space Σ of our Seifert fibered 3-manifold X,

$$\begin{array}{ccc} \mathrm{U}(1) & \hookrightarrow & X \\ & & \downarrow \\ & & \Sigma \end{array}$$

and $n = c_1(X)$ is the Chern number of the U(1)-bundle X. The generator h arises from the generic fiber over Σ. Observe that in the abelianization of $\pi_1(X)$ the following relation is satisfied:

$$\prod_{p=1}^{g}[a_p,b_p] = h^n. \tag{8}$$

We have

$$[\pi_1(X),\pi_1(X)] = \left\langle [a_p,b_p] \,|\, \prod_{p=1}^{g}[a_p,b_p] = h^n \right\rangle \tag{9}$$

and therefore

$$\begin{aligned} \frac{\pi_1(X)}{[\pi_1(X),\pi_1(X)]} &= \langle a_p, b_p, h \,|\, [a_p,b_p] = h^n = 1\rangle \\ &= \bigoplus_{p=1}^{g}\langle a_p\rangle \bigoplus_{p=1}^{g}\langle b_p\rangle \bigoplus \left(\frac{\langle h\rangle}{\langle h^n\rangle}\right), \end{aligned} \tag{10}$$

where a_p, b_p, h now represent equivalence classes in the abelianization and $\langle a_p \rangle \simeq \mathbb{Z}$, $\langle b_p \rangle \simeq \mathbb{Z}$, $\frac{\langle h \rangle}{\langle h^n \rangle} \simeq \mathbb{Z}/n\mathbb{Z} \simeq \mathbb{Z}_n$. Thus,

$$\frac{\pi_1(X)}{[\pi_1(X), \pi_1(X)]} \simeq \begin{cases} \mathbb{Z}^{2g} \times \mathbb{Z}_n, & n \geq 1, \\ \mathbb{Z}^{2g+1}, & n = 0. \end{cases} \tag{11}$$

Finally, we have

$$H^1(X, \mathbb{R}) \simeq \mathrm{Hom}(H_1(X, \mathbb{Z}), \mathbb{R}) \simeq \mathrm{Hom}\left(\frac{\pi_1(X)}{[\pi_1(X), \pi_1(X)]}, \mathbb{R}\right)$$

$$\simeq \begin{cases} \mathrm{Hom}(\mathbb{Z}^{2g} \times \mathbb{Z}_n, \mathbb{R}), & n \geq 1, \\ \mathrm{Hom}(\mathbb{Z}^{2g+1}, \mathbb{R}), & n = 0, \end{cases} \tag{12}$$

$$\simeq \begin{cases} \mathbb{R}^{2g}, & n \geq 1, \\ \mathbb{R}^{2g+1}, & n = 0. \end{cases} \tag{13}$$

In conclusion,

$$\dim H^1(X, \mathbb{R}) = \begin{cases} 2g, & n \geq 1, \\ 2g + 1, & n = 0. \end{cases} \tag{14}$$

We will restrict ourselves to the case $n \neq 0$ because this is assumed in [BW]. Thus we obtain the k dependence

$$Z_X \sim k^{\frac{2g-1}{2}}.$$

Now consider the k dependence for (5.172) of [BW]. For this computation we assume that all work leading to (5.172) is relevant to the case of a trivial principal U(1)-bundle over X. We note that the main difference between our case (with a U(1) gauge group) and the case studied in [BW] is that [BW] assume that their gauge group G is compact, connected, simply connected, and simple. In particular, they can conclude that their principal G-bundle

$$\begin{array}{ccc} G & \hookrightarrow & P \\ & & \downarrow \\ & & X \end{array}$$

is trivial. However, U(1) is neither simply connected nor simple, and there exist non-trivial principal U(1)-bundles over X. It is surprising that our results show that [BW] (5.172) is still valid in the U(1) case, although we could not infer this from Beasley and Witten's calculation because our situation does not satisfy the hypotheses of [BW].

Consider eq. 5.172 of [BW]:

$$Z_X := Z(\epsilon)|_{\widetilde{\mathcal{M}}_0} \tag{15}$$
$$= \frac{1}{|\Gamma|}\exp\left(-\frac{\imath\pi}{2}\eta_0\right)\int_{\widetilde{\mathcal{M}}_0}\widehat{A}(\widetilde{\mathcal{M}}_0)\exp\left[\frac{1}{2\pi\epsilon}\Omega+\frac{1}{2}c_1(T\widetilde{\mathcal{M}}_0)+\frac{\imath n}{4\pi^2\epsilon_r}\Theta\right].$$

Here

- $\widetilde{\mathcal{M}}_0$ is a smooth component of the moduli space of irreducible flat connections on a Seifert manifold X (we assume that our Seifert manifold X is a smooth line bundle of degree n over Σ);
- $\Gamma = Z(G)$ is the center of G;
- $\eta_0 = -\frac{n\dim G}{6}$;
- $\widehat{A}(\widetilde{\mathcal{M}}_0) = \prod_{j=1}^{\dim\widetilde{\mathcal{M}}_0}\frac{x_j/2}{\sinh(x_j/2)}$, where x_j $(1 \le j \le n)$ are the Chern roots of $T\widetilde{\mathcal{M}}_0$. so that $c(T\widetilde{\mathcal{M}}_0) = \prod_{j=1}^{n}(1+x_j),\ x_j \in H^2(\widetilde{\mathcal{M}}_0, \mathbb{Z})$;
- Ω is the symplectic form on $\widetilde{\mathcal{M}}_0$;
- $\epsilon_r = \frac{2\pi}{k+\widehat{c}_{\mathfrak{g}}}$, where $\widehat{c}_{\mathfrak{g}}$ is the dual Coxeter number of G;
- $\Theta \in H^4(\widetilde{\mathcal{M}}_0)$ is the cohomology class corresponding to the degree-4 element $-(\phi,\phi)/2$ in the equivariant cohomology $H^4_G(\mathrm{pt})$ (for $\phi \in \mathfrak{g}$ using the Cartan model of equivariant cohomology); Θ can also be described in terms of the universal bundle $\mathbb{U}$:

$$\begin{array}{ccc} \mathbb{C} & \hookrightarrow & \mathbb{U} \\ & & \downarrow \\ & & \mathrm{Jac}(\Sigma)\times\Sigma \end{array}$$

  In other words,
$$\Theta = -\frac{1}{2}c_1(\mathbb{U})^2|_{\mathrm{pt.}\in\Sigma},$$
  where $\mathrm{Jac}(\Sigma)$ is the Jacobian of Σ.

The overall constant Γ does not make sense for $G = \mathrm{U}(1)$, since it is infinite. We disregard the overall constant in front of the integrand in $Z(\epsilon)|_{\widetilde{\mathcal{M}}_0}$. Looking only at the k dependence, we consider $Z(\epsilon)|_{\widetilde{\mathcal{M}}_0}$, ignoring overall multiplicative constants:

$$Z(\epsilon)|_{\widetilde{\mathcal{M}}_0} \sim \int_{\widetilde{\mathcal{M}}_0}\widehat{A}(\widetilde{\mathcal{M}}_0)\exp\left[\frac{1}{2\pi\epsilon}\Omega+\frac{1}{2}c_1(T\widetilde{\mathcal{M}}_0)+\frac{\imath n}{4\pi^2\epsilon_r}\Theta\right]. \tag{16}$$

Note that $\widetilde{\mathcal{M}}_0 \simeq \mathrm{U}(1)^{2g}\times\mathbb{Z}_n$ by Proposition 2.2 of [M], and since we are restricting to the trivial bundle case, we identify $\widetilde{\mathcal{M}}_0 \simeq \mathrm{U}(1)^{2g}$ as the connected component corresponding to $p = 0$.

The first thing we observe is that $\Theta = 0$ in our case. This follows, since the universal bundle $\mathbb{U}$ for U(1)-bundles is the classical Poincaré line bundle, and the Poincaré line bundle is normalized to have degree $d = 0$ when restricted to the Jacobian of Σ. Since $c_1(\mathbb{U}) = d[\Sigma] \in H^2(\Sigma)$, this implies $c_1(\mathbb{U}) = 0$, and hence $\Theta = 0$. Also, since $\widetilde{\mathcal{M}}_0 \simeq \mathrm{U}(1)^{2g}$, we know that

$$c(\widetilde{\mathcal{M}}_0) = c(T\widetilde{\mathcal{M}}_0) = \prod_{i=1}^{g} c(L_i) = \prod_{i=1}^{g}(1 + x_i),$$

where

$$L_i = T\Sigma_i, \quad \text{and} \quad x_i = c_1(L_i) \in H^2(\Sigma_i, \mathbb{Z}),$$

where $\Sigma_i \simeq (\mathrm{U}(1))^2$. Then the tangent bundles $T\Sigma_i$ are trivial, and hence

$$x_i = c_1(T\Sigma_i) = 0.$$

Thus

$$\widehat{A}(\widetilde{\mathcal{M}}_0) = \prod_{j=1}^{\dim \widetilde{\mathcal{M}}_0} \frac{x_j/2}{\sinh(x_j/2)} = 1. \tag{17}$$

Clearly, $c_1(T\widetilde{\mathcal{M}}_0) = 0$ as well, and we arrive at

$$Z(\epsilon)|_{\widetilde{\mathcal{M}}_0} \sim \int_{\widetilde{\mathcal{M}}_0} \exp\left[\frac{1}{2\pi\epsilon}\Omega\right]. \tag{18}$$

Recalling that $\epsilon = \frac{2\pi}{k}$, we have

$$Z(\epsilon)|_{\widetilde{\mathcal{M}}_0} \sim \int_{\widetilde{\mathcal{M}}_0} \exp[k\Omega] = \int_{\widetilde{\mathcal{M}}_0} k^g \Omega^g/g! = k^g \mathrm{Vol}_\Omega(\widetilde{\mathcal{M}}_0).$$

Thus the U(1) Chern–Simons partition function computed from eq. 5.172 of [BW] is

$$Z_X := Z(\epsilon)|_{\widetilde{\mathcal{M}}_0} \sim k^g. \tag{19}$$

There is a difference of $k^{1/2}$ between the two cases, since $Z_X \sim k^g$ for Beasley–Witten, whereas $Z_X \sim k^{\frac{2g-1}{2}}$ for Manoliu. Let us analyze this difference further. In the case of [M], this extra factor of $k^{-1/2}$ appears because the dimension of the stabilizer of the gauge group action (for U(1) gauge groups) is $\dim(H^0(X, \mathbb{R})) = 1$.

A similar phenomenon occurs in Yang–Mills theory at the higher nonflat critical points of the Yang–Mills action. As observed in Section 4.3 of [BW] (for example equation (4.45)), there is a factor of $k^{1/2}$ in the Yang–Mills partition function coming from the fact that the gauge group $\mathcal{G}$ does not act locally freely on the locus of nonflat Yang–Mills solutions. This $k^{1/2}$ factor comes from the U(1) stabilizer at a nonflat Yang–Mills solution. U(1)-Chern–Simons theory also has a U(1) stabilizer at all points, the subgroup of constant gauge transformations with values in U(1). This accounts for the extra factor of $k^{-1/2}$ in the Chern–Simons partition function in Manoliu's paper.

In fact, Beasley and Witten recast the Chern–Simons partition function as a Yang–Mills partition function (see (3.61) in [BW]). In the computation of (5.172) of [BW] it is assumed that one is localizing at an irreducible flat connection, and therefore the isotropy group of A, $\Gamma_A = \{u \in \mathcal{G} | u(A) = A\}$, is finite. Hence there is no factor of $k^{1/2}$ in [BW] (5.172) because the dimension of the stabilizer is zero.

The extra factor of $k^{-1/2}$ also appears in [JKKW] (see Section 9, Example 46). This article treats integrals of the same form as [BW] (3.61) over symplectic manifolds equipped with Hamiltonian group actions. When the group acts locally freely at the zero locus of the moment map, it is shown in [JK] that the integral is a polynomial in $\epsilon = 2\pi/k$. When the group acts with nontrivial stabilizer at points in the zero locus of the moment map, the partition function is a polynomial in $\sqrt{\epsilon}$ but not in ϵ.

3 Reidemeister torsion and symplectic volume

We would like to show that the remainder of the calculation in [M] involving the Reidemeister torsion yields the symplectic volume as above. In this section we review Reidemeister torsion (R-torsion) and provide some relevant examples.

The R-torsion is an invariant for a CW-complex and a representation of its fundamental group. Before we define the R-torsion, we recall the definition of the torsion of a chain complex. Let

$$C_* = (0 \to C_n \xrightarrow{d_n} C_{n-1} \xrightarrow{d_{n-1}} \cdots C_1 \xrightarrow{d_1} C_0 \to 0)$$

be a chain complex over $\mathbb{F}$ (either $\mathbb{R}$ or $\mathbb{C}$). Let Z_i denote the cycles of this complex, B_i denote the boundaries, and H_i the homology. Let $\{c^i\}$ be a basis of C_i and let c be the collection $\{c^i\}_{i \geq 0}$. We call a pair (C_*, c) a *based chain complex*, c the preferred basis of C_*, and c^i the preferred basis of C_i. Let h^i be a basis of H_i.

We construct another basis as follows. By the definitions of Z_i, B_i, and H_i, the following two split exact sequences exist:

$$0 \to Z_i \to C_i \xrightarrow{d_i} B_{i-1} \to 0,$$

$$0 \to B_i \to Z_i \to H_i \to 0.$$

Let $\widetilde{B}_{i-1}$ be a lift of B_{i-1} to C_i and $\widetilde{H}_i$ a lift of H_i to Z_i. Then we can decompose C_i as follows:

$$C_i = Z_i \oplus \widetilde{B}_{i-1} = B_i \oplus \widetilde{H}_i \oplus \widetilde{B}_{i-1} = d_{i+1}\widetilde{B}_i \oplus \widetilde{H}_i \oplus .\widetilde{B}_{i-1}.$$

Choose a basis b^i for B_i. We write $\widetilde{b}^{i+1} = \{\widetilde{b}^{i+1}_j\}_{j=1}^{n_i}$ for a lift of b^i and $\widetilde{h}^i = \{\widetilde{h}^i_j\}_{j=1}^{m_i}$ for a lift of h^i. By construction, the set $\{\widetilde{b}^i \cup d_{i+1}(\widetilde{b}^{i+1}) \cup \widetilde{h}^i\}$ forms another ordered basis of C_i. Denote this basis by $\{\widetilde{b}^i d_{i+1}(\widetilde{b}^{i+1})\widetilde{h}^i\}$. The definition of the R-torsion, $\text{Tor}(C_*, c, h)$, is as follows:

$$\mathrm{Tor}(C_*, c)\{h\} = (-1)^{|C_*|} \cdot \prod_{i=1}^{n} [\widetilde{b^i} d_{i+1}(\widetilde{b^{i+1}})\widetilde{h^i}/c^i]^{(-1)^{i+1}} \in \mathbb{F}^*, \tag{20}$$

where $[\widetilde{b^i} d_{i+1}(\widetilde{b^{i+1}})\widetilde{h^i}/c^i]$ denotes the determinant of the change of basis matrix from the basis $\{c^i\}$ to the basis $\{\widetilde{b^i} d_{i+1}(\widetilde{b^{i+1}})\widetilde{h^i}\}$. An alternative definition is to equip our complex C_* with volumes $\mu_i \in (\wedge^{\max} C_i)^*$, one for each i, and then to define

$$\mathrm{Tor}(C_*, \mu)\{h\} = \frac{\bigwedge_{i \text{ even}} \mu_i[\widetilde{b^i} \wedge d_{i+1}(\widetilde{b^{i+1}}) \wedge \widetilde{h^i}]}{\bigwedge_{i \text{ odd}} \mu_i[\widetilde{b^i} \wedge d_{i+1}(\widetilde{b^{i+1}}) \wedge \widetilde{h^i}]}, \tag{21}$$

where we take $\widetilde{b^i} = \wedge_{j=1}^{n_i} \widetilde{b^i_j}$, $d_{i+1}\widetilde{b^{i+1}} = \wedge_{j=1}^{n_i} d_{i+1}\widetilde{b^{i+1}_j}$, and $\widetilde{h^i} = \wedge_{j=1}^{m_i} \widetilde{h^i_j}$. The torsion is an element

$$\mathrm{Tor}(C_*, \mu) \in \otimes_{2i+1}[\wedge^{\max} H_{2i+1}(C_*)] \otimes_{2i} [\wedge^{\max} H_{2i}(C_*)]^*.$$

The latter definition specializes to the former definition when we choose the canonical volumes associated to a choice of preferred basis c for C.

It is well known (see, e.g., [F]) that the torsion is independent of the choices of b^i and of the choices of lifts $\widetilde{b^i}$, $\widetilde{h^i}$. In the case that our complex C_* is acyclic, we define

$$\mathrm{Tor}(C_*, \mu) = \frac{\bigwedge_{i \text{ even}} \mu_i[\widetilde{b^i} \wedge d_{i+1}(\widetilde{b^{i+1}})]}{\bigwedge_{i \text{ odd}} \mu_i[\widetilde{b^i} \wedge d_{i+1}(\widetilde{b^{i+1}})]} \in \mathbb{R}^*. \tag{22}$$

We will be interested in a specific chain complex C_*. In particular, let N be a cell complex, and ρ a representation of $\pi_1(N)$ in G. The Lie algebra $\mathfrak{g}$ is acted on by $\pi_1(N)$ under the composition of the adjoint action of G and the representation ρ. Let $\mathfrak{g}_\rho$ denote $\mathfrak{g}$ with the $\pi_1(N)$-module structure from ρ. Let $\widetilde{N}$ denote the universal cover of N. Since the fundamental group $\pi_1(N)$ acts on $\widetilde{N}$ by covering transformations, the chain complex $C_*(\widetilde{N})$ also has a natural $\pi_1(N)$-module structure. The chain complex of interest is then $C_*(N, \mathfrak{g}_\rho)$, defined as the quotient of $C_*(\widetilde{N}) \otimes \mathfrak{g}$ under the equivalence

$$\sigma \otimes X \sim \sigma a \otimes \mathrm{Ad}(\rho(a))^{-1} X, \tag{23}$$

where $a \in \pi_1(N)$, $\sigma \in C_*(\widetilde{N})$, and $X \in \mathfrak{g}$. The usual differential on $C_*(\widetilde{N})$ is compatible with the equivalence relation, and thus descends to a differential δ_ρ on $C_*(N, \mathfrak{g}_\rho)$. By dualizing one obtains the corresponding cochain complex $C^*(N, \mathfrak{g}_\rho)$ with differential $d_\rho = \delta_\rho^*$. We have the following lemma.

Lemma 3.1 ([JW]). *Suppose $h \in G$. If ρ and $h\rho h^{-1}$ are conjugate representations of $\pi_1(N)$ in G, then the map $\mathrm{Ad}(h) : \mathfrak{g} \to \mathfrak{g}$ induces an isomorphism of the chain complexes $C_*(N, \mathfrak{g}_\rho)$ and $C_*(N, \mathfrak{g}_{h\rho h^{-1}})$. Hence one obtains a natural isomorphism between the cohomology groups*

$$H^i(C_*(N, \mathfrak{g}_\rho))$$

and

$$H^i(C_*(N, \mathfrak{g}_{h\rho h^{-1}})).$$

We will mainly be interested in the zeroth and first cohomology groups of this complex. We recall the following result.

Proposition 3.2 ([JW]). *Let* $[\rho] \in \mathrm{Hom}(\pi_1(N), G)/G$. *The choice of a particular* $\rho \in \mathrm{Hom}(\pi_1(N), G)$ *in the conjugacy class* $[\rho]$ *identifies the Zariski tangent space at* ρ *of the space* $\mathrm{Hom}(\pi_1(N), G)/G$ *with the first cohomology group* $H^1(N, \mathfrak{g}_\rho)$.

Furthermore, the Lie algebra of the isotropy group of ρ *(the subgroup of* G *fixing the representation* ρ *under conjugation) is* $H^0(N, \mathfrak{g}_\rho)$.

Using the definition of the R-torsion in (21) above, we may define volumes on $C_*(N, \mathfrak{g}_\rho)$ using the metric on $\mathfrak{g}$. We take $\{\sigma_j^i \otimes X_k\}$ to be an orthonormal basis of $C_*(N, \mathfrak{g}_\rho)$, where σ_j^i are the i-cells in the universal cover $\widetilde{N}$ and the X_k are an orthonormal basis of $\mathfrak{g}$. This volume is well defined, since the adjoint representation is an orthogonal representation of G, and hence compatible with the equivalence relation (23). The torsion is then an element of

$$\mathrm{Tor}(C_*(N, \mathfrak{g}_\rho), \mu) \in \otimes_{2i+1}[\wedge^{\max} H_{2i+1}(N, \mathfrak{g}_\rho)] \otimes_{2i} [\wedge^{\max} H_{2i}(N, \mathfrak{g}_\rho)]^*.$$

Since $H^i(N, \mathfrak{g}_\rho) \simeq H_i(N, \mathfrak{g}_\rho)^*$, the torsion may be identified with an element

$$\mathrm{Tor}(C_*(N, \mathfrak{g}_\rho), \mu) \in \otimes_{2i+1}[\wedge^{\max} H^{2i+1}(N, \mathfrak{g}_\rho)]^* \otimes_{2i} [\wedge^{\max} H^{2i}(N, \mathfrak{g}_\rho)]. \quad (24)$$

The isomorphisms in Lemma 3.1 identify the torsion $\mathrm{Tor}(C_*(N, \mathfrak{g}_\rho))$ with $\mathrm{Tor}(C_*(N, \mathfrak{g}_{h\rho h^{-1}})$, so the torsion descends to an equivalence class $\tau(N, \rho)$ depending only on the conjugacy class $[\rho] \in \mathrm{Hom}(\pi_1(N), G)/G$.

It is instructive to consider the case that N is a genus-g surface Σ^g.

Example 3.1. From the relation (24) above, we see that the torsion $\tau(\Sigma^g; \rho)$ of a surface Σ^g takes values in

$$\wedge^{\max} H^1(\Sigma^g, \mathfrak{g}_\rho)^* \otimes \wedge^{\max} H^2(\Sigma^g, \mathfrak{g}_\rho) \otimes \wedge^{\max} H^0(\Sigma^g, \mathfrak{g}_\rho).$$

By Poincaré duality, $H^2(\Sigma^g, \mathfrak{g}_\rho)$ is canonically dual to $H^0(\Sigma^g, \mathfrak{g}_\rho)$, so we have

$$\tau(\Sigma^g; \rho) \in \wedge^{\max} H^1(\Sigma^g, \mathfrak{g}_\rho)^*.$$

Observe that $H^1(\Sigma^g, \mathfrak{g}_\rho)$ is a symplectic vector space (the tangent space to the moduli space of gauge equivalence classes of flat connections on Σ^g) with the symplectic form given by the cup product $H^1(\Sigma^g, \mathfrak{g}_\rho) \otimes H^1(\Sigma^g, \mathfrak{g}_\rho) \to H^2(\Sigma^g, \mathbb{R}) \simeq \mathbb{R}$. One can show that in fact, the torsion may be identified with the symplectic volume on $H^1(\Sigma^g, \mathfrak{g}_\rho)$. A rigorous proof of this is given in [W].

The case of interest to us here is that of N a Seifert manifold and $G = \mathrm{U}(1)$. In this case, the torsion $\tau(N; \rho)$ takes values in

$$\wedge^{\max} H^1(N, \mathfrak{g}_\rho)^* \wedge^{\max} H^3(N, \mathfrak{g}_\rho)^* \otimes \wedge^{\max} H^2(N, \mathfrak{g}_\rho) \otimes \wedge^{\max} H^0(N, \mathfrak{g}_\rho),$$

where by Poincaré duality $H^3(N, \mathfrak{g}_\rho)$ is canonically dual to

$$H^0(N, \mathfrak{g}_\rho) \simeq \mathbb{R}$$

and $H^1(N, \mathfrak{g}_\rho)$ is canonically dual to $H^2(N, \mathfrak{g}_\rho)$. Note that

$$H^0(N, \mathfrak{g}_\rho) \simeq \mathbb{R}$$

once we choose a basis, because we are working with U(1). Thus,

$$\tau(N; \rho) \in (\wedge^{\max} H^1(N, \mathfrak{g}_\rho)^*)^{\otimes 2},$$

or

$$\sqrt{\tau(N; \rho)} \in \wedge^{\max} H^1(N, \mathfrak{g}_\rho)^*.$$

When N is two-dimensional we observed above (see [W]) that the torsion can be identified with the symplectic volume on $H^1(\Sigma, \mathfrak{g}_\rho)$.

We recall our previous results in equation (18), where we observed that for the gauge group U(1) the Chern–Simons partition function Z_X was proportional to the symplectic volume. In the case $G = \mathrm{U}(1)$, for a Seifert manifold $\mathrm{U}(1) \to N \to \Sigma$, we would also like to see that $\sqrt{\tau(N)}$ is proportional to the symplectic volume on the moduli space $\mathrm{U}(1)^{2g} \times \mathbb{Z}_n$.

Writing this more concisely, we want to see that

$$\sqrt{\tau(N)} = C \cdot \omega^g, \tag{25}$$

where $C \in \mathbb{R}^*$ is some nonzero constant. Here $\omega \in \Omega^2(\mathrm{U}(1)^{2g} \times \mathbb{Z}_n); \mathbb{R})$ is the symplectic form on $\mathrm{U}(1)^{2g} \times \mathbb{Z}_n$, i.e., the symplectic form on each of the n disjoint copies of $\mathrm{U}(1)^{2g}$. We introduce the notation $a = (\rho, m) \in \mathrm{U}(1)^{2g} \times \mathbb{Z}_n$. Here $\omega_a(\alpha, \beta) := \int_\Sigma \alpha \wedge \beta$, for $\alpha, \beta \in H^1(N, \mathfrak{g}_a) \simeq H^1(\Sigma, \mathfrak{g}_\rho) \simeq T_a(\mathrm{U}(1)^{2g} \times \mathbb{Z}_n)$. Here the last equality follows from Proposition 3.2. Also $\sqrt{\tau(N)} \in \Omega^{2g}(\mathrm{U}(1)^{2g} \times \mathbb{Z}_n; \mathbb{R})$, i.e., we define $\sqrt{\tau(N)}_a := \sqrt{\tau(N; a)}$, $\forall a \in \mathrm{U}(1)^{2g} \times \mathbb{Z}_n$ as a section of the top exterior power of $T^*(\mathrm{U}(1)^{2g} \times \mathbb{Z}_n)$]. To summarize, both the torsion and the symplectic volume are volume elements on $\mathrm{U}(1)^{2g} \times \mathbb{Z}_n$ and the Lie group structure of U(1) means that the tangent bundle is trivial, and there is a natural basis vector given by a generator of the Lie algebra of U(1). In terms of this basis vector, we note that the definition of the torsion does not depend on the choice of a point in $\mathrm{U}(1)^{2g} \times \mathbb{Z}_n$, since the differential of the chain complex is simply the exterior derivative. For nonabelian groups the differential is the twisted differential $d_A = d + \mathrm{Ad}(A)$, which does depend on the choice of a flat connection A. If the group is abelian, d_A reduces to the exterior derivative d, and it does not depend on A.

It will be sufficient to identify $\sqrt{\tau(N)}$ and ω^g at a single point of the moduli space $\mathrm{U}(1)^{2g} \times \mathbb{Z}_n$, since we will show that $\sqrt{\tau(N)}$ and ω are invariant under left multiplication, i.e., invariant under the action,

$$L_a : \mathrm{U}(1)^{2g} \times \mathbb{Z}_n \to \mathrm{U}(1)^{2g} \times \mathbb{Z}_n$$

defined by $L_a : b \mapsto a \cdot b$ for $a \in \mathrm{U}(1)^{2g} \times \mathbb{Z}_n$.

First, we show that $\sqrt{\tau(N)}$ is invariant under this action. As discussed above, the torsion does not depend on the choice of point $a = [A] \in \mathrm{U}(1)^{2g} \times \mathbb{Z}_n$ corresponding to a flat connection A. By construction, the torsion is independent of a.

We identify $H^1(N, \mathbb{R}) \simeq H^1(N, d)$, viewed as differential 1-forms, with $T_a(\mathrm{U}(1)^{2g} \times \mathbb{Z}_n)$ in the following way. The generators of $H^1(N, \mathbb{R})$ come from generators of the cohomology for U(1), one for each generating loop of the fundamental group $\pi_1(N)$. The tangent space to the moduli space $\mathrm{U}(1)^{2g} \times \mathbb{Z}_n$ at a then has $2g$ generators $\{\frac{\partial}{\partial \phi^i}{}_a\}_{i=1}^{2g}$. Since the definition of $\tau(N)$ is independent of a, where $\frac{\partial}{\partial \phi^i}$ is the vector field on the ith copy of U(1) in $\mathrm{U}(1)^{2g} \times \mathbb{Z}_n$, we have that

$$(dL_a)_b \frac{\partial}{\partial \phi^i}\bigg|_a = \frac{\partial}{\partial \phi^i}\bigg|_{a \cdot b}, \quad \forall b \in \mathrm{U}(1)^{2g} \times \mathbb{Z}_n,\ 1 \le i \le 2g. \tag{26}$$

We conclude that $L_a^* \tau(N) = \tau(N)$, $\forall a \in \mathrm{U}(1)^{2g} \times \mathbb{Z}_n$, i.e., $\tau(N)$ is invariant under the action of $\mathrm{U}(1)^{2g} \times \mathbb{Z}_n$ on itself.

The next thing that we will show is that ω, the Goldman-Atiyah-Bott symplectic form on the moduli space, is invariant under the action of $\mathrm{U}(1)^{2g} \times \mathbb{Z}_n$ on itself. This can be seen directly. Consider

$$\begin{aligned}
(L_a^* \omega)_a \left(\frac{\partial}{\partial \phi^i}, \frac{\partial}{\partial \theta^j} \right) &= \omega_{a \cdot a} \left((dL_a)_a \left(\frac{\partial}{\partial \phi^i} \right), (dL_a)_a \left(\frac{\partial}{\partial \phi^j} \right) \right) \\
&= \omega_{a \cdot a} \left(\frac{\partial}{\partial \phi^i}, \frac{\partial}{\partial \phi^j} \right) \\
&= \omega_a \left(\frac{\partial}{\partial \phi^i}, \frac{\partial}{\partial \phi^j} \right).
\end{aligned}$$

Thus,

$$L_a^* \omega = \omega, \quad \forall a \in \mathrm{U}(1)^{2g} \times \mathbb{Z}_n,$$

i.e., ω is invariant under the action of $\mathrm{U}(1)^{2g} \times \mathbb{Z}_n$ on itself.

Now we can prove our original claim. Let e denote the identity element of $\mathrm{U}(1)^{2g} \times \mathbb{Z}_n$. Then at the point e, $\sqrt{\tau(N)}$ and ω^g must agree up to a nonzero multiplicative constant:

$$\sqrt{\tau(N)}|_e = C \cdot \omega^g|_e$$

for some $C \in \mathbb{R}^*$. By left invariance, we therefore have

$$\sqrt{\tau(N)}|_a = C \cdot \omega^g|_a, \quad \forall a \in \mathrm{U}(1)^{2g} \times \mathbb{Z}_n.$$

Thus,

$$\sqrt{\tau(N)} = C \cdot \omega^g.$$

References

[BW] Chris Beasley and Edward Witten, Non-abelian localization for Chern–Simons theory. *J. Diff. Geom.* **70** (2005), no. 2, 183–323.

[F] D.S. Freed, Reidemeister torsion, spectral sequences and Brieskorn spheres. *J. Reine Angew. Math.* **429** (1992) 75–89.

[FS] M. Furuta, B. Steer, Seifert fibred homology 3-spheres and the Yang–Mills equations on Riemann surfaces with marked points. *Adv. Math.* **96** (1992) 38–102.

[J1] L.C. Jeffrey, Chern–Simons–Witten invariants of lens spaces and torus bundles, and the semiclassical approximation. *Commun. Math. Phys.* **147** (1992), 563–604.

[JK] L. Jeffrey, F. Kirwan, Localization for nonabelian group actions. *Topology* **34** (1995) 291–327.

[JKKW] L. Jeffrey, Y. Kiem, F. Kirwan, J. Woolf, Cohomology pairings on singular quotients in geometric invariant theory. singular quotients in geometric invariant theory. *Transformation Groups* **8** (2003) 217–259.

[JW] L.C. Jeffrey, J. Weitsman, Half density quantization of the moduli space of flat connections and Witten's semiclassical manifold invariants. *Topology* **32** (1993), 509–529.

[M] M. Manoliu, Abelian Chern–Simons theory. I. A topological quantum field theory. *J. Math. Phys.* **39** (1998), no. 1, 170–206.

[MPR] J. Mattes, M. Polyak, N. Reshetikhin, On invariants of 3-manifolds derived from abelian groups. In *Quantum Topology*, Ser. *Knots Everything* **3**, World Scientific (1993) 324–338.

[O1] P. Orlik, *Seifert Manifolds*. Lecture Notes in Mathematics **291**, Springer (1972).

[W] E. Witten, On quantum gauge theories in two dimensions. *Commun. Math. Phys.* **141** (1991), 153–209.

Symplectic implosion and nonreductive quotients

Frances Kirwan

This paper was originally written to help celebrate the 65th birthday of Hans Duistermaat; it is now dedicated to the memory of a wonderful mathematician

Abstract To a Hamiltonian action of a compact Lie group K on a symplectic manifold X, the symplectic implosion construction of Guillemin, Jeffrey and Sjamaar associates a stratified symplectic space X_{impl} with a Hamiltonian action of the maximal torus T of K such that, if ζ lies in a fixed positive Weyl chamber in the dual of the Lie algebra of T, then the symplectic reduction of X by K at level ζ can be canonically identified with the symplectic reduction of X_{impl} by T at level ζ. Moreover X_{impl} can be obtained as the symplectic quotient by K of the product of X and the universal symplectic implosion $(T^*K)_{\mathrm{impl}}$, and $(T^*K)_{\mathrm{impl}}$ can be naturally identified with the canonical affine completion of $G/U_{\max}$ where G is the complexification of K and $U_{\max}$ is a maximal unipotent subgroup of G (or equivalently the unipotent radical of a Borel subgroup of G). Thus if X is a projective variety with a linear G-action, the symplectic implosion X_{impl} can be identified with the nonreductive GIT quotient $G/\!/U_{\max}$.

In this paper the symplectic implosion construction is generalised so that if U is the unipotent radical of any parabolic subgroup P of G then the associated generalised symplectic implosion of T^*K can be naturally identified with the canonical affine completion of G/U, and hence when G acts linearly on a projective variety X the nonreductive GIT quotient $X/\!/U$ can be identified with the associated generalised symplectic implosion of X.

Key words: Symplectic reduction, symplectic implosion, moment map, geometric invariant theory, non-reductive algebraic group action

Mathematics Subject Classification (2010): 14L24, 53D20

Mathematical Institute, Oxford OX1 3BJ, UK
e-mail: frances.kirwan@balliol.ox.ac.uk

1 Introduction

There is a close relationship between Mumford's geometric invariant theory (GIT) in (complex) algebraic geometry and the process of reduction in symplectic geometry. GIT was developed to construct quotients of algebraic varieties by reductive group actions and thus to construct and study moduli spaces [27, 28]. When a moduli space (or a compactification of a moduli space) over $\mathbb{C}$ can be constructed as a GIT quotient of a complex projective variety by the action of a complex reductive group G, then it can be identified with a symplectic reduction by a maximal compact subgroup K of G, and techniques from symplectic geometry can be used to study its topology (for example [2, 15, 16, 19, 20, 21, 22]). Many moduli spaces arise as quotients of algebraic group actions, but the groups concerned are not necessarily reductive, so that classical GIT does not apply and different methods need to be used to construct the quotients (cf., e.g., [18, 24]). Nonetheless, in suitable situations GIT can be generalised to allow us to construct GIT-like quotients (and compactified quotients) for these actions [7, 23]. This paper describes some ways in which such nonreductive compactified quotients can be studied using symplectic techniques closely related to the "symplectic implosion" construction of Guillemin, Jeffrey, and Sjamaar [14].

More precisely, suppose that U is a maximal unipotent subgroup of a complex reductive group G acting linearly (with respect to an ample line bundle L) on a complex projective variety X, and suppose that the linear action of U on X extends to a linear action of G. Then the ring of invariants $\bigoplus_{k\geq 0} H^0(X, L^{\otimes k})^U$ is finitely generated and the enveloping quotient $X/\!/U$ (in the sense of [7]) is the projective variety $\mathrm{Proj}(\bigoplus_{k\geq 0} H^0(X, L^{\otimes k})^U)$ associated to the ring of invariants. Moreover, if K is a maximal compact subgroup of G, and X is given a suitable K-invariant Kähler form, then $X/\!/U$ can be identified with the imploded cross-section X_{impl} of X by K in the sense of the symplectic implosion construction of Guillemin, Jeffrey, and Sjamaar [14]. Note that here U is the unipotent radical of a Borel subgroup of G. The aim of this paper is to generalise symplectic implosion to give a symplectic construction for GIT-like (compactified) quotients by the unipotent radical U of any parabolic subgroup P of a complex reductive group G, when the action extends to an action of G. Hence we obtain a "moment map" description of such compactified quotients of projective varieties by unipotent radicals of parabolics which is analogous to the description of a reductive GIT quotient $Y/\!/G$ as a symplectic quotient $\mu^{-1}(0)/K$, where K is a maximal compact subgroup of G and μ is a moment map.

The layout of the paper is as follows. Section 2 reviews classical GIT and its relationship with symplectic geometry, while Section 3 reviews symplectic implosion from [14] and extends its construction to cover quotients by unipotent radicals of parabolics. Section 4 gives a brief description of the results of [7, 23] on nonreductive actions and the construction of compactified quotients (more details and a much more leisurely introduction to nonreductive GIT can be found in [7]) and finally relates them to symplectic implosion. A simple example when $G = \mathrm{SL}(2; \mathbb{C})$ is worked out in detail at the very end of the paper in Example 4.8.

1.1 Index of notation

Notation is introduced in this paper as follows:

$\mu, \mathfrak{k}, K_\zeta$	§2.1
$\hat{\mathcal{O}}_L(X), X/\!/G, X^{\mathrm{ss}}, X^{\mathrm{s}}$ (for reductive actions)	§2.2
$X_{\mathrm{impl}}, T, \mathfrak{t}, \mathfrak{t}^*_+, W, [K_\zeta, K_\zeta], \Sigma, B, U_{\max}, \overline{G/U_{\max}}^{\mathrm{aff}}, \Lambda, \Lambda^*_+, V_\lambda, \Pi, \iota, w_0$	§3.1
$B^{\mathrm{op}}, U^{\mathrm{op}}_{\max}, V_\lambda^{(T)}, \upsilon_\varpi, \mathcal{F}, \alpha^\vee, S, \widetilde{X_{\mathrm{impl}}}$	§3.1
$U, P, L^{(P)}, K^{(P)}, S_P, R^+, R(S_P), Q^{(P)}, \mathfrak{k}^{(P)}, \mathfrak{z}^{(P)}, \overline{G/U}^{\mathrm{aff}}, X/\!/U, E^{(P)}$	§3.2
$V_\varpi^{(P)}, V_\varpi^{K^{(P)}}, \upsilon_\varpi^{(P)}, \upsilon_{\varpi,\lambda}^{(P)}, \pi^{K^{(P)}}, \mathfrak{t}^*_{(P)+}, \mathcal{F}^{(P)}, \mathfrak{k}_+^{(P)*}, K_\zeta(P), \upsilon_\sigma^{(P)}, X_{\mathrm{impl}}^{K,K^{(P)}}$	§3.2
$\widetilde{X_{\mathrm{impl}}^{K,K^{(P)}}}, \widetilde{G/U}^{\mathrm{aff}}$	§3.3
$X^{\mathrm{ss}}, X^{\mathrm{s}}, X^{\mathrm{nss}}, X^{\mathrm{ns}}, X/\!/U, \overline{G \times_U X}, X^{\bar{\mathrm{s}}}, X^{\overline{\mathrm{ss}}}$	§4.1
$\hat{U}, \widehat{X/\!/U}, \hat{L}_\epsilon = \hat{L}_\epsilon^{(N)}, \mathcal{X}, \tilde{\mathcal{X}}, \widetilde{X/\!/U}$	§4.2

2 Symplectic reduction and geometric invariant theory

The GIT quotient construction in complex algebraic geometry is closely related to the process of reduction in symplectic geometry.

2.1 Symplectic reduction

Suppose that a compact, connected Lie group K with Lie algebra $\mathfrak{k}$ acts smoothly on a symplectic manifold X and preserves the symplectic form ω. Let us denote the vector field on X defined by the infinitesimal action of $a \in \mathfrak{k}$ by $x \mapsto a_x$. Recall that a moment map for the action of K on X is then a smooth map $\mu : X \to \mathfrak{k}^*$ which satisfies

$$d\mu(x)(\xi) \cdot a = \omega_x(\xi, a_x)$$

for all $x \in X$, $\xi \in T_x X$, and $a \in \mathfrak{k}$. Equivalently, if $\mu_a : X \to \mathbb{R}$ denotes the component of μ along $a \in \mathfrak{k}$ defined for all $x \in X$ by the pairing $\mu_a(x) = \mu(x) \cdot a$ between $\mu(x) \in \mathfrak{k}^*$ and $a \in \mathfrak{k}$, then μ_a is a Hamiltonian function for the vector field on X induced by a. We shall assume that any moment map $\mu : X \to \mathfrak{k}^*$ is K-equivariant with respect to the given action of K on X and the coadjoint action of K on $\mathfrak{k}^*$. If the stabiliser K_ζ of $\zeta \in \mathfrak{k}^*$ acts freely on $\mu^{-1}(\zeta)$ then $\mu^{-1}(\zeta)$ is a submanifold of X and the symplectic form ω induces a symplectic structure on the quotient $\mu^{-1}(\zeta)/K_\zeta$, which is the Marsden–Weinstein reduction, or symplectic reduction, at ζ of the action of K on X. The quotient $\mu^{-1}(\zeta)/K_\zeta$ also inherits a symplectic structure when the action of K_ζ on $\mu^{-1}(\zeta)$ is not free, but in this case it is likely to have singularities (although these will be only orbifold singularities if ζ is a regular value of μ, or equivalently if K_ζ acts on $\mu^{-1}(\zeta)$ with finite stabilisers). The case $\zeta = 0$ is of particular importance; $\mu^{-1}(0)/K$ is often called the symplectic quotient of X by the action of K.

Now let X be a nonsingular connected complex projective variety embedded in complex projective space $\mathbb{P}^n$, and let G be a complex Lie group acting on X via a complex linear representation $\rho : G \to \mathrm{GL}(n+1; \mathbb{C})$. By an appropriate choice of coordinates on $\mathbb{P}^n$ we may assume that ρ maps a maximal compact subgroup K of G into the unitary group $\mathrm{U}(n+1)$. Then the Fubini–Study form ω on $\mathbb{P}^n$ restricts to a K-invariant Kähler form on X, and there is a moment map $\mu : X \to \mathfrak{k}^*$ defined (up to multiplication by a constant scalar factor depending on the convention chosen for the normalisation of the Fubini–Study form) by

$$\mu(x) \cdot a = \frac{\bar{\hat{x}}^t \rho_*(a) \hat{x}}{2\pi i \|\hat{x}\|^2} \tag{1}$$

for all $a \in \mathfrak{k}$, where $\hat{x} \in \mathbb{C}^{n+1} - \{0\}$ is a representative vector for $x \in \mathbb{P}^n$ and the representation $\rho : K \to \mathrm{U}(n+1)$ induces $\rho_* : \mathfrak{k} \to \mathfrak{u}(n+1)$ and dually $\rho^* : \mathfrak{u}(n+1)^* \to \mathfrak{k}^*$.

In this situation there are two possible quotient constructions: the symplectic reduction $\mu^{-1}(0)/K$ in symplectic geometry and the GIT quotient $X//G$ in algebraic geometry described below. In fact these give us the same space, at least up to homeomorphism (and diffeomorphism away from the singularities).

2.2 Mumford's geometric invariant theory

Let X be a complex projective variety and let G be a complex reductive group acting on X. Recall that over $\mathbb{C}$ a linear algebraic group G is reductive if and only if it is the complexification of a maximal compact subgroup K. The simplest nontrivial example is the complexification $\mathbb{C}^*$ of the circle S^1, and more generally $\mathrm{GL}(n;\mathbb{C})$ is the complexification of the unitary group $U(n)$ and thus is reductive. In contrast the additive group of complex numbers $\mathbb{C}^+$ has no nontrivial compact subgroups and so is not reductive; the same is true of any complex linear algebraic group U which is unipotent (that is, U is isomorphic to a closed subgroup of the group of strictly upper triangular matrices in $\mathrm{GL}(n;\mathbb{C})$ for some n). In some sense reductive and unipotent groups sit at the opposite extremes of a spectrum, and any linear algebraic group H has a unique maximal unipotent normal subgroup U (its unipotent radical) such that the quotient group H/U is reductive.

Geometric invariant theory needs an extra ingredient in addition to the action of G on X, which is a linearisation of the action, that is, a line bundle L on X and a lift of the action of G to L. The line bundle L is usually taken to be ample, and then very little generality is lost by assuming that for some projective embedding

$$X \subseteq \mathbb{P}^n$$

the action of G on X extends to an action on $\mathbb{P}^n$ given by a representation

$$\rho : G \to \mathrm{GL}(n+1),$$

and taking for L the hyperplane line bundle $\mathcal{O}_{\mathbb{P}^n}(1)$ on $\mathbb{P}^n$.

A *categorical quotient* of a variety X under an action of G is a G-invariant morphism $\phi : X \to Y$ from X to a variety Y such that any other G-invariant morphism $\tilde{\phi} : X \to \tilde{Y}$ factors as $\tilde{\phi} = \chi \circ \phi$ for a unique morphism $\chi : Y \to \tilde{Y}$ [28, Chapter 2, §4]. An *orbit space* for the action is a categorical quotient $\phi : X \to Y$ such that each fibre $\phi^{-1}(y)$ is a single G-orbit, and a *geometric quotient* is an orbit space $\phi : X \to Y$ which is an affine morphism such that

(i) if U is open in Y then

$$\phi^* : \mathcal{O}(U) \to \mathcal{O}(\phi^{-1}(U))$$

induces an isomorphism of $\mathcal{O}(U)$ onto $\mathcal{O}(\phi^{-1}(U))^G$, and

(ii) if W_1 and W_2 are disjoint closed G-invariant subvarieties of X then their images $\phi(W_1)$ and $\phi(W_2)$ in Y are disjoint closed subvarieties of Y.

When G acts linearly on X as above there is an induced action of G on the homogeneous coordinate ring

$$\hat{\mathcal{O}}_L(X) = \bigoplus_{k \geq 0} H^0(X, L^{\otimes k}) \cong \mathbb{C}[x_0, \ldots, x_n]/\mathcal{I}_X, \tag{2}$$

where $\mathcal{I}_X$ is the ideal in $\mathbb{C}[x_0, \ldots, x_n]$ generated by the homogeneous polynomials vanishing on X. The subring $\hat{\mathcal{O}}_L(X)^G$ consisting of the elements of $\hat{\mathcal{O}}_L(X)$ left invariant by G is a finitely generated graded complex algebra because G is reductive, and so we can define the GIT quotient $X/\!/G$ to be the projective variety $\mathrm{Proj}(\hat{\mathcal{O}}_L(X)^G)$ associated to $\hat{\mathcal{O}}_L(X)^G$ [27]. The inclusion of $\hat{\mathcal{O}}_L(X)^G$ in $\hat{\mathcal{O}}_L(X)$ determines a rational map q from X to $X/\!/G$, but in general there will be points of $X \subseteq \mathbb{P}^n$ where every G-invariant polynomial vanishes and so this map will not be well defined everywhere on X. Hence we define the set X^{ss} of *semistable* points in X to be the set of those $x \in X$ for which there exists some $f \in \hat{\mathcal{O}}_L(X)^G$ not vanishing at x, and then the rational map q restricts to a surjective G-invariant morphism from the open subset X^{ss} of X to the quotient variety $X/\!/G$, which is a categorical quotient for the action of G on X^{ss}. This restriction $q : X^{\mathrm{ss}} \to X/\!/G$ is not necessarily an orbit space: when x and y are semistable points of X we have $q(x) = q(y)$ if and only if the closures $\overline{O_G(x)}$ and $\overline{O_G(y)}$ of the G-orbits of x and y meet in X^{ss}. Topologically $X/\!/G$ is the quotient of X^{ss} by the equivalence relation $\sim$ such that if x and y lie in X^{ss} then $x \sim y$ if and only if $\overline{O_G(x)}$ and $\overline{O_G(y)}$ meet in X^{ss}.

A *stable* point of X ("properly stable" in the terminology of [27]) is a point x of X^{ss} with a G-invariant neighbourhood in X^{ss} such that every G-orbit in this neighbourhood is closed in X^{ss} and has dimension $\dim G$. If U is any G-invariant open subset of the set X^{s} of stable points of X, then $q(U)$ is an open subset of $X/\!/G$ and the restriction $q|_U : U \to q(U)$ of q to U is an orbit space for the action of

G on U, so that it makes sense to write U/G for $q(U)$; in fact U/G is a geometric quotient for the action of G on U. In particular there is a geometric quotient X^s/G for the action of G on X^s, and $X//G$ can be thought of as a compactification of X^s/G:

$$\begin{array}{ccccc} X^s & \underset{\text{open}}{\subseteq} & X^{ss} & \underset{\text{open}}{\subseteq} & X \\ \downarrow & & \downarrow & & \\ X^s/G & \underset{\text{open}}{\subseteq} & X//G = X^{ss}/\sim & & \end{array} \tag{3}$$

Remark 2.1. X^s, X^{ss}, and $X//G$ are unaltered if for any $k > 0$ the line bundle L is replaced by $L^{\otimes k}$ with the induced action of G, so it is sometimes convenient to allow *fractional* linearisations $L^{\otimes \ell/m}$.

The subsets X^{ss} and X^s of X are characterised by the following properties (see Chapter 2 of [27] or [28]).

Proposition 2.2 (Hilbert–Mumford criteria).

(i) A point $x \in X$ is semistable (respectively stable) for the action of G on X if and only if for every $g \in G$ the point gx is semistable (respectively stable) for the action of a fixed maximal (complex) torus of G.

(ii) A point $x \in X$ with homogeneous coordinates $[x_0 : \ldots : x_n]$ in some coordinate system on $\mathbb{P}^n$ is semistable (respectively stable) for the action of a maximal (complex) torus of G acting diagonally on $\mathbb{P}^n$ with weights $\alpha_0, \ldots, \alpha_n$ if and only if the convex hull

$$\mathrm{Conv}\{\alpha_i : x_i \neq 0\}$$

contains 0 (respectively contains 0 in its interior).

The GIT quotient $X//G$ is homeomorphic to the symplectic quotient $\mu^{-1}(0)/K$, and the subsets X^{ss} and X^s of X can be described using the moment map μ at (1) above. More precisely [19], any $x \in X$ is semistable if and only if the closure of its G-orbit meets $\mu^{-1}(0)$, while x is stable if and only if its G-orbit meets

$$\mu^{-1}(0)_{\mathrm{reg}} = \{x \in \mu^{-1}(0) | d\mu(x) : T_xX \to \mathfrak{k}^* \text{ is surjective}\},$$

and the inclusions of $\mu^{-1}(0)$ into X^{ss} and of $\mu^{-1}(0)_{\mathrm{reg}}$ into X^s induce homeomorphisms

$$\mu^{-1}(0)/K \to X//G$$

and

$$\mu^{-1}(0)_{\mathrm{reg}}/K \to X^s/G.$$

Thus the moment map picks out a unique K-orbit in each stable G-orbit, and also in each equivalence class of strictly semistable G-orbits, where x and y in X^{ss} are equivalent if the closures of their G-orbits meet in X^{ss} (that is, if their images under the natural surjection $q : X^{ss} \to X//G$ agree).

Remark 2.3. It follows from the formula (1) that if we change the linearisation of the G-action of X by multiplying by a character $\chi : G \to \mathbb{C}^*$ of G, then the moment map is modified by the addition of a central constant c_χ in $\mathfrak{k}^*$, which we can identify with the restriction to $\mathfrak{k}$ of the derivative of χ.

Example 2.4. Let $G = \mathrm{SL}(2;\mathbb{C})$ act on $X = (\mathbb{P}^1)^4$ via Möbius transformations and let K be the maximal compact subgroup $\mathrm{SU}(2)$ of G. If we identify $\mathbb{P}^1$ with the unit sphere S^2 in $\mathbb{R}^3$ then there is a moment map

$$\mu : X = (S^2)^4 \to \mathfrak{k}^* \cong \mathbb{R}^3$$

given by $\mu(x_1, x_2, x_3, x_4) = x_1 + x_2 + x_3 + x_4$. Thus $\mu^{-1}(0)$ consists of configurations of four points on S^2 which are balanced in the sense that their centre of gravity lies at the origin, while $\mu^{-1}(0)\backslash\mu^{-1}(0)_{\mathrm{reg}}$ consists of the configurations in which two points coincide at some $p \in S^2$ and the other two points coincide at the antipodal point $-p$. The open subset

$$X^{\mathrm{s}} = \{(x_1, x_2, x_3, x_4) \in (\mathbb{P}^1)^4 : x_1, x_2, x_3, x_4 \text{ distinct}\}$$

of $X = (\mathbb{P}^1)^4$ has a geometric quotient which, using the cross-ratio, can be identified with

$$\mathbb{P}^1 - \{0, 1, \infty\},$$

and this in turn can be identified with $\mu^{-1}(0)_{\mathrm{reg}}/K$. In addition,

$$X^{\mathrm{ss}} = \{(x_1, x_2, x_3, x_4) \in (\mathbb{P}^1)^4 : \text{ at most two of } x_1, x_2, x_3, x_4 \text{ coincide}\}$$

has a categorical quotient $X/\!/G \cong X^{\mathrm{ss}}/\sim \cong \mathbb{P}^1$ in which the points $0, 1, \infty$ each represent three strictly semistable G-orbits in X: one G-orbit consisting of configurations in which two points x_i and x_j coincide at some $p \in \mathbb{P}^1$ and the other two points x_k and x_m coincide at a distinct point $q \in \mathbb{P}^1$, a second consisting of configurations in which x_i and x_j coincide at some $p \in \mathbb{P}^1$ and the other two points x_k and x_m are distinct from each other and from p, and the third consisting of configurations in which x_k and x_m coincide at a some point $q \in \mathbb{P}^1$ while x_i and x_j are distinct from each other and from q. The first of these orbits is closed in X^{ss} and lies in the closure of each of the other two orbits.

3 Symplectic implosion and quotients by nonreductive groups

Ways in which classical GIT might be generalised to actions of nonreductive affine algebraic groups on algebraic varieties were studied in [7] (see also [23]) building on earlier work such as [8, 9, 10, 11, 35]. Every affine algebraic group H has a unipotent radical $U \trianglelefteq H$ such that H/U is reductive, so we can concentrate on unipotent actions. It is shown in [7] that when a unipotent group U acts linearly (with respect to an ample line bundle L) on a complex projective variety X, then

X has invariant open subsets $X^{\mathrm{s}} \subseteq X^{\mathrm{ss}}$, consisting of the "stable" and "semistable" points for the action, such that X^{s} has a geometric quotient X^{s}/U and X^{ss} has a canonical "enveloping quotient" $X^{\mathrm{ss}} \to X/\!/U$, which restricts to $X^{\mathrm{s}} \to X^{\mathrm{s}}/U$, where X^{s}/U is an open subset of $X/\!/U$. However, in contrast to the reductive case, the natural map from X^{ss} to $X/\!/U$ is not necessarily surjective, and indeed its image is not necessarily a subvariety of $X/\!/U$, so this does not in general give us a categorical quotient of X^{ss}. Furthermore, $X/\!/U$ is in general only quasiprojective, not projective, though when the ring of invariants $\hat{\mathcal{O}}_L(X)^U = \bigoplus_{k\geq 0} H^0(X, L^{\otimes k})^U$ is finitely generated as a $\mathbb{C}$-algebra then $X/\!/U$ is the projective variety $\mathrm{Proj}(\hat{\mathcal{O}}_L(X)^U)$.

In order to obtain a compactification $\overline{X/\!/U}$ of the enveloping quotient $X/\!/U$ when the ring of invariants $\hat{\mathcal{O}}_L(X)^U$ is not finitely generated, and to understand its geometry even when $X/\!/U = \overline{X/\!/U}$ is itself projective, we can transfer the problem of constructing a quotient for the U-action to the construction of a quotient for an action of a reductive group G which contains U as a subgroup, by finding a reductive envelope. This is a projective completion

$$\overline{G \times_U X}$$

of the quasiprojective variety $G \times_U X$ (which is the quotient of $G \times X$ by the free action of U acting diagonally on the left on X and by right multiplication on G), with a linear G-action on $\overline{G \times_U X}$ extending the induced G-action on $G \times_U X$, such that the U-invariants on X lying in a suitable set (see Definition 4.3 below) extend to G-invariants on $\overline{G \times_U X}$. If the linearisation on $\overline{G \times_U X}$ is ample, then the classical GIT quotient

$$\overline{G \times_U X}/\!/G$$

is a compactification $\overline{X/\!/U}$ of $X/\!/U$, and hence also of its open subset X^{s}/U if $X^{\mathrm{s}} \neq \emptyset$. Moreover, if $X^{\bar{\mathrm{s}}}$ and $X^{\bar{\mathrm{ss}}}$ denote the open subsets of X consisting of points of X which are stable and semistable for the G-action on $\overline{G \times_U X}$ under the inclusion

$$X \hookrightarrow G \times_U X \hookrightarrow \overline{G \times_U X},$$

then

$$X^{\bar{\mathrm{s}}} \subseteq X^{\mathrm{s}} \subseteq X^{\mathrm{ss}} \subseteq X^{\bar{\mathrm{ss}}}.$$

Note, however, that $X^{\bar{\mathrm{s}}}$, $X^{\bar{\mathrm{ss}}}$, and $\overline{X/\!/U}$ depend in general on the choice of reductive envelope $\overline{G \times_U X}$ with its linear G-action, whereas X^{s}, X^{ss}, and $X/\!/U$ depend only on the linear action of U on X.

Just as GIT quotients by complex reductive groups are closely related to symplectic reduction, so quotients by suitable unipotent groups (in particular maximal unipotent subgroups of complex reductive groups) are closely related to the construction called symplectic implosion [14], which we will discuss below.

3.1 Symplectic implosion for a maximal unipotent subgroup

Let (X, ω) be a symplectic manifold on which a compact connected Lie group K acts with a moment map $\mu : X \to \mathfrak{k}^*$, where $\mathfrak{k}$ is the Lie algebra of K. Let us choose an invariant inner product on $\mathfrak{k}$ and use it to identify $\mathfrak{k}^*$ with $\mathfrak{k}$. Let T be a maximal torus of K with Lie algebra $\mathfrak{t} \subseteq \mathfrak{k}$ and Weyl group $W = N_K(T)/T$, and let $\mathfrak{t}^*_+ \cong \mathfrak{t}^*/W \cong \mathfrak{k}^*/\mathrm{Ad}^*(K)$ be a positive Weyl chamber in $\mathfrak{k}^*$. The *imploded cross-section* [14] of X is then

$$X_{\mathrm{impl}} = \mu^{-1}(\mathfrak{t}^*_+)/\approx, \tag{4}$$

where $x \approx y$ if and only if $\mu(x) = \mu(y) = \zeta \in \mathfrak{t}^*_+$ and $x = ky$ for some $k \in [K_\zeta, K_\zeta]$. Here K_ζ denotes the stabiliser $K_\zeta = \{k \in K : (\mathrm{Ad}^*k)\zeta = \zeta\}$ of ζ under the coadjoint action of K on $\mathfrak{k}^*$, and $[K_\zeta, K_\zeta]$ is its commutator subgroup. If Σ is the set of faces of $\mathfrak{t}^*_+$ then

$$X_{\mathrm{impl}} = \coprod_{\sigma \in \Sigma} \frac{\mu^{-1}(\sigma)}{[K_\sigma, K_\sigma]} = \mu^{-1}((\mathfrak{t}^*_+)^\circ) \sqcup \coprod_{\substack{\sigma \in \Sigma \\ \sigma \neq (\mathfrak{t}^*_+)^\circ}} \frac{\mu^{-1}(\sigma)}{[K_\sigma, K_\sigma]}, \tag{5}$$

where $K_\sigma = K_\zeta$ for any $\zeta \in \sigma$. The topology on X_{impl} is the quotient topology induced from $\mu^{-1}(\mathfrak{t}^*_+)$, and X_{impl} also inherits a symplectic structure. More precisely, it is stratified by the locally closed subsets $\mu^{-1}(\sigma)/[K_\sigma, K_\sigma]$, each of which is the symplectic reduction by the action of $[K_\sigma, K_\sigma]$ of a locally closed symplectic submanifold

$$X_\sigma = K_\sigma \mu^{-1}\left(\bigcup_{\tau \in \Sigma, \bar{\tau} \supseteq \sigma} \tau \right)$$

of X (and locally near every point, X_{impl} can be identified symplectically with the product of the stratum and a suitable cone in the normal direction). The induced action of T on X_{impl} preserves this symplectic structure and has a moment map

$$\mu_{X_{\mathrm{impl}}} : X_{\mathrm{impl}} \to \mathfrak{t}^*_+ \subseteq \mathfrak{t}^*$$

inherited from the restriction of μ to $\mu^{-1}(\mathfrak{t}^*_+)$. If $\zeta \in \mathfrak{t}^*_+$, the symplectic reduction of X_{impl} at ζ for this action of T is the symplectic reduction of X at ζ for the action of K:

$$\frac{\mu^{-1}_{X_{\mathrm{impl}}}(\zeta)}{T} = \frac{\mu^{-1}(\zeta)}{T \cdot [K_\zeta, K_\zeta]} = \frac{\mu^{-1}(\zeta)}{K_\zeta}. \tag{6}$$

The *universal imploded cross-section* is the imploded cross-section

$$(T^*K)_{\mathrm{impl}} = K \times \mathfrak{t}^*_+/\approx \tag{7}$$

of the cotangent bundle $T^*K \cong K \times \mathfrak{k}^*$ with respect to the K-action induced from the right action of K on itself; it inherits an action of $K \times T$ from the left action of K on itself and the right action of T on K. Any other imploded cross-section X_{impl} can be constructed as the symplectic quotient of the product $X \times (T^*K)_{\text{impl}}$ by the diagonal action of K [14, Theorem 4.9].

In fact $(T^*K)_{\text{impl}}$ is always a complex affine variety, and its symplectic structure is given by a Kähler form. Indeed, let $G = K_c$ be the complexification of K and let B be a Borel subgroup of G with $G = KB$ and $K \cap B = T$. If $U_{\max} \leq B$ is the unipotent radical of B (and hence a maximal unipotent subgroup of G), then $U_{\max}$ is a Grosshans subgroup of G [12]: that is, the quasiaffine variety $G/U_{\max}$ can be embedded as an open subset of an affine variety in such a way that its complement has (complex) codimension at least two. This means that the ring of invariants $\mathcal{O}(G)^{U_{\max}}$ is finitely generated (see for example [12]), and by [14, Proposition 6.8], there is a natural $K \times T$-equivariant identification

$$(T^*K)_{\text{impl}} \cong \operatorname{Spec}(\mathcal{O}(G)^{U_{\max}})$$

of the canonical affine completion $\operatorname{Spec}(\mathcal{O}(G)^{U_{\max}})$ of $G/U_{\max}$ with $(T^*K)_{\text{impl}}$. It follows that if X is a complex projective variety on which G acts linearly with respect to a very ample line bundle L, and ω is an associated K-invariant Kähler form on X, then the symplectic quotient X_{impl} of $X \times (T^*K)_{\text{impl}}$ by K can be identified with the GIT quotient $(X \times \operatorname{Spec}(\mathcal{O}(G)^{U_{\max}}))//G$. Moreover,

$$\hat{\mathcal{O}}_L(X)^{U_{\max}} \cong (\hat{\mathcal{O}}_L(X) \otimes \mathcal{O}(G)^{U_{\max}})^G$$

is finitely generated, and if we define the GIT quotient $X//U_{\max}$ to be the projective variety $\operatorname{Proj}(\hat{\mathcal{O}}_L(X)^{U_{\max}})$ associated to the ring of invariants $\hat{\mathcal{O}}_L(X)^{U_{\max}}$ then

$$X//U_{\max} = \operatorname{Proj}(\hat{\mathcal{O}}_L(X)^{U_{\max}}) \cong (X \times \operatorname{Spec}(\mathcal{O}(G)^{U_{\max}}))//G \cong X_{\text{impl}}. \quad (8)$$

The proof in [14, Section 6] that $(T^*K)_{\text{impl}}$ is homeomorphic to the canonical affine completion

$$\overline{G/U_{\max}}^{\text{aff}} = \operatorname{Spec}(\mathcal{O}(G)^{U_{\max}})$$

of $G/U_{\max}$ runs as follows. First it is possible to reduce to the case that K is semisimple and simply connected, by regarding K as the quotient by a finite central subgroup of $Z(K) \times \widetilde{[K,K]}$, where $Z(K)$ is the centre of K and $\widetilde{[K,K]}$ is the universal cover of the commutator subgroup $[K,K]$ of K.

Following [14, Section 6], if K is a semisimple, connected, and simply connected compact group let $\Lambda = \ker(\exp|_{\mathfrak{t}})$ be the exponential lattice in $\mathfrak{t}$, and let $\Lambda^* = \operatorname{Hom}_{\mathbb{Z}}(\Lambda, \mathbb{Z})$ be the weight lattice in $\mathfrak{t}^*$, so that $\Lambda^*_+ = \Lambda^* \cap \mathfrak{t}^*_+$ is the monoid of dominant weights. For $\lambda \in \Lambda^*_+$ let V_λ be the irreducible G-module with highest weight λ, and let

$$\Pi = \{\varpi_1, \ldots, \varpi_r\}$$

be the set of fundamental weights, which forms a $\mathbb{Z}$-basis of Λ^* and a minimal set of generators for Λ^*_+. Recall that $V^*_\lambda = V_{\iota\lambda}$ is the irreducible G-module with highest weight $\iota\lambda$, where $\iota : \mathfrak{t}^* \to \mathfrak{t}^*$ is the involution given by $\iota\lambda = -w_0\lambda$ and w_0 denotes the element of the Weyl group W of G such that $w_0 U_{\max} w_0^{-1} = U_{\max}^{\mathrm{op}}$ is the unipotent radical of the Borel subgroup B^{op} of G which is opposite to $B \geq U$ in the sense that $B \cap B^{\mathrm{op}}$ is the complexification T_c of T and $U_{\max} \cap U_{\max}^{\mathrm{op}}$ is the identity subgroup. We have an isomorphism of G-modules

$$\mathcal{O}(G)^{U_{\max}} \cong \bigoplus_{\lambda \in \Lambda_+^*} V_\lambda^* \cong \bigoplus_{\lambda \in \Lambda_+^*} V_{\iota\lambda}, \tag{9}$$

where G acts on itself on the left and $U_{\max}$ acts on G on the right. Note that T_c normalises $U_{\max}$ and this isomorphism (9) becomes an isomorphism of $G \times T_c$-modules if we let T_c act on V_λ with weight $-\lambda$ so that it acts on V_λ^* with weight λ (see [12, Section 12]). Equivalently we have an isomorphism of $G \times T_c$-modules

$$\mathcal{O}(G)^{U_{\max}} \cong \bigoplus_{\lambda \in \Lambda_+^*} V_\lambda^{(T)} \otimes V_\lambda^*, \tag{10}$$

where $V_\lambda^{(T)}$ is the irreducible T_c-module with weight λ, and by [12, Theorem 12.9] this isomorphism extends to an isomorphism of $G \times G$-modules

$$\mathcal{O}(G) \cong \bigoplus_{\lambda \in \Lambda_+^*} V_\lambda \otimes V_\lambda^*. \tag{11}$$

In particular, the algebra $\mathcal{O}(G)^{U_{\max}}$ is generated by its finite-dimensional vector subspace

$$\bigoplus_{\varpi \in \Pi} V_\varpi^* \cong \bigoplus_{\varpi \in \Pi} V_\varpi^{(T)} \otimes V_\varpi^*.$$

The inclusion of this finite-dimensional subspace into $\mathcal{O}(G)^{U_{\max}}$ induces a closed $G \times T_c$-equivariant embedding of $\overline{G/U_{\max}}^{\mathrm{aff}} = \mathrm{Spec}(\mathcal{O}(G)^{U_{\max}})$ into the affine space

$$E = \bigoplus_{\varpi \in \Pi} V_\varpi \cong \bigoplus_{\varpi \in \Pi} (V_\varpi^{(T)})^* \otimes V_\varpi,$$

sending the identity coset $U_{\max}$ in $G/U_{\max} \subseteq \overline{G/U_{\max}}^{\mathrm{aff}}$ to a sum

$$\sum_{\varpi \in \Pi} v_\varpi$$

of highest-weight vectors $v_\varpi \in V_\varpi \cong (V_\varpi^{(T)})^* \otimes V_\varpi$. Under this embedding $G/U_{\max}$ is identified with $G\,E^{U_{\max}}$, where $E^{U_{\max}}$ is the subspace of E consisting of vectors fixed by $U_{\max}$. We give E a flat Kähler structure ω_E via the unique $K \times T$-invariant Hermitian inner product on E which satisfies $\|v_\varpi\| = 1$ for each $\varpi \in \Pi$. Then by [14, Proposition 6.8] there is a $K \times T$-equivariant map $\mathcal{F} : K \times \mathfrak{t}_+^* \to E$ defined on $\mathfrak{t}_+^*$ by

$$\mathcal{F}(1,\lambda) = \frac{1}{\sqrt{\pi}} \sum_{j=1}^{r} \sqrt{\lambda(\alpha_j^\vee)} v_{\varpi_j}, \tag{12}$$

where $\alpha^\vee = 2\alpha/(\alpha \cdot \alpha)$ and

$$S = \{\alpha_1, \dots, \alpha_r\}$$

is the set of simple roots corresponding to the fundamental weights $\{\varpi_1, \dots, \varpi_r\}$ (so that $\varpi_i \cdot \alpha_j^\vee = \delta_{ij}$ for $i, j \in \{1, \dots, r\}$); moreover, $\mathcal{F}$ induces a homeomorphism from $(T^*K)_{\mathrm{impl}}$ to $\overline{G/U_{\max}}^{\mathrm{aff}}$ whose restriction to each stratum $\mu^{-1}(\sigma)/[K_\sigma, K_\sigma]$ of $(T^*K)_{\mathrm{impl}}$ is a symplectic isomorphism onto its image.

Remark 3.1. Let M be any compact Kähler manifold on which the complexified torus T_c acts in such a way that T preserves the Kähler structure and has a moment map $\mu_T : M \to \mathfrak{t}^*$. In [1, Theorem 2] Atiyah shows

(a) that the image $\mu_T(\bar{Y})$ under the torus moment map μ_T of the closure $\bar{Y}$ in M of the T_c-orbit $Y = T_c m$ of any $m \in M$ is a convex polytope $\mathcal{P}$ whose vertices are the images under μ_T of the connected components of $\bar{Y} \cap M^T$, where M^T is the T-fixed-point set in M,
(b) that the inverse image in $\bar{Y}$ of each open face of $\mathcal{P}$ consists of a single T_c-orbit, and
(c) that μ_T induces a homeomorphism of $\bar{Y}/T$ onto $\mathcal{P}$.

In fact Atiyah's proof shows that if $\mathcal{Y} = \exp(i\mathfrak{t})$ is the orbit of $m \in M$ under the subgroup $\exp(i\mathfrak{t})$ of T_c then μ_T restricts to a homeomorphism from $\bar{\mathcal{Y}}$ onto $\mathcal{P}$, and the inverse image in $\bar{\mathcal{Y}}$ of each open face of $\mathcal{P}$ consists of a single $\exp(i\mathfrak{t})$-orbit.

We can apply this to the compactification $M = \mathbb{P}(\mathbb{C} \oplus E)$ of the affine space E. The moment map $\mu_T^E : E \to \mathfrak{t}^*$ for the T-action on E with its chosen flat Kähler structure is given (up to multiplication by a positive constant) by

$$\sum_{\varpi} u_\varpi \mapsto \sum_{\varpi} \|u_\varpi\|^2 \varpi$$

when $u_\varpi \in V_\varpi$ for $\varpi \in \Pi$, while the moment map $\mu_T^{\mathbb{P}(\mathbb{C}\oplus E)} : \mathbb{P}(\mathbb{C} \oplus E) \to \mathfrak{t}^*$ for the T-action on $\mathbb{P}(\mathbb{C} \oplus E)$ with the induced Fubini–Study Kähler structure is given (up to multiplication by a positive constant) by

$$\left[z : \sum_{\varpi \in \Pi} u_\varpi \right] \mapsto \frac{\sum_\varpi \|u_\varpi\|^2 \varpi}{|z|^2 + \sum_{\varpi \in \Pi} \|u_\varpi\|^2}$$

when $z \in \mathbb{C}$ and $u_\varpi \in V_\varpi$ for $\varpi \in \Pi$ are not all zero. Comparing these two moment maps on E (regarded as an open subset of $\mathbb{P}(\mathbb{C} \oplus E)$ in the usual way) we see that the image under μ_T^E of the closure $\bar{\mathcal{Y}}$ in E of the $\exp(i\mathfrak{t})$-orbit $\mathcal{Y}$ in E of the vector $\sum_{\varpi \in \Pi} v_\varpi$ corresponding to the identity coset $U_{\max}$ in $G/U_{\max}$ is the cone in

$\mathfrak{t}^*$ spanned by the half-lines $\mathbb{R}_+\varpi$ for $\varpi \in \Pi$, which is of course the positive Weyl chamber $\mathfrak{t}^*_+$. We also find that the restriction

$$\mu_T^E|_{\bar{\mathcal{Y}}} : \bar{\mathcal{Y}} \to \mathfrak{t}^*_+ \tag{13}$$

is a homeomorphism, and it is easy to check that the map $\mathcal{F} : \mathfrak{t}^*_+ \to E$ of [14, Proposition 6.8] defined at (12) above can be identified with the composition of the inverse $(\mu_T^E|_{\bar{\mathcal{Y}}})^{-1} : \mathfrak{t}^*_+ \to \bar{\mathcal{Y}}$ of (13) and the inclusion of $\bar{\mathcal{Y}}$ in E. From this it can be deduced that its $K \times T$-equivariant extension $\mathcal{F} : K \times \mathfrak{t}^*_+ \to E$ induces a bijection from $(T^*K)_{\mathrm{impl}}$ onto the closure $\overline{G/U_{\max}}^{\mathrm{aff}}$ of $G(\sum_{\varpi\in\Pi} v_\varpi) \cong G/U_{\max}$ in E using

(i) the Iwasawa decomposition

$$G = K\exp(i\mathfrak{t})U_{\max}$$

of G, which tells us that $\overline{G/U_{\max}}^{\mathrm{aff}} = K\bar{\mathcal{Y}} = \mathcal{F}(K \times \mathfrak{t}^*_+)$, and

(ii) Lemma 6.2 of [14], which shows that for each face σ of $\mathfrak{t}^*_+$ the stabiliser in K of

$$\sum_{\varpi\in\sigma} v_\varpi$$

is $[K_\sigma, K_\sigma]$.

Guillemin, Jeffrey, and Sjamaar also construct a $K \times T$-equivariant desingularisation $\widetilde{(T^*K)}_{\mathrm{impl}}$ for the universal imploded cross-section $(T^*K)_{\mathrm{impl}} \cong \overline{G/U_{\max}}^{\mathrm{aff}}$ and a partial desingularisation $\widetilde{X}_{\mathrm{impl}}$ for X_{impl}. In [14, Section 7] they show that if the action of K on X has principal face the interior $(\mathfrak{t}^*_+)^\circ$ of $\mathfrak{t}^*_+$ (where the principal face is the minimal open face σ of $\mathfrak{t}^*_+$ such that $\mu(X) \cap \mathfrak{t}^*_+$ is contained in $\bar{\sigma}$), then $\widetilde{X}_{\mathrm{impl}}$ can be identified with the symplectic quotient of $X \times \widetilde{(T^*K)}_{\mathrm{impl}}$ by the induced action of K (and they observe without proof that the same is true for any principal face). Moreover, $\widetilde{(T^*K)}_{\mathrm{impl}}$ can be identified as a Hamiltonian K-manifold with the homogeneous complex vector bundle

$$\widetilde{G/U_{\max}}^{\mathrm{aff}} = G \times_B E^{U_{\max}} \tag{14}$$

over the flag manifold G/B, where the restriction to $G \times E^{U_{\max}}$ of the multiplication map $G \times E \to E$ induces a birational G-equivariant morphism

$$p_{U_{\max}} : \widetilde{G/U_{\max}}^{\mathrm{aff}} \to \overline{G/U_{\max}}^{\mathrm{aff}} = (T^*K)_{\mathrm{impl}} \subseteq E.$$

Note that the fixed-point set $E^{U_{\max}}$ of $U_{\max}$ in E is the closure in E of the T_c-orbit of $\sum_{\varpi\in\Pi} v_\varpi$. If $\lambda_0 \in \mathfrak{t}^*$ is regular dominant and $\epsilon > 0$ is sufficiently close to 0, and if ω_0 is the Kähler form on G/B given by regarding G/B as the coadjoint

K-orbit through $\epsilon\lambda_0$, then $p_{U_{\max}}^* \omega_E + q^*\omega_0$ is a Kähler form on $\widetilde{G/U_{\max}}^{\mathrm{aff}}$, where $q : G \times_B E \to G/B$ is the projection.

It is also shown in [14, Section 7] that the partial desingularisation $\widetilde{X_{\mathrm{impl}}}$ can alternatively be obtained from X_{impl} via a symplectic cut with respect to the T-action and the polyhedral cone $\epsilon\lambda_0 + \bar{\tau}$, where τ is the principal face of X, $\lambda_0 \in \tau$, and $\epsilon > 0$ is sufficiently close to 0; that is, $\widetilde{X_{\mathrm{impl}}}$ is the symplectic reduction at $\epsilon\lambda_0$ for the diagonal T-action on the product of X_{impl} and the symplectic toric manifold associated to the polyhedron $-\bar{\tau}$ (see [25, 26]).

3.2 Symplectic implosion for the unipotent radical of a parabolic subgroup

Now suppose that U is the unipotent radical of a parabolic subgroup P of the complex reductive group G. Recall (see, e.g., [4, 33]) that a parabolic subgroup of G is a closed subgroup which contains some Borel subgroup, and its unipotent radical is its unique maximal normal unipotent subgroup; thus by replacing P with a suitable conjugate in G if necessary, we can assume that P contains the Borel subgroup B of G and $U \leq U_{\max}$. Then $P = UL^{(P)} \cong U \rtimes L^{(P)}$, where the Levi subgroup $L^{(P)}$ of P contains the complex maximal torus T_c of G, and we can assume in addition that $L^{(P)}$ is the complexification of its intersection

$$K^{(P)} = L^{(P)} \cap K = P \cap K$$

with K. For some subset S_P of the set S of simple roots, P is the unique parabolic subgroup of G which contains B such that the root space $\mathfrak{g}_{-\alpha}$ for $\alpha \in S$ is contained in the Lie algebra of P if and only if $\alpha \in S_P$. The Lie algebra of $L^{(P)}$ is generated by the root spaces $\mathfrak{g}_\alpha$ and $\mathfrak{g}_{-\alpha}$ for $\alpha \in S_P$ together with the Lie algebra $\mathfrak{t}_c = \mathfrak{t} \otimes_{\mathbb{R}} \mathbb{C}$ of the complexification T_c of T. In addition, the Lie algebra of U is

$$\mathfrak{u} = \bigoplus_{\substack{\alpha \in R^+ \\ \mathfrak{g}_\alpha \not\subseteq \mathrm{Lie}(L^{(P)})}} \mathfrak{g}_\alpha, \tag{15}$$

where R^+ is the set of positive roots for G, while the Lie algebra of P is

$$\mathfrak{p} = \mathfrak{t}_c \oplus \bigoplus_{\alpha \in R(S_P)} \mathfrak{g}_\alpha, \tag{16}$$

where $R(S_P)$ is the union of R^+ with the set of all roots which can be written as sums of negatives of the simple roots in S_P. If we identify S with the set of vertices of the Dynkin diagram of K then the Dynkin diagram of the semisimple part $Q^{(P)} = [K^{(P)}, K^{(P)}]$ of $K^{(P)}$ is the subdiagram given by leaving out the vertices which do not belong to S_P. We can decompose $\mathfrak{k}^{(P)} = \mathrm{Lie}K^{(P)}$ and $\mathfrak{t}$ as

$$\mathfrak{k}^{(P)} = [\mathfrak{k}^{(P)}, \mathfrak{k}^{(P)}] \oplus \mathfrak{z}^{(P)} \quad \text{and} \quad \mathfrak{t} = \mathfrak{t}^{(P)} \oplus \mathfrak{z}^{(P)},$$

where $[\mathfrak{k}^{(P)}, \mathfrak{k}^{(P)}]$ is the Lie algebra of $Q^{(P)} = [K^{(P)}, K^{(P)}]$, while $\mathfrak{t}^{(P)}$ is the Lie algebra of the maximal torus $T^{(P)} = T \cap [K^{(P)}, K^{(P)}]$ of $Q^{(P)}$, and $\mathfrak{z}^{(P)}$ is the Lie algebra of the centre $Z(K^{(P)})$ of $K^{(P)}$. As before let $B^{\mathrm{op}} = T_c U_{\max}^{\mathrm{op}}$ be the Borel subgroup of G, with unipotent radical $U_{\max}^{\mathrm{op}}$, which is opposite to B in the sense that $B \cap B^{\mathrm{op}} = T_c$ and $U_{\max} \cap U_{\max}^{\mathrm{op}} = \{1\}$, and let $\iota : \mathfrak{t}^* \to \mathfrak{t}^*$ be the involution given by $\iota\lambda = -w_0\lambda$, where w_0 denotes the element of the Weyl group W of G such that $w_0 U_{\max} w_0^{-1} = U_{\max}^{\mathrm{op}}$.

By [13, Theorem 2.2] U is a Grosshans subgroup of G, and so, just as in the case $U = U_{\max}$, the ring of invariants $\mathcal{O}(G)^U$ is finitely generated and G/U has a canonical affine completion

$$G/U \subseteq \overline{G/U}^{\mathrm{aff}} = \mathrm{Spec}(\mathcal{O}(G)^U) \tag{17}$$

such that the complement of G/U in $\overline{G/U}^{\mathrm{aff}}$ has codimension two.

Remark 3.2. When $U = U_{\max}$ the Iwasawa decomposition

$$G = K \exp(i\mathfrak{t}) U_{\max}$$

enables us to identify $G/U_{\max}$ with $K \exp(i\mathfrak{t})$. More generally we have an analogous decomposition

$$\begin{aligned} G = K \times_{K^{(P)}} P = K \times_{K^{(P)}} L^{(P)} U = K \times_{K^{(P)}} K^{(P)} \exp(i\mathfrak{k}^{(P)}) U \\ = K \exp(i\mathfrak{k}^{(P)}) U, \end{aligned} \tag{18}$$

which enables us to identify G/U with $K \exp(i\mathfrak{k}^{(P)})$.

Let X be a complex projective variety on which G acts linearly with respect to a very ample line bundle L, and let ω be an associated K-invariant Kähler form on X. Then it follows by the Borel transfer theorem (see, e.g., [5] Lemma 4.1) that

$$\hat{\mathcal{O}}_L(X)^U \cong (\hat{\mathcal{O}}_L(X) \otimes \mathcal{O}(G)^U)^G$$

is finitely generated, and the associated projective variety

$$X/\!/U = \mathrm{Proj}(\hat{\mathcal{O}}_L(X)^U)$$

is isomorphic to the GIT quotient $(\overline{G/U}^{\mathrm{aff}} \times X)/\!/G$. Just as in the case $U = U_{\max}$, if we have a suitable K-invariant Kähler form on $\overline{G/U}^{\mathrm{aff}}$, then we will be able to identify $X/\!/U$ with a symplectic quotient of $\overline{G/U}^{\mathrm{aff}} \times X$ by K, and obtain a symplectic description of $X/\!/U$ analogous to symplectic implosion, with $\overline{G/U}^{\mathrm{aff}}$ playing the role of the universal imploded cross-section $(T^*K)_{\mathrm{impl}}$. As is observed in [14, Section 6], the easiest case is that in which K is semisimple and simply connected (for example when $K = \mathrm{SU}(r+1)$); for general compact connected K

one can reduce to this case by considering the product $\tilde{K}$ of the centre of K and the universal cover of its commutator subgroup $[K, K]$, and expressing K as $\tilde{K}/\Upsilon$, where Υ is a finite central subgroup of $\tilde{K}$.

Therefore, as in the previous subsection, let K be a semisimple, connected, and simply connected compact group, let $\Lambda = \ker(\exp|_{\mathfrak{t}})$ be the exponential lattice in $\mathfrak{t}$, and let $\Lambda^* = \mathrm{Hom}_{\mathbb{Z}}(\Lambda, \mathbb{Z})$ be the weight lattice in $\mathfrak{t}^*$, so that $\Lambda^*_+ = \Lambda^* \cap \mathfrak{t}^*_+$ is the monoid of dominant weights. For $\lambda \in \Lambda^*_+$ let V_λ be the irreducible G-module with highest weight λ, and let $\Pi = \{\varpi_1, \dots, \varpi_r\}$ be the set of fundamental weights, forming a $\mathbb{Z}$-basis of Λ^* and a minimal set of generators for Λ^*_+. Recall that we have an isomorphism of $G \times G$-modules

$$\mathcal{O}(G) \cong \bigoplus_{\lambda \in \Lambda^*_+} V_\lambda \otimes V_\lambda^* \cong \bigoplus_{\lambda \in \Lambda^*_+} V_\lambda \otimes V_{\iota\lambda} \tag{19}$$

which restricts to an isomorphism of $G \times T_c$-modules

$$\mathcal{O}(G)^{U_{\max}} \cong \bigoplus_{\lambda \in \Lambda^*_+} V_\lambda^{(T)} \otimes V_\lambda^* \cong \bigoplus_{\lambda \in \Lambda^*_+} V_\lambda^*, \tag{20}$$

which is generated as an algebra by its finite-dimensional vector subspace

$$E^* = \bigoplus_{\varpi \in \Pi} V_\varpi^*,$$

giving us a closed $G \times T_c$-equivariant embedding of $\overline{G/U_{\max}}^{\mathrm{aff}} = \mathrm{Spec}(\mathcal{O}(G)^{U_{\max}})$ into the affine space E equipped with a flat Kähler structure. We have seen how Guillemin, Jeffrey, and Sjamaar identify $(T^*K)_{\mathrm{impl}}$ with $\overline{G/U_{\max}}^{\mathrm{aff}}$ equipped with the Kähler structure obtained from this embedding in E. To extend their construction to $\overline{G/U}^{\mathrm{aff}}$ when U is the unipotent radical of a parabolic subgroup $P \geq B$ as above, we first observe from the proof of [13, Theorem 2.2] that $\mathcal{O}(G)^U$ is generated by any finite-dimensional $L^{(P)}$-invariant (or equivalently $K^{(P)}$-invariant) vector subspace of

$$\mathcal{O}(G) \cong \bigoplus_{\lambda \in \Lambda^*_+} V_\lambda \otimes V_\lambda^* \cong \bigoplus_{\lambda \in \Lambda^*_+} V_\lambda \otimes V_{\iota\lambda}$$

which contains

$$E^* = \bigoplus_{\varpi \in \Pi} V_\varpi^* \cong \bigoplus_{\varpi \in \Pi} V_\varpi^{(T)} \otimes V_\varpi^*.$$

Here as above $V_\varpi^{(T)}$ is the irreducible T_c-module with weight ϖ, while $K^{(P)} = K \cap L^{(P)} = K \cap P$ is a maximal compact subgroup of the Levi subgroup $L^{(P)} = {K^{(P)}}_c$ of P, and $K^{(P)}$ acts on $\mathcal{O}(G)$ via left multiplication on G.

Let $E^{(P)}$ be the dual of the smallest $K^{(P)}$-invariant subspace $(E^{(P)})^*$ of $\mathcal{O}(G)$ containing E^*; then $(E^{(P)})^*$ is fixed pointwise by U, since $K^{(P)}$ normalises U and U is a subgroup of $U_{\max}$, which fixes E pointwise. The inclusion of $(E^{(P)})^*$ in $\mathcal{O}(G)^U \subseteq \mathcal{O}(G)$ induces a closed $L^{(P)} \times G$-equivariant embedding of $\overline{G/U}^{\mathrm{aff}} =$

Spec($\mathcal{O}(G)^U$) into the affine space $E^{(P)}$, whose projection to E induces the embedding of $\overline{G/U}^{\mathrm{aff}}_{\max}$ described in the previous subsection.

The subspace $(E^{(P)})^*$ decomposes under the action of $K \times K^{(P)}$ as a direct sum of irreducible $K \times K^{(P)}$-modules

$$(E^{(P)})^* = \bigoplus_{\varpi \in \Pi} (V_\varpi^{(P)})^*,$$

where $(V_\varpi^{(P)})^*$ is the smallest $K \times K^{(P)}$-invariant subspace of $\mathcal{O}(G)$ containing V_ϖ^*. As in [12, Section 12] we have

$$(V_\varpi^{(P)})^* \cong V_\varpi^{K^{(P)}} \otimes V_\varpi^*,$$

where $V_\varpi^{K^{(P)}}$ is the irreducible $K^{(P)}$-module with highest weight ϖ, so

$$E^{(P)} = \bigoplus_{\varpi \in \Pi} V_\varpi^{(P)} = \bigoplus_{\varpi \in \Pi} (V_\varpi^{K^{(P)}})^* \otimes V_\varpi. \tag{21}$$

Moreover, if $v_\varpi^{(P)}$ is the vector in $V_\varpi^{(P)} \cong (V_\varpi^{K^{(P)}})^* \otimes V_\varpi$ representing the inclusion of $V_\varpi^{K^{(P)}}$ in V_ϖ then the embedding of $G/U \subseteq \overline{G/U}^{\mathrm{aff}}$ in $E^{(P)}$ induced by the inclusion of $(E^{(P)})^*$ in $\mathcal{O}(G)^U$ takes the identity coset U to $\sum_{\varpi \in \Pi} v_\varpi^{(P)}$. Let

$$V_\varpi^{K^{(P)}} = \bigoplus_{\lambda \in \Lambda_\varpi^*} V_{\varpi,\lambda}^{K^{(P)}}$$

be the decomposition of $V_\varpi^{K^{(P)}}$ into weight spaces with weights $\lambda \in \mathfrak{t}^*$ under the action of the maximal torus T of $K^{(P)}$. Then $V_\varpi^{(P)}$ decomposes as a $K \times T$-module into a sum of irreducible $K \times T$-modules

$$V_\varpi^{(P)} \cong \bigoplus_{\lambda} V_\varpi \otimes (V_{\varpi,\lambda}^{K^{(P)}})^* \tag{22}$$

and $v_\varpi^{(P)} = \sum_\lambda v_{\varpi,\lambda}^{(P)}$, where $v_{\varpi,\lambda}^{(P)} \in V_\varpi \otimes (V_{\varpi,\lambda}^{K^{(P)}})^*$ represents the inclusion of $V_{\varpi,\lambda}^{K^{(P)}}$ in V_ϖ. In particular, $v_{\varpi,\varpi}^{(P)}$ is a highest-weight vector for the action of $K \times K^{(P)}$ on $V_\varpi^{(P)}$.

Remark 3.3. The embedding of $G/U \subseteq \overline{G/U}^{\mathrm{aff}}$ in $E^{(P)}$ induced by the inclusion of $(E^{(P)})^*$ in $\mathcal{O}(G)^U$ takes the identity coset to $\sum_{\varpi \in \Pi} v_\varpi^{(P)}$. From the decomposition $G = K \exp(i\mathfrak{k}^{(P)})U$ (see Remark 3.2 above) and the compactness of K it follows that the closure $\overline{G/U}^{\mathrm{aff}}$ of the G-orbit of $\sum_{\varpi \in \Pi} v_\varpi^{(P)}$ in $E^{(P)}$ is given by the K-sweep

$$\overline{G/U}^{\mathrm{aff}} = K\left(\overline{\exp(i\mathfrak{k}^{(P)}) \sum_{\varpi \in \Pi} v_\varpi^{(P)}}\right)$$

of the closure in $E^{(P)}$ of the $\exp(i\mathfrak{k}^{(P)})$-orbit of $\sum_{\varpi\in\Pi} v_\varpi^{(P)}$. Similarly the closure in $E^{(P)}$ (or equivalently in the linear subspace $\bigoplus_{\varpi\in\Pi}(V_\varpi^{K^{(P)}})^*\otimes V_\varpi^{K^{(P)}}$ of $E^{(P)}$) of the $L^{(P)}$-orbit of $\sum_{\varpi\in\Pi} v_\varpi^{(P)}$ (which is a free orbit since $U\cap L^{(P)}=\{1\}$) is given by $K^{(P)}\overline{\left(\exp(i\mathfrak{k}^{(P)})\sum_{\varpi\in\Pi} v_\varpi^{(P)}\right)}$. Note alse that $\mathfrak{k}^{(P)}=\bigcup_{k\in K^{(P)}}\mathrm{Ad}(k)\mathfrak{t}$ and so

$$\exp(i\mathfrak{k}^{(P)})=\bigcup_{k\in K^{(P)}} k\exp(i\mathfrak{t})k^{-1}. \tag{23}$$

Let $S_P=\{\alpha_1,\dots,\alpha_{r(P)}\}\subseteq S=\{\alpha_1,\dots,\alpha_r\}$ be the set of simple roots for the root system of $(K^{(P)},T)$ with corresponding positive Weyl chamber

$$\mathfrak{t}^*_{+,P}=\{\zeta\in\mathfrak{t}^*:\zeta\cdot\alpha\geq 0 \text{ for all } \alpha\in S_P\}=\mathfrak{t}_+^{(P)*}\oplus\mathfrak{z}^{(P)*},$$

where $\mathfrak{z}^{(P)}$ is the Lie algebra of the centre $Z(K^{(P)})\leq T$ of $K^{(P)}$ and $\mathfrak{t}_+^{(P)*}$ is the positive Weyl chamber for the semisimple part

$$Q^{(P)}=[K^{(P)},K^{(P)}]$$

of $K^{(P)}$ with respect to the maximal torus $T^{(P)}=T\cap[K^{(P)},K^{(P)}]$ of $Q^{(P)}$ and simple roots given by restricting S_P to $T^{(P)}$. If $\varpi\cdot\alpha=0$ for all $\alpha\in S_P$ (or equivalently if $\varpi=\varpi_j$ for $j>r(P)$) then $\varpi\in\mathfrak{z}^{(P)*}$ and $V_\varpi^{K^{(P)}}$ is one-dimensional; in this situation $Q^{(P)}$ acts trivially on $V_\varpi^{K^{(P)}}$ and we have $V_\varpi^{K^{(P)}}=V_{\varpi,\varpi}^{K^{(P)}}$ with $V_{\varpi,\lambda}^{K^{(P)}}=0$ if $\lambda\neq\varpi$, and $v_\varpi^{K^{(P)}}=v_{\varpi,\varpi}^{K^{(P)}}$, while $v_{\varpi,\lambda}^{K^{(P)}}=0$ if $\lambda\neq\varpi$. On the other hand, if $j\leq r(P)$ then $\varpi=\varpi_j$ restricts to a fundamental weight for $Q^{(P)}$ and $V_\varpi^{K^{(P)}}=V_\varpi^{Q^{(P)}}$ is the irreducible $Q^{(P)}$-module with highest weight $\varpi|_{Q^{(P)}}$ on which $Z(K^{(P)})$ acts as scalar multiplication by $\varpi|_{Z(K^{(P)})}$.

There is a unique $K\times K^{(P)}$-invariant Hermitian inner product on $E^{(P)}=\bigoplus_{\varpi\in\Pi}V_\varpi^{(P)}$ satisfying $\|v_{\varpi,\varpi}^{(P)}\|=1$ for each $\varpi\in\Pi$, which is obtained from K-invariant Hermitian inner products on the irreducible K-modules V_ϖ and their restrictions to $K^{(P)}$-invariant Hermitian inner products on the irreducible $K^{(P)}$-modules $V_\varpi^{(P)}$. This gives $E^{(P)}$ a flat Kähler structure which is $K\times K^{(P)}$-invariant.

Remark 3.4. Recall that

$$E^{(P)}=\bigoplus_{\varpi\in\Pi}V_\varpi^{(P)}=\bigoplus_{\varpi\in\Pi}(V_\varpi^{K^{(P)}})^*\otimes V_\varpi,$$

where $V_\varpi^{K^{(P)}}\subseteq V_\varpi$, and the embedding of $G/U\subseteq\overline{G/U}^{\mathrm{aff}}$ in $E^{(P)}$ induced by the inclusion of $(E^{(P)})^*$ in $\mathcal{O}(G)^U$ takes the identity coset U to $\sum_{\varpi\in\Pi}v_\varpi^{(P)}$, where $v_\varpi^{(P)}\in V_\varpi^{(P)}\cong(V_\varpi^{K^{(P)}})^*\otimes V_\varpi$ represents the inclusion of $V_\varpi^{K^{(P)}}$ in V_ϖ. Thus

$$\sum_{\varpi\in\Pi}v_\varpi^{(P)}\in\bigoplus_{\varpi\in\Pi}(V_\varpi^{K^{(P)}})^*\otimes V_\varpi^{K^{(P)}}\subseteq E^{(P)},$$

where $\bigoplus_{\varpi\in\Pi}(V_\varpi^{K^{(P)}})^*\otimes V_\varpi^{K^{(P)}}$ is invariant under the action of the subgroup $K^{(P)}\times K^{(P)}$ of $K\times K^{(P)}$ on $E^{(P)}$, and indeed is invariant under the action of $L^{(P)}\times L^{(P)}$. If we identify $(V_\varpi^{K^{(P)}})^*\otimes V_\varpi^{K^{(P)}}$ with $\mathrm{End}(V_\varpi^{K^{(P)}})$ equipped with the Hermitian structure

$$\langle A, B\rangle = \mathrm{Trace}(AB^*)$$

in the standard way, then $v_\varpi^{(P)}$ is identified with the identity map in $\mathrm{End}(V_\varpi^{K^{(P)}})$. If V is any Hermitian vector space then the moment map for the action of the product of unitary groups $U(V)\times U(V)$ on $\mathrm{End}(V)$ by left and right multiplication is given (up to a nonzero real scalar) by

$$A\mapsto (iAA^*, iA^*A)$$

(cf. [29, Section 3.3]). Thus the moment map for the action of $K^{(P)}\times K^{(P)}$ on

$$\bigoplus_{\varpi\in\Pi}(V_\varpi^{K^{(P)}})^*\otimes V_\varpi^{K^{(P)}}\cong\bigoplus_{\varpi\in\Pi}\mathrm{End}(V_\varpi^{K^{(P)}})$$

is given (up to multiplication by a nonzero real scalar) by

$$\sum_{\varpi\in\Pi}A_\varpi\mapsto\left(\pi^{K^{(P)}}\left(\sum_{\varpi\in\Pi}iA_\varpi A_\varpi^*\right),\pi^{K^{(P)}}\left(\sum_{\varpi\in\Pi}iA_\varpi^*A_\varpi\right)\right), \tag{24}$$

where $\pi^{K^{(P)}}:\mathfrak{u}\big(\bigoplus_{\varpi\in\Pi}V_\varpi^{K^{(P)}}\big)^*\to(\mathfrak{k}^{(P)})^*$ is the projection induced by the inclusion of $K^{(P)}$ as a subgroup of the unitary group $\mathrm{U}\big(\bigoplus_{\varpi\in\Pi}V_\varpi^{K^{(P)}}\big)$. In particular, if g belongs to the complexification $L^{(P)}$ of $K^{(P)}$ and $g_\varpi : V_\varpi^{K^{(P)}}\to V_\varpi^{K^{(P)}}$ is the action of g on $V_\varpi^{K^{(P)}}$, then

$$g\sum_{\varpi\in\Pi}v_\varpi^{(P)}=\sum_{\varpi\in\Pi}g_\varpi$$

and the moment map for the left $K^{(P)}$-action sends this to

$$\pi^{K^{(P)}}\left(\sum_{\varpi\in\Pi}ig_\varpi g_\varpi^*\right)\in\mathfrak{k}^{(P)}.$$

Using the decomposition $\mathfrak{k}^{(P)}=[\mathfrak{k}^{(P)},\mathfrak{k}^{(P)}]\oplus\mathfrak{z}^{(P)}$ we can decompose $\pi^{K^{(P)}}:\mathfrak{u}\big(\bigoplus_{\varpi\in\Pi}V_\varpi^{K^{(P)}}\big)^*\to(\mathfrak{k}^{(P)})^*$ as

$$\pi^{K^{(P)}}=\pi^{[K^{(P)},K^{(P)}]}\oplus\pi^{Z(K^{(P)})}:\mathfrak{u}\left(\bigoplus_{\varpi\in\Pi}V_\varpi^{K^{(P)}}\right)^*\to[\mathfrak{k}^{(P)},\mathfrak{k}^{(P)}]^*\oplus\mathfrak{z}^{(P)*}. \tag{25}$$

If $g=yz$ with $y\in[L^{(P)},L^{(P)}]=Q_c^{(P)}$ and $z\in Z(L^{(P)})=Z(K^{(P)})_c$, then the $K^{(P)}$-moment map above sends $g\sum_{\varpi\in\Pi}v_\varpi^{(P)}$ to

$$\pi^{[K^{(P)},K^{(P)}]}\left(\sum_{1\le j\le r(P)} iy_{\varpi_j}y^*_{\varpi_j}\right)+\pi^{Z(K^{(P)})}\left(\sum_{1\le j\le r} iz_{\varpi_j}z^*_{\varpi_j}\right)$$
$$\in [\mathfrak{k}^{(P)},\mathfrak{k}^{(P)}]^*\oplus\mathfrak{z}^{(P)*}.$$

It follows by the arguments of [29, Section 3] (in particular Proposition 3.10) that the $T_c^{(P)}$-orbit of $\sum_{\varpi\in\Pi} v_{\varpi}^{(P)}$ is mapped diffeomorphically onto $\mathfrak{t}^{(P)}$ by the moment map

$$y\sum_{\varpi\in\Pi} v_{\varpi}^{(P)} \mapsto \pi^{T^{(P)}}\left(\sum_{1\le j\le r(P)} iy_{\varpi_j}y^*_{\varpi_j}\right) \tag{26}$$

for the action of $T^{(P)}$ on $E^{(P)}$, since its image in the projective space $\mathbb{P}(E^{(P)})$ is mapped diffeomorphically by the associated moment map onto the convex hull of the set $\{w\varpi : \varpi\in\Pi, w\in W^{(P)}\}$, where $W^{(P)}$ is the Weyl group of $Q^{(P)}=[K^{(P)},K^{(P)}]$ (cf. Remark 3.1).

Now consider the moment map $\mu_T^{E^{(P)}}$ for the restriction to T of the $K^{(P)}$-action on $E^{(P)}$. This is given (up to multiplication by a positive constant) by

$$\sum_{\varpi,\lambda} u_{\varpi,\lambda}\mapsto\sum_{\varpi,\lambda}\|u_{\varpi,\lambda}\|^2\lambda$$

when $u_{\varpi,\lambda}\in V_{\varpi,\lambda}^{K^{(P)}}$ for $\varpi\in\Pi$ and $\lambda\in\Lambda^*_{\varpi}\subseteq\Lambda^*$. The embedding of $G/U\subseteq\overline{G/U}^{\text{aff}}$ in $E^{(P)}$ induced by the inclusion of $(E^{(P)})^*$ in $\mathcal{O}(G)^U$ takes the coset of $t\in T_c$ to

$$\sum_{\varpi,\lambda}\lambda(t)^{-1}v_{\varpi,\lambda}^{(P)},$$

so the value taken by this moment map on the coset tU of $t=t_1t_2\in T_c$, where $t_1\in T_c^{(P)}$ and $t_2\in Z(L^{(P)})=Z(K^{(P)})_c$, is given by

$$\sum_{\varpi,\lambda}|\lambda(t)|^{-2}\|v_{\varpi,\lambda}^{(P)}\|^2\lambda=\sum_{j=1}^{r(P)}|\varpi_j(t_2)|^{-2}\sum_{\lambda}|\lambda(t_1)|^{-2}\|v_{\varpi_j,\lambda}^{(P)}\|^2\lambda$$
$$+\sum_{j=r(P)+1}^{r}|\varpi_j(t_2)|^{-2}\|v_{\varpi_j,\varpi_j}^{(P)}\|^2\varpi_j, \tag{27}$$

where the jth sum over λ runs over all the weights of the irreducible $K^{(P)}$-module $V_{\varpi_j}^{K^{(P)}}$ with highest weight ϖ_j. When we decompose $\mathfrak{t}^*$ as $\mathfrak{t}^{(P)*}\oplus\mathfrak{z}^{(P)*}$ this has component

$$\sum_{j=1}^{r}|\varpi_j(t_2)|^{-2}\|v_{\varpi_j}^{(P)}\|^2\varpi_j|_{Z(K^{(P)}}\text{ in }\mathfrak{z}^{(P)*}$$

and

$$\sum_{j=1}^{r(P)} |\varpi_j(t_2)|^{-2} \sum_{\lambda} |\lambda(t_1)|^{-2} \|v_{\varpi_j,\lambda}^{(P)}\|^2 \lambda|_{T^{(P)}} \text{ in } \mathfrak{t}^{(P)*}.$$

Definition 3.5. Let $\mathfrak{t}^*_{(P)+}$ be the cone

$$\mathfrak{t}^*_{(P)+} = \bigcup_{w \in W^{(P)}} \mathrm{Ad}^*(w)\mathfrak{t}^*_+$$

in $\mathfrak{t}^*$, where $W^{(P)}$ is the Weyl group of $Q^{(P)} = [K^{(P)}, K^{(P)}]$ (which is a subgroup of the Weyl group W of K).

Lemma 3.6. *The restriction to the closure $\overline{\exp(i\mathfrak{t}) \sum_{\varpi \in \Pi} v_\varpi^{(P)}}$ of the $\exp(i\mathfrak{t})$-orbit in $E^{(P)}$ of $\sum_{\varpi \in \Pi} v_\varpi^{(P)}$ of the moment map $\mu_T^{E^{(P)}}$ for the action of T on $E^{(P)}$ is a homeomorphism onto the cone $\mathfrak{t}^*_{(P)+}$ in $\mathfrak{t}^*$. Its inverse provides a continuous injection*

$$\mathcal{F}^{(P)} : \mathfrak{t}^*_{(P)+} \to \overline{G/U}^{\mathrm{aff}} \subseteq E^{(P)} \tag{28}$$

such that $\mu_T^{E^{(P)}} \circ \mathcal{F}^{(P)}$ is the identity on $\mathfrak{t}^*_{(P)+}$. Moreover, $\overline{\exp(i\mathfrak{t}) \sum_{\varpi \in \Pi} v_\varpi^{(P)}}$ is the union of finitely many $\exp(i\mathfrak{t})$-orbits, each of the form

$$\mathcal{F}^{(P)}(\sigma) = \exp(i\mathfrak{t}) \sum_{\varpi \in \Pi, \lambda \in \Lambda_\varpi^* \cap \bar{\sigma}} v_{\varpi,\lambda}^{(P)},$$

where σ is an open face of $\mathfrak{t}^*_{(P)+}$.

Proof. This follows by applying the results of [1] to the compactification $\mathbb{P}(\mathbb{C} \oplus E^{(P)})$ of the affine space $E^{(P)}$, as in Remark 3.1, and observing that the convex hull of the weights λ of the T-action on the $K^{(P)}$-module $V_\varpi^{K^{(P)}}$ is the convex hull of $\{w\varpi : w \in W^{(P)}\}$, and thus the convex hull of the half-lines $\mathbb{R}_+\lambda$ for $\lambda \in \Lambda_\varpi$ with $\varpi \in \Pi$ is the cone $\mathfrak{t}^*_{(P)+}$.

Lemma 3.7 (cf. [29, Lemma 3.12]). *The image of the closure $\overline{T_c \sum_{\varpi \in \Pi} v_\varpi^{(P)}}$ of the T_c-orbit in $E^{(P)}$ of $\sum_{\varpi \in \Pi} v_\varpi^{(P)}$ under the $K^{(P)}$-moment map $\mu^{E^{(P)}} : E^{(P)} \to (\mathfrak{k}^{(P)})^* \cong \mathfrak{k}^{(P)}$ is contained in $\mathfrak{t}$.*

Proof. The orthogonal complement to $\mathfrak{t}$ in $\mathfrak{k}^{(P)}$ is $[\mathfrak{k}^{(P)}, \mathfrak{t}]$, and if $\zeta \in \mathfrak{t}$ and $\xi \in \mathfrak{k}^{(P)}$ and $t \in T_c$ then by Remark 3.4,

$$\begin{aligned} \mu^{E^{(P)}}\left(t \sum_{\varpi \in \Pi} v_\varpi^{(P)}\right) \cdot [\xi, \zeta] &= \sum_{\varpi \in \Pi} \mathrm{Trace}(i[\xi, \zeta] t_\varpi t_\varpi^*) \\ &= \sum_{\varpi \in \Pi} \mathrm{Trace}(i\xi[\zeta, t_\varpi t_\varpi^*]) = 0, \end{aligned}$$

since $[\zeta, t_\varpi t_\varpi^*] = 0$.

Corollary 3.8. *The restriction of the $K^{(P)}$-moment map $\mu^{E^{(P)}} : E^{(P)} \to (\mathfrak{k}^{(P)})^*$ to the closure*

$$\overline{\exp(i\mathfrak{k}^{(P)}) \sum_{\varpi \in \Pi} v_\varpi^{(P)}}$$

of the $\exp(i\mathfrak{k}^{(P)})$-orbit in $E^{(P)}$ of $\sum_{\varpi \in \Pi} v_\varpi^{(P)}$ is a homeomorphism from $\overline{\exp(i\mathfrak{k}^{(P)}) \sum_{\varpi \in \Pi} v_\varpi^{(P)}}$ onto the closed subset

$$\mathfrak{k}_+^{(P)*} = \mathrm{Ad}^*(K^{(P)})\mathfrak{t}^*_{(P)+}$$

of $\mathfrak{k}^{(P)*}$. Moreover, $\overline{\exp(i\mathfrak{k}^{(P)}) \sum_{\varpi \in \Pi} v_\varpi^{(P)}}$ is the union of finitely many $\exp(i\mathfrak{k}^{(P)})$-orbits which correspond under this homeomorphism to the open faces of $\mathfrak{k}_+^{(P)*}$.

Proof. We have already observed that the restriction of the T-moment map $\mu_T^{E^{(P)}}$: $E^{(P)} \to \mathfrak{t}^*$ to the closure

$$\overline{\exp(i\mathfrak{t}) \sum_{\varpi \in \Pi} v_\varpi^{(P)}}$$

of the $\exp(i\mathfrak{t})$-orbit of the image $\sum_{\varpi \in \Pi} v_\varpi^{(P)}$ of the identity coset U under the embedding of G/U in $E^{(P)}$ is a homeomorphism from this closure onto the cone $\mathfrak{t}^*_{(P)+}$. Since $\mu_T^{E^{(P)}}$ is the projection of $\mu^{E^{(P)}}$ onto $\mathfrak{t}^*$, it follows immediately from Lemma 3.7 above that the restriction of $\mu^{E^{(P)}} : E^{(P)} \to (\mathfrak{k}^{(P)})^* \cong \mathfrak{k}^{(P)}$ to this closure $\overline{\exp(i\mathfrak{t}) \sum_{\varpi \in \Pi} v_\varpi^{(P)}}$ is a homeomorphism onto the cone $\mathfrak{t}^*_{(P)+}$ when $\mathfrak{t}^*$ is identified with $\mathfrak{t} \subseteq \mathfrak{k}^{(P)}$ via the restriction of the fixed invariant inner product on $\mathfrak{k}$. Replacing the maximal torus T with kTk^{-1} for any $k \in K^{(P)}$, it follows that the restriction of $\mu^{E^{(P)}} : E^{(P)} \to (\mathfrak{k}^{(P)})^*$ to the closure $\overline{k \exp(i\mathfrak{t})k^{-1} \sum_{\varpi \in \Pi} v_\varpi^{(P)}}$ of the $\exp(i\,\mathrm{Ad}(k)\mathfrak{t})$-orbit of the image $\sum_{\varpi \in \Pi} v_\varpi^{(P)}$ of the identity coset U under the embedding of G/U in $E^{(P)}$ is a homeomorphism onto the cone $\mathrm{Ad}^*(k)\mathfrak{t}^*_{(P)+}$. Putting these homeomorphisms together for $k \in K^{(P)}$, we get a homeomorphism $\mathcal{M}$ from

$$\mathcal{Z} = \left\{ (kN_T^{(P)}, x) \in K^{(P)}/N_T^{(P)} \times E^{(P)} : x \in \overline{k \exp(i\mathfrak{t})k^{-1} \sum_{\varpi \in \Pi} v_\varpi^{(P)}} \right\},$$

where $N_T^{(P)}$ is the normaliser of T in $K^{(P)}$, to

$$K^{(P)} \times_{N_T^{(P)}} \mathfrak{t}^*_{(P)+} = \{(kN_T^{(P)}, \xi) \in K^{(P)}/N_T^{(P)} \times \mathfrak{k}^{(P)*} | \xi \in \mathrm{Ad}^*(k)\mathfrak{t}^*_{(P)+}\},$$

which fits into a diagram

$$\begin{array}{ccc} \mathcal{Z} & \rightarrow & K^{(P)} \times_{N_T^{(P)}} \mathfrak{t}^*_{(P)+} \\ \alpha \downarrow & & \downarrow \beta \\ \overline{\exp(i\mathfrak{k}^{(P)}) \sum_{\varpi \in \Pi} v_{\varpi}^{(P)}} & \rightarrow & \bigcup_{k \in K^{(P)}} \mathrm{Ad}^*(k)\mathfrak{t}^*_{(P)+} = \mathfrak{k}_+^{(P)*} \end{array} \tag{29}$$

where the first horizontal map is the homeomorphism $\mathcal{M}$ and the second is $\mu^{E^{(P)}}$. Since the image of α is dense and $K^{(P)}$ is compact, it follows that α is surjective. Moreover, β is surjective, and $\beta(kN_T^{(P)}, \xi) = \beta(k'N_T^{(P)}, \xi')$ if and only if $\mathrm{Ad}^*(k^{-1})\xi$ lies in an open face σ of $\mathfrak{t}_+$ such that $k'k^{-1} \in K_\sigma^{(P)}$, in which case $\alpha(\mathcal{M}^{-1}(kN_T^{(P)}, \xi)) = \alpha(\mathcal{M}^{-1}(kN_T^{(P)}, \xi))$. Thus

$$\mu^{E^{(P)}} : \overline{\exp(i\mathfrak{k}^{(P)}) \sum_{\varpi \in \Pi} v_{\varpi}^{(P)}} \rightarrow \mathfrak{k}_+^{(P)*}$$

is a continuous bijection, which is a homeomorphism since K is compact and $\mathcal{M}$ is a homeomorphism.

The inverse of $\mu^{E^{(P)}} : \overline{\exp(i\mathfrak{k}^{(P)}) \sum_{\varpi \in \Pi} v_{\varpi}^{(P)}} \rightarrow \mathfrak{k}_+^{(P)*}$ gives us a continuous $K^{(P)}$-equivariant map

$$\mathcal{F}^{(P)} : \mathfrak{k}_+^{(P)*} \rightarrow \overline{G/U}^{\mathrm{aff}} \subseteq E^{(P)}$$

extending (28) such that $\mu_T^{E^{(P)}} \circ \mathcal{F}^{(P)}$ is the identity on $\mathfrak{k}_+^{(P)*}$. This in turn extends to a continuous $K \times K^{(P)}$-equivariant map

$$\mathcal{F}^{(P)} : K \times \mathfrak{k}_+^{(P)*} \rightarrow \overline{G/U}^{\mathrm{aff}}, \tag{30}$$

which is surjective since $\overline{G/U}^{\mathrm{aff}} = K(\overline{\exp(i\mathfrak{k}^{(P)}) \sum_{\varpi \in \Pi} v_{\varpi}^{(P)}})$ by Remark 3.3.

Definition 3.9. If $\zeta \in \mathfrak{k}_+^{(P)*} = \mathrm{Ad}^*(K^{(P)})\mathfrak{t}^*_{(P)+} = \mathrm{Ad}^*(K^{(P)})\mathfrak{t}^*_+$ let $\zeta = \mathrm{Ad}^*(k)\xi$ with $k \in K^{(P)}$ and $\xi \in \mathfrak{t}^*_+$, and let σ_0 be the open face of $\mathfrak{t}^*_+$ containing ξ. Let $\sigma_0(P)$ be the open face of $\mathfrak{t}^*_+$ whose closure is

$$\overline{\sigma_0(P)} = \{\zeta \in \mathfrak{t}^* : \zeta \cdot \alpha = 0 \text{ for all } \alpha \in R_{\sigma_0} \backslash R^{(P)}\},$$

where R and $R^{(P)}$ are the sets of roots of K and $K^{(P)}$, and

$$R_{\sigma_0} = \{\alpha \in R : \zeta \cdot \alpha = 0 \text{ for all } \zeta \in \sigma_0\},$$

so that $\sigma_0(P)$ is an open subset of the open face containing σ_0 of the cone $\mathfrak{t}^*_{(P)+}$. Finally, let $K_\zeta(P) = kK_\xi k^{-1}$, where $K_\xi(P) = K_{\sigma_0(P)}$ is the stabiliser under the adjoint action of K of any element of $\sigma_0(P)$.

Note that $K_\zeta(P) \leq K_\zeta$ for any $\zeta \in \mathfrak{k}_+^{(P)*}$.

Lemma 3.10 (cf. [14, Lemma 6.2]). *Let σ be an open face of $\mathfrak{t}^*_{(P)+}$ and let*

$$v_\sigma^{(P)} = \sum_{\substack{\varpi \in \Pi \\ \lambda \in \mathrm{Ad}^*(W^{(P)})\varpi \cap \bar{\sigma}}} v_{\varpi,\lambda}^{(P)}.$$

If $\zeta \in \sigma$ then the stabiliser of $v_\sigma^{(P)}$ in K is $[K_\zeta(P), K_\zeta(P)]$.

Proof. Recall that $\mathfrak{t}^*_{(P)+} = \bigcup_{w \in W^{(P)}} \mathrm{Ad}^*(w)\mathfrak{t}^*_+$, so there is an element w_0 of the Weyl group $W^{(P)}$ of $Q^{(P)} = [K^{(P)}, K^{(P)}]$ such that $\mathrm{Ad}^*(w_0)\zeta \in \mathfrak{t}^*_+$ and $\mathrm{Ad}^*(w_0)\sigma$ contains an open face σ_0 of $\mathfrak{t}^*_+$ with σ_0 an open subset of σ. First assume that $\xi = \mathrm{Ad}^*(w_0)\zeta$ lies in σ_0. Then if $\varpi \in \Pi$ and $w \in W^{(P)}$ we have $\mathrm{Ad}^*(w)\varpi \in \bar{\sigma}$ if and only if $\mathrm{Ad}^*(w)\varpi$ lies in the linear subspace of $\mathfrak{t}^*$ spanned by $\Pi \cap \bar{\sigma} = \Pi \cap \overline{\sigma_0}$, and since $\xi \in \sigma_0$, this happens if and only if $K_\xi \leq w K_\varpi w^{-1}$, so that

$$\bigcap_{\substack{\varpi \in \Pi \\ \lambda = w\varpi \in W^{(P)}\varpi \cap \bar{\sigma}}} w K_\varpi w^{-1} = K_\xi.$$

As in the proof of [14, Lemma 6.2] we find that if $w \in W^{(P)}$, the stabiliser in $G = K_c$ of $[v_{\varpi,w\varpi}^{(P)}] \in \mathbb{P}((V_\varpi^{K^{(P)}})^* \otimes V_\varpi)$ is $wP_\varpi w^{-1}$, where P_ϖ is the parabolic subgroup of G associated to ϖ, and thus the stabiliser in K of $v_\sigma^{(P)}$ is the conjugate by w_0 of

$$\left\{ k \in \bigcap_{\substack{\varpi \in \Pi \\ \lambda = w\varpi \in W^{(P)}\varpi \cap \bar{\sigma}}} w K_\varpi w^{-1} : \tilde{\lambda}(g) = 1 \text{ for all } \tilde{\lambda} \in \Lambda^* \cap \bar{\sigma} \right\}$$

$$= \{k \in K_\xi : \tilde{\lambda}(g) = 1 \text{ for all } \tilde{\lambda} \in \Lambda^* \cap \bar{\sigma}\} = [K_\xi, K_\xi] = [K_{\sigma_0}, K_{\sigma_0}]. \quad (31)$$

In general, if $\xi = \mathrm{Ad}^*(w_0)\zeta$ lies in $\mathfrak{t}^*_+ \cap \mathrm{Ad}^*(w_0)\sigma$ then there is a unique open face σ_0 of $\mathfrak{t}^*_+$ containing ξ. Let $\sigma_0(P)$ be as in Definition 3.9; then $\bar{\sigma} \cap \mathfrak{t}^*_+ = \overline{\sigma_0(P)}$, and so by the previous paragraph the stabiliser of $v_\sigma^{(P)}$ in K is

$$w_0[K_{\sigma_0(P)}, K_{\sigma_0(P)}]w_0^{-1} = [K_\zeta(P), K_\zeta(P)].$$

Thus we extend the definition of the imploded cross-section X_{impl} to a $K^{(P)}$-imploded cross-section $X_{\mathrm{impl}}^{K,K^{(P)}}$ as follows.

Definition 3.11. Let (X, ω) be a symplectic manifold on which K acts with a moment map $\mu : X \to \mathfrak{k}^*$. As before let

$$\mathfrak{k}_+^{(P)*} = \mathrm{Ad}^*(K^{(P)})\mathfrak{t}^*_{(P)+} = \mathrm{Ad}^*(K^{(P)})\mathfrak{t}^*_+ = \mathrm{Ad}^*(Q^{(P)})\mathfrak{t}^*_+ \subseteq \mathfrak{k}^{(P)*} \quad (32)$$

be the sweep of $\mathfrak{t}_+^*$ under the coadjoint action of $K^{(P)}$ on $\mathfrak{k}^*$, and let $\Sigma^{(P)}$ be the set of open faces of $\mathfrak{k}_+^{(P)*}$. If $\zeta \in \mathfrak{k}^{(P)*}$ let $K_\zeta(P)$ be defined as in Definition 3.9. The $K^{(P)}$*-imploded cross-section* of X is

$$X_{\text{impl}}^{K,K^{(P)}} = \mu^{-1}(\mathfrak{k}_+^{(P)*})/\approx_{K^{(P)}},$$

where $x \approx_{K^{(P)}} y$ if and only if $\mu(x) = \mu(y) = \zeta \in \mathfrak{k}_+^{(P)*}$ and $x = \kappa y$ for some $\kappa \in [K_\zeta(P), K_\zeta(P)]$.

The *universal* $K^{(P)}$*-imploded cross-section* is the $K^{(P)}$-imploded cross-section

$$(T^*K)_{\text{impl}}^{K,K^{(P)}} = K \times \mathfrak{k}_+^{(P)*}/\approx_{K^{(P)}}$$

for the cotangent bundle $T^*K \cong K \times \mathfrak{k}^*$ with respect to the K-action induced from the right action of K on itself.

Theorem 3.12. *The map* $\mathcal{F}^{(P)} : K \times \mathfrak{k}_+^{(P)*} \to \overline{G/U}^{\text{aff}}$ *of (30) induces a* $K \times K^{(P)}$*-equivariant homeomorphism*

$$(T^*K)_{\text{impl}}^{K,K^{(P)}} = K \times \mathfrak{k}_+^{(P)*}/\approx_{K^{(P)}} \to \overline{G/U}^{\text{aff}} \subseteq E^{(P)}.$$

Moreover, under this identification of $K \times \mathfrak{k}_+^{(P)*}/\approx_{K^{(P)}}$ *with* $\overline{G/U}^{\text{aff}} \subseteq E^{(P)}$*, the moment map for the action of* $K \times K^{(P)}$ *on* $E^{(P)}$ *is induced by the map* $(K \times \mathfrak{k}_+^{(P)*})/\approx_{K^{(P)}} \to \mathfrak{k}^* \times \mathfrak{k}^{(P)*}$ *given by*

$$(k, \zeta) \mapsto (\text{Ad}^*(k)(\zeta), \zeta)).$$

Proof. By Lemma 3.10, $\mathcal{F}^{(P)}$ induces a continuous map $(T^*K)_{\text{impl}}^{K,K^{(P)}} \to \overline{G/U}^{\text{aff}} \subseteq E^{(P)}$, which is surjective since $\overline{G/U}^{\text{aff}} = \overline{K(\exp(i\mathfrak{k}^{(P)})\sum_{\varpi\in\Pi} v_\varpi^{(P)})}$ by Remark 3.3. The map $(K \times \mathfrak{k}_+^{(P)*})/\approx_{K^{(P)}} \to \mathfrak{k}^* \times \mathfrak{k}^{(P)*}$ given by

$$(k, \zeta) \mapsto (\text{Ad}^*(k)(\zeta), \zeta))$$

is the composition of $\mathcal{F}^{(P)} : K \times \mathfrak{k}_+^{(P)*} \to \overline{G/U}^{\text{aff}}$ with the restriction to $\overline{G/U}^{\text{aff}} \subseteq E^{(P)}$ of the moment map $\mu^{E^{(P)}}$ for the action of $K \times K^{(P)}$ on $E^{(P)}$. Moreover, $\mathcal{F}^{(P)}$ is continuous and surjective and restricts to a homeomorphism from $\overline{\exp(i\mathfrak{k}^{(P)})\sum_{\varpi\in\Pi} v_\varpi^{(P)}}$ to $\mathfrak{k}_+^{(P)*}$ by Corollary 3.8. If $\mathcal{F}^{(P)}(k_1, \zeta_1) = \mathcal{F}^{(P)}(k_2, \zeta_2)$ then it follows by applying $\mu^{E^{(P)}}$ that $(\text{Ad}^*(k_1)(\zeta_1), \zeta_1) = (\text{Ad}^*(k_2)(\zeta_2), \zeta_2)$ and therefore $\zeta_1 = \zeta_2$ and $k_1 k_2^{-1} \in K_{\zeta_1} = K_{\zeta_2}$. Thus

$$\begin{aligned}\mathcal{F}^{(P)}(1, \zeta_1) &= (k_1^{-1}, 1)\mathcal{F}^{(P)}(k_1, \zeta_1) = (k_1^{-1}, 1)\mathcal{F}^{(P)}(k_2, \zeta_2)\\ &= (k_1^{-1}k_2, 1)\mathcal{F}^{(P)}(1, \zeta_2) = (k_1^{-1}k_2, 1)\mathcal{F}^{(P)}(1, \zeta_1).\end{aligned}$$

Since $\zeta_1 = \zeta_2 \in \mathfrak{k}_+^{(P)*} = \mathrm{Ad}^*(K^{(P)})\mathfrak{t}^*_{(P)+}$, we can write $\zeta_1 = \zeta_2 = \mathrm{Ad}^*(k_0)(\zeta)$, where $\zeta \in \mathfrak{t}^*_{(P)+}$ and $k_0 \in K^{(P)}$, so

$$\begin{aligned}\mathcal{F}^{(P)}(1,\zeta) = (1,k_0^{-1})\mathcal{F}^{(P)}(1,\zeta_1) &= (k_1^{-1}k_2, k_0^{-1})\mathcal{F}^{(P)}(1,\zeta_1)\\ &= (k_1^{-1}k_2, 1)\mathcal{F}^{(P)}(1,\zeta).\end{aligned}$$

By Lemma 3.6, $\mathcal{F}^{(P)}(1,\zeta)$ lies in the $\exp(i\mathfrak{t})$-orbit of $\sum_{\varpi\in\Pi,\lambda\in\Lambda^*_{\varpi}\cap\bar{\sigma}} v^{(P)}_{\varpi,\lambda}$, where σ is the open face of $\mathfrak{t}^*_{(P)+}$ containing ζ. Hence by Lemma 3.10, $k_1^{-1}k_2 \in [K_\zeta(P), K_\zeta(P)]$, and thus $\mathcal{F}^{(P)}$ induces a continuous bijection $(T^*K)^{K,K^{(P)}}_{\mathrm{impl}} \to \overline{G/U}^{\mathrm{aff}} \subseteq E^{(P)}$. Since K is compact and so the map $(K\times\mathfrak{k}_+^{(P)*})/\approx_{(P)} \to \mathfrak{k}^*\times\mathfrak{k}^{(P)*}$ given by $(k,\zeta)\mapsto(Ad^*(k)(\zeta),\zeta))$ is proper, this continuous bijection is a homeomorphism.

Remark 3.13. If $K^{(P)} = T$ and $\zeta \in \mathfrak{k}_+^{(P)*}$ then $K_\zeta(P) = K_\zeta$, and so X_{KimplT} is the standard imploded cross-section X_{impl} of [14]. On the other hand, if $K^{(P)} = K$ then $K_\zeta(P)$ is conjugate to T and $[K_\zeta(P), K_\zeta(P)]$ is trivial for all $\zeta \in \mathfrak{k}_+^{(P)*}$, so $X_{\mathrm{KimplK}} = T^*K$.

Of course $\overline{G/U}^{\mathrm{aff}}$ inherits a $K\times K^{(P)}$-invariant Kähler structure as a complex subvariety of $E^{(P)}$. The subvariety $\overline{G/U}^{\mathrm{aff}}$ (which is in general singular) is stratified by the (finitely many) G-orbits in $\overline{G/U}^{\mathrm{aff}}$, and the $K\times K^{(P)}$-invariant Kähler structure on $E^{(P)}$ restricts to a $K\times K^{(P)}$-invariant symplectic structure on each stratum, which gives $\overline{G/U}^{\mathrm{aff}}$ a stratified symplectic structure. Under the homeomorphism $(T^*K)^{K,K^{(P)}}_{\mathrm{impl}} \to \overline{G/U}^{\mathrm{aff}}$ of Theorem 3.12 these strata correspond to the locally closed subsets

$$\begin{aligned}\frac{K\times\mathrm{Ad}^*(K^{(P)})\sigma}{\approx_{K^{(P)}}} &\cong K^{(P)}\times_{K_\sigma\cap K^{(P)}}\left(\frac{K\times\sigma}{\approx_{K^{(P)}}}\right)\\ &\cong K^{(P)}\times_{K_\sigma\cap K^{(P)}}\left(\frac{K\times\sigma}{[K_{\sigma(P)},K_{\sigma(P)}]}\right)\end{aligned}$$

of $(T^*K)^{K,K^{(P)}}_{\mathrm{impl}}$, where $\sigma\in\Sigma$ runs over the open faces of $\mathfrak{t}^*_+$. So the homeomorphism $(T^*K)^{K,K^{(P)}}_{\mathrm{impl}} \to \overline{G/U}^{\mathrm{aff}}$ of Theorem 3.12 induces a stratified $K\times K^{(P)}$-invariant symplectic structure on the universal $K^{(P)}$-imploded cross-section $(T^*K)^{K,K^{(P)}}_{\mathrm{impl}}$. As in [14] the induced symplectic structure on

$$K^{(P)}\times_{K_\sigma\cap K^{(P)}}\left(\frac{K\times\sigma}{[K_{\sigma(P)},K_{\sigma(P)}]}\right)$$

can be described directly, and can be expressed in terms of the symplectic reduction by the action of the subgroup $[K_{\sigma(P)}, K_{\sigma(P)}]$ of K on a locally closed symplectic submanifold of T^*K (cf. [14, Section 2]).

Using this symplectic structure on $(T^*K)_{\mathrm{impl}}^{K,K^{(P)}}$ we obtain the following corollary.

Corollary 3.14. *Let K act on a symplectic manifold X with moment map $\mu : X \to \mathfrak{k}^*$. Then the symplectic quotient of $\overline{G/U}^{\mathrm{aff}} \times X = (T^*K)_{\mathrm{impl}}^{K,K^{(P)}} \times X$ by the diagonal action of K can be identified via $\mathcal{F}^{(P)}$ with $X_{\mathrm{impl}}^{K,K^{(P)}}$.*

Remark 3.15. In particular, if X is a projective variety with a linear action of the complexification G of K, then $X_{\mathrm{impl}}^{K,K^{(P)}}$ can be identified with the GIT quotient of $\overline{G/U}^{\mathrm{aff}} \times X$ by the diagonal action of G.

It follows from Corollary 3.14 that if (X, ω) is any symplectic manifold on which K acts with moment map $\mu : X \to \mathfrak{k}^*$ then $X_{\mathrm{impl}}^{K,K^{(P)}}$ inherits a stratified $K \times K^{(P)}$-invariant symplectic structure

$$X_{\mathrm{impl}}^{K,K^{(P)}} = \bigsqcup_{\sigma \in \Sigma} \frac{\mu^{-1}(\sigma)}{\approx K^{(P)}}$$

$$= \mu^{-1}((\mathfrak{k}_+^{(P)*})^\circ) \sqcup \bigsqcup_{\substack{\sigma \in \Sigma \\ \sigma \neq (\mathfrak{t}_+^*)^\circ}} K^{(P)} \times_{K_\sigma \cap K^{(P)}} \left(\frac{\mu^{-1}(\sigma)}{[K_{\sigma(P)}, K_{\sigma(P)}]} \right) \tag{33}$$

with strata indexed by the set Σ of open faces of $\mathfrak{t}_+^*$, which are locally closed symplectic submanifolds of $X_{\mathrm{impl}}^{K,K^{(P)}}$. The induced action of $K^{(P)}$ on $X_{\mathrm{impl}}^{K,K^{(P)}}$ preserves this symplectic structure and has a moment map

$$\mu_{X_{\mathrm{impl}}^{K,K^{(P)}}} : X_{\mathrm{impl}}^{K,K^{(P)}} \to \mathfrak{k}_+^{(P)*} \subseteq \mathfrak{k}^{(P)*}$$

inherited from the restriction of μ to $\mu^{-1}(\mathfrak{k}_+^{(P)})$.

Remark 3.16. In order to identify $\overline{G/U}^{\mathrm{aff}}$ with $(T^*K)_{\mathrm{impl}}^{K,K^{(P)}}$ we made the assumption that K is semisimple and simply connected. However, the construction of $X_{\mathrm{impl}}^{K,K^{(P)}}$ makes sense whenever K is a compact connected Lie group with a Hamiltonian action on the symplectic manifold X, and as in [14] we can identify $\overline{G/U}^{\mathrm{aff}}$ with $(T^*K)_{\mathrm{impl}}^{K,K^{(P)}}$ in this more general situation by expressing K as the quotient of the product of its centre $Z(K)$ and the universal cover of $[K, K]$ by a finite central subgroup. We then get an identification of $X_{\mathrm{impl}}^{K,K^{(P)}}$ with the symplectic quotient of $\overline{G/U}^{\mathrm{aff}} \times X$ by K in the general case.

3.3 Wonderful compactifications, symplectic cuts, and partial desingularisations

Recently, Paradan [29] introduced a generalisation of the technique of symplectic cutting (originally due to Lerman [25]) which is valid for a (not necessarily abelian) compact connected group K and is motivated by the wonderful compactifications of De Concini and Procesi. He defines a K-*adapted* polytope in $\mathfrak{t}^*$ to be a W-invariant Delzant polytope $\mathcal{P}$ in $\mathfrak{t}^*$ whose vertices are regular elements of the weight lattice Λ^*. If $\{\lambda_1, \dots, \lambda_N\}$ are the dominant weights lying in the union of all the closed one-dimensional faces of $\mathcal{P}$, then there is a $G \times G$-equivariant embedding of $G = K_c$ into

$$\mathbb{P}\left(\bigoplus_{i=1}^{N} V_{\lambda_i}^* \otimes V_{\lambda_i}\right)$$

associating to $g \in G$ its representation on $\bigoplus_{i=1}^{N} V_{\lambda_i}$. The closure $\mathcal{X}_{(\mathcal{P},K)}$ of the image of G in this projective space is smooth and has moment map

$$\mu^{\mathcal{P}}_{K\times K} : \mathcal{X}_{(\mathcal{P},K)} \to \mathfrak{k}^* \times \mathfrak{k}^*,$$

whose image is

$$\mu^{\mathcal{P}}_{K\times K}(\mathcal{X}_{(\mathcal{P},K)}) = \{(\mathrm{Ad}^*(k_1)\xi, -\mathrm{Ad}^*(k_2)\xi) : \xi \in \mathcal{P}, k_1, k_2 \in K\}.$$

The symplectic cut $X_{(\mathcal{P},K)}$ defined by Paradan of a symplectic manifold X under a Hamiltonian K-action with respect to such a K-adapted polytope $\mathcal{P}$ is given by the symplectic quotient of $\mathcal{X}_{(\mathcal{P},K)} \times X$ by K, so that if X is a complex projective variety with a linear K-action then $X_{(\mathcal{P},K)}$ is the GIT quotient

$$X_{(\mathcal{P},K)} = (\mathcal{X}_{(\mathcal{P},K)} \times X)//G,$$

where $G = K_c$. Then $X_{(\mathcal{P},K)}$ inherits a Hamiltonian K-action with moment map $\mu^{X_{(\mathcal{P},K)}} : X_{(\mathcal{P},K)} \to \mathfrak{k}^*$ whose image is

$$\mu^{X_{(\mathcal{P},K)}}(X_{(\mathcal{P},K)}) = \mu(X) \cap \mathrm{Ad}^*(K)(\mathcal{P}).$$

Moreover, if $U_{\mathcal{P}} = \mathrm{Ad}^*(K)(\mathcal{P}^\circ)$, where $\mathcal{P}^\circ$ is the interior of $\mathcal{P}$, then $(\mu^{X_{(\mathcal{P},K)}})^{-1}(U_{\mathcal{P}})$ is an open dense subset of $X_{(\mathcal{P},K)}$ which is K-equivariantly diffeomorphic to the open subset $\mu^{-1}(U_{\mathcal{P}})$ of X. This diffeomorphism is a quasisymplectomorphism in the sense that there is a homotopy of symplectic forms taking the symplectic form on $(\mu^{X_{(\mathcal{P},K)}})^{-1}(U_{\mathcal{P}})$ to the pullback of the symplectic form on $\mu^{-1}(U_{\mathcal{P}})$.

Recall from [14, Section 7] that if $\mathcal{P}_\epsilon$ is the polyhedral cone $-(\epsilon\lambda_0 + \mathfrak{t}^*_+)$, where λ_0 is a generic element of $\mu(X) \cap \mathfrak{t}^*_+$ and $0 < \epsilon \ll 1$, then the imploded cross-section $X_{\mathrm{impl}} = X_{\mathrm{KimplT}}$ has a partial desingularisation

$$\widetilde{X_{\mathrm{impl}}} = (X_{\mathrm{impl}})_{(\mathcal{P}_\epsilon, T)},$$

which is the symplectic reduction of $\mathcal{X}_{(-\mathfrak{t}_+^*, T)} \times X_{\mathrm{impl}}$ at $\epsilon\lambda_0$. Similarly, just as in [14], if $P \geq B$ is a parabolic subgroup of $G = K_c$ with maximal compact subgroup $K^{(P)} = K \cap P$ and unipotent radical U, then we can construct a $K \times K^{(P)}$-equivariant desingularisation $\widetilde{(T^*K)_{\mathrm{impl}}^{K,K^{(P)}}}$ for the universal imploded cross-section $(T^*K)_{\mathrm{impl}}^{K,K^{(P)}} \cong \overline{G/U}^{\mathrm{aff}}$ and a partial desingularisation $\widetilde{X_{\mathrm{impl}}^{K,K^{(P)}}}$ for $X_{\mathrm{impl}}^{K,K^{(P)}}$, which can be identified with the symplectic quotient of $X \times \widetilde{(T^*K)_{\mathrm{impl}}^{K,K^{(P)}}}$ by the induced action of K. Moreover, $\widetilde{(T^*K)_{\mathrm{impl}}^{K,K^{(P)}}}$ can be identified as a Hamiltonian K-manifold with

$$\widetilde{G/U}^{\mathrm{aff}} = G \times_P \left(\overline{L^{(P)} \sum_{\varpi \in \Pi} v_\varpi^{(P)}} \right) = K \times_{K^{(P)}} \left(\overline{L^{(P)} \sum_{\varpi \in \Pi} v_\varpi^{(P)}} \right), \tag{34}$$

where $\overline{L^{(P)} \sum_{\varpi \in \Pi} v_\varpi^{(P)}}$ is the closure in $E^{(P)}$ (or equivalently in the linear subspace $\bigoplus_{\varpi \in \Pi} (V_\varpi^{K^{(P)}})^* \otimes V_\varpi^{K^{(P)}}$ of $E^{(P)}$) of the $L^{(P)}$-orbit (or equivalently the P-orbit) of $\sum_{\varpi \in \Pi} v_\varpi^{(P)}$, and the restriction to $G \times \overline{L^{(P)} \sum_{\varpi \in \Pi} v_\varpi^{(P)}}$ of the multiplication map $G \times E^{(P)} \to E^{(P)}$ induces a birational G-equivariant morphism

$$p_U : \widetilde{G/U}^{\mathrm{aff}} \to \overline{G/U}^{\mathrm{aff}} = (T^*K)_{\mathrm{impl}}^{K,K^{(P)}} \subseteq E^{(P)}.$$

It follows from Theorem 3.5 of [29] that $\overline{L^{(P)} \sum_{\varpi \in \Pi} v_\varpi^{(P)}})$ is a nonsingular subvariety of

$$\bigoplus_{\varpi \in \Pi} (V_\varpi^{K^{(P)}})^* \otimes V_\varpi^{K^{(P)}} \subseteq \mathbb{P}\left(\mathbb{C} \oplus \bigoplus_{\varpi \in \Pi} (V_\varpi^{K^{(P)}})^* \otimes V_\varpi^{K^{(P)}} \right).$$

If $\lambda_0 \in \mu(X) \cap \mathfrak{t}_+^* \cap \mathfrak{z}^{(P)*}$ is generic and $\epsilon > 0$ is sufficiently close to 0, and if ω_ϵ is the Kähler form on G/P given by regarding G/P as the coadjoint K-orbit through $\epsilon\lambda_0$, then $p_U^* \omega_{E^{(P)}} + q_P^* \omega_\epsilon$ is a Kähler form on $\widetilde{G/U}^{\mathrm{aff}}$, where $q_P : G \times_P E^{(P)} \to G/P$ is the projection.

The partial desingularisation $\widetilde{X_{\mathrm{impl}}^{K,K^{(P)}}}$ can alternatively be obtained from $X_{\mathrm{impl}}^{K,K^{(P)}}$ via a symplectic cut following Paradan [29]. Let $W^{(P)}$ be the Weyl group of the compact subgroup $K^{(P)}$ of K; then we have an identification

$$\widetilde{X_{\mathrm{impl}}^{K,K^{(P)}}} = (X_{\mathrm{impl}}^{K,K^{(P)}})_{(\mathcal{P}_\epsilon, K^{(P)})}, \tag{35}$$

where the cut is with respect to the $K^{(P)}$-action and the polyhedral cone $\mathcal{P}_\epsilon = -(\epsilon\lambda_0 + \mathfrak{t}_{(P)+}^*)$. If we wish we can cut with respect to a suitable $W^{(P)}$-invariant

Delzant polytope $\mathcal{P}_\epsilon$ in this cone which is large enough that its complement does not meet the compact subset $\mu(X)$, but then the identification (35) is not quite symplectic according to Paradan's construction; as in Remark 3.1 we have to distinguish between the flat Kähler metric on

$$\bigoplus_{\varpi\in\Pi} (V_\varpi^{K^{(P)}})^* \otimes V_\varpi^{K^{(P)}} \subseteq E^{(P)}$$

and the Fubini–Study metric on

$$\bigoplus_{\varpi\in\Pi} (V_\varpi^{K^{(P)}})^* \otimes V_\varpi^{K^{(P)}} \subseteq \mathbb{P}\left(\mathbb{C} \oplus \bigoplus_{\varpi\in\Pi} (V_\varpi^{K^{(P)}})^* \otimes V_\varpi^{K^{(P)}}\right) \subseteq \mathbb{P}(\mathbb{C} \oplus E^{(P)}).$$

4 Nonreductive geometric invariant theory

The last section discussed a generalisation of symplectic implosion which is closely related to a GIT-like quotient construction for a linear action of the unipotent radical U of a parabolic subgroup P of a complex reductive group G on a complex variety X. This section will recall from [7] a version of GIT for nonreductive group actions and then relate it to symplectic implosion.

4.1 Background

Let H be an affine algebraic group, with unipotent radical U (that is, U is the unique maximal normal unipotent subgroup of H), acting linearly on a complex projective variety X with respect to an ample line bundle L. If we wish to generalise Mumford's GIT to this nonreductive situation, the first problem to be faced is that the ring of invariants

$$\hat{\mathcal{O}}_L(X)^H = \bigoplus_{k\geq 0} H^0(X, L^{\otimes k})^H$$

is not necessarily finitely generated as a graded complex algebra, so that $\mathrm{Proj}(\hat{\mathcal{O}}_L(X)^H)$ is not well defined as a projective variety. Note, however, that in the case considered in Section 3 in which the unipotent radical U of a parabolic subgroup of a reductive group G acts linearly on X and the linear action extends to G, then the ring of invariants is finitely generated. Even when $\hat{\mathcal{O}}_L(X)^H$ is not finitely generated $\mathrm{Proj}(\hat{\mathcal{O}}_L(X)^H)$ does make sense as a scheme, and the inclusion of $\hat{\mathcal{O}}_L(X)^H$ in $\hat{\mathcal{O}}_L(X)$ gives us a rational map of schemes q from X to $\mathrm{Proj}(\hat{\mathcal{O}}_L(X)^H)$, whose image in $\mathrm{Proj}(\hat{\mathcal{O}}_L(X)^H)$ is constructible (that is, a finite union of locally closed subschemes).

We will consider only the case that $H = U$ is unipotent, since H/U is always reductive and classical GIT allows us to deal with quotients by reductive groups.

A more leisurely introduction to nonreductive GIT and details and proofs of the results quoted below can be found in [7].

Definition 4.1 (See [7]). Let $I = \bigcup_{m>0} H^0(X, L^{\otimes m})^U$ and for $f \in I$ let X_f be the U-invariant affine open subset of X where f does not vanish, with $\mathcal{O}(X_f)$ its coordinate ring. The (finitely generated) *semistable set* of X is

$$X^{\mathrm{ss}} = X^{\mathrm{ss,fg}} = \bigcup_{f \in I^{\mathrm{fg}}} X_f$$

where I^{fg} consists of $f \in I$ such that $\mathcal{O}(X_f)^U$ is finitely generated. The set of (locally trivial) *stable* points is

$$X^{\mathrm{s}} = X^{\mathrm{lts}} = \bigcup_{f \in I^{\mathrm{lts}}} X_f,$$

where I^{lts} is the set of $f \in I$ such that $\mathcal{O}(X_f)^U$ is finitely generated, and $q : X_f \longrightarrow \mathrm{Spec}(\mathcal{O}(X_f)^U)$ is a locally trivial geometric quotient. The set of *naively semistable* points of X is the domain of definition

$$X^{\mathrm{nss}} = \bigcup_{f \in I} X_f$$

of the rational map q, and the set of *naively stable* points of X is

$$X^{\mathrm{ns}} = \bigcup_{f \in I^{\mathrm{ns}}} X_f,$$

where I^{ns} consists of those $f \in I$ such that $\mathcal{O}(X_f)^U$ is finitely generated, and $q : X_f \longrightarrow \mathrm{Spec}(\mathcal{O}(X_f)^U)$ is a geometric quotient.

The *enveloped quotient* of X^{ss} is $q : X^{\mathrm{ss}} \to q(X^{\mathrm{ss}})$, where $q(X^{\mathrm{ss}})$ is a dense constructible subset (but not necessarily a subvariety) of the *enveloping quotient*

$$X/\!/U = \bigcup_{f \in I^{\mathrm{ss,fg}}} \mathrm{Spec}(\mathcal{O}(X_f)^U)$$

of X^{ss}.

Lemma 4.2 ([7, 4.2.9 and 4.2.10]). *The enveloping quotient $X/\!/U$ is a quasi-projective variety, and if $\hat{\mathcal{O}}_L(X)^U$ is finitely generated then it is the projective variety* $\mathrm{Proj}(\hat{\mathcal{O}}_L(X)^U)$.

Let G be a complex reductive group with U as a closed subgroup, and let $G \times_U X$ denote the quotient of $G \times X$ by the free action of U defined by $h(g, x) = (gh^{-1}, hx)$, which is a quasiprojective variety by [30, Theorem 4.19]. There is an induced G-action on $G \times_U X$ given by left multiplication of G on itself. If the action of U on X extends to an action of G, there is an isomorphism of G-varieties

$$G \times_U X \cong (G/U) \times X \tag{36}$$

given by $[g,x] \mapsto (gH, gx)$. When U acts linearly on X with respect to a very ample line bundle L inducing an embedding of X in $\mathbb{P}^n$, and G is a subgroup of $\mathrm{SL}(n+1;\mathbb{C})$, then there is a very ample G-linearisation (which we will also denote by L) on $G \times_U X$ via the embedding

$$G \times_U X \hookrightarrow G \times_U \mathbb{P}^n \cong (G/U) \times \mathbb{P}^n,$$

and using the trivial bundle on the variety G/U, which is quasiaffine by [12, Corollary 2.8]. For large enough m we can choose a G-equivariant embedding of G/U in $\mathbb{C}^m$ with a linear G-action to get a G-equivariant embedding of $G \times_U X$ in $\mathbb{C}^m \times \mathbb{P}^n \subseteq \mathbb{P}^m \times \mathbb{P}^n \subseteq \mathbb{P}^{nm+m+n}$, and the G-invariants on $G \times_U X$ are given by

$$\bigoplus_{m\geq 0} H^0(G \times_U X, L^{\otimes m})^G \cong \bigoplus_{m \geq 0} H^0(X, L^{\otimes m})^U = \hat{\mathcal{O}}_L(X)^U. \tag{37}$$

Definition 4.3 ([7, Section 5.2]). A *finite separating set of invariants* for the linear action of U on X is a collection of invariant sections $\{f_1, \ldots, f_n\}$ of positive tensor powers of L such that if x, y are any two points of X then $f(x) = f(y)$ for all invariant sections f of $L^{\otimes k}$ and all $k > 0$ if and only if

$$f_i(x) = f_i(y) \qquad \forall i = 1, \ldots, n.$$

If G is any reductive group containing U, a finite separating set S of invariant sections of positive tensor powers of L is a *finite fully separating set of invariants* for the linear U-action on X if

(i) for every $x \in X^s$ there exists $f \in S$ with associated G-invariant F over $G \times_U X$ (under the isomorphism (37)) such that $x \in (G \times_U X)_F$ and $(G \times_U X)_F$ is affine; and
(ii) for every $x \in X^{\mathrm{ss}}$ there exists $f \in S$ such that $x \in X_f$ and S is a generating set for $\mathcal{O}(X_f)^U$.

By [7, Remark 5.2.3] this definition is in fact independent of the choice of G.

A G-equivariant projective completion $\overline{G \times_U X}$ of $G \times_U X$, together with a G-linearisation with respect to a line bundle L which restricts to the given U-linearisation on X, is a *reductive envelope* of the linear U-action on X if every U-invariant f in some finite fully separating set of invariants S for the U-action on X extends to a G-invariant section of a tensor power of L over $\overline{G \times_U X}$. If L is ample on $(\overline{G \times_U X})$ it is an *ample reductive envelope*.

There always exists an ample reductive envelope for any linear U-action on a projective variety X, at least if we replace the line bundle L with a suitable positive tensor power of itself (see [7, Proposition 5.2.8]).

Definition 4.4. Let X be a projective variety with a linear U-action and a reductive envelope $\overline{G \times_U X}$. Let $i : X \hookrightarrow G \times_U X$ and $j : G \times_U X \hookrightarrow \overline{G \times_U X}$ be the inclusions, and $\overline{G \times_U X}^{\mathrm{s}}$ and $\overline{G \times_U X}^{\mathrm{ss}}$ the stable and semistable sets for the linear G-action on $\overline{G \times_U X}$. Then the set of *completely stable points* of X with respect to the reductive envelope is

$$X^{\bar{\mathrm{s}}} = (j \circ i)^{-1}(\overline{G \times_U X}^{\mathrm{s}})$$

and the set of *completely semistable points* is

$$X^{\overline{\mathrm{ss}}} = (j \circ i)^{-1}(\overline{G \times_U X}^{\mathrm{ss}}).$$

Theorem 4.5 ([7, 5.3.1]). *Let X be a normal projective variety with a linear U-action, for U a connected unipotent group, and let $(\overline{G \times_U X}, L)$ be any ample reductive envelope. Then there is a diagram*

$$\begin{array}{ccccccccc} X^{\bar{\mathrm{s}}} & \subseteq & X^{\mathrm{s}} & \subseteq & X^{\mathrm{ns}} & \subseteq & X^{\mathrm{ss}} & \subseteq & X^{\overline{\mathrm{ss}}} = X^{\mathrm{nss}} \\ \downarrow & & \downarrow & & \downarrow & & \downarrow & & \downarrow \\ X^{\bar{\mathrm{s}}}/U & \subseteq & X^{\mathrm{s}}/U & \subseteq & X^{\mathrm{ns}}/U & \subseteq & X/\!/U & \subseteq & \overline{G \times_U X}/\!/G \end{array}$$

where all the inclusions are open and the first three vertical maps provide quasi-projective geometric quotients of the stable sets $X^{\bar{\mathrm{s}}}$, X^{s}, and X^{ns} by the action of U. The fourth vertical map is the enveloping quotient $q : X^{\mathrm{ss}} \to X/\!/U$ defined in Definition 4.1 and $X/\!/U$ is an open subvariety of the projective variety $\overline{G \times_U X}/\!/G$.

Note, however, that even when $\hat{\mathcal{O}}_L(X)^U$ is finitely generated so that

$$X/\!/U = \mathrm{Proj}(\hat{\mathcal{O}}_L(X)^U) = \overline{G \times_U X}/\!/G,$$

the maps $q : X^{\mathrm{ss}} \to X/\!/U$ and $X^{\overline{\mathrm{ss}}} \to \overline{G \times_U X}/\!/G$ are *not* necessarily surjective, and their images are in general only constructible subsets and not subvarieties.

4.2 Some examples of reductive envelopes

Now let us assume that $U = (\mathbb{C}^+)^r$, where $\mathbb{C}^+$ is the additive group of complex numbers and r is any positive integer.

Remark 4.6. Each affine algebraic group H over $\mathbb{C}$ has a unipotent radical U, which is the unique maximal normal unipotent subgroup of H and has a reductive quotient group $R = H/U$ (see, e.g., [4, 33] for more details). Given a linear action of H on a projective variety X with respect to an ample line bundle L, we can hope to quotient first by the action of U, and then by the induced action of the reductive group H/U, provided that the unipotent quotient (or compactified quotient) is sufficiently canonical to inherit an induced linear action of H/U. For example, if the algebra of invariants $\hat{\mathcal{O}}_L(X)^U$ is finitely generated then the enveloping quotient $X/\!/U = \mathrm{Proj}(\hat{\mathcal{O}}_L(X)^U)$ is a projective variety with an induced linear action of H/U on an induced ample line bundle on $X/\!/U$, and then classical GIT allows us to construct $X/\!/H = \mathrm{Proj}(\hat{\mathcal{O}}_L(X)^H)$ as a GIT quotient $(X/\!/U)/\!/(H/U)$ of $X/\!/U$ by the reductive group H/U; even when $\hat{\mathcal{O}}_L(X)^U$ is not finitely generated, the same is true for $\mathrm{Proj}(\hat{\mathcal{O}}_L(X)^U_m)$, where m is a sufficiently large positive integer and $\hat{\mathcal{O}}_L(X)^U_m$ is the subalgebra of $\hat{\mathcal{O}}_L(X)^U$ generated by invariant sections of $L^{\otimes j}$ for $1 \leq j \leq m$. Moreover, the unipotent radical U has canonical sequences of normal subgroups such that each successive subquotient is isomorphic to $(\mathbb{C}^+)^r$ for some r (for example by taking the ascending or descending central series of U), so we can hope to quotient successively by unipotent groups of the form $(\mathbb{C}^+)^r$, and then finally by the reductive group R. Therefore the case $U \cong (\mathbb{C}^+)^r$ for some r is less special than it might appear at first sight.

Note that when $U = (\mathbb{C}^+)^r$ we have $\mathrm{Aut}(U) \cong \mathrm{GL}(r;\mathbb{C})$; let

$$\hat{U} = \mathbb{C}^* \ltimes U$$

be the semidirect product, where $\mathbb{C}^*$ is the centre of $\mathrm{Aut}(U)$. The centre of $\hat{U}$ is finite and meets U in the trivial subgroup, so U is isomorphic to a closed subgroup of the reductive group $G = \mathrm{SL}(\mathbb{C} \oplus \mathfrak{u})$ via the inclusion

$$U \hookrightarrow \hat{U} \to \mathrm{Aut}(\hat{U}) \to \mathrm{GL}(\mathrm{Lie}\hat{U}) = \mathrm{GL}(\mathbb{C} \oplus \mathfrak{u}),$$

where $\mathfrak{u}$ is the Lie algebra of U and $\hat{U}$ is identified with its group of inner automorphisms. Then U is the unipotent radical of a parabolic subgroup P of $G = \mathrm{SL}(r+1;\mathbb{C})$, where P is the stabiliser of the r-dimensional linear subspace $\mathfrak{u}$ of $\mathbb{C} \oplus \mathfrak{u}$, so we are in the situation of Section 3.2 above. The parabolic $P = U \rtimes \mathrm{GL}(r;\mathbb{C})$ in $G = \mathrm{SL}(r+1;\mathbb{C})$ has Levi subgroup $\mathrm{GL}(r;\mathbb{C})$ embedded in $\mathrm{SL}(r+1;\mathbb{C})$, since

$$g \mapsto \begin{pmatrix} g & 0 \\ 0 & \det g^{-1} \end{pmatrix}.$$

Note that

$$G/U \cong \{\alpha \in (\mathbb{C}^r)^* \otimes \mathbb{C}^{r+1} | \alpha : \mathbb{C}^r \to \mathbb{C}^{r+1} \text{ is injective}\}$$

with the natural G-action $g\alpha = g \circ \alpha$. Since the injective linear maps from $\mathbb{C}^r$ to $\mathbb{C}^{r+1}$ form an open subset in the affine space $(\mathbb{C}^r)^* \otimes \mathbb{C}^{r+1}$, whose complement has codimension two, we see directly in this case that $U = (\mathbb{C}^+)^r$ is a Grosshans subgroup of $G = \mathrm{SL}(r+1;\mathbb{C})$ and hence that

$$\mathcal{O}(G)^U \cong \mathcal{O}(G/U) \cong \mathcal{O}((\mathbb{C}^r)^* \otimes \mathbb{C}^{r+1})$$

is finitely generated [12] with

$$\overline{G/U}^{\mathrm{aff}} = \mathrm{Spec}\mathcal{O}(G)^U = (\mathbb{C}^r)^* \otimes \mathbb{C}^{r+1}.$$

Now suppose that the linear action of $U = (\mathbb{C}^+)^r$ on X extends to a linear action of $G = \mathrm{SL}(r+1;\mathbb{C})$, giving us an identification of G-spaces

$$G \times_U X \cong (G/U) \times X$$

as at (36) via $[g, x] \mapsto (gH, gx)$. Then (as in the Borel transfer theorem [5, Lemma 4.1])

$$\hat{\mathcal{O}}_L(X)^U \cong \hat{\mathcal{O}}_L(G \times_U X)^G \cong [\mathcal{O}(G/U) \otimes \hat{\mathcal{O}}_L(X)]^G \tag{38}$$

is finitely generated [13] and we have a reductive envelope

$$\overline{G \times_U X} = \mathbb{P}(\mathbb{C} \oplus ((\mathbb{C}^r)^* \otimes \mathbb{C}^{r+1})) \times X$$

with

$$\overline{G \times_U X}/\!/G \cong X/\!/U = \mathrm{Proj}(\hat{\mathcal{O}}_L(X)^U),$$

where we choose for our linearisation on $\overline{G \times_U X}$ the line bundle

$$L^{(N)} = \mathcal{O}_{\mathbb{P}(\mathbb{C}\oplus((\mathbb{C}^r)^*\otimes\mathbb{C}^{r+1}))}(N) \otimes L$$

with $N > 0$ sufficiently large (see [23, Section 4.1]). This reductive envelope is ample and so satisfies Theorem 4.5; in addition, by [23, Section 4.1(6)] we have

$$X^{\bar{\mathrm{s}}} = X^{\mathrm{s}} \quad \text{and} \quad X^{\overline{\mathrm{ss}}} = X^{\mathrm{ss}}. \tag{39}$$

Thus we have a diagram

$$\begin{array}{ccc} X^{\mathrm{s}} & \subseteq & X^{\mathrm{ss}} \\ \downarrow & & \downarrow \\ X^{\mathrm{s}}/U & \subseteq & X/\!/U = \overline{G \times_U X}/\!/G \end{array}$$

but the enveloping quotient map $q : X^{\mathrm{ss}} \to X/\!/U = \overline{G \times_U X}/\!/G$ is not necessarily surjective, so in contrast to the reductive situation we cannot describe $X/\!/U$ topologically as the quotient of X^{ss} by an equivalence relation.

In order to describe $X/\!/U$ topologically (and geometrically) it is useful to consider the linear action of the Levi subgroup $\mathrm{GL}(r;\mathbb{C}) \leq P$ on the closure $\overline{P \times_U X} = \mathbb{P}(\mathbb{C} \oplus ((\mathbb{C}^r)^* \otimes \mathbb{C}^r)) \times X$ of $P \times_U X \cong L^{(P)} \times X$ in $\overline{G \times_U X} = \mathbb{P}(\mathbb{C} \oplus ((\mathbb{C}^r)^* \otimes \mathbb{C}^{r+1})) \times X$. We have

$$G \times_U X \cong G \times_P (P \times_U X),$$

where $P/U \cong \mathrm{GL}(r;\mathbb{C})$ and $G/P \cong \mathbb{P}^r$ is projective, so $G \times_P (\overline{P \times_U X})$ is a projective completion of $G \times_U X$. The induced linearisation of the action of G on $G \times_P (\overline{P \times_U X})$ is not ample: if we regard $G \times_P (\overline{P \times_U X})$ as a subvariety in the obvious way of

$$
\begin{aligned}
G \times_P (\overline{G \times_U X}) &= G \times_P (\mathbb{P}(\mathbb{C} \oplus ((\mathbb{C}^r)^* \otimes \mathbb{C}^{r+1})) \times X) \\
&\cong (G/P) \times \mathbb{P}(\mathbb{C} \oplus ((\mathbb{C}^r)^* \otimes \mathbb{C}^{r+1})) \times X \\
&\cong \mathbb{P}^r \times \mathbb{P}(\mathbb{C} \oplus ((\mathbb{C}^r)^* \otimes \mathbb{C}^{r+1})) \times X
\end{aligned}
$$

then the birational morphism

$$G \times_P (\overline{P \times_U X}) \to \overline{G \times_U X} \cong \mathbb{P}(\mathbb{C} \oplus ((\mathbb{C}^r)^* \otimes \mathbb{C}^r)) \times X$$

given by $[g, y] \mapsto gy$ extends to the projection

$$\mathbb{P}^r \times \mathbb{P}(\mathbb{C} \oplus ((\mathbb{C}^r)^* \otimes \mathbb{C}^{r+1})) \times X \to \mathbb{P}(\mathbb{C} \oplus ((\mathbb{C}^r)^* \otimes \mathbb{C}^{r+1})) \times X$$

and the induced line bundle is the restriction to $G \times_P (\overline{P \times_U X})$ of $\mathcal{O}_{\mathbb{P}(\mathbb{C} \oplus ((\mathbb{C}^r)^* \otimes \mathbb{C}^{r+1}))}(N) \otimes L$. However, if $\epsilon \in \mathbb{Q} \cap (0, \infty)$, the tensor product $\hat{L}_\epsilon = \hat{L}_\epsilon^{(N)}$ of this line bundle with the pullback via the morphism

$$G \times_P (\overline{P \times_U X}) \to G/P \cong \mathbb{P}^r$$

of the fractional line bundle $\mathcal{O}_{\mathbb{P}^r}(\epsilon)$ provides an ample fractional linearisation for the action of G on $G \times_P (\overline{P \times_U X})$ with, when ϵ is sufficiently small, an induced surjective birational morphism

$$\widehat{X/\!/U} =_{\text{def}} G \times_P (\overline{P \times_U X})/\!/_{\hat{L}_\epsilon} G \to \overline{G \times_U X}/\!/G = X/\!/U, \qquad (40)$$

which is an isomorphism over

$$(G \times_U X^{\bar{s}})/G \cong X^{\bar{s}}/U = X^{s}/U.$$

This line bundle $\hat{L}_\epsilon$ can be thought of as the bundle $G \times_P (\mathcal{O}_{\mathbb{P}(\mathbb{C} \oplus ((\mathbb{C}^r)^* \otimes \mathbb{C}^r))}(N) \otimes L)$ on $G \times_P (\overline{P \times_U X})$, where now the P-action on $\mathcal{O}_{\mathbb{P}(\mathbb{C} \oplus ((\mathbb{C}^r)^* \otimes \mathbb{C}^r))}(N) \otimes L$ is no longer the restriction of the G-action on the line bundle $\mathcal{O}_{\mathbb{P}(\mathbb{C} \oplus ((\mathbb{C}^r)^* \otimes \mathbb{C}^{r+1}))}(N) \otimes L$ but has been twisted by ϵ times the character of P which restricts to the determinant on $\mathrm{GL}(r; \mathbb{C})$.

Since $\mathrm{GL}(r; \mathbb{C}) = P/U$ has a central one-parameter subgroup $\mathbb{C}^*$, we can modify the linearisation of any linear actions of P and $\mathrm{GL}(r; \mathbb{C})$ by multiplying by ϵ times the standard character det of $\mathrm{GL}(r; \mathbb{C})$ for any $\epsilon \in \mathbb{Q}$. By the Hilbert–Mumford criteria (Proposition 2.2 above) we have

$$\overline{P \times_U X}^{\mathrm{ss}, P, \epsilon} \subseteq \overline{P \times_U X}^{\mathrm{ss}, \mathrm{GL}(r;\mathbb{C}), \epsilon} \subseteq \overline{P \times_U X}^{\mathrm{ss}, \mathrm{SL}(r;\mathbb{C})}, \qquad (41)$$

where $\overline{P \times_U X}^{\mathrm{ss}, \mathrm{GL}(r;\mathbb{C}), \epsilon}$ and $\overline{P \times_U X}^{\mathrm{ss}, \mathrm{SL}(r;\mathbb{C})}$ (independent of ϵ) denote the $\mathrm{GL}(r; \mathbb{C})$-semistable and $\mathrm{SL}(r; \mathbb{C})$-semistable sets of $\overline{P \times_U X}$ after twisting the linearisation by ϵ times the character det of $\mathrm{GL}(r; \mathbb{C})$; this character is of course trivial on $\mathrm{SL}(r; \mathbb{C})$. It turns out (see [23, Section 4.1(11)]) that if ϵ is chosen appropriately (close to $-N/2$, where N is as in the choice of linearisation above) then

$$\overline{P \times_U X}^{\mathrm{ss},\mathrm{GL}(r;\mathbb{C}),\epsilon} = (\mathbb{P}(\mathbb{C} \oplus ((\mathbb{C}^r)^* \otimes \mathbb{C}^r)) \times X)^{\mathrm{ss},\mathrm{GL}(r;\mathbb{C}),\epsilon} = \mathrm{GL}(r;\mathbb{C}) \times X, \tag{42}$$

and so quotienting we get

$$\overline{P \times_U X}//_{\hat{L}^{(N)}_{-N/2}} \mathrm{GL}(r;\mathbb{C}) \cong X. \tag{43}$$

Therefore

$$\mathcal{X} =_{\mathrm{def}} \overline{P \times_U X}//\mathrm{SL}(r;\mathbb{C}) = (\mathbb{P}(\mathbb{C} \oplus ((\mathbb{C}^r)^* \otimes \mathbb{C}^r)) \times X)//\mathrm{SL}(r;\mathbb{C}) \tag{44}$$

is a projective variety with a linear action of $\mathbb{C}^* = \mathrm{GL}(r;\mathbb{C})/\mathrm{SL}(r;\mathbb{C})$, which we can twist by ϵ times the standard character of $\mathbb{C}^*$, such that when $\epsilon = -N/2$ we get

$$\mathcal{X}//_{-N/2}\mathbb{C}^* \cong X, \tag{45}$$

while for $\epsilon > 0$ sufficiently small we have a surjection from an open subset $(\mathcal{X}//_\epsilon\mathbb{C}^*)^{\hat{\mathrm{ss}}}$ of $\mathcal{X}//_\epsilon\mathbb{C}^*$ onto $\widehat{X//U}$, and hence onto $X//U$ (see [23, Proposition 4.6]). More precisely, let $(\mathcal{X}//_\epsilon\mathbb{C}^*)^{\hat{\mathrm{s}}}$ be the open subset $\overline{P \times_U X}^{\mathrm{s},P,\epsilon}/\mathrm{GL}(r;\mathbb{C})$ of

$$\begin{aligned}\overline{P \times_U X}^{\mathrm{s},\mathrm{GL}(r;\mathbb{C}),\epsilon}/\mathrm{GL}(r;\mathbb{C}) &= (\overline{P \times_U X}^{\mathrm{s},\mathrm{GL}(r;\mathbb{C}),\epsilon}/\mathrm{SL}(r;\mathbb{C})/\mathbb{C}^* \\ &= \mathcal{X}^{\mathrm{s},\epsilon}/\mathbb{C}^* \subseteq \mathcal{X}//_\epsilon\mathbb{C}^*\end{aligned}$$

and let $\mathcal{X}^{\hat{\mathrm{ss}},\epsilon} = \pi^{-1}((\mathcal{X}//_\epsilon\mathbb{C}^*)^{\hat{\mathrm{ss}}})$ and $\mathcal{X}^{\hat{\mathrm{s}},\epsilon} = \pi^{-1}((\mathcal{X}//_\epsilon\mathbb{C}^*)^{\hat{\mathrm{s}}})$, where $\pi : \mathcal{X}^{\mathrm{ss},\epsilon} \to \mathcal{X}//_\epsilon\mathbb{C}^*$ is the quotient map, so that

$$(\mathcal{X}//_\epsilon\mathbb{C}^*)^{\hat{\mathrm{s}}} = \mathcal{X}^{\hat{\mathrm{s}},\epsilon}/\mathbb{C}^*. \tag{46}$$

In this construction we can replace the compactification $\mathbb{P}(\mathbb{C} \oplus ((\mathbb{C}^r)^* \otimes \mathbb{C}^r))$ of $\mathrm{GL}(r;\mathbb{C})$ by its wonderful compactification $\mathbb{P}(\mathbb{C} \oplus \widetilde{((\mathbb{C}^r)^*} \otimes \mathbb{C}^r))$ given by blowing up $\mathbb{P}(\mathbb{C} \oplus ((\mathbb{C}^r)^* \otimes \mathbb{C}^r)) = \{[z : (z_{ij})_{i,j=1}^r]\}$ along the (proper transforms of the) subvarieties defined by

$$z = 0 \quad \text{and} \quad \mathrm{rank}(z_{ij}) \leq \ell$$

for $\ell = 0, 1, \ldots, r$ and by

$$\mathrm{rank}(z_{ij}) \leq \ell$$

for $\ell = 0, 1, \ldots, r-1$ [17]. The action of $\mathrm{SL}(r;\mathbb{C})$ on $\mathbb{P}(\mathbb{C} \oplus \widetilde{((\mathbb{C}^r)^*} \otimes \mathbb{C}^r))$, linearised with respect to a small perturbation of the pullback of $\mathcal{O}_{\mathbb{P}(\mathbb{C}\oplus((\mathbb{C}^r)^*\otimes\mathbb{C}^r))}(1)$, satisfies

$$\mathbb{P}(\mathbb{C} \oplus \widetilde{((\mathbb{C}^r)^*} \otimes \mathbb{C}^r))^{\mathrm{ss}} = \mathbb{P}(\mathbb{C} \oplus \widetilde{((\mathbb{C}^r)^*} \otimes \mathbb{C}^r))^{\mathrm{s}}$$

and

$$\mathbb{P}(\mathbb{C} \oplus \widetilde{((\mathbb{C}^r)^* \otimes \mathbb{C}^r)})//\mathrm{SL}(r;\mathbb{C}) \cong \mathbb{P}^1.$$

If we take $\overline{P \times_U X}$ to be

$$\mathbb{P}(\mathbb{C} \oplus \widetilde{((\mathbb{C}^r)^* \otimes \mathbb{C}^r)}) \times X$$

instead of $\mathbb{P}(\mathbb{C} \oplus ((\mathbb{C}^r)^* \otimes \mathbb{C}^r)) \times X$, and define

$$\tilde{\mathcal{X}} = \mathbb{P}(\mathbb{C} \oplus \widetilde{((\mathbb{C}^r)^* \otimes \mathbb{C}^r)}) \times X//\mathrm{SL}(r;\mathbb{C}), \tag{47}$$

then the properties of $\mathcal{X}$ given above are satisfied by $\tilde{\mathcal{X}}$, and if X is nonsingular then

$$\widetilde{X//U} =_{\mathrm{def}} G \times_P (\mathbb{P}(\mathbb{C} \oplus \widetilde{((\mathbb{C}^r)^* \otimes \mathbb{C}^r)}) \times X)//G$$

is a partial desingularisation of $X//U$ and a compactification of X^{s}/U. Indeed, it is shown in [23, Proposition 4.6] (combined with [23, Remark 4.8]) that if $\epsilon > 0$ is sufficiently small then the natural rational map from $\tilde{\mathcal{X}}//_{\epsilon}\mathbb{C}^*$ to $\widetilde{X//U}$ restricts to surjective morphisms

$$(\tilde{\mathcal{X}}//_{\epsilon}\mathbb{C}^*)^{\tilde{\mathrm{s}}\tilde{\mathrm{s}}} \to \widetilde{X//U} \to X//U$$

and

$$(\tilde{\mathcal{X}}//_{\epsilon}\mathbb{C}^*)^{\tilde{\mathrm{s}}} \to X^{\mathrm{s}}/U.$$

Using the theory of variation of GIT [6, 32, 34], we can relate the quotient $\hat{\mathcal{X}}//_{\epsilon}\mathbb{C}^*$ to $\tilde{\mathcal{X}}//_{N/2}\mathbb{C}^* \cong X$ via a sequence of flips which occur as walls are crossed between the linearisations corresponding to ϵ and to $-N/2$. Thus we have a diagram

$$\begin{array}{ccccc}
(\tilde{\mathcal{X}}//_{\epsilon}\mathbb{C}^*)^{\tilde{\mathrm{s}}} \subseteq & (\tilde{\mathcal{X}}//_{\epsilon}\mathbb{C}^*)^{\tilde{\mathrm{s}}\tilde{\mathrm{s}}} & \subseteq \tilde{\mathcal{X}}//_{\epsilon}\mathbb{C}^* & \overset{}{\leftarrow - - \rightarrow} & X = \tilde{\mathcal{X}}//_{-N/2}\mathbb{C}^* \\
\downarrow & \downarrow & & \text{flips} & \\
X^{\mathrm{s}}/U \quad \subseteq & \widetilde{X//U} & & & \\
\| & \downarrow & & & \\
X^{\mathrm{s}}/U \quad \subseteq & X//U & & &
\end{array} \tag{48}$$

where the vertical maps are all surjective, and the inclusions are all open.

Remark 4.7. The construction of a reductive envelope described here is valid only if the action of $U = (\mathbb{C}^+)^r$ on X extends to an action of $G = \mathrm{SL}(\mathbb{C} \oplus \mathfrak{u})$ (which is a rather special situation when the ring of invariants $\hat{\mathcal{O}}_L(X)^U$ is always finitely generated). Moreover, at least a priori this construction may depend on the choice of the extension of the U-action to a G-action, although $\overline{G \times_U X}//G = X//U = \mathrm{Proj}(\hat{\mathcal{O}}_L(X)^U)$ depends only on the linearisation of the U-action on X. However, it is shown in [23] that we can associate to a linear U-action on X a family of projective varieties Y_m (one for every sufficiently large positive integer m), each of which contains X and has an action of $G = \mathrm{SL}(\mathbb{C} \oplus \mathfrak{u})$ and a G-linearisation on an ample line bundle L_{Y_m}, which restricts to the given linearisation of the U-action on

X and is such that every U-invariant in a finite fully separating set of U-invariants on X extends to a U-invariant on Y_m. Then we can embed X in the G-variety

$$\mathbb{P}(\mathbb{C} \oplus ((\mathbb{C}^r)^* \otimes \mathbb{C}^{r+1})) \times Y_m$$

as $\{\iota\} \times X$, where $\iota \in (\mathbb{C}^r)^* \otimes \mathbb{C}^{r+1} \subseteq \mathbb{P}(\mathbb{C} \oplus ((\mathbb{C}^r)^* \otimes \mathbb{C}^{r+1}))$ is the standard embedding of $\mathbb{C}^r$ in $\mathbb{C}^{r+1}$. The closure of $GX \cong G \times_U X$ in $\mathbb{P}(\mathbb{C} \oplus ((\mathbb{C}^r)^* \otimes \mathbb{C}^{r+1})) \times Y_m$ will provide us with a reductive envelope $\overline{G \times_U X}$ (which is, however, not necessarily ample), and we can study the closures of the images of X^s/U in $Y_m//U = \overline{Y_m//U}$ and its partial desingularisation $\widetilde{Y_m//U}$ constructed as above.

4.3 Symplectic implosion for $U = (\mathbb{C}^+)^r \le \mathrm{SL}(r+1;\mathbb{C})$ actions

Let X be a complex projective variety on which the complexification $G = \mathrm{SL}(r+1;\mathbb{C})$ of $K = \mathrm{SU}(r+1)$ acts linearly with respect to a very ample line bundle L, and let $U = (\mathbb{C}^+)^r$ be the unipotent radical of the parabolic $P = \mathrm{GL}(r;\mathbb{C})U$ as in the previous subsection. As before, let T be the maximal torus of K consisting of the diagonal matrices in K, and let B be the upper triangular Borel subgroup of G. In the notation of Section 3.2 we have $L^{(P)} = \mathrm{GL}(r;\mathbb{C})$ and $K^{(P)} = \mathrm{U}(r)$. We can identify the Lie algebra $\mathfrak{k}^{(P)} = \mathfrak{u}(r)$ of $K^{(P)}$ with the product $[\mathfrak{k}^{(P)}, \mathfrak{k}^{(P)}] \oplus \mathfrak{z}^{(P)}$ of the Lie algebras of its semisimple part $Q^{(P)} = [K^{(P)}, K^{(P)}] = \mathrm{SU}(r)$ and its centre $Z(K^{(P)}) \cong S^1$. If we identify $\mathfrak{t}^*$ with

$$\{\zeta = (\zeta_1, \dots, \zeta_{r+1}) \in \mathbb{R}^{r+1} : \zeta_1 + \dots + \zeta_{r+1} = 0\}$$

in the usual way so that

$$\mathfrak{t}^*_+ = \{\zeta = (\zeta_1, \dots, \zeta_{r+1}) \in \mathfrak{t}^* : \zeta_1 \ge \zeta_2 \ge \dots \ge \zeta_{r+1}\},$$

then

$$\mathfrak{t}^*_{(P)+} = \{\zeta = (\zeta_1, \dots, \zeta_{r+1}) \in \mathfrak{t}^* : \zeta_j \ge \zeta_{r+1} \text{ for } j = 1, \dots, r\} \tag{49}$$

and

$$\mathfrak{z}^{(P)*} = \{\zeta = (\zeta_1, \dots, \zeta_{r+1}) \in \mathfrak{t}^* : \zeta_1 = \dots = \zeta_r\}. \tag{50}$$

Moreover, $\mathfrak{k}^{(P)*}_+$ can be identified with the set of skew-Hermitian matrices in $\mathfrak{su}(r+1)^*$ of the form

$$\zeta = \begin{pmatrix} \xi & 0 \\ 0 & i\lambda_{r+1} \end{pmatrix}, \tag{51}$$

where ξ is a skew-Hermitian $r \times r$ matrix with all its eigenvalues of the form $i\lambda$ with $\lambda \in \mathbb{R}$ and $\lambda \ge \lambda_{r+1}$. If all the eigenvalues $i\lambda$ of ξ satisfy $\lambda > \lambda_{r+1}$ then $K_\zeta(P)$ is conjugate to T and $[K_\zeta(P), K_\zeta(P)]$ is trivial. In general, $\mathrm{Ad}^*(K^{(P)})\zeta$ contains a matrix of the form

$$\begin{pmatrix} \xi & 0 \\ 0 & i\lambda_{r+1} I_j \end{pmatrix} \tag{52}$$

for some $j \in \{0, 1, \dots, r\}$, where ξ is a skew-Hermitian $(r-j) \times (r-j)$ matrix with all its eigenvalues of the form $i\lambda$ with $\lambda \in \mathbb{R}$ and $\lambda > \lambda_{r+1}$, and I_j is the $j \times j$ identity matrix. Then $K_\zeta(P)$ is conjugate in $K^{(P)} = \mathrm{U}(r)$ to the product of a torus and the unitary group $U(j)$ embedded in $K = \mathrm{SU}(r+1)$ as

$$A \mapsto \begin{pmatrix} I_{r-j} & 0 & 0 \\ 0 & A & 0 \\ 0 & 0 & \det A^{-1} \end{pmatrix},$$

and the face σ of $\mathfrak{k}_+^{(P)*}$ to which ζ belongs is determined by j and the partition $\pi \in \Pi_{r-j}$ given by the eigenvalues $i\lambda$ of ζ with $\lambda > \lambda_{r+1}$. Thus $[K_\zeta(P), K_\zeta(P)] \cong \mathrm{SU}(j)$ and the universal $K^{(P)}$-imploded cross-section is

$$\begin{aligned} (T^*K)_{\mathrm{impl}}^{K,K^{(P)}} &= \bigsqcup_{j=0}^{r} (K \times \mathfrak{k}_{+,j,\pi}^{(P)*})/\approx^{K^{(P)}} \\ &= (K \times \mathfrak{k}_+^{(P)*})^\circ \sqcup \bigsqcup_{j=1}^{r} \bigsqcup_{\pi \in \Pi_j} (K \times \mathfrak{k}_{+,j}^{(P)*})/\approx^{K^{(P)}} \\ &= (K \times \mathfrak{k}_+^{(P)*})^\circ \sqcup \bigsqcup_{j=1}^{r} \bigsqcup_{\pi = (\pi_1, \dots, \pi_\ell) \in \Pi_j} \mathrm{U}(r) \times_{(\mathrm{U}(\pi_1) \times \cdots \times \mathrm{U}(\pi_\ell) \times \mathrm{U}(j))} \\ &\quad \times ((K \times \mathfrak{k}_{+,j,\pi}^{(P)*})/\mathrm{SU}(j)). \end{aligned} \tag{53}$$

Here $\mathfrak{k}_{+,j}^{(P)*}$ consists of all $\zeta \in \mathfrak{su}(r+1)^*$ of the form (51) with ξ a skew-Hermitian $r \times r$ matrix with all its eigenvalues of the form $i\lambda$ with $\lambda \in \mathbb{R}$ and $\lambda \geq \lambda_{r+1}$ and exactly j of its eigenvalues equal to $i\lambda_{r+1}$, and $\mathfrak{k}_{+,j,\pi}^{(P)*}$ consists of all $\zeta \in \mathfrak{k}_{+,j}^{(P)*}$ of the form (51) such that the partition of $r-j$ determined by the eigenvalues of ζ of the form $i\lambda$ with $\lambda > \lambda_{r+1}$ is π. Moreover, if (k_1, ζ_1) and (k_2, ζ_2) lie in $K \times \mathfrak{k}_{+,j}^{(P)*}$ then $(k_1, \zeta_1) \approx^{K^{(P)}} (k_2, \zeta_2)$ if and only if there is some $\kappa \in K^{(P)}$ such that

$$\zeta_1 = \zeta_2 = \kappa \begin{pmatrix} \xi & 0 \\ 0 & i\lambda_{r+1} I_j \end{pmatrix} \kappa^{-1}$$

and $\kappa^{-1} k_1 k_2^{-1} \kappa \in [K_\zeta(P), K_\zeta(P)] \cong \mathrm{SU}(j)$. Thus $(T^*K)_{\mathrm{impl}}^{K,K^{(P)}}$ is isomorphic to $\overline{G/U}^{\mathrm{aff}} = (\mathbb{C}^r)^* \otimes \mathbb{C}^{r+1}$ via

$$(k, \zeta) \mapsto k \circ \mathcal{F}(\zeta),$$

where if ζ is at (51) then $\mathcal{F}(\zeta) : \mathbb{C}^r \to \mathbb{C}^r \subseteq \mathbb{C}^{r+1}$ is the linear map represented by the unique $r \times r$ Hermitian positive definite matrix α satisfying $i\alpha^*\alpha = \xi - i\lambda_{r+1} I_r$.

Let ω be a K-invariant Kähler form on X, given in some choice of coordinates by the Fubini–Study form on the projective space into which the very ample line bundle L embeds X. Then we know that

$$\hat{\mathcal{O}}_L(X)^U \cong (\hat{\mathcal{O}}_L(X) \otimes \mathcal{O}(G)^U)^G$$

is finitely generated, and the associated projective variety

$$X/\!/U = \operatorname{Proj}(\hat{\mathcal{O}}_L(X)^U)$$

is isomorphic to the GIT quotient $(\overline{G/U}^{\text{aff}} \times X)/\!/G$, which as in Section 3.2 can be identified with a symplectic quotient of $\overline{G/U}^{\text{aff}} \times X$ by K, and thus with the $K^{(P)}$-imploded cross-section

$$X_{\text{impl}}^{K,K^{(P)}} = \mu^{-1}(\mathfrak{k}_+^{(P)*})/\approx_{K^{(P)}}$$

of X, where $x \approx_{K^{(P)}} y$ if and only if $\mu(x) = \mu(y) = \zeta \in \mathfrak{k}_+^{(P)*}$ and $x = \kappa y$ for some $\kappa \in [K_\zeta(P), K_\zeta(P)]$. Equivalently

$$\begin{aligned} X_{\text{impl}}^{K,K^{(P)}} &= \mu^{-1}((\mathfrak{k}_+^{(P)*})^\circ) \sqcup \bigsqcup_{j=1}^{r} \mu^{-1}(\mathfrak{k}_{+,j}^{(P)*})/\approx^{K^{(P)}} \\ &= \mu^{-1}((\mathfrak{k}_+^{(P)*})^\circ) \sqcup \bigsqcup_{j=1}^{r} \bigsqcup_{\pi=(\pi_1,\dots,\pi_\ell)\in\Pi_j} \mathrm{U}(r) \times_{(\mathrm{U}(\pi_1)\times\cdots\times\mathrm{U}(\pi_\ell)\times\mathrm{U}(j))} \\ &\quad \times (\mu^{-1}(\mathfrak{k}_{+,j,\pi}^{(P)*} \cap \mathfrak{t}_+^*)/\mathrm{SU}(j)), \end{aligned} \tag{54}$$

since $[K_\zeta(P), K_\zeta(P)] \cong \mathrm{SU}(j)$ if $\zeta \in \mathfrak{k}_{+,j}^{(P)*}$.

The desingularisation $\widetilde{(T^*K)}_{\text{impl}}^{K,K^{(P)}}$ of $(T^*K)_{\text{impl}}^{K,K^{(P)}}$ is given by

$$\widetilde{(T^*K)}_{\text{impl}}^{K,K^{(P)}} = (K \times \mathfrak{k}_+^{(P)*,\epsilon})/\approx_\epsilon^{K^{(P)}}, \tag{55}$$

where $\mathfrak{k}_+^{(P)*,\epsilon} = \mathrm{Ad}^*(K^{(P)})(\epsilon\lambda_0 + \mathfrak{t}_{(P)+}^*)$ for $0 < \epsilon \ll 1$ and $\lambda_0 = \operatorname{diag}(1, 1, \dots, 1, -r) \in \mathfrak{t}_{(P)+}^* \cap \mathfrak{z}^{(P)*}$, and if (k_1, ζ_1) and (k_2, ζ_2) lie in $K \times \mathfrak{k}_{+,j}^{(P)*,\epsilon}$, then $(k_1, \zeta_1) \approx_\epsilon^{K^{(P)}} (k_2, \zeta_2)$ if and only if there is some $\kappa \in K^{(P)} \cong \mathrm{U}(r)$ such that

$$\zeta_1 = \zeta_2 = \kappa \begin{pmatrix} \xi & 0 \\ 0 & i\lambda_{r+1} I_j \end{pmatrix} \kappa^{-1}$$

and $\kappa^{-1}k_1k_2^{-1}\kappa$ lies in the maximal torus T_j of $[K_\zeta(P), \widetilde{K_\zeta(P)}] \cong \mathrm{SU}(j)$ which is its intersection with T. The partial desingularisation $\widetilde{X_{\mathrm{impl}}^{K,K^{(P)}}}$ of $X_{\mathrm{impl}}^{K,K^{(P)}}$ is the symplectic quotient of $\widetilde{(T^*K)_{\mathrm{impl}}^{K,K^{(P)}}} \times X$ by the diagonal action of K; as a stratified symplectic space, it is given by

$$\widetilde{X_{\mathrm{impl}}^{K,K^{(P)}}} = \mu^{-1}((\mathfrak{k}_+^{(P)*,\epsilon})^\circ) \sqcup \bigsqcup_{j=1}^{r} \bigsqcup_{\pi=(\pi_1,\ldots,\pi_\ell)\in\Pi_j} \mathrm{U}(r)\times_{(\mathrm{U}(\pi_1)\times\cdots\times\mathrm{U}(\pi_\ell)\times\mathrm{U}(j))}$$
$$\times\,(\mu^{-1}(\epsilon\lambda_0 + \mathfrak{k}_{+,j,\pi}^{(P)*} \cap \mathfrak{t}_+^*)/T_j),$$

and it can also be identified with the partial desingularisation $\widetilde{X/\!/U}$ described in Section 4.2.

Example 4.8. Let $U = \mathbb{C}^+$ act linearly on a projective space $\mathbb{P}^n$, and suppose that coordinates have been chosen so that the natural generator of $\mathrm{Lie}(\mathbb{C}^+) = \mathbb{C}$ has Jordan normal form with blocks of sizes $k_1+1, \ldots, k_s+1$, where $\sum_{j=1}^{s}(k_j+1) = n+1$. The $\mathbb{C}^+$ action extends to an action of $G = \mathrm{SL}(2;\mathbb{C})$ by identifying $\mathbb{C}^+$ with the group of upper triangular matrices

$$\left\{\begin{pmatrix}1 & a\\ 0 & 1\end{pmatrix} : a \in \mathbb{C}\right\} \le \mathrm{SL}(2;\mathbb{C})$$

and $\mathbb{C}^{n+1}$ with $\bigoplus_{j=1}^{s}\mathrm{Sym}^{k_j}(\mathbb{C}^2)$, where $\mathrm{Sym}^k(\mathbb{C}^2)$ is the kth symmetric power of the standard representation $\mathbb{C}^2$ of $G = \mathrm{SL}(2;\mathbb{C})$. We have

$$G/\mathbb{C}^+ \cong \mathbb{C}^2\backslash\{0\} \subseteq \mathbb{C}^2 \subseteq \mathbb{P}^2 = \overline{G/\mathbb{C}^+}$$

and thus $\mathbb{P}^n/\!/\mathbb{C}^+$ is the GIT quotient $\mathrm{Proj}(\mathbb{C}[x_0,\ldots,x_n]^{\mathbb{C}^+}) \cong (\mathbb{P}^2\times\mathbb{P}^n)/\!/G$ with respect to the linearisation $\mathcal{O}_{\mathbb{P}^2}(N)\otimes\mathcal{O}_{\mathbb{P}^n}(1)$ on $\mathbb{P}^2\times\mathbb{P}^n$ for N a sufficiently large positive integer. Since $(\mathbb{P}^2)^{\mathrm{ss},G} = \mathbb{C}^2$ and N is large, we have

$$(\mathbb{P}^2\times\mathbb{P}^n)^{\mathrm{ss},G} \subseteq \mathbb{C}^2\times\mathbb{P}^n = (G\times_{\mathbb{C}^+}\mathbb{P}^n)\sqcup(\{0\}\times\mathbb{P}^n),$$

and if semistability implies stability then

$$\mathbb{P}^n/\!/\mathbb{C}^+ = (\mathbb{P}^n)^{\mathrm{s},U}/\mathbb{C}^+ \sqcup (\{0\}\times\mathbb{P}^n)/\!/\mathrm{SL}(2;\mathbb{C}).$$

In this example the parabolic subgroup P of $G = \mathrm{SL}(2;\mathbb{C})$ is its standard (upper triangular) Borel subgroup with $\overline{B/\mathbb{C}^+} = \overline{\mathbb{C}^*} = \mathbb{P}^1$ and

$$\overline{B\times_{\mathbb{C}^+}\mathbb{P}^n} = \mathbb{P}^1\times\mathbb{P}^n,$$

while $G\times_B \overline{B/\mathbb{C}^+} = G\times_B\mathbb{P}^1$ is the blowup of $\mathbb{P}^2$ at the origin $0\in\mathbb{C}^2\subseteq\mathbb{P}^2$. Similarly $G\times_B(\overline{B\times_{\mathbb{C}^+}\mathbb{P}^n})$ is the blowup of $\overline{G\times_{\mathbb{C}^+}\mathbb{P}^n}\cong\mathbb{P}^2\times\mathbb{P}^n$ along $\{0\}\times\mathbb{P}^n$, and its quotient $\widetilde{X//U}$ is the blowup of $\mathbb{P}^n//\mathbb{C}^+$ along its "boundary"

$$\mathbb{P}^n//\mathrm{SL}(2;\mathbb{C}) \cong (\{0\} \times \mathbb{P}^n)//\mathrm{SL}(2;\mathbb{C}) \subseteq (\mathbb{P}^2 \times \mathbb{P}^n)//\mathrm{SL}(2;\mathbb{C}) = \mathbb{P}^n//\mathbb{C}^+.$$

From the point of view of symplectic geometry we have

$$\mathbb{P}^n//\mathbb{C}^+ \cong (\mathbb{P}^n)_{\mathrm{impl}} = \mu^{-1}((\mathfrak{t}_+^*)^\circ) \sqcup \frac{\mu^{-1}(0)}{\mathrm{SU}(2)} = \mu^{-1}(0,\infty) \sqcup \frac{\mu^{-1}(0)}{\mathrm{SU}(2)},$$

where $\mathfrak{t}_+^*$ is identified with $(0,\infty)$ in the usual way, and

$$\widetilde{\mathbb{P}^n//\mathbb{C}^+} \cong \widetilde{(\mathbb{P}^n)}_{\mathrm{impl}} = \mu^{-1}(\epsilon,\infty) \sqcup \frac{\mu^{-1}(\epsilon)}{S^1}$$

for $0 < \epsilon \ll 1$.

References

1. M. F. Atiyah, Convexity and commuting Hamiltonians. Bull. London Math. Soc. **14** (1982), 1–15.
2. M. F. Atiyah and R. Bott, The Yang–Mills equations over Riemann surfaces. Phil. Trans. Roy. Soc. London **308** (1982), 523–615.
3. D. Birkes, Orbits of linear algebraic groups. Annals of Math. **93** (1971), 459–475.
4. A. Borel, Linear algebraic groups. Graduate Texts in Mathematics **126**, Springer, 1991.
5. I. Dolgachev, Lectures on invariant theory. London Mathematical Society Lecture Note Series 296, Cambridge University Press, 2003.
6. I. Dolgachev and Y. Hu, Variation of Geometric Invariant Theory quotients. Publ. Math. I.H.E.S. **87** (1998), 5–56.
7. B. Doran and F. Kirwan, Towards non-reductive geometric invariant theory. Pure Appl. Math. Q. **3** (2007), 61–105.
8. A. Fauntleroy, Categorical quotients of certain algebraic group actions. Illinois Journal Math. **27** (1983), 115–124.
9. A. Fauntleroy, Geometric invariant theory for general algebraic groups. Compositio Mathematica **55** (1985), 63–87.
10. G.-M. Greuel and G. Pfister, Geometric quotients of unipotent group actions. Proc. London Math. Soc. (3) **67** (1993), 75–105.
11. G.-M. Greuel and G. Pfister, Geometric quotients of unipotent group actions II. Singularities (Oberwolfach 1996), 27–36, Progress in Math. 162, Birkhäuser, Basel 1998.
12. F. Grosshans, Algebraic homogeneous spaces and invariant theory. Lecture Notes in Mathematics, 1673, Springer-Verlag, Berlin, 1997.
13. F. Grosshans, The invariants of unipotent radicals of parabolic subgroups. Invent. Math. **73** (1983), 1–9.
14. V. Guillemin, L. Jeffrey, and R. Sjamaar, Symplectic implosion. Transformation Groups **7** (2002), 155–184.
15. L. Jeffrey and F. Kirwan, Localization for nonabelian group actions. Topology **34** (1995), no. 2, 291–327.
16. L. Jeffrey, Y.-H. Kiem, F. Kirwan, and J. Woolf, Cohomology pairings on singular quotients in geometric invariant theory. Transform. Groups **8** (2003), no. 3, 217–259.
17. I. Kausz, A modular compactification of the general linear group. Doc. Math. **5** (2000), 553–594.

18. S. Keel and S. Mori, Quotients by groupoids. Annals of Math. (2) **145** (1997), 193–213.
19. F. Kirwan, Cohomology of quotients in symplectic and algebraic geometry. Mathematical Notes, 31. Princeton University Press, Princeton, NJ, 1984.
20. F. Kirwan, Partial desingularisations of quotients of nonsingular varieties and their Betti numbers. Ann. of Math. (2) **122** (1985), no. 1, 41–85.
21. F. Kirwan, Rational intersection cohomology of quotient varieties. Invent. Math. **86** (1986), no. 3, 471–505.
22. F. Kirwan, Rational intersection cohomology of quotient varieties II. Invent. Math. **90** (1987), no. 1, 153–167.
23. F. Kirwan, Quotients by non-reductive algebraic group actions. Moduli spaces and vector bundles, 311–366, London Math. Soc. Lecture Note Ser., 359, Cambridge Univ. Press, Cambridge, 2009.
24. J. Kollár, Quotient spaces modulo algebraic groups. Ann. Math. (2) **145** (1997), 33–79.
25. E. Lerman, Symplectic cuts. Math. Res. Lett. **2** (1995), 247–258.
26. E. Meinrenken and R. Sjamaar, Singular reduction and quantization. Topology **38** (1999), 699–762.
27. D. Mumford, J. Fogarty, and F. Kirwan, Geometric invariant theory. 3rd edition, Springer, 1994.
28. P. E. Newstead, Introduction to moduli problems and orbit spaces. Tata Institute Lecture Notes, Springer, 1978.
29. P.-E. Paradan, Formal geometric quantization. arXiv.math/0702224.
30. V. Popov and E. Vinberg, Invariant theory. Algebraic geometry IV, Encyclopaedia of Mathematical Sciences v. 55, 1994.
31. Z. Reichstein, Stability and equivariant maps. Invent. Math. **96** (1989), 349–383.
32. N. Ressayre, The GIT-equivalence for G-line bundles. Geom. Dedicata **81** (2000), 295–324.
33. T. A. Springer, Linear algebraic groups. Second edition, Birkhäuser, 1998.
34. M. Thaddeus, Geometric invariant theory and flips. Journal of Amer. Math. Soc. 1996.
35. J. Winkelmann, Invariant rings and quasiaffine quotients. Math. Z. **244** (2003), 163–174.

Quantization of q-Hamiltonian SU(2)-spaces

Eckhard Meinrenken

Dedicated to Hans Duistermaat on the occasion of his 65th birthday

Abstract We will explain how to define the quantization of q-Hamiltonian SU(2)-spaces as push-forwards in twisted equivariant K-homology, and prove the "quantization commutes with reduction" theorem for this setting. As applications, we show how the Verlinde formulas for flat SU(2)- or SO(3)-bundles are obtained via localization in twisted K-homology.

Key words: Moment maps, moduli spaces, twisted K-theory

Mathematics Subject Classification (2010): 53D30, 19L50

1 Introduction

The theory of q-Hamiltonian G-spaces was introduced ten years ago in the paper *Lie group valued moment maps* [1]. The motivation was to treat Hamiltonian loop group actions with proper moment maps in a purely finite-dimensional framework, obtaining for instance a finite-dimensional construction of the moduli space of flat G-bundles over a surface. Many of the standard constructions for ordinary Hamiltonian group actions on symplectic manifolds carried over to the new setting, but often with nontrivial "twists." For example, all q-Hamiltonian G-spaces M carry a natural volume form [5], which may be viewed informally as a push-forward of the (ill-defined) Liouville form on the associated infinite-dimensional loop group space. This volume form admits an equivariant extension (but for a nonstandard equivariant cohomology theory) [3], and the total volume may be computed by localization techniques, just as in the usual Duistermaat–Heckman theory [18].

University of Toronto, Department of Mathematics, 40 St. George Street, Toronto, Ontario M5S2E4, Canada
e-mail: mein@math.toronto.edu

One problem that had remained open until recently is how to define a "quantization" of q-Hamiltonian spaces. In contrast to the Hamiltonian theory, the 2-form on a q-Hamiltonian space is usually degenerate. Hence, there is no obvious notion of a compatible almost complex structure, and the usual quantization as the equivariant index of a Spin_c-Dirac operator [16] is no longer possible. In [31], rather than trying to construct such an operator, we define the quantization more abstractly as the push-forward of a K-homology fundamental class $[M]$. This fundamental class is canonically defined as an element in *twisted* equivariant K-homology of M. Our construction defines a push-forward of this element to the twisted equivariant K-homology of a Lie group. The Freed–Hopkins–Teleman theorem [20, 19] identifies the latter with the fusion ring $R_k(G)$ (Verlinde algebra), at an appropriate level-k. We take the resulting element $\mathcal{Q}(M) \in R_k(G)$ to be the "quantization" of our q-Hamiltonian space. As in the usual Hamiltonian theory [22, 21, 30], the quantization procedure satisfies a "quantization commutes with reduction" principle.

In the present paper, we will preview this quantization of q-Hamiltonian G-spaces for the simplest of simple compact Lie groups $G = \mathrm{SU}(2)$. Much of the general theory simplifies in this special case—for example, there is a fairly simple proof of the q-Hamiltonian "quantization commutes with reduction" theorem. As an application, we explain, following [4], how the SU(2)-Verlinde formulas are obtained in our theory. In the last section, we will show how to derive Verlinde-type formulas for moduli spaces of flat SO(3)-bundles. The paper will be largely self-contained, except for certain details that are better handled with the techniques from [31].

Notation. We fix the following notation and conventions for the Lie group SU(2). The group unit will be denoted by e, and the nontrivial central element by $c = \mathrm{diag}(-1,-1)$. We define an open cover by contractible subsets

$$\mathrm{SU}(2)_+ = \mathrm{SU}(2)\backslash\{c\}, \quad \mathrm{SU}(2)_- = \mathrm{SU}(2)\backslash\{e\} \tag{1}$$

with intersection the set $\mathrm{SU}(2)_{\mathrm{reg}}$ of regular elements. We take the maximal torus T to consist of the diagonal matrices, isomorphic to U(1) by the homomorphism

$$j\colon \mathrm{U}(1) \to T, \quad z \mapsto \mathrm{diag}(z, z^{-1}).$$

The Weyl group $W = \mathbb{Z}_2$ acts on T by permutation of the diagonal entries, or equivalently on U(1) by $z \mapsto z^{-1}$. We let $\Lambda \subset \mathfrak{t}$ be the integral lattice (kernel of $\exp|_{\mathfrak{t}}$) and $\Lambda^* \subset \mathfrak{t}^*$ its dual, the (real) weight lattice. For any $\mu \in \Lambda^*$ we denote by $t \mapsto t^\mu$ the corresponding homomorphism $T \to \mathrm{U}(1)$; the resulting 1-dimensional representation of T is denoted by $\mathbb{C}_\mu$. The weight lattice is generated by the element $\rho \in \Lambda^*$ such that $\mathbb{C}_\rho$ is the defining representation of U(1). The corresponding positive root is $\alpha = 2\rho$. We will identify $\mathfrak{su}(2)^* \cong \mathfrak{su}(2)$ using the *basic inner product*

$$\xi \cdot \xi' = \frac{1}{4\pi^2}\,\mathrm{tr}(\xi^\dagger \xi'), \quad \xi, \xi' \in \mathfrak{su}(2).$$

Similarly we identify $\mathfrak{t} \cong \mathfrak{t}^*$ using the induced inner product. Under this identification, $\Lambda = 2\Lambda^*$, with generators $\alpha = 2\pi i \operatorname{diag}(1, -1)$ and $\rho = i\pi \operatorname{diag}(1, -1)$.

For any subset $A \subset \mathfrak{t}$, we define $T_A = \exp(A) = \{\exp\xi \mid \xi \in A\}$. Any conjugacy class in SU(2) passes through a unique point in $T_{[0,\rho]}$, so that $[0, \rho]$ labels the conjugacy classes. We will frequently use the equivariant diffeomorphism,

$$T_{(0,\rho)} \times \mathrm{SU}(2)/T \to \mathrm{SU}(2)_{\mathrm{reg}}, \quad (t, gT) \mapsto \mathrm{Ad}_g(t). \tag{2}$$

2 The fusion ring $R_k(\mathrm{SU}(2))$

In this section, we review three simple descriptions of the level-k fusion ring (Verlinde algebra) $R_k(G)$ for the case $G = \mathrm{SU}(2)$. The fusion ring may be identified with the set of irreducible projective representations of the loop group $L\,\mathrm{SU}(2)$ at level-k [36], but we will not need that interpretation here.

2.1 *First description*

Let $R(\mathrm{SU}(2))$ be the representation ring of SU(2), viewed as the ring of virtual characters. For $m = 0, 1, 2, \ldots$ let $\chi_m \in R(\mathrm{SU}(2))$ be the character of the $(m+1)$-dimensional irreducible representation of SU(2). These form a basis of $R(\mathrm{SU}(2))$ as a $\mathbb{Z}$-module, and the ring structure is given by

$$\chi_m \chi_{m'} = \chi_{m+m'} + \chi_{m+m'-2} + \cdots + \chi_{|m-m'|}.$$

For $k = 0, 1, 2, \ldots$, the *level-k fusion ring* (or *Verlinde algebra*) is a quotient

$$R_k(\mathrm{SU}(2)) = R(\mathrm{SU}(2))/I_k(\mathrm{SU}(2))$$

by the ideal $I_k(\mathrm{SU}(2))$ generated by the character χ_{k+1}. Additively, the ideal is spanned by the characters $\chi_{k+1}, \chi_{2k+3}, \chi_{3k+5}, \ldots$, together with all characters of the form $\chi_{l'} - (-1)^r \chi_l$, where $l \in \{0, \ldots, k\}$, and l' is obtained from l by r reflections across the set of elements $k+1, 2k+3, 3k+5, \ldots$. It follows that as an abelian group, $R_k(\mathrm{SU}(2))$ is free with generators $\tau_0, \ldots, \tau_k$ the images of $\chi_0, \ldots, \chi_k$. For example, if $k = 4$, $m = 3$, $m' = 4$ we have

$$\chi_3\chi_4 = \chi_1 + \chi_3 + \chi_5 + \chi_7 \Rightarrow \tau_3\tau_4 = \tau_1 + \tau_3 + 0 - \tau_3 = \tau_1.$$

For any given level-k, the element $\tau_k \in R_k(\mathrm{SU}(2))$ defines an involution of the group $R_k(\mathrm{SU}(2))$,

$$\tau_l \mapsto \tau_l \tau_k = \tau_{k-l}.$$

2.2 Second description

Let q be the $(2k+4)$th root of unity

$$q = e^{\frac{i\pi}{k+2}}.$$

Then $I_k(\mathrm{SU}(2)) \subset R(\mathrm{SU}(2))$ may be described as the ideal of all characters vanishing at all points $j(q^s)$, for $s = 1, \ldots, k+1$. Put differently, letting

$$T_{k+2} = \{t \in T \mid t^{2k+4} = e\}$$

be the cyclic subgroup generated by $j(q)$, $I_k(\mathrm{SU}(2))$ is the vanishing ideal of $T_{k+2} \cap \mathrm{SU}(2)_{\mathrm{reg}} = T^{\mathrm{reg}}_{k+2}$. Hence, for any $t \in T^{\mathrm{reg}}_{k+2}$ the evaluation map $\mathrm{ev}_t : R(\mathrm{SU}(2)) \to \mathbb{C}$ descends to an evaluation map

$$\mathrm{ev}_t : R_k(\mathrm{SU}(2)) \to \mathbb{C}, \quad \tau \mapsto \tau(t) = \mathrm{ev}_t(\tau).$$

For the basis elements one obtains, by the Weyl character formula,

$$\tau_l(j(q^s)) = \frac{q^{(l+1)s} - q^{-(l+1)s}}{q^s - q^{-s}}.$$

The orthogonality relations

$$\sum_{s=1}^{k+1} \frac{|q^s - q^{-s}|^2}{2k+4} \tau_l(j(q^s))\tau_{l'}(j(q^s)) = \delta_{l,l'} \tag{3}$$

allow us to recover $\tau \in R_k(\mathrm{SU}(2))$ from the values $\tau(j(q^s))$ for $s = 1, \ldots, k$. The coefficients in this sum may alternatively be written as

$$\frac{|q^s - q^{-s}|^2}{2k+4} = \left(\frac{k}{2} + 1\right)^{-1} \sin^2\left(\frac{\pi s}{k+2}\right).$$

2.3 Third description

The third way of describing the fusion ring is to write down the structure constants relative to the basis $\tau_0, \ldots, \tau_k$. The *level-k fusion coefficient* $N^{(k)}_{l_1,l_2,l_3}$ for $0 \le l_i \le k$ is the multiplicity of τ_0 in the triple product $\tau_{l_1}\tau_{l_2}\tau_{l_3}$. The fusion coefficients are invariant under permutations of the l_i, and have the additional symmetry property $N^{(k)}_{l_1,l_2,l_3} = N^{(k)}_{l_1,k-l_2,k-l_3}$ (coming from $\tau_{k-l} = \tau_k \tau_l$). One has

$$\tau_{l_1}\tau_{l_2} = \sum_{l_3=0}^{k} N^{(k)}_{l_1,l_2,l_3} \tau_{l_3}.$$

Let $\Delta \subset [0,1]^3$ be the *Jeffrey–Weitsman polytope*, cut out by the inequalities

$$u_3 \leq u_1 + u_2, \quad u_1 \leq u_2 + u_3, \quad u_2 \leq u_3 + u_1, \quad u_1 + u_2 + u_3 \leq 2.$$

Suppose $\mathcal{C}_i, i = 1, 2, 3$, are conjugacy classes of elements $\exp(u_i \rho)$. As shown by Jeffrey–Weitsman [27, Proposition 3.1], the set $\{g_1 g_2 g_3 | \ g_i \in \mathcal{C}_i\}$ contains e if and only if $(u_1, u_2, u_3) \in \Delta$. Similarly,

$$N^{(k)}_{l_1,l_2,l_3} = \begin{cases} 1 & \text{if } \ l_1 + l_2 + l_3 \text{ even}, \ \left(\frac{l_1}{k}, \frac{l_2}{k}, \frac{l_3}{k}\right) \in \Delta, \\ 0 & \text{otherwise}. \end{cases}$$

3 The twisted equivariant K-homology of SU(2)

We will follow the approach to twisted K-homology via Dixmier–Douady bundles.

3.1 G-Dixmier–Douady bundles

Suppose G is a compact Lie group, acting on a (reasonable) topological space X. A G-Dixmier–Douady bundle over X is a G-equivariant bundle $\mathcal{A} \to X$ of $*$-algebras, with typical fiber $\mathbb{K}(\mathcal{H})$ the compact operators on a separable Hilbert space $\mathcal{H}$, and structure group $\mathrm{Aut}(\mathbb{K}(\mathcal{H})) = \mathrm{PU}(\mathcal{H})$ the projective unitary group. Here $\mathcal{H}$ is allowed to be finite-dimensional. A *Morita isomorphism* between two such bundles $\mathcal{A}_1, \mathcal{A}_2 \to X$ is a G-equivariant bundle of $(\mathcal{A}_2 - \mathcal{A}_1)$-bimodules $\mathcal{E} \to X$ such that $\mathcal{E}$ is locally modeled on the $(\mathbb{K}(\mathcal{H}_2) - \mathbb{K}(\mathcal{H}_1))$-bimodule $\mathbb{K}(\mathcal{H}_1, \mathcal{H}_2)$ of compact operators from $\mathcal{H}_1$ to $\mathcal{H}_2$. We write

$$\mathcal{A}_1 \simeq_{\mathcal{E}} \mathcal{A}_2.$$

One then also has $\mathcal{A}_2 \simeq_{\mathcal{E}^{\mathrm{op}}} \mathcal{A}_1$, where the opposite bimodule $\mathcal{E}^{\mathrm{op}}$ is modeled on $\mathbb{K}(\mathcal{H}_2, \mathcal{H}_1)$. Any two Morita isomorphisms $\mathcal{E}, \mathcal{E}'$ between $\mathcal{A}_1, \mathcal{A}_2$ differ by a G-equivariant line bundle J, given as the bundle of bimodule homomorphisms:

$$J = \mathrm{Hom}_{\mathcal{A}_2 - \mathcal{A}_1}(\mathcal{E}, \mathcal{E}'), \quad \mathcal{E}' = \mathcal{E} \otimes J.$$

Two equivariant Morita isomorphisms $\mathcal{E}, \mathcal{E}'$ will be called *equivalent* if this line bundle is equivariantly trivial. By the Dixmier–Douady theorem [14] (extended to the equivariant case by Atiyah–Segal [6]), the Morita isomorphism classes of G-Dixmier–Douady bundles $\mathcal{A} \to X$ are classified by an equivariant *Dixmier–Douady class* $\mathrm{DD}_G(\mathcal{A}) \in H^3_G(X, \mathbb{Z})$. Put differently, the Dixmier–Douady class is the obstruction to an equivariant Morita trivialization $\mathbb{C} \simeq_{\mathcal{E}} \mathcal{A}$, i.e., an equivariant Hilbert space bundle $\mathcal{E}$ with an isomorphism $\mathcal{A} \cong \mathbb{K}(\mathcal{E})$.

Remark 3.1. For $G = \{e\}$ the Dixmier–Douady class is realized as a Čech cohomology class, as follows: Choose a cover $\{U_a\}$ of M with Morita trivialization $\mathbb{C} \simeq_{\mathcal{E}_a} \mathcal{A}|_{U_a}$. On overlaps, the $\mathcal{E}_a$ are related by "transition line bundles":

$$J_{ab} = \mathrm{Hom}_{\mathcal{A}}(\mathcal{E}_a, \mathcal{E}_b), \quad \mathcal{E}_b = \mathcal{E}_a \otimes J_{ab}.$$

On triple overlaps, one has a trivializing section θ_{abc} of $J_{ab} \otimes J_{bc} \otimes J_{ca}$. Taking U_a sufficiently fine, the J_{ab} are all trivial, and a choice of trivialization makes θ_{abc} into a collection of U(1)-valued functions defining a Čech cocycle. A different choice of trivialization of the J_{ab} changes the cocycle by a coboundary. The class $\mathrm{DD}(\mathcal{A})$ equals the cohomology class of θ, under the isomorphism $H^2(X, \underline{U(1)}) = H^3(X, \mathbb{Z})$.

3.2 The Dixmier–Douady bundle over SU(2)

We will now give a fairly explicit construction of an equivariant Dixmier–Douady bundle representing the generator of $H^3_{\mathrm{SU}(2)}(\mathrm{SU}(2), \mathbb{Z}) = \mathbb{Z}$, using the cover (1). Let $\mathcal{H}$ be any SU(2)-Hilbert space with the property that $\mathcal{H}$ contains all T-weights with infinite multiplicity. (A possible choice is $\mathcal{H} = L^2(\mathrm{SU}(2))$ with the left regular representation.) As a consequence, there exists a T-equivariant unitary isomorphism

$$\mathcal{H} \to \mathcal{H} \otimes \mathbb{C}_\rho \tag{4}$$

(given by a collection of isomorphisms of the μ-weight spaces with the $(\mu - \rho)$-weight spaces). Let

$$\mathcal{E}_\pm = \mathrm{SU}(2)_\pm \times \mathcal{H}$$

with the diagonal SU(2)-action. By (2), any SU(2)-equivariant bundle over $\mathrm{SU}(2)_{\mathrm{reg}}$ is uniquely determined by its restriction to a T-equivariant bundle over $T_{(0,\rho)}$. Let $J \to \mathrm{SU}(2)_{\mathrm{reg}}$ be the equivariant line bundle such that $J|_{T_{(0,\rho)}} = T_{(0,\rho)} \times \mathbb{C}_\rho$. The isomorphism (4) defines a T-equivariant isomorphism

$$\mathcal{E}_-|_{T_{(0,\rho)}} \to \mathcal{E}_+|_{T_{(0,\rho)}} \otimes J|_{T_{(0,\rho)}},$$

which extends to an SU(2)-equivariant isomorphism $\mathcal{E}_-|_{\mathrm{SU}(2)_{\mathrm{reg}}} \to \mathcal{E}_+|_{\mathrm{SU}(2)_{\mathrm{reg}}} \otimes J$. This then defines an isomorphism $\mathbb{K}(\mathcal{E}_-)|_{\mathrm{SU}(2)_{\mathrm{reg}}} \to \mathbb{K}(\mathcal{E}_+)|_{\mathrm{SU}(2)_{\mathrm{reg}}}$, which we use to glue $\mathbb{K}(\mathcal{E}_\pm)$ to a global bundle $\mathcal{A}$. The bundle $\mathcal{A}$ represents the generator of $H^3_{\mathrm{SU}(2)}(\mathrm{SU}(2), \mathbb{Z}) = \mathbb{Z}$. Since $H^2_{\mathrm{SU}(2)}(\mathrm{SU}(2), \mathbb{Z}) = 0$, any other Dixmier–Douady bundle $\mathcal{A}'$ representing the generator is related to $\mathcal{A}$ by a *unique* (up to equivalence) Morita isomorphism. Again, this can be made quite explicit: Let $\mathcal{E}'_\pm$ be Morita trivializations of $\mathcal{A}'$, with transition line bundle J'. Then the Morita $\mathcal{A} - \mathcal{A}'$ bimodule is obtained by gluing $\mathbb{K}(\mathcal{E}'_+, \mathcal{E}_+)$ with $\mathbb{K}(\mathcal{E}'_-, \mathcal{E}_-)$, where the isomorphism over $\mathrm{SU}(2)_{\mathrm{reg}}$ is defined by the choice of an equivariant isomorphism $J' \cong J$ (the latter is unique up to homotopy).

3.3 The equivariant Cartan 3-form on SU(2)

The equivariant Dixmier–Douady bundle $\mathcal{A} \to \mathrm{SU}(2)$ may be viewed as a "prequantization" of the generator of equivariant Cartan 3-form on SU(2). To explain this viewpoint, we need some notation. For any manifold M with an action of a Lie group G, we denote by $\xi_M \in \mathfrak{X}(M), \xi \in \mathfrak{g}$, the generating vector fields for the infinitesimal $\mathfrak{g}$-action. That is, $\xi_M(f) = \frac{\partial}{\partial u}|_{u=0}(\exp(-u\xi))^* f$ for $f \in C^\infty(M)$. We let $(\Omega_G^\bullet(M), \mathrm{d}_G)$ denote the complex of equivariant differential forms

$$\Omega_G^k(M) = \bigoplus_{2i+j=k} (S^i\mathfrak{g}^* \otimes \Omega^j(M))^G,$$

with equivariant differential $(\mathrm{d}_G\gamma)(\xi) = \mathrm{d}\gamma(\xi) - \iota(\xi_M)\gamma(\xi)$. For G compact, its cohomology is identified with Borel's equivariant cohomology $H_G^k(M, \mathbb{R})$.

Let $\theta^L, \theta^R \in \Omega^1(\mathrm{SU}(2), \mathfrak{su}(2))$ be the Maurer–Cartan forms on SU(2). The *Cartan 3-form* $\eta \in \Omega^3(\mathrm{SU}(2))$ is given in terms of the basic inner product $\cdot$ on $\mathfrak{su}(2)$ by

$$\eta = \frac{1}{12}\theta^L \cdot [\theta^L, \theta^L].$$

It is d-closed, and has an equivariantly closed extension $\eta_{\mathrm{SU}(2)} \in \Omega^3_{\mathrm{SU}(2)}(\mathrm{SU}(2))$,

$$\eta_{\mathrm{SU}(2)}(\xi) = \eta - \frac{1}{2}(\theta^L + \theta^R) \cdot \xi.$$

Let $\varpi \in \Omega^2(\mathfrak{su}(2))$ be the invariant primitive of $\exp^* \eta$ defined by the de Rham homotopy operator for the radial homotopy. The image of the (nonclosed) 2-form $\mathrm{d}\mu - \frac{1}{2}\exp^*(\theta^L + \theta^R)$ under the homotopy operator is zero, since its pull-back to any line through the origin vanishes. Hence

$$\exp^* \eta_{\mathrm{SU}(2)} = \mathrm{d}_{\mathrm{SU}(2)}(\varpi - \mu), \tag{5}$$

where the "identity function" $\mu\colon \mathfrak{g} \to \mathfrak{g}$ is viewed as an element of $\mathfrak{su}(2)^* \otimes \Omega^0(\mathfrak{su}(2))$.

Lemma 3.2. *For any G-manifold with a closed equivariant 3-form $\gamma \in \Omega_G^3(M)$, all G-orbits $S \subset M$ acquire* unique *invariant 2-forms $\omega_S \in \Omega^2(S)^G$ such that $d_G\omega_S = i_S^*\gamma$.*

The straightforward proof is left to the reader. As special cases, we obtain 2-forms $\omega_{\mathcal{C}}$ on the conjugacy classes $\mathcal{C} \subset \mathrm{SU}(2)$ and $\omega_{\mathcal{O}}$ on the adjoint orbits $\mathcal{O} \subset \mathfrak{su}(2)$ such that

$$\mathrm{d}_{\mathrm{SU}(2)}\omega_{\mathcal{C}} = -\iota_{\mathcal{C}}^*\eta_{\mathrm{SU}(2)}, \quad \mathrm{d}_{\mathrm{SU}(2)}\omega_{\mathcal{O}} = \iota_{\mathcal{O}}^*(\mathrm{d}\mu).$$

Under the identification of $\mathfrak{su}(2)$ with its dual, $\omega_{\mathcal{O}}$ is just the usual symplectic form on coadjoint orbits. Suppose $\mathcal{C} = \exp(\mathcal{O})$. Then (5) and the uniqueness part of the lemma imply

$$i_{\mathcal{O}}^*\varpi = \omega_{\mathcal{O}} - (\exp|_{\mathcal{O}})^*\omega_{\mathcal{C}}. \tag{6}$$

Let $V \subset \mathfrak{su}(2)$ be the open ball of radius $\frac{1}{\sqrt{2}}$. We have diffeomorphisms

$$\exp_\pm \colon V \cong \mathrm{SU}(2)_\pm,$$

where $\exp_+$ is the restriction of the exponential map, and $\exp_- = l_c \circ \exp_+$ is its left translate by the central element c. The inverse maps will be denoted by

$$\log_\pm \colon \mathrm{SU}(2)_\pm \to V \subset \mathfrak{su}(2).$$

Let $\varpi_\pm = \log_\pm^* \varpi \in \mathrm{SU}(2)_\pm$. Then $\mathrm{d}\varpi_\pm = \eta$ over $\mathrm{SU}(2)_\pm$. Furthermore, by equation (5) we have, over $\mathrm{SU}(2)_\pm$,

$$\mathrm{d}_{\mathrm{SU}(2)}(\varpi_\pm - \log_\pm) = \eta_{\mathrm{SU}(2)}. \tag{7}$$

Over $\mathrm{SU}(2)_{\mathrm{reg}}$, both $\varpi_\pm$ are primitives of η; hence their difference is closed. To determine this closed 2-form, recall (cf. equation (2)) that $\mathrm{SU}(2)_{\mathrm{reg}} \cong T_{(0,\rho)} \times \mathrm{SU}(2)/T$. Let

$$\Psi\colon \mathrm{SU}(2)_{\mathrm{reg}} \to \mathrm{SU}(2)/T$$

be the projection to the second factor, and identify $\mathrm{SU}(2)/T$ with the (co)adjoint orbit $\mathcal{O} = \mathrm{SU}(2)\cdot\rho$.

Lemma 3.3. *One has* $\varpi_- - \varpi_+ = \Psi^*\omega_{\mathcal{O}}$ *over* $\mathrm{SU}(2)_{\mathrm{reg}}$, *where* $\mathcal{O}$ *is the adjoint orbit of the element* ρ.

Proof. By (7) we have

$$\mathrm{d}_{\mathrm{SU}(2)}(\varpi_- - \varpi_+ - (\log_- - \log_+)) = 0$$

over $\mathrm{SU}(2)_{\mathrm{reg}}$. Thus, $\log_+ - \log_-$ serves as a moment map for the closed invariant 2-form $\varpi_- - \varpi_+$. We claim that

$$\log_+ - \log_- = \iota_{\mathcal{O}} \circ \Psi.$$

Since both sides are SU(2)-equivariant, it suffices to compare the restrictions to $T_{(0,\rho)} \subset \mathrm{SU}(2)_{\mathrm{reg}}$. Indeed, $\log_+(\exp(u\rho)) = u\rho$ and $\log_-\exp(u\rho) = \log(\exp(u-1)\rho) = (u-1)\rho$, so the difference is $(\log_+ - \log_-)(\exp(u\rho)) = \rho$ as needed. This gives

$$0 = \mathrm{d}_{\mathrm{SU}(2)}(\varpi_- - \varpi_+ + \iota_{\mathcal{O}} \circ \Psi) = \mathrm{d}_{\mathrm{SU}(2)}(\varpi_- - \varpi_+ - \Psi^*\omega_{\mathcal{O}}).$$

In particular, $\varpi_- - \varpi_+ - \Psi^*\omega_{\mathcal{O}}$ is annihilated by all contractions with generating vector fields for the conjugation action. It is hence enough to show that its pull-back to $T_{(0,\rho)}$ is zero. Indeed, by applying the homotopy operator to $\exp_T^* \iota_T^* \eta_{\mathrm{SU}(2)} = 0$, we see that $\iota_{\mathfrak{t}}^*\varpi = 0$, which implies that $\varpi_\pm$ pull back to 0 on T.

The 2-form $\omega_{\mathcal{O}}$ is the curvature form $\mathrm{curv}(\nabla)$ of the line bundle $\mathrm{SU}(2) \times_T \mathbb{C}_\rho$, for the unique invariant connection ∇ on this bundle. Let $J = \Psi^*(\mathrm{SU}(2) \times_T \mathbb{C}_\rho)$ carry the pull-back connection ∇_J. The identities

$$\varpi_- - \varpi_+ = \mathrm{curv}(\nabla_J), \quad \mathrm{d}\varpi_\pm = \eta$$

say that $(\nabla_J, \varpi_\pm)$ is a "gerbe connection" in the sense of Chatterjee–Hitchin [12, 24], with η as its 3-curvature. Similarly, $(\nabla_J, \varpi_\pm - \log_\pm)$ is an equivariant gerbe connection, with equivariant 3-curvature $\eta_{\mathrm{SU}(2)}$.

We conclude this section with an easy proof of the fact that η integrates to 1. Observe that $\partial V = \overline{V}\backslash V$ is the (co)adjoint orbit $\mathcal{O}$ of the element ρ. It has symplectic volume $\int_{\mathcal{O}} \omega_{\mathcal{O}} = 1$ by the well-known formula for volume of coadjoint orbits [11, Corollary 7.27]. Since $\mathcal{C} := \exp \mathcal{O} = \{c\}$, we have $\omega_{\mathcal{C}} = 0$. Hence equation (6) together with Stokes's theorem gives

$$\int_{\mathrm{SU}(2)} \eta = \int_V \mathrm{d}\varpi = \int_{\mathcal{O}} \iota_{\mathcal{O}}^* \varpi = \int_{\mathcal{O}} \omega_{\mathcal{O}} = 1.$$

3.4 Twisted K-homology

Let G be a compact Lie group acting on a compact G-space X. Given a G-Dixmier–Douady bundle $\mathcal{A} \to X$, one defines (following J. Rosenberg [37]) the twisted K-homology group

$$K_0^G(X, \mathcal{A}) = K_G^0(\Gamma(X, \mathcal{A})),$$

where the right-hand side denotes the K-homology group of the $(G - C^*)$-algebra of sections of $\mathcal{A}$. (For K-homology of C^*-algebras, see [23, 28].) The twisted K-homology is a covariant functor: If $\Phi\colon X_1 \to X_2$ is an equivariant map of compact G-spaces, together with an equivariant Morita isomorphism $\mathcal{A}_1 \simeq_{\mathcal{E}} \Phi^* \mathcal{A}_2$, one obtains a push-forward map

$$\Phi_*\colon K_0^G(X_1, \mathcal{A}_1) \to K_0^G(X_2, \mathcal{A}_2).$$

It is possible to work out many examples of twisted equivariant K-homology groups simply from its formal properties such as excision, Poincaré duality, and so on. For $\mathcal{A} = \mathbb{C}$ one obtains the untwisted K-homology groups. One has a ring isomorphism

$$K_0^G(\mathrm{pt}) = R(G),$$

where the ring structure on the left-hand side is realized as push-forward under $\mathrm{pt} \times \mathrm{pt} \to \mathrm{pt}$. The following is the simplest nontrivial case of the Freed–Hopkins–Teleman theorem [20]. This special case may be proved by an elementary Mayer–Vietoris argument; see Freed [19].

Theorem 3.4. *Let* $\mathrm{SU}(2)$ *act on itself by conjugation, and let* $\mathcal{A} \to \mathrm{SU}(2)$ *be the basic Dixmier–Douady bundle. For all levels* $k = 0, 1, 2, \ldots$*, the* $R(\mathrm{SU}(2))$*-module homomorphism*

$$R(SU(2)) \cong K_0^{\mathrm{SU}(2)}(\mathrm{pt}) \to K_0^{\mathrm{SU}(2)}(\mathrm{SU}(2), \mathcal{A}^{k+2})$$

given as push-forward under the inclusion of the group unit pt $\to$ SU(2) *is onto, with kernel the level-k fusion ideal* $I_k(\mathrm{SU}(2))$. *It hence defines a ring isomorphism,*

$$R_k(\mathrm{SU}(2)) \cong K_0^{\mathrm{SU}(2)}(\mathrm{SU}(2), \mathcal{A}^{k+2}).$$

3.5 The K-homology fundamental class

Recall that for n even, the complex Clifford algebra $\mathbb{C}\mathrm{l}(n) = \mathbb{C}\mathrm{l}(\mathbb{R}^n)$ admits a unique (up to isomorphism) irreducible $*$-representation. Concretely, the identification $\mathbb{R}^n \cong \mathbb{C}^{n/2}$ gives a Clifford action on the standard spinor module $\mathsf{S} = \wedge\mathbb{C}^{n/2}$. This realizes the Clifford algebra as a matrix algebra, $\mathbb{C}\mathrm{l}(n) = \mathrm{End}(\mathsf{S})$. Given $A \in \mathrm{SO}(n)$ there exists a unitary transformation $U \in \mathrm{U}(\mathsf{S})$, unique up to a scalar, such that $A(v) \cdot U(z) = U(v \cdot z)$ for $v \in \mathbb{R}^n, z \in \mathsf{S}$. The set of such *implementers* U forms a closed subgroup of $\mathrm{U}(\mathsf{S})$, denoted by $\mathrm{Spin}_c(n)$, and the map taking U to A makes this group into a central extension

$$1 \to \mathrm{U}(1) \to \mathrm{Spin}_c(n) \to \mathrm{SO}(n) \to 1.$$

If M is an oriented Riemannian G-manifold of even dimension n, then its Clifford algebra bundle $\mathbb{C}\mathrm{l}(TM)$ is a G-equivariant bundle of complex matrix algebras. It is thus a G-Dixmier–Douady bundle. Its Dixmier–Douady class is the third integral equivariant[1] Stiefel–Whitney class, $W_G^3(M) \in H_G^3(M, \mathbb{Z})$. As pointed out by Connes [13] and Plymen [35], an equivariant Spin_c-structure on M is exactly the same thing as an equivariant Morita trivialization of $\mathbb{C}\mathrm{l}(TM)$. Indeed, given an equivariant lift $P_{\mathrm{Spin}_c}(M) \to P_{\mathrm{SO}}(M)$ of the $\mathrm{SO}(n)$-frame bundle to the group $\mathrm{Spin}_c(n)$, the Morita trivialization is defined by the bundle of spinors $\mathcal{S} = P_{\mathrm{Spin}_c}(M) \times_{\mathrm{Spin}_c(n)} \mathsf{S}$. Conversely, given an equivariant Morita trivialization $\mathbb{C}\mathrm{l}(TM) \simeq_{\mathcal{S}} \mathbb{C}$, one obtains a lift of the structure group: The fiber of the bundle $P_{\mathrm{Spin}_c}(M)$ at $m \in M$ is the set of pairs (A, U), where $A\colon T_mM \to \mathbb{R}^n$ is an oriented orthonormal frame, and $U\colon \mathcal{S}_m \to \mathsf{S}$ is a unitary isomorphism intertwining the Clifford actions of $v \in T_mM$ and $A(v) \in \mathbb{R}^n$.

The Clifford bundle $\mathbb{C}\mathrm{l}(TM)$ is naturally a $\mathbb{C}\mathrm{l}(TM) - \mathbb{C}\mathrm{l}(TM)$ bimodule. Using the canonical antiautomorphism of $\mathbb{C}\mathrm{l}(TM)$, it may also be viewed as a module over $\mathbb{C}\mathrm{l}(TM)\otimes\mathbb{C}\mathrm{l}(TM)$, defining a Morita trivialization of the latter. Given any Spin_c-structure $\mathcal{S}$, one obtains a Hermitian line bundle

$$\mathcal{L} := \mathcal{L}(\mathcal{S}) = \mathrm{Hom}_{\mathbb{C}\mathrm{l}(TM)\otimes\mathbb{C}\mathrm{l}(TM)}(\mathbb{C}\mathrm{l}(TM), \mathcal{S}\otimes\mathcal{S})$$

[1] We remark that for G compact and simply connected, the vanishing of $W_G^3(M)$ is equivalent to the vanishing of the nonequivariant Stiefel–Whitney class $W^3(M)$, since the map $H_G^3(M, \mathbb{Z}) \to H^3(M, \mathbb{Z})$ is injective (cf. [29]).

called the Spin$_c$-*line bundle*. Twisting $\mathcal{S}$ by a line bundle L changes the Spin$_c$-line bundle as follows:

$$\mathcal{L}(\mathcal{S} \otimes L) = \mathcal{L}(\mathcal{S}) \otimes L^2.$$

For any equivariant Spin$_c$-structure on an even-dimensional manifold, the class of the Spin$_c$-Dirac operator defines a *fundamental class* in equivariant K-homology. In the absence of a Spin$_c$-structure, there is still a fundamental class, but as an element

$$[M] \in K_0^G(M, \mathbb{C}\mathrm{l}(TM))$$

in twisted K-homology.[2] For an explicit construction of $[M]$, see Kasparov [28]. Below, we will construct elements of $R_k(\mathrm{SU}(2)) = K_0^{\mathrm{SU}(2)}(\mathrm{SU}(2), \mathcal{A}^{k+2})$ as push-forwards of $[M]$ under SU(2)-equivariant maps $\Phi\colon M \to \mathrm{SU}(2)$. In order to define such a push-forward, we need an equivariant Morita isomorphism

$$\mathbb{C}\mathrm{l}(TM) \simeq_{\mathcal{E}} \Phi^* \mathcal{A}^{k+2}.$$

We will explain how such a "twisted Spin$_c$-structure" arises for prequantized q-Hamiltonian SU(2)-spaces. The counterpart to the Spin$_c$-line bundle is the Morita isomorphism $\Phi^*\mathcal{A}^{2k+4} \simeq_{\mathcal{K}} \mathbb{C}$ given by

$$\mathcal{K} = \mathrm{Hom}_{\mathbb{C}\mathrm{l}(TM)\otimes\mathbb{C}\mathrm{l}(TM)}(\mathbb{C}\mathrm{l}(TM), (\mathcal{E} \otimes \mathcal{E})^{\mathrm{op}}).$$

4 q-Hamiltonian SU(2)-spaces

4.1 Basic definitions

Let G be a compact Lie group, with Lie algebra $\mathfrak{g}$. Given an invariant inner product B on its Lie algebra, define the equivariant Cartan 3-form

$$\eta_G^{(B)}(\xi) = \frac{1}{12} B(\theta^L, [\theta^L, \theta^L]) - \frac{1}{2} B(\theta^L + \theta^R, \xi).$$

A *q-Hamiltonian G-space* (relative to the inner product B) is a triple (M, ω, Φ), where M is a G-manifold, ω is an invariant 2-form, and $\Phi\colon M \to G$ is an equivariant smooth map, called the *moment map*, such that

(i) $\mathrm{d}_G\omega = -\Phi^*\eta_G^{(B)}$,
(ii) $\ker\omega \cap \ker(\mathrm{d}\Phi) = 0$ everywhere.

Remark 4.1. If $G = T$ is a torus, this is just the usual definition of a symplectic T-space with torus-valued moment map. Indeed, Condition (i) in this case says that $\mathrm{d}\omega = 0$ and $\omega_m(\xi_M(m), v) = -B(\theta_T(\mathrm{d}_m\Phi(v)), \xi)$ for all $\xi \in \mathfrak{g}$, $v \in T_mM$.

[2] More precisely, one has to view $\mathbb{C}\mathrm{l}(TM)$ as a $\mathbb{Z}_2$-graded Dixmier–Douady bundle, and work with the twisted K-homology for such $\mathbb{Z}_2$-graded bundles.

Hence it implies $\ker(\omega) \subset \ker(\mathrm{d}\Phi)$, whence (ii) simplifies to $\ker(\omega) = \{0\}$. For general G, a similar argument shows that $\ker(\omega_m)$ is spanned by all $\xi_M(m)$ such that $\mathrm{Ad}_{\Phi(m)}\xi + \xi = 0$.

Basic examples of q-Hamiltonian G-spaces are the conjugacy classes $\mathcal{C} \subset G$, with moment map the embedding. The *double* $D(G) = G \times G$, with G acting by conjugation and with moment $\Phi(a, b) = aba^{-1}b^{-1}$, is another example. The 2-form is

$$\omega = \frac{1}{2}a^*\theta^L \cdot b^*\theta^R + \frac{1}{2}a^*\theta^R \cdot b^*\theta^L + \frac{1}{2}(ab)^*\theta^L \cdot (a^{-1}b^{-1})^*\theta^R,$$

where, for example, $a^{-1}b^{-1}$ denotes the map $(a, b) \mapsto a^{-1}b^{-1}$. If G' is the quotient of G by a finite subgroup of $Z(G)$, then the moment map, action, and 2-form on $D(G)$ descends to $D(G')$, so that $D(G')$ is again a q-Hamiltonian G-space.

Given two q-Hamiltonian G-spaces (M_i, ω_i, Φ_i), $i = 1, 2$, their product $M_1 \times M_2$ with the diagonal G-action, moment map $\Phi_1\Phi_2$, and 2-form $\omega_1 + \omega_2 + \frac{1}{2}B(\Phi_1^*\theta^L, \Phi_2^*\theta^R)$ is again a q-Hamiltonian G-space. This is called the *fusion product* of M_1, M_2. The *symplectic quotient* of a q-Hamiltonian G-space is $M /\!\!/ G = \Phi^{-1}(e)/G$. Similarly to the Hamiltonian theory, e is a regular value of Φ if and only if G acts locally freely on $\Phi^{-1}(e)$, and in this case $M /\!\!/ G$ is a symplectic orbifold. (If e is a singular value, then $M /\!\!/ G$ is a singular symplectic space as defined in [39].) More generally, given a conjugacy class $\mathcal{C}$ one can define a symplectic quotient

$$M /\!\!/_{\mathcal{C}} G = (M \times \mathcal{C}) /\!\!/ G.$$

It was shown in [1] that moduli spaces of flat G-bundles over compact oriented surfaces Σ_h^r of genus h with r boundary circles, with boundary holonomies in prescribed conjugacy classes $\mathcal{C}_j$, are symplectic quotients

$$M(\Sigma_h^r, \mathcal{C}_1, \ldots, \mathcal{C}_r) = (\underbrace{D(G) \times \cdots \times D(G)}_{h \text{ times}} \times \mathcal{C}_1 \times \cdots \times \mathcal{C}_r) /\!\!/ G.$$

We now specialize to q-Hamiltonian SU(2)-spaces (M, ω, Φ), with B the basic inner product. Put $M_\pm = \Phi^{-1}(\mathrm{SU}(2)_\pm)$, and let

$$\omega_{0,\pm} = \omega + \Phi^*\varpi_\pm,$$
$$\Phi_{0,\pm} = \log_\pm \circ \Phi.$$

Then

$$\mathrm{d}_{\mathrm{SU}(2)}(\omega_{0,\pm} - \Phi_{0,\pm}) = \mathrm{d}_{\mathrm{SU}(2)}(\omega + \Phi^*(\varpi_\pm - \log_\pm)) = 0.$$

That is, $\omega_{0,\pm}$ is closed, with $\Phi_{0,\pm}$ as a moment map. Using condition (ii) above one can show [1] that $\omega_{0,\pm}$ are nondegenerate, i.e., *symplectic*. Thus, $(M_\pm, \omega_{0,\pm}, \Phi_{0,\pm})$ are ordinary (symplectic) Hamiltonian SU(2)-spaces. In particular, $M_\pm$ are even-dimensional, with a natural orientation. If M is compact and connected, then the spaces $M_\pm$ are connected. (This follows from the convexity properties and the fiber connectivity of group-valued moment maps [1].)

Conversely, (M, ω, Φ) is determined by the pair of Hamiltonian SU(2)-spaces $(M_\pm, \omega_{0,\pm}, \Phi_{0,\pm})$. This correspondence reduces many properties of q-Hamiltonian spaces to standard facts about ordinary Hamiltonian spaces. It is also used to construct q-Hamiltonian spaces, as in the following example.

4.2 Example: The 4-sphere

The following construction of a q-Hamiltonian structure of S^4 is taken from [5]. An independent construction due to Hurtubise–Jeffrey [26] was later generalized by Hurtubise–Jeffrey–Sjamaar [25] to define the structure of a q-Hamiltonian SU(n)-space on S^{2n}, for any n.

Let $\mathbb{C}^2$ carry the standard SU(2)-action and the standard symplectic structure $\omega_0 = \frac{i}{2}(\mathrm{d}z_1 \wedge \mathrm{d}\overline{z}_1 + \mathrm{d}z_2 \wedge \mathrm{d}\overline{z}_2)$. The moment map for the SU(2)-action can be written, for $z \neq 0$, as

$$\Phi_0(z) = -i\pi^2\|z\|^2 P(z) + i\pi^2\|z\|^2(I - P(z)),$$

where $P(z)$ is the projection operator,

$$P(z) = \|z\|^{-2}\begin{pmatrix} z_1 \\ z_2 \end{pmatrix}\begin{pmatrix} z_1 \\ z_2 \end{pmatrix}^\dagger = \frac{1}{\|z\|^2}\begin{pmatrix} |z_1|^2 & z_1\overline{z}_2 \\ \overline{z}_1 z_2 & |z_2|^2 \end{pmatrix}.$$

Hence,

$$\exp(\Phi_0(z)) = e^{-i\pi^2\|z\|^2} P(z) + e^{i\pi^2\|z\|^2}(I - P(z)).$$

Let $V \subset \mathfrak{su}(2)$ be the open ball of radius $\frac{1}{\sqrt{2}}$ (cf. Section 3.3). We have $\|\Phi_0(z)\| = \frac{1}{\sqrt{2}}\pi\,\|z\|^2$, so that

$$S^4_\pm := \Phi_0^{-1}(V) = \{z \in \mathbb{C}^2 | \pi\,\|z\|^2 < 1\}.$$

Define a diffeomorphism F of the annulus $0 < \pi\,\|z\|^2 < 1$ by

$$F(z_1, z_2) = (-\overline{z_2}, \overline{z}_1)\sqrt{\frac{1}{\pi\|z\|^2} - 1}.$$

Then F is equivariant, with $\pi\,\|F(z)\|^2 = 1 - \pi\,\|z\|^2$. Gluing the charts $S^4_\pm$ under F, one obtains a 4-sphere S^4 with an action of SU(2).

Put $\Phi_+ = \exp\Phi_0$ and $\Phi_- = l_c \circ \exp\Phi_0 = -\exp\Phi_0$. The diffeomorphism F satisfies $P(F(z)) = I - P(z)$, and therefore,

$$\Phi_+(F(z)) = \exp(\Phi_0(F(z))) = -\exp(\Phi_0(z)) = \Phi_-(z).$$

Hence $\Phi_\pm$ glue to a global equivariant map $\Phi\colon S^4 \to \mathrm{SU}(2)$. Similarly, the 2-forms $\omega_\pm = \omega_0 + \Phi_0^*\varpi$ glue[3] to a global invariant 2-form $\omega \in \Omega^2(S^4)$, defining a q-Hamiltonian SU(2)-space (S^4, ω, Φ).

Remark 4.2. The space S^4 carries an involution $I\colon S^4 \to S^4$, given in charts by complex conjugation. It has the equivariance property $I(g\cdot x) = I(g)\cdot I(x)$ relative to the involution of SU(2) given by complex conjugation of matrices, $I(A) = \overline{A}$. The involution satisfies $I^*\omega = -\omega$ and $I^*\Phi = \overline{\Phi}$. The fixed-point set of the involution is a 2-sphere $S^2 \subset S^4$. The theory of anti-involutions of q-Hamiltonian G-spaces was developed in recent work of Schaffhauser [38], who established an analogue of the convexity results of Duistermaat [15] and O'Shea–Sjamaar [33] in this context.

Remark 4.3. It is well known that the complement of the zero section in $T^*(S^2)$ is SU(2)-equivariantly symplectomorphic to the complement of the origin in $\mathbb{C}^2$. One may thus modify the construction above, and obtain examples in which the fiber over e or over c (or both) is a 2-sphere rather than a point. The four examples obtained in this way are the complete list of 4-dimensional q-Hamiltonian SU(2)-spaces with surjective moment map.

5 Cross-sections

Let (M, ω, Φ) be a q-Hamiltonian SU(2)-space. By the q-Hamiltonian cross-section theorem [1], the preimage

$$Y = \Phi^{-1}(T_{(0,\rho)}) \tag{8}$$

is a q-Hamiltonian T-space $(Y, \omega_Y, \Phi|_Y)$, with 2-form $\omega_Y = i_Y^*\omega$. In particular, ω_Y is symplectic. Letting $\Phi_Y\colon Y \to (0,\rho) \subset \mathfrak{t}$ with $\exp\Phi_Y = \Phi|_Y$, it is immediate that (Y, ω_Y, Φ_Y) is an ordinary Hamiltonian T-space. We have

$$M_{\mathrm{reg}} = M_+ \cap M_- = \mathrm{SU}(2) \times_T Y$$

and

$$TM|_Y = TY \oplus \mathfrak{t}^\perp,$$

where the second summand is embedded by the generating vector fields. This splitting is ω-orthogonal, and the 2-form on $Y \times \mathfrak{t}^\perp$ is given at $y \in Y$, with $g = \Phi(y) \in T_{(0,\rho)}$, by $(\xi_1, \xi_2) \mapsto \frac{1}{2}((\mathrm{Ad}_g - \mathrm{Ad}_{g^{-1}})\xi_1, \xi_2)$. Note that since the pull-back of $\varpi_\pm$ to $T_{(0,\rho)}$ is zero, the 2-forms $\omega_{0,\pm}$ both pull back to ω_Y. Similarly

$$\Phi_{0,+}|_Y = \Phi_Y = \Phi_{0,-}|_Y + \rho.$$

---

[3] To check that these 2-forms agree on the overlap $S^4_{\mathrm{reg}} = S^4_+ \cap S^4_-$, it suffices to consider their pull-back to symplectic cross-sections as in Section 5.

That is, (Y, ω_Y, Φ_Y) may also be viewed as the symplectic cross-section of $M_\pm$. (To be precise, in the case of M_-, it is the *opposite* cross-section, given as the preimage of $(-\infty, 0) \subset \mathfrak{t}$ under $\Phi_{0,-}$.) The 2-forms on the bundles $Y \times \mathfrak{t}^\perp$ induced by $\omega_{0,\pm}$ are

$$(\xi_1, \xi_2) \mapsto \mathrm{ad}_{\mu_\pm} \xi_1 \cdot \xi_2,$$

where $\mu_+ = \Phi_{0,+}(y)$ and $\mu_- = \Phi_{0,-}(y)$.

The space Y is only a "partial" cross-section for M, since it leaves out the subsets $\Phi^{-1}(e)$, $\Phi^{-1}(c)$. On the other hand, the "full" cross-section $\tilde{Y} = \Phi^{-1}(T_{[0,\rho]})$ is usually not a manifold, let alone symplectic. However, following Hurtubise–Jeffrey–Sjamaar [25] one can "implode" $\tilde{Y}$ to obtain a symplectic T-space X, which is a symplectic orbifold under regularity conditions. As a topological space, the *imploded cross-section* is a quotient space

$$X = \Phi^{-1}(T_{[0,\rho]})/\sim,$$

where the equivalence relation divides out the SU(2)-action on both $\Phi^{-1}(e)$ and $\Phi^{-1}(c)$. We have a decomposition of X into three symplectic spaces,

$$X = (M /\!/ \mathrm{SU}(2)) \cup Y \cup (M /\!/_c \mathrm{SU}(2)). \tag{9}$$

The action of $T \subset \mathrm{SU}(2)$ on $\Phi^{-1}(T_{[0,\rho]})$ descends to an action on X, and the map $\Phi^{-1}(T_{[0,\rho]}) \to [0, \rho] \subset \mathfrak{t}$ descends to a T-equivariant map

$$\Phi_X \colon X \to \mathfrak{t}.$$

Let

$$X_+ = (M /\!/ \mathrm{SU}(2)) \cup Y, \quad X_- = Y \cup (M /\!/_c \mathrm{SU}(2)),$$

so that $X_\pm$ are the imploded cross-sections of $M_\pm$. View $M_\pm$ as Hamiltonian SU(2)-spaces with 2-forms $\omega_{0,\pm}$, and let $\mathbb{C}^2$ carry the standard structure as a Hamiltonian SU(2)-space.

Proposition 5.1. *Suppose* SU(2) *acts locally freely (respectively freely) on* $\Phi^{-1}(e)$, $\Phi^{-1}(c)$. *Then the imploded cross-section X admits a unique structure of a symplectic orbifold (respectively symplectic manifold) such that the open subsets* $X_\pm$ *are symplectic quotients,*

$$X_\pm = (M_\pm \times \mathbb{C}^2) /\!/ \mathrm{SU}(2).$$

Furthermore,

a. The restriction of Φ_X *to* $X_\pm$ *is smooth, and is a moment map for the action of* $T \cong \mathrm{U}(1)$.

b. The Hamiltonian T-space (Y, ω_Y, Φ_Y) *is embedded as an open symplectic submanifold of X.*

c. $M /\!/ \mathrm{SU}(2)$ *is a symplectic suborbifold (respectively submanifold), with normal bundle* $\Phi^{-1}(e) \times_{\mathrm{SU}(2)} \mathbb{C}^2$. *The* U(1) *action on the normal bundle is with weights* $(-1, -1)$.

d. $M/\!\!/_c \mathrm{SU}(2)$ *is a symplectic suborbifold (respectively submanifold), with normal bundle* $\Phi^{-1}(c) \times_{\mathrm{SU}(2)} \mathbb{C}^2$*. The* $\mathrm{U}(1)$*-action on the normal bundle is with weights* $(1,1)$.

Thus, X is obtained by gluing the Hamiltonian imploded cross-sections for $(M_\pm, \omega_{0,\pm}, \Phi_{0,\pm})$. For the case $G = \mathrm{SU}(2)$, the imploded cross-sections construction was introduced by Eugene Lerman as an SU(2)-counterpart of symplectic cutting. Its basis properties for Hamiltonian SU(2)-spaces are described in [30, Appendix], and directly imply the properties for q-Hamiltonian SU(2)-spaces.

Remark 5.2. More intrinsically, the imploded cross-section can directly be constructed as a q-Hamiltonian symplectic quotient $X = (M \times S^4)/\!\!/ \mathrm{SU}(2)$. This is the approach taken in [26, 25]. However, in this paper we will have more use for the construction in terms of ordinary Hamiltonian quotients.

6 The canonical "twisted Spin$_c$-structure"

Choose invariant almost complex structures on $M_\pm$, which are compatible with $\omega_{0,\pm}$ in the sense that each tangent space is isomorphic to $\mathbb{C}^{n/2}$ with the standard complex structure and standard symplectic form. The almost complex structure defines spinor modules

$$\mathcal{S}_{0,\pm} = \wedge_{\mathbb{C}} TM_\pm \to M_\pm$$

for the Clifford bundles $\mathbb{C}\mathrm{l}(TM)|_{M_\pm}$, where the notation $\wedge_{\mathbb{C}}$ denotes the complex exterior powers of $TM_\pm$ relative to the given complex structure. On the overlap $M_+ \cap M_- = M_{\mathrm{reg}}$, the two spinor bundles differ by $\mathrm{Hom}_{\mathbb{C}\mathrm{l}(TM)}(\mathcal{S}_{0,+}, \mathcal{S}_{0,-})$.

Proposition 6.1. *The line bundle* $\mathrm{Hom}_{\mathbb{C}\mathrm{l}(TM)}(\mathcal{S}_{0,+}, \mathcal{S}_{0,-})$ *is equivariantly isomorphic to the pull-back* $\Phi^*(J^{\otimes 2})$.

Proof. An SU(2)-invariant almost complex structure on $M_{\mathrm{reg}} = \mathrm{SU}(2) \times_T Y$ is equivalent to a T-invariant complex structure on the bundle $TM|_Y = TY \oplus \mathfrak{t}^\perp$. This bundle carries two symplectic structures, defined by the 2-forms $\omega_{0,\pm}$ on $M_\pm$. Pick a T-invariant compatible structure on the bundle TY. Its sum with the complex structure on $\mathfrak{t}^\perp$, coming from the identification $\mathfrak{t}^\perp \cong \mathbb{C}_\alpha$, is compatible with $\omega_{0,+}$. Similarly its sum with the complex structure on $\mathfrak{t}^\perp$, coming from the identification $\mathfrak{t}^\perp \cong \mathbb{C}_{-\alpha}$, is compatible with $\omega_{0,-}$. The corresponding spinor bundles $\tilde{\mathcal{S}}_{0,\pm}|_Y \to Y$ are related by a twist by a T-equivariant line bundle, corresponding to the change of the complex structure on $\mathfrak{t}^\perp$ to its opposite. Clearly, this is the line bundle $Y \times \mathbb{C}_\alpha = Y \times (\mathbb{C}_\rho)^2$:

$$\tilde{\mathcal{S}}_{0,-}|_Y = \tilde{\mathcal{S}}_{0,+}|_Y \otimes (Y \times (\mathbb{C}_\rho)^2).$$

Extending to M_{reg}, and using the definition of $J \to \mathrm{SU}(2)_{\mathrm{reg}}$ we obtain

$$\tilde{\mathcal{S}}_{0,-} = \tilde{\mathcal{S}}_{0,+} \otimes \Phi^* J^2.$$

But $\tilde{S}_{0,\pm}$ are equivariantly isotopic to $\mathcal{S}_{0,\pm}$, since any two choices of equivariant compatible almost complex structures are isotopic. Hence we also have $\mathcal{S}_{0,-} \cong \mathcal{S}_{0,+} \otimes \Phi^* J^2$, or equivalently $\mathrm{Hom}_{\mathbb{C}\mathrm{l}(TM)}(\mathcal{S}_{0,+}, \mathcal{S}_{0,-}) \cong \Phi^* J^2$.

Equivalently, we can express this result as follows:

Proposition 6.2. *For any q-Hamiltonian* SU(2)*-space* (M, ω, Φ)*, there is a distinguished (up to equivalence)* SU(2)*-equivariant Morita isomorphism*

$$\Phi^* \mathcal{A}^2 \simeq_{\mathcal{S}} \mathbb{C}\mathrm{l}(TM). \tag{10}$$

Proof. Let $\mathcal{F}_\pm \to \mathrm{SU}(2)_\pm$ define Morita trivializations $\mathbb{C} \simeq_{\mathcal{F}_\pm} \mathcal{A}^2$. Fix isomorphisms $\mathcal{F}_- \cong \mathcal{F}_+ \otimes J^2$ and $\mathcal{S}_{0,-} \cong \mathcal{S}_{0,+} \otimes \Phi^* J^2$ on intersections. The desired Morita $\mathbb{C}\mathrm{l}(TM) - \Phi^*\mathcal{A}^2$ bimodule $\mathcal{S}$ is then obtained by gluing the bundles $\mathcal{S}_\pm = \mathrm{Hom}_{\mathbb{C}}(\Phi^*\mathcal{F}_\pm, \mathcal{S}_{0,\pm})$, using that

$$\begin{aligned} \mathrm{Hom}_{\mathbb{C}}(\Phi^*\mathcal{F}_-, \mathcal{S}_{0,-}) &\cong \mathrm{Hom}_{\mathbb{C}}(\Phi^*(\mathcal{F}_+ \otimes J^2), \\ \mathcal{S}_{0,+} \otimes \Phi^* J^2) &= \mathrm{Hom}_{\mathbb{C}}(\Phi^*\mathcal{F}_+, \mathcal{S}_{0,+}) \end{aligned}$$

on the intersection.

We refer to the Morita isomorphism (10) as the canonical twisted Spin_c-structure of a q-Hamiltonian manifold.

Remark 6.3. In particular, we see that the third integral Stiefel–Whitney class of any q-Hamiltonian SU(2)-space satisfies

$$W^3(M) = 2\Phi^* x,$$

where $x \in H^3(\mathrm{SU}(2), \mathbb{Z})$ is the generator. Since this is a 2-torsion class, it follows that $4\Phi^* x = 0$. The fact that $\Phi^* x$ is torsion is a consequence of the condition $\mathrm{d}\omega = -\Phi^*\eta$. The more precise statement relies on the minimal degeneracy condition $\ker(\omega) \cap \ker(\mathrm{d}\Phi) = 0$.

7 Prequantization of q-Hamiltonian SU(2)-spaces

Suppose (M, ω, Φ) is a q-Hamiltonian SU(2)-space. The conditions $\mathrm{d}\omega = -\Phi^*\eta$ and $\mathrm{d}\eta = 0$ mean that the pair $(\omega, -\eta)$ defines a cocycle for the relative de Rham complex[4] $\Omega^\bullet(\Phi)$. For $k > 0$, we define a *level-k prequantization* of (M, ω, Φ) to be a lift of the class $k[(\omega, -\eta)] \in H^3(\Phi, \mathbb{R})$ to a class in $H^3(\Phi, \mathbb{Z})$.

[4] Recall that for any morphism of cochain complexes $F^\bullet \colon C^\bullet \to \tilde{C}^\bullet$, the relative cohomology $H^\bullet(F)$ is the cohomology of the algebraic mapping cone $(\tilde{C}^{k-1} \oplus C^k, \mathrm{d})$, with differential $\mathrm{d}(x, y) = (F(y) - \mathrm{d}x, \mathrm{d}y)$. In our case $F = \Phi^*$, acting on differential forms or on singular cochains, and we write $H(\Phi, \cdot)$ for the relative cohomology.

Remark 7.1. One can similarly define an *equivariant* level-k prequantization to be an integral lift of $k[(\omega, -\eta_{\mathrm{SU}(2)})] \in H^3_{\mathrm{SU}(2)}(\Phi, \mathbb{R})$. However, the equivariance is automatic: Indeed, for any simply connected compact Lie group G and any G-space M one has $H^p_G(M, \mathbb{Z}) = H^p(M, \mathbb{Z})$ for $p \leq 2$, and if $\Phi\colon M \to G$ is an equivariant map one has $H^p_G(\Phi, \mathbb{Z}) = H^p(\Phi, \mathbb{Z})$ for $p \leq 3$. See, e.g., [29].

Lemma 7.2. *If* (M, ω, Φ) *admits a level-k prequantization, then the set of such prequantizations is a principal homogeneous space under the group* $\mathrm{Tor}(H^2(M, \mathbb{Z}))$ *of flat line bundles over* M.

Proof. Clearly, the set of prequantizations is a principal homogeneous space under $\mathrm{Tor}(H^3(\Phi, \mathbb{Z}))$. Since $H^3(\mathrm{SU}(2), \mathbb{Z}) = \mathbb{Z}$ has no torsion, $\mathrm{Tor}(H^3(\Phi, \mathbb{Z}))$ lies in the image of the map $H^2(M, \mathbb{Z}) \to H^3(\Phi, \mathbb{Z})$ in the long exact sequence for relative cohomology. But this map is injective since $H^2(\mathrm{SU}(2), \mathbb{Z}) = 0$, and hence restricts to an isomorphism of the torsion subgroups.

The class $k[(\omega, -\eta)]$ is integral if and only if it takes integer values on all relative 3-cycles; that is, for every smooth singular 2-cycle $\Sigma \in C_2(M)$, and every smooth singular 3-chain $\Gamma \in C_3(\mathrm{SU}(2))$ bounding $\Phi(\Sigma)$, we must have

$$k\left(\int_\Gamma \eta + \int_\Sigma \omega\right) \in \mathbb{Z}. \tag{11}$$

(Given Σ, it is actually enough, by the integrality of η, to check the condition for *some* Γ bounding $\Phi(\Sigma)$.) If $H^2(M, \mathbb{R}) = 0$, there is a much simpler criterion [29]: Let $x \in H^3(\mathrm{SU}(2), \mathbb{Z})$ be the generator. Since $\Phi^*[\eta] = 0$, the class $\Phi^* x$ is torsion. If $H^2(M, \mathbb{R}) = 0$, then (M, ω, Φ) is prequantizable at level-k if and only if

$$k\Phi^* x = 0. \tag{12}$$

Proposition 7.3. *The conjugacy class* $\mathcal{C}$ *of* $t \in T_{[0,\rho]} \subset \mathrm{SU}(2)$ *is prequantizable at level-k if and only if* $t = \exp(\frac{n}{k}\rho)$ *for some* $n \in \{0, 1, \ldots, k\}$.

Proof. It is enough to check criterion (11) for $\Sigma = \mathcal{C}$. Write $t = \exp(u\rho)$ with $u \in [0, 1]$. Let $\mathcal{O}$ be the adjoint orbit of $u\rho$, so that $\mathcal{C} = \Phi(\mathcal{O})$. As above, let $V \subset \mathfrak{su}(2)$ be the open ball of radius $\frac{1}{\sqrt{2}}$. Then $\mathcal{O}$ is the boundary of $V_u = uV$, and we compute, with $\Gamma = \Phi(\overline{V_u})$,

$$\int_\Gamma \eta = \int_{V_u} \exp^* \eta = \int_{V_u} \mathrm{d}\varpi = \int_{\mathcal{O}} i_{\mathcal{O}}^* \varpi = \int_{\mathcal{O}} \omega_{\mathcal{O}} - \int_{\mathcal{C}} \omega_{\mathcal{C}}.$$

Hence

$$k\left(\int_\Gamma \eta + \int_{\mathcal{C}} \omega_{\mathcal{C}}\right) = k \int_{\mathcal{O}} \omega_{\mathcal{O}},$$

which is an integer if and only if the orbit through $ku\rho$ is integral, i.e., $ku \in \mathbb{Z}$.

Proposition 7.4. *The 4-sphere* S^4 *and the double* $D(\mathrm{SU}(2))$ *are prequantizable at any integer level-k. More generally, this is the case for any q-Hamiltonian*

SU(2)-*space* (M, ω, Φ) *with vanishing second homology. The double* $D(\mathrm{SO}(3))$ *(viewed as a q-Hamiltonian* SU(2)*-space) is prequantizable at level-k if and only if k is even.*

The condition for $D(\mathrm{SO}(3))$ was first obtained by Derek Krepski [29].

Proof. In each of these examples we have $H^2(M, \mathbb{R}) = 0$; hence it suffices to find all k such that $k\Phi^* x = 0$. For $M = S^4$, one has $\Phi^* x = 0$, since $H^3(S^4, \mathbb{Z}) = 0$. For $M = D(\mathrm{SU}(2))$, one again has $\Phi^* x = 0$, by the properties of x under group multiplication and inversion ($\mathrm{Mult}^* x = \mathrm{pr}_1^* x + \mathrm{pr}_2^* x$, $\mathrm{Inv}^* x = -x$.) For $M = D(\mathrm{SO}(3))$, one checks that the torsion subgroup of $H^3(M, \mathbb{Z})$ is $\mathbb{Z}_2$, so that M is prequantizable at either all levels or at all even levels. We claim that M is *not* prequantizable at level 1. To see this, consider the symplectic submanifold $T' \times T' \subset D(\mathrm{SO}(3))$, where T' is the maximal torus in SO(3) given as the image of T. For the symplectic volume one finds (see Section 11.1 below)

$$\mathrm{vol}(T' \times T') = \frac{1}{4}\,\mathrm{vol}(T \times T) = \frac{2}{4} = \frac{1}{2}.$$

By criterion (11), with $\Sigma = T' \times T'$ and $\Gamma = \emptyset$, the prequantized levels k must satisfy $k \int_\Sigma \omega \in \mathbb{Z}$; hence they must be even.

Finally, we remark that if (M_i, ω_i, Φ_i) are prequantized at level-k, then their fusion product $M_1 \times M_2$ inherits a prequantization at level-k.

For an ordinary Hamiltonian SU(2)-space (M, ω_0, Φ_0), a prequantization is an integral lift of the class of the equivariant symplectic form. More generally, by a *level-k prequantization* of such a space we mean a prequantization of $(M, k\omega_0, k\Phi_0)$. Geometrically, the lift is realized as the equivariant Chern class of an equivariant *prequantum line bundle* over M.

Proposition 7.5. *A level-k prequantization of a q-Hamiltonian* SU(2)*-space* (M, ω, Φ) *is equivalent to a pair of level-k prequantizations of the Hamiltonian* SU(2)*-spaces* $(M_\pm, \omega_{0,\pm}, \Phi_{0,+})$, *with the property that the prequantum line bundles* $L_\pm \to M_\pm$ *satisfy*

$$L_- \cong L_+ \otimes \Phi^* J^k$$

on the overlap $M_{\mathrm{reg}} = M_+ \cap M_-$.

Proof. Let $\Phi_\pm \colon M_\pm \to \mathrm{SU}(2)_\pm$ be the restrictions of Φ. Since $\mathrm{SU}(2)_+, \mathrm{SU}(2)_-$ retract onto e, c respectively, the long exact sequences in relative cohomology give isomorphisms $H^2(M_\pm, \cdot) \xrightarrow{\cong} H^3(\Phi_\pm, \cdot)$, and a commutative diagram

$$\begin{array}{ccc} H^3(\Phi, \mathbb{Z}) & \longrightarrow & H^3(\Phi_\pm, \mathbb{Z}) \cong H^2(M_\pm, \mathbb{Z}) \\ \downarrow & & \downarrow \\ H^3(\Phi, \mathbb{R}) & \longrightarrow & H^3(\Phi_\pm, \mathbb{R}) \cong H^2(M_\pm, \mathbb{R}) \end{array}$$

The lower horizontal map is given on $k[(\omega, -\eta)]$ by

$$k[(\omega, -\eta)] \mapsto k[\omega_\pm + \Phi_\pm^* \varpi_\pm] = k[\omega_{0,\pm}].$$

To give a parallel discussion of the upper horizontal map, let $C^k(\cdot, R) = \mathrm{Hom}(C_k(\cdot), R)$ denote the complex of smooth singular cochains, with coefficient in the ring R. We have two natural cochain maps,

$$C^k(\cdot, \mathbb{Z}) \to C^k(\cdot, \mathbb{R}) \leftarrow \Omega^k(\cdot).$$

Let $\eta^{\mathbb{Z}} \in C^3(\mathrm{SU}(2), \mathbb{Z})$ be a smooth singular cocycle whose image in $C^3(\mathrm{SU}(2), \mathbb{R})$ is cohomologous to the image of η, and let $\varpi_\pm^{\mathbb{Z}} \in C^2(\mathrm{SU}(2)_\pm, \mathbb{Z})$ be primitives of the restriction of $\eta^{\mathbb{Z}}$ to $\mathrm{SU}(2)_\pm$. Let $\sigma^{\mathbb{Z}} \in C^2(M, \mathbb{Z})$ be such that $\mathrm{d}\sigma^{\mathbb{Z}} = -k\Phi^*\eta^{\mathbb{Z}}$, and such that $[(\sigma^{\mathbb{Z}}, k\eta^{\mathbb{Z}})] \in H^3(\Phi, \mathbb{Z})$ represents the lift of $k[(\omega, -\eta)]$ given by the prequantization. The upper map in the commutative diagram above is given on $[(\sigma^{\mathbb{Z}}, k\eta^{\mathbb{Z}})]$ by

$$[(\sigma^{\mathbb{Z}}, k\eta^{\mathbb{Z}})] \mapsto [\sigma_\pm^{\mathbb{Z}} + k\Phi^*\varpi_\pm^{\mathbb{Z}}].$$

Hence $[\sigma_\pm^{\mathbb{Z}} + k\Phi^*\varpi_\pm^{\mathbb{Z}}] \in H^2(M_\pm, \mathbb{Z})$ are integral lifts of $k[\omega_{0,\pm}]$. Let $L_\pm \to M_\pm$ be the corresponding $\mathrm{SU}(2)$-equivariant prequantum line bundles, so that

$$c_1(L_\pm) = [\sigma_\pm^{\mathbb{Z}} + k\Phi^*\varpi_\pm^{\mathbb{Z}}].$$

On the overlap $M_{\mathrm{reg}} = M_+ \cap M_-$, the difference between the 2-cocycles $\sigma_\pm^{\mathbb{Z}} + \Phi^*\varpi_\pm^{\mathbb{Z}}$ is $k\Phi^*(\varpi_-^{\mathbb{Z}}|_{\mathrm{SU}(2)_{\mathrm{reg}}} - \varpi_{\mathbb{Z}}^+|_{\mathrm{SU}(2)_{\mathrm{reg}}})$. The 2-cochain $\varpi_-^{\mathbb{Z}}|_{\mathrm{SU}(2)_{\mathrm{reg}}} - \varpi_+^{\mathbb{Z}}|_{\mathrm{SU}(2)_{\mathrm{reg}}} \in C^2(\mathrm{SU}(2)_{\mathrm{reg}}, \mathbb{Z})$ is closed, and its cohomology class is an integral lift of $[\varpi_-|_{\mathrm{SU}(2)_{\mathrm{reg}}} - \varpi_+|_{\mathrm{SU}(2)_{\mathrm{reg}}}] = \Psi^*[\omega_{\mathcal{O}}] \in H^2(\mathrm{SU}(2)_{\mathrm{reg}}, \mathbb{R})$. Hence it represents the Chern class $c_1(J)$. We have shown that

$$c_1(L_-|_{M_{\mathrm{reg}}}) - c_1(L_+|_{M_{\mathrm{reg}}}) = k\Phi^* c_1(J)$$

and consequently $L_-|_{M_{\mathrm{reg}}} \cong L_+|_{M_{\mathrm{reg}}} \otimes \Phi^* J^k$. Conversely, given a pair of prequantum line bundles $L_\pm$ with this property, we may retrace the steps of this proof to obtain an integral lift of $[k(\omega, -\eta)]$.

In particular, we see that if (M, ω, Φ) is prequantized at level-k, and e is a regular value of Φ, then the symplectic quotient $M /\!\!/ \mathrm{SU}(2)$ inherits a level-k prequantization. The corresponding prequantum line bundle over $M /\!\!/ \mathrm{SU}(2)$ is $L_+ /\!\!/ \mathrm{SU}(2) = L_+|_{\Phi^{-1}(e)} / \mathrm{SU}(2)$.

The prequantization result may be expressed in terms of Morita trivializations:

Proposition 7.6. *A level-k prequantization of a q-Hamiltonian* $\mathrm{SU}(2)$*-space* (M, ω, Φ) *gives rise to a Morita isomorphism*

$$\mathbb{C} \simeq_{\mathcal{E}} \Phi^* \mathcal{A}^k.$$

Proof. Pick Morita trivializations $\mathbb{C} \simeq_{\mathcal{F}_\pm} \mathcal{A}^k$ over $\mathrm{SU}(2)_\pm$, with $\mathcal{F}_- \cong \mathcal{F}_+ \otimes J^k$ on the overlap. The prequantum line bundles $L_\pm \to M_\pm$ defined by the level-k

prequantization satisfy $L_- \cong L_+ \otimes \Phi^* J^k$ on the overlap. Hence the Hilbert space bundles

$$\mathcal{E}_\pm := \mathrm{Hom}_{\mathbb{C}}(L_\pm, \Phi^* \mathcal{F}_\pm)$$

(where $\mathrm{Hom}_{\mathbb{C}}$ denotes continuous bundle homomorphisms) glue to give the desired Morita isomorphism.

Proposition 7.7. *Suppose (M, ω, Φ) is a level-k prequantized q-Hamiltonian* SU(2)*-space. Assume that e, c are regular values of Φ. Then the imploded cross-section (X, ω_X, Φ_X) inherits a level-k prequantization.*

Proof. Let $(M_\pm, \omega_{0,\pm}, \Phi_{0,\pm})$ carry the corresponding prequantum line bundles with $L_- = L_+ \otimes \Phi^* J^k$ on the overlap. Since $X_\pm = (M_\pm \times \mathbb{C}^2) /\!\!/ \mathrm{SU}(2)$ are ordinary Hamiltonian quotients, we obtain prequantizations of the Hamiltonian T-spaces $(X_\pm, \omega_X, \Phi_X)$. The prequantum line bundles $L_{X_\pm}$ satisfy $L_{X_\pm}|_Y \cong L_\pm|_Y$; hence

$$L_{X_-}|_Y = L_{X_+}|_Y \otimes \Phi_Y^* J^k = L_{X_+}|_Y \otimes \mathbb{C}_{k\rho}.$$

We conclude that L_{X_+} and $L_{X_-} \otimes \mathbb{C}_{-k\rho}$ patch to define a global T-equivariant prequantum line bundle $L_X \to X$.

8 Quantization of q-Hamiltonian SU(2)-spaces

We are now in a position to define the quantization of prequantized q-Hamiltonian SU(2)-spaces. We begin with a quick overview of the quantization of ordinary *Hamiltonian* G-spaces (M, ω, Φ). Choose an invariant almost complex structure on M, compatible with the symplectic form. Such an almost complex structure is unique up to equivariant homotopy, and hence the isomorphism class of the resulting equivariant Spin_c-structure given by a G-equivariant spinor bundle $\mathcal{S}$ is independent of this choice. We obtain a Morita isomorphism $\mathbb{C}\mathrm{l}(TM) \simeq_{\mathcal{S}^{\mathrm{op}}} \mathbb{C}$. Given a prequantum line bundle $L \to M$, one can twist by L to obtain a new Spin_c-structure $\mathcal{S} \otimes L^{-1}$, hence a Morita isomorphism

$$\mathbb{C}\mathrm{l}(TM) \simeq_{\mathcal{S}^{\mathrm{op}} \otimes L} \mathbb{C}.$$

This allows us to define a push-forward map relative to $p\colon M \to \mathrm{pt}$,

$$p_*\colon K_0^G(M, \mathbb{C}\mathrm{l}(TM)) \to K_0^G(\mathrm{pt}) = R(G),$$

and to set $\mathcal{Q}(M) = p_*([M]) \in R(G)$. (For $G = \{e\}$, this is just an integer.) Equivalently, $\mathcal{Q}(M)$ may be viewed as the equivariant index of the Spin_c-Dirac operator for the Spin_c-structure $\mathcal{S} \otimes L^{-1}$. The quantization procedure for Hamiltonian G-spaces is compatible with products:

$$\mathcal{Q}(M_1 \times M_2) = \mathcal{Q}(M_1)\mathcal{Q}(M_2). \tag{13}$$

For any $g \in G$, the value of the equivariant index $\mathcal{Q}(M)$ at g may be computed by Atiyah–Segal's localization theorem. On the other hand, one has the Guillemin–Sternberg *quantization commutes with reduction* property: Let $\mathcal{Q}(M)^G \in \mathbb{Z}$ be the multiplicity with which the trivial representation occurs in $\mathcal{Q}(M)$. Then [30, 32]

$$\mathcal{Q}(M)^G = \mathcal{Q}(M /\!\!/ G).$$

Here the index $\mathcal{Q}(M /\!\!/ G)$ is well defined if 0 is a regular value of Φ and the G-action on $\Phi^{-1}(0)$ is free. If the action is only locally free, then $M /\!\!/ G$ is an orbifold and the quantization is defined by the index theorem for orbifolds. In the general case, if 0 is not a regular value and $M /\!\!/ G$ is a singular space, $\mathcal{Q}(M /\!\!/ G)$ may be defined by partial desingularization of the singular symplectic quotient [32].

Suppose now that (M, ω, Φ) is a compact q-Hamiltonian SU(2)-space, prequantized at level-k. By combining the Morita isomorphisms $\Phi^*\mathcal{A}^2 \simeq_{\mathcal{S}} \mathbb{C}\mathrm{l}(TM)$ from Proposition 6.2 and $\mathbb{C} \simeq_{\mathcal{E}} \Phi^*\mathcal{A}^k$ from Proposition 7.6 we obtain a Morita isomorphism

$$\mathbb{C}\mathrm{l}(TM) \simeq_{\mathcal{S}^{\mathrm{op}} \otimes \mathcal{E}} \Phi^*\mathcal{A}^{k+2}.$$

This defines a push-forward map in K-homology,

$$K_0^{\mathrm{SU}(2)}(M, \mathbb{C}\mathrm{l}(TM)) \to K_0^{\mathrm{SU}(2)}(\mathrm{SU}(2), \mathcal{A}^{k+2}) \cong R_k(\mathrm{SU}(2)).$$

Definition 8.1. Let (M, ω, Φ) be a compact q-Hamiltonian SU(2)-space, prequantized at level-k. We define the *quantization* $\mathcal{Q}(M) \in R_k(\mathrm{SU}(2))$ to be the push-forward of the K-homology fundamental class $[M] \in K_0^{\mathrm{SU}(2)}(M, \mathbb{C}\mathrm{l}(TM))$,

$$\mathcal{Q}(M) = \Phi_*([M]).$$

The properties of this quantization procedure for q-Hamiltonian spaces are very similar to those for the Hamiltonian case: In particular, the analogue to the "quantization commutes with products" property (13) holds, with the left-hand side involving the fusion product of q-Hamiltonian spaces, and the right-hand side the product in $R_k(\mathrm{SU}(2))$. However, while (13) is rather obvious in the Hamiltonian theory, its q-Hamiltonian counterpart is a nontrivial fact (proved in [31]). In what follows, we will focus on "localization" and "quantization commutes with reduction" for q-Hamiltonian SU(2)-spaces.

9 Localization

We mentioned in Section 2.2 that any $\tau \in R_k(\mathrm{SU}(2))$ is determined by its values $\tau(t)$ at elements $t \in T_{k+2}^{\mathrm{reg}}$. For a level-$k$ prequantized q-Hamiltonian SU(2)-space (M, ω, Φ), the number $\mathcal{Q}(M)(t)$ may be computed by localization to the fixed-point set M^t of t. By equivariance, and since t is regular, the moment map takes the fixed-point set to the maximal torus $T = \mathrm{SU}(2)^t$.

Proposition 9.1. *The restriction $\mathcal{A}^{k+2}|_T$ admits a T_{k+2}-equivariant Morita trivialization,*

$$\mathbb{C} \simeq_{\mathcal{G}} \mathcal{A}^{k+2}|_T.$$

This Morita trivialization is uniquely determined (up to equivalence) by requiring that $\mathcal{G}|_e$ extend to an SU(2)*-equivariant Morita trivialization of $\mathcal{A}^{k+2}|_e$.*

Proof. Choose SU(2)-equivariant Morita trivializations $\mathbb{C} \simeq_{\mathcal{F}_\pm} \mathcal{A}^{k+2}|_{\mathrm{SU}(2)_\pm}$ such that on the overlap, $\mathcal{F}_- \cong \mathcal{F}_+ \otimes J^{k+2}$. Restrict to T-equivariant Morita trivializations over

$$T \cap \mathrm{SU}(2)_+ = T_{(-\rho,\rho)}, \quad T \cap \mathrm{SU}(2)_- = T_{(0,2\rho)}.$$

The intersection $T_{(-\rho,\rho)} \cap T_{(0,2\rho)}$ has two connected components, $T_{(0,\rho)}$ and $T_{(\rho,2\rho)}$. The restrictions of J^{k+2} to the two components are

$$J^{k+2}|_{T_{(0,\rho)}} = T_{(0,\rho)} \times \mathbb{C}_{(k+2)\rho},$$

$$J^{k+2}|_{T_{(\rho,2\rho)}} = T_{(\rho,2\rho)} \times \mathbb{C}_{-(k+2)\rho}.$$

Let

$$\mathcal{G}_+ = \mathcal{F}_+|_{T_{(-\rho,\rho)}}, \quad \mathcal{G}_- = \mathcal{F}_-|_{T_{0,2\rho}} \otimes \mathbb{C}_{(k+2)\rho}.$$

Then $\mathcal{G}_- \cong \mathcal{G}_+$ over $T_{(0,\rho)}$, while $\mathcal{G}_- = \mathcal{G}_+ \otimes \mathbb{C}_{2(k+2)\rho}$ over $T_{(\rho,2\rho)}$. But T_{k+2} is exactly the subgroup of T acting trivially on $\mathbb{C}_{2(k+2)\rho}$. That is, the bundles $\mathcal{G}_\pm$ glue to define a T_{k+2}-equivariant Morita trivialization

$$\mathbb{C} \simeq_{\mathcal{G}} \mathcal{A}^{k+2}|_T.$$

By construction, $\mathcal{G}|_e$ extends to the unique (up to equivalence) SU(2)-equivariant trivialization $\mathcal{F}_+|_e$ of $\mathcal{A}|_e$. Any other T_{k+2}-equivariant Morita trivialization differs from $\mathcal{G}$ by twist with a T_{k+2}-equivariant line bundle. Since $\dim T = 1$, we have $H^2_{T_{k+2}}(T) = H^2_{T_{k+2}}(\mathrm{pt})$; hence such a line bundle is detected by its restriction to e. Since only the trivial T_{k+2}-representation extends to an SU(2)-representation, the proof is complete.

Remark 9.2. The last part of the proof relied on $\dim T = 1$. Indeed, the corresponding statement for higher rank groups is trickier [31].

Proposition 9.3. *Suppose $\Phi\colon M \to$ SU(2) is an equivariant map, and that we are given an equivariant Morita isomorphism $\mathbb{C}\mathrm{l}(TM) \simeq_{\mathcal{E}} \Phi^*\mathcal{A}^{k+2}$. Then, for all regular elements $t \in T \cap \mathrm{SU}(2)_{\mathrm{reg}}$, and any component of the fixed-point set $F \subset M^t$, the restriction $TM|_F$ inherits a distinguished T_{k+2}-equivariant* Spin_c*-structure.*

Proof. By equivariance, and since t is regular, Φ restricts to a map $\Phi_F\colon F \to \mathrm{SU}(2)^t = T$. Hence we have T_{k+2}-equivariant Morita isomorphisms

$$\mathbb{C} \simeq_{\Phi^*\mathcal{G}} \Phi^*(\mathcal{A}^{k+2}|_T) \simeq_{\mathcal{E}^{\mathrm{op}}|_F} \mathbb{C}\mathrm{l}(TM|_F).$$

But a Morita trivialization of a Clifford algebra bundle is equivalent to a Spin_c-structure.

Let $\mathcal{L}_F \to F$ be the Spin_c-line bundle associated to this Spin_c-structure on $TM|_F$.

Remark 9.4. The line bundle $\mathcal{L}_F$ may be described as follows. From $\mathbb{C}\mathrm{l}(TM) \simeq_{\mathcal{E}} \Phi^*\mathcal{A}^{k+2}$ we obtain a Morita trivialization,

$$\mathbb{C} \simeq \mathbb{C}\mathrm{l}(TM) \otimes \mathbb{C}\mathrm{l}(TM) \simeq_{\mathcal{E}\otimes\mathcal{E}} \Phi^*\mathcal{A}^{2k+4}.$$

Over $M_\pm$, we have another Morita trvialization of $\Phi^*\mathcal{A}^{2k+4}$ coming from the defining Morita trivializations of $\mathcal{A}$ over $U_\pm$. The two Morita trivializations are related by line bundles $\mathcal{L}_\pm \to M_\pm$, with $\mathcal{L}_- = \mathcal{L}_+ \otimes \Phi^* J^{-(2k+4)}$ on the overlap. The restriction of J^{2k+4} to T is T_{k+2}-equivariantly trivial, and $\mathcal{L}_F$ is the T_{k+2}-equivariant line bundle obtained by gluing $\mathcal{L}_\pm|_{F\cap M_\pm}$.

Using Proposition 9.3 we see that even though M does not come with a Spin_c-structure, the fixed-point contributions from the usual Atiyah–Segal–Singer theorem [8, 7, 9] are well defined. Indeed one has, the following result.

Theorem 9.5 (Localization). *Suppose (M, ω, Φ) is a compact q-Hamiltonian* SU(2)*-space, prequantized at level-k. For all $t \in T_{k+2}^{\mathrm{reg}}$, the number $\mathcal{Q}(M)(t)$ is given as a sum of fixed-point contributions,*

$$\mathcal{Q}(M)(t) = \sum_{F\subset M^t} \mathcal{Q}(\nu_F)(t),$$

where $\mathcal{Q}(\nu_F)(t)$ is defined using the T_{k+2}-equivariant Spin_c*-structure on $TM|_F$.*

The proof of Theorem 9.5 is parallel to the proof of the localization formula in Atiyah–Segal [7]; details will be given in [31]. In the cohomological form of the index theorem, the fixed-point contributions $\mathcal{Q}(\nu_F)$ are given as integrals of certain characteristic classes over F (cf. [16, 4]),

$$\mathcal{Q}(\nu_F)(t) = (\sigma(\mathcal{L}_F)(t))^{1/2} \int_F \frac{\widehat{A}(F)\exp(\frac{1}{2}c_1(\mathcal{L}_F))}{D_{\mathbb{R}}(\nu_F, t)}.$$

Here $\widehat{A}(F)$ is the $\widehat{A}$-class, and $D_{\mathbb{R}}(\nu_F, t)$ is given on the level of differential forms by

$$D_{\mathbb{R}}(\nu_F, t) = e^{\frac{i\pi}{4}\mathrm{rank}_{\mathbb{R}}(\nu_F)} \det{}^{1/2}_{\mathbb{R}}\left(1 - t^{-1} e^{\frac{1}{2\pi}\mathrm{curv}_{\mathbb{R}}(\nu_F)}\right),$$

with $\mathrm{curv}_{\mathbb{R}}(\nu_F) \in \Omega^2(F, \mathfrak{o}(\nu_F))$ the curvature form for an invariant Riemannian connection. The expression in parentheses lies in $\Omega(F, \mathrm{End}(\nu_F))$, with zeroth-order term the identity, and the (positive) square root of its determinant is well defined. Finally, $\mathcal{L}_F$ is the line bundle associated to the Spin_c-structure on $TM|_F$, the phase factor $\sigma(\mathcal{L}_\mathcal{F})(t) \in \mathrm{U}(1)$ is given by the action of t on $\mathcal{L}|_F$, and $\sigma(\mathcal{L}_\mathcal{F})(t)^{1/2}$ is a suitable choice of square root.[5] If $F \subset M_+$, the Spin_c-structure on $TM|_F$ is defined by the almost complex structure on M_+, twisted by the line bundle L_+. Hence, the fixed-point contribution can be written in "Riemann–Roch" form as

$$\mathcal{Q}(\nu_F)(t) = \sigma(L_+|_F)(t) \int_F \frac{\mathrm{Td}(F)\,\mathrm{ch}(L_+|_F)}{D(\nu_{F,+}, t)},$$

where $D(\nu_{F,+}, t)$ is the equivariant characteristic class

$$D(\nu_{F,+}, t) = \det_{\mathbb{C}}\left(1 - t^{-1} e^{\frac{i}{2\pi}\operatorname{curv}_{\mathbb{C}}(\nu_{F,+})}\right),$$

with $\operatorname{curv}_{\mathbb{C}}(\nu_{F,+})$ the curvature form for an invariant Hermitian connection, and $\sigma(L_+|_F)(t)$ the phase factor defined by the action of t on $L_+|_F$. There is a similar formula for the case $F \subset M_-$:

$$\mathcal{Q}(\nu_F)(t) = -t^{(k+2)\rho}\sigma(L_-|_F)(t) \int_F \frac{\mathrm{Td}(F)\,\mathrm{ch}(L_-|_F)}{D(\nu_{F,-}, t)}.$$

If $t = j(q^s)$ with $s = 1, \ldots, k+1$, we have

$$-t^{(k+2)\rho} = (-1)^{s-1}.$$

This sign factor may be traced back to our choice of Morita trivialization of $\mathcal{A}^{k+2}|_T$, which was chosen to be compatible with the SU(2)-equivariant Morita trivialization of $\mathcal{A}^{k+2}|_e$ (rather than that of $\mathcal{A}^{k+2}|_c$).

Remark 9.6. A detailed check of the equivalence of the "Spin_c" and "Riemann–Roch" forms of the fixed-point contribution may be found in [4, Section 2.3]. In general, it is quite possible that F is contained neither in M_+ nor in M_-: this happens for instance for $M = D(\mathrm{SO}(3))$, as discussed in the final section of this paper.

Remark 9.7. The right-hand side of the localization formula appears in [4], as a "working definition" of the quantization of a q-Hamiltonian space. However, in [4] it was not understood how to view this expression as the localization of an appropriate equivariant object on M.

---

[5] The square root is determined as follows. Let $\mathcal{S}_x$ be the fiber of the spinor module at any given $x \in F$. Choose a T_{k+2}-invariant complex structure on T_xM, compatible with the orientation. Let $c_1, \ldots, c_{n/2} \in \mathrm{U}(1)$ be the eigenvalues (with multiplicities) for the action of t on T_xM, and $u \in \mathrm{U}(1)$ the action of t on the line $\mathrm{Hom}_{\mathbb{C}\mathrm{l}(T_xM)}(\wedge_{\mathbb{C}} T_xM, \mathcal{S}_x)$. Then

$$\sigma(\mathcal{L}_F)(t)^{1/2} = u \prod_{c_r \neq 1} c_r^{1/2},$$

using the square roots of $c_r \neq 1$ with positive imaginary part.

10 Quantization commutes with reduction

Suppose (M, ω, Φ) is a compact q-Hamiltonian SU(2)-space, with a prequantization at level-k. For each $l = 0, \ldots, k$, let $\mathcal{C}_l$ be the conjugacy class of the element $\exp(\frac{l}{k}\rho)$. If SU(2) acts freely (respectively locally freely) on $\Phi^{-1}(\mathcal{C}_l)$, then

$$M/\!\!/_{\mathcal{C}_l}\,\mathrm{SU}(2) = (M \times \mathcal{C}_l)/\!\!/\,\mathrm{SU}(2) \cong \Phi^{-1}(\mathcal{C}_l)/\,\mathrm{SU}(2)$$

is a smooth symplectic manifold (respectively orbifold), with a level-k prequantization from M. The Riemann–Roch numbers

$$\mathcal{Q}(M/\!\!/_{\mathcal{C}_l}\,\mathrm{SU}(2)) \in \mathbb{Z}$$

are thus defined. If SU(2) does not act locally freely, it is still possible to define the Riemann–Roch numbers using a partial desingularization, as in [32].

Theorem 10.1 (q-Hamiltonian quantization commutes with reduction). *Let (M, ω, Φ) be a level-k prequantized q-Hamiltonian* SU(2)*-manifold, and $\mathcal{Q}(M) \in R_k(\mathrm{SU}(2))$ its quantization. Let $N(l) \in \mathbb{Z}$ be the multiplicity of τ_l in $\mathcal{Q}(M)$. Then*

$$N(l) = \mathcal{Q}(M/\!\!/_{\mathcal{C}_l}\,\mathrm{SU}(2)),$$

where the right-hand side denotes the level-k quantization of the symplectic quotient.

A general proof of this result, for arbitrary simply connected groups, can be found in [4]. Here we will present a much simpler approach for the rank-1 case. It is modeled after a similar proof for the Hamiltonian case [30, Appendix].

Proposition 10.2. *Let (M, ω, Φ) be a level-k prequantized q-Hamiltonian* SU(2)*-space. Suppose* SU(2) *acts (locally) freely on $\Phi^{-1}(e)$, $\Phi^{-1}(c)$, so that the imploded cross-section (X, ω_X, Φ_X) is a smooth Hamiltonian T-space, with a prequantization at level-k. Let $N_X(l), l \in \mathbb{Z}$, be the multiplicity function for the Hamiltonian T-space X, and $N(l), 0 \le l \le k$, that for the q-Hamiltonian* SU(2)*-space M. Then*

$$N_X(l) = \begin{cases} N(l) & \text{if } 0 \le l \le k, \\ 0 & \text{otherwise.} \end{cases}$$

Proof. We will consider only the case that SU(2) acts freely on $\Phi^{-1}(e), \Phi^{-1}(c)$. The fact that $N_X(l)$ vanishes unless $0 \le l \le k$ is an easy special case of the Hamiltonian "quantization commutes with reduction" theorem; see, e.g., [17]. The statement is thus equivalent to showing that $\mathcal{Q}(M)$ is the image, under the induction map $R_k(T) \to R_k(\mathrm{SU}(2))$, of $t^\rho \mathcal{Q}(M)(t) \in R(T)$ (restricted to T_{k+2}). That is, we have to show that for all $t = j(z)$, with $z \in \{q, q^2, \ldots, q^{k+1}\}$,

$$\mathcal{Q}(M)(t) = \frac{t^\rho\, \mathcal{Q}(X)(t) - t^{-\rho}\, \mathcal{Q}(X)(t^{-1})}{t^\rho - t^{-\rho}} = \frac{\mathcal{Q}(X)(t)}{1 - t^{-2\rho}} + \frac{\mathcal{Q}(X)(t^{-1})}{1 - t^{2\rho}}.$$

The equivariant index theorem expresses $\mathcal{Q}(M)(t)$ as a sum of fixed-point contributions, $\mathcal{Q}(\nu_F)(t)$, as explained above. Since SU(2) acts freely on $\Phi^{-1}(e)$, $\Phi^{-1}(c)$, the fixed-point manifolds F are all contained in M_{reg}; hence we may work with the Riemann–Roch form of the fixed-point contributions. By regularity, $\Phi(F) \subset T^{\mathrm{reg}}$. Thus, either $F \subset Y$, or the image of F under the Weyl group action lies in Y. That is, all fixed-point manifolds come in pairs F, F', with $F \in Y$ and F' its image under the action of the nontrivial Weyl group element. We have

$$\mathcal{Q}(\nu_{F'})(t) = \mathcal{Q}(\nu_F)(t^{-1}).$$

Now, since $F \subset Y$, it also appears as a fixed-point set in X. The normal bundle of F in M splits as a direct sum of its normal bundle ν_F^X in X and the normal bundle of Y in M, the latter being T-equivariantly isomorphic to $\mathbb{C}_\alpha = \mathbb{C}_{2\rho}$. Hence, the fixed-point contributions are related by

$$\mathcal{Q}(\nu_F)(t) = \frac{\mathcal{Q}(\nu_F^X)(t)}{1 - t^{-2\rho}}, \quad \mathcal{Q}(\nu_{F'})(t) = \frac{\mathcal{Q}(\nu_F^X)(t^{-1})}{1 - t^{2\rho}}.$$

Summing over all fixed-point components $F \subset Y^t$, one obtains all contributions to the fixed-point formula for X, *except* the contributions from $F = M/\!\!/ \mathrm{SU}(2)$ and $F = M/\!\!/_c \mathrm{SU}(2)$. From the explicit description of the normal bundle of $M/\!\!/ \mathrm{SU}(2)$ as $\Phi^{-1}(0) \times_{\mathrm{SU}(2)} \mathbb{C}^2$, and the identity, for $\xi \in \mathfrak{su}(2)$,

$$\det(1 - z^{-1}e^{-\xi}) = z^{-2}\det(1 - ze^{\xi}) = z^{-2}\det(1 - ze^{-\xi}),$$

we obtain

$$D(\nu^X_{M/\mathrm{SU}(2)}, z^{-1}) = z^{-2} D(\nu^X_{M/\mathrm{SU}(2)}, z).$$

Hence, the two terms for $F = M/\!\!/ \mathrm{SU}(2)$ cancel in the fixed-point formula for X. Similarly, the two contributions from $F = M/\!\!/_c \mathrm{SU}(2)$ cancel.

Proof of Theorem 10.1. We have seen that $N(l) = N_X(l)$. From the "quantization commutes with reduction theorem" for Hamiltonian U(1)-spaces [17], we know that $N_X(l)$ is the Riemann–Roch number of the level-k quantization of a symplectic quotient of X:

$$N_X(l) = \mathcal{Q}\left(\Phi_X^{-1}\left(\frac{i\pi l}{k}\right) / \mathrm{U}(1)\right) = \mathcal{Q}(M/\!\!/_{\mathcal{C}_l} \mathrm{SU}(2)).$$

One obtains the multiplicities $N(l)$ by the orthogonality relations (3). Writing $N(l) = \mathcal{Q}(M/\!\!/_{\mathcal{C}_l} \mathrm{SU}(2))$ we obtain

$$\mathcal{Q}(M/\!\!/_{\mathcal{C}_l} \mathrm{SU}(2)) = \sum_{s=1}^{k+1} \frac{|q^s - q^{-s}|^2}{2k+4} \tau_l(j(q^s))\, \mathcal{Q}(M)(j(q^s)).$$

11 Examples

Using the localization formula, we can compute the quantizations $\mathcal{Q}(M) \in R_k(\mathrm{SU}(2))$ for our basic examples. Recall that τ_n, $n = 0, \ldots, k$, are the basis elements of $R_k(\mathrm{SU}(2))$.

11.1 The double

We begin with the q-Hamiltonian SU(2)-space $D(\mathrm{SU}(2))$. Recall that this space is prequantizable at any integer level $k \geq 1$.

Proposition 11.1. *The level-k quantization of the double* $D(\mathrm{SU}(2))$ *is given by*

$$\mathcal{Q}(D(\mathrm{SU}(2))=\sum_{j=0}^{[\frac{k}{2}]}(k+1-2j)\tau_{2j}.$$

Here $[x]$ *denotes the largest integer less than or equal to* x. *Equivalently,*

$$\mathcal{Q}(D(\mathrm{SU}(2))(j(q^s))=\frac{2k+4}{|q^s-q^{-s}|^2}$$

for $s = 1, \ldots, k+1$.

Proof. We first verify the equivalence of the two formulas. Using the known formulas for products of τ_n's, one finds that

$$\sum_{j=0}^{[\frac{k}{2}]}(k+1-2j)\tau_{2j}=\sum_{n=0}^{k}(\tau_n)^2.$$

Write $z = q^s$. Then

$$\sum_{n=0}^{k}(\tau_n(j(z)))^2=-\frac{1}{|z-z^{-1}|^2}\sum_{n=0}^{k}(z^{n+1}-z^{-(n+1)})^2$$

$$=\frac{1}{|z-z^{-1}|^2}\sum_{n=0}^{k}(2-z^{2(n+1)}-z^{-2(n+1)})=\frac{2k+4}{|z-z^{-1}|^2},$$

where the sum is evaluated as a geometric series (using $z^{k+2} = (-1)^s$). We next compare this result to the fixed-point computation for $M = D(\mathrm{SU}(2))$ (the following computation may be found in [4]). Since the action of SU(2) on $M = \mathrm{SU}(2)\times\mathrm{SU}(2)$ is by conjugation on each factor, and $j(z)$ is a regular element, its fixed-point set is

$$M^{j(z)} = T \times T =: F.$$

Note that $\Phi(F) = \{e\}$, in particular $F \subset M_+$. The induced symplectic structure on F is the standard symplectic structure on $T \times T$, defined by the inner product:

$$\omega_F = \mathrm{pr}_1^* \theta_T \cdot \mathrm{pr}_2^* \theta_T,$$

where $\mathrm{pr}_i : T \times T \to T$ are the two projections. The symplectic volume of F is

$$\mathrm{vol}(F) = \int_{T\times T} \omega_F = \Big(\int_T \theta_T\Big) \cdot \Big(\int_T \theta_T\Big) = \alpha \cdot \alpha = 2.$$

The Spin_c-line bundle $\mathcal{L}_F$ comes from the level-$(k+2)$ Morita isomorphism $\mathbb{C}\mathrm{l}(TM) \simeq \Phi^* \mathcal{A}^{k+2}$,

$$\mathbb{C} \simeq \mathbb{C}\mathrm{l}(TM) \otimes \mathbb{C}\mathrm{l}(TM) \simeq \Phi^* \mathcal{A}^{2k+4};$$

hence it is isomorphic to the $(2k+4)$th power of the level-1 prequantum line bundle over F. (We are using that $H^2(M, \mathbb{Z}) = 0$.) Hence $\frac{1}{2} c_1(\mathcal{L}_F) = (k+2)\omega_F$. By considering the action at $x = (e, e) \in F$, one checks that $\zeta(\mathcal{L}_F)(t) = 1$. Indeed, the Spin_c-structure on $T_x M$ extends to an SU(2)-equivariant Spin_c-structure, and the corresponding representation of SU(2) on $\mathcal{L}_F|_x$ is necessarily trivial. The normal bundle to F in M is a trivial bundle

$$\nu_F = \mathfrak{su}(2)/\mathfrak{t} \oplus \mathfrak{su}(2)/\mathfrak{t} = \mathbb{C} \oplus \mathbb{C}^-,$$

with T acting by weight 2 on the first summand and -2 on the second summand. Hence

$$\frac{\zeta_F(t)^{1/2}}{D_{\mathbb{R}}(\nu_F, t)} = \frac{1}{|(1-z^2)(1-z^{-2})|} = \frac{1}{|z - z^{-1}|^2}.$$

Since finally $\widehat{A}(F) = 1$, the fixed-point contribution is

$$\chi(\nu_F, j(z)) = \int_F \frac{e^{(k+2)\omega_F}}{|z - z^{-1}|^2} = \frac{2k+4}{|z - z^{-1}|^2},$$

as claimed.

Recall now that $M(\Sigma_h) = D(\mathrm{SU}(2))^h /\!\!/ \mathrm{SU}(2)$ is the moduli space of flat SU(2)-bundles over a surface of genus h. Using that quantization commutes with products, we have $\mathcal{Q}(D(\mathrm{SU}(2))^h) = \mathcal{Q}(D(\mathrm{SU}(2)))^h$. Together with the quantization commutes with reduction principle we hence obtain the Verlinde formula for this moduli space (cf. [40]):

$$\mathcal{Q}(M(\Sigma_h)) = \sum_{s=1}^{k+1} \left(\frac{|q^s - q^{-s}|^2}{2k+4} \right)^{1-h} = \sum_{s=1}^{k+1} \left(\frac{2\sin^2(\frac{s\pi}{k+2})}{k+2} \right)^{1-h}.$$

11.2 Conjugacy classes

We have seen that the conjugacy classes $\mathcal{C} \subset \mathrm{SU}(2)$ admitting a level-k prequantization are precisely those of elements $\exp(\frac{n}{k}\rho)$ with $0 \leq n \leq k$.

Proposition 11.2. *The level-k quantization of the conjugacy class* $\mathcal{C} = \mathrm{SU}(2) \cdot \exp(\frac{n}{k}\rho)$ *is given by*

$$\mathcal{Q}(\mathcal{C}) = \tau_n. \tag{14}$$

Equivalently, for $s = 1, \ldots, k+1$,

$$\mathcal{Q}(\mathcal{C})(j(q^s)) = \frac{q^{s(n+1)} - q^{-s(n+1)}}{q^s - q^{-s}}. \tag{15}$$

Proof. The equivalence of the two formulations follows from the discussion in Section 2.2. Write $z = q^s$. If $n < k$, then $\Phi(\mathcal{C}) \subset \mathrm{SU}(2)_+$. The symplectic form on $\mathcal{C} = \mathcal{C}_+$ identifies $\mathcal{C}$ with the coadjoint orbit of $\frac{n}{k}\rho$, and the level-k prequantization corresponds to the usual (level-1) prequantization of the orbit through $n\rho$. Written in Riemann–Roch form, the fixed-point contributions for the conjugacy class are just the same as those for the coadjoint orbit, given by (15). If $n = k$, the conjugacy class $\mathcal{C}$ coincides with the central element $\{c\}$. Since $z^{k+2} = (-1)^s$ we have

$$\chi_k(z) = \frac{z^{k+1} - z^{-(k+1)}}{z - z^{-1}} = \frac{z^{k+2} - z^{-(k+2)} z^2}{z^2 - 1} = -(-1)^s,$$

which on the other hand is also the fixed-point contribution for $\mathcal{Q}(\mathcal{C})(j(z))$, for $\mathcal{C} \in \Phi^{-1}(\mathrm{SU}(2)_-)$. This gives (15) for $n = k$.

As a consequence, we may compute the level-k quantization of

$$M(\Sigma_h^r; \mathcal{C}_1, \ldots, \mathcal{C}_r) = D(\mathrm{SU}(2))^h \times \mathcal{C}_1 \times \cdots \times \mathcal{C}_r,$$

where $\mathcal{C}_i, i = 1, \ldots, r$, are conjugacy classes of elements $\exp(\frac{l_i}{k}\rho)$ with $0 \leq l_i \leq k$. One obtains

$$\mathcal{Q}(M(\Sigma_h^r; \mathcal{C}_1, \ldots, \mathcal{C}_r)) = \sum_{s=1}^{k+1} \left(\frac{|q^s - q^{-s}|^2}{2k+4} \right)^{1-h} \tau_{l_1}(q^s) \cdots \tau_{l_r}(q^s).$$

For $h = 0$ and $r = 3$, the right-hand side of this formula consists of the fusion coefficients. That is,

$$\mathcal{Q}(M(\Sigma_0^3 : \mathcal{C}_1, \mathcal{C}_2, \mathcal{C}_3)) = N_{l_1,l_2,l_3}^{(k)}.$$

11.3 The 4-sphere

Recall that the q-Hamiltonian space S^4 admits a unique prequantization for all k.

Proposition 11.3. *The level-k quantization of the 4-sphere is given by*

$$\mathcal{Q}(S^4) = \sum_{n=0}^{k} \tau_n.$$

Equivalently, for $s = 1, \dots, k+1$,

$$\mathcal{Q}(S^4)(j(q^s)) = \begin{cases} 2\,|1 - q^{-s}|^{-2}, & s \text{ odd}, \\ 0, & s \text{ even}. \end{cases}$$

Proof. Write $z = q^s$. We first verify the equivalence of the two formulas:

$$\begin{aligned} \sum_{n=0}^{k} \tau_n(j(q(z))) &= \frac{1}{z - z^{-1}} \sum_{n=0}^{k} (z^{n+1} - z^{-(n+1)}) \\ &= \frac{1}{z - z^{-1}} \left(\frac{z - z^{k+2}}{1 - z} - \frac{z^{-1} - z^{-(k+2)}}{1 - z^{-1}} \right). \end{aligned}$$

If s is even, then $z^{k+2} = 1$ and the two terms cancel. If s is odd, then $z^{k+2} = -1$ and we obtain, writing $(z - z^{-1}) = (1 - z^{-1})(z + 1)$, that

$$\sum_{n=0}^{k} \tau_n(j(z)) = \frac{2}{(1 - z^{-1})(1 - z)} = \frac{2}{|1 - z^{-1}|^2}.$$

The fixed-point set of t consists of the "north pole" $\Phi^{-1}(e)$ and the "south pole" $\Phi^{-1}(c)$. By construction, $S^4_{\pm}$ are identified with open balls in $\mathbb{C}^2$, with the standard SU(2)-action. Hence the weights for the $T \subset \mathrm{SU}(2)$-action are $+1, -1$ respectively, and the fixed-point formulas give (using $j(z)^{(k+2)\rho} = z^{k+2} = (-1)^s$)

$$\mathcal{Q}(S^4)(j(z)) = \frac{1}{(1 - z)(1 - z^{-1})} - (-1)^s \frac{1}{(1 - z)(1 - z^{-1})},$$

as needed.

11.4 Moduli spaces of flat SO(3)*-bundles*

The symplectic quotient

$$D(\mathrm{SO}(3))^h /\!\!/ \mathrm{SO}(3)$$

of an h-fold product of $D(\mathrm{SO}(3))$'s (viewed as q-Hamiltonian SO(3)-spaces) is the moduli space of flat SO(3)-bundles over a surface of genus h. It has two connected components, given as symplectic quotients of $D(\mathrm{SO}(3))^h$, where $D(\mathrm{SO}(3))$ is now viewed as a q-Hamiltonian SU(2)-space:

$$D(\mathrm{SO}(3))^h /\!\!/ \mathrm{SO}(3) = D(\mathrm{SO}(3))^h /\!\!/ \mathrm{SU}(2) \cup D(\mathrm{SO}(3))^h /\!\!/_c \mathrm{SU}(2). \tag{16}$$

The two components correspond to the trivial and the nontrivial SO(3)-bundles over the surface. To obtain Verlinde numbers for these moduli spaces, we need to work out the quantization of the q-Hamiltonian SU(2)-space $D(\mathrm{SO}(3))$.

We have seen that $D(\mathrm{SO}(3))$ is prequantizable at level-k if and only if k is even. The different prequantizations form a principal homogeneous space under the torsion subgroup of $H^2(D(\mathrm{SO}(3)), \mathbb{Z})$. In fact, this group is all torsion, and

$$\begin{aligned} H^2(D(\mathrm{SO}(3)), \mathbb{Z}) &= H^2_{\mathbb{Z}_2\times\mathbb{Z}_2}(D(\mathrm{SU}(2)), \mathbb{Z}) \\ &= H^2_{\mathbb{Z}_2\times\mathbb{Z}_2}(\mathrm{pt}, \mathbb{Z}) \\ &= \mathrm{Hom}(\mathbb{Z}_2 \times \mathbb{Z}_2, \mathrm{U}(1)). \end{aligned}$$

Letting $\mathbb{C}_\phi$ denote the 1-dimensional representation given by $\phi \in \mathrm{Hom}(\mathbb{Z}_2 \times \mathbb{Z}_2, \mathrm{U}(1))$, this group acts by tensoring with the flat line bundle

$$D(\mathrm{SU}(2)) \times_{\mathbb{Z}_2\times\mathbb{Z}_2} \mathbb{C}_\phi.$$

Let $T' = T/\mathbb{Z}_2$ be the maximal torus in SO(3), and $N(T) \subset \mathrm{SU}(2)$, $N(T') \subset \mathrm{SO}(3)$ the normalizers. Similarly, for elements $a, b, \ldots$ of SU(2) we denote by $a', b', \ldots$ their images in SO(3).

Lemma 11.4. *For any $t \in T_{\mathrm{reg}} \subset \mathrm{SU}(2)$, the fixed-point set of its action on $\mathrm{SO}(3) = \mathrm{SU}(2)/\mathbb{Z}_2$ is $T' = T/\mathbb{Z}_2$ unless $t^2 = c$, in which case it is $N(T') = N(T)/\mathbb{Z}_2$.*

Proof. For $a \in \mathrm{SU}(2)$, the element a' is fixed under Ad_t if and only if a is fixed up to a central element, i.e., $tat^{-1}a^{-1} \in Z(\mathrm{SU}(2))$. If this central element is e, this just means that $a \in T$. If the central element is c, then $at^{-1}a^{-1} = t^{-1}c$ shows that $a \in N(T)$ represents the nontrivial Weyl element w, and $c = tw(t^{-1}) = t^2$. We have thus shown that the fixed-point set of a regular element t is the image of T in SO(3), unless $t^2 = c$, in which case it is the image of the normalizer $N(T)$.

Let us consider the fixed contributions of any $t = j(q^s)$, $s = 1, 2, \ldots, k+1$, for the q-Hamiltonian space $D(\mathrm{SO}(3))$, for k even. Note that $t^2 = c \Leftrightarrow s = k/2 + 1$, and so we have to consider two cases:

Case 1. $s \neq 1 + \frac{k}{2}$, i.e., $t^2 \neq c$. Then $D(\mathrm{SO}(3))^t = T' \times T' =: F$ is connected, and its moment map image is $\{e\}$. Since SU(2) acts trivially on the fiber of L_+ at $(e', e') \subset F$, the action of t on $L_+|_F$ is trivial. Hence the fixed-point contribution is just $1/4$ that of the corresponding fixed-point manifold in $D(\mathrm{SU}(2))$:

$$\chi(\nu_F,t)=\frac{1}{4}\frac{2k+4}{|q^s-q^{-s}|^2}=\frac{1}{4\sin^2(\frac{\pi s}{k+2})}\left(\frac{k}{2}+1\right).$$

Case 2. $s=1+\frac{k}{2}$, i.e., $t^2=c$ and $q^s=i$. Then $D(\mathrm{SO}(3))^t=N(T')\times N(T')$ has four connected components, indexed by the elements of $u=(u_1,u_2)\in W\times W=\mathbb{Z}_2\times\mathbb{Z}_2$. Choose

$$n=\begin{pmatrix}0&1\\-1&0\end{pmatrix}\in N(T)$$

as a lift of the nontrivial Weyl group element, and let $n'\in N(T')$ be its image. Then each fixed-point component F_u has a base point

$$x_u\in\{(e',e'),(n',e'),(e',n'),(n',n')\}$$

with the property $\Phi(x_u)=e$. For any given choice of the prequantization, one finds that the contribution of the component labeled by $u=(u_1,u_2)$ is of the form[6]

$$\chi(\nu_{F_u},t)=\frac{\lambda(u)}{4}\frac{2k+4}{|q^s-q^{-s}|^2}=\frac{\lambda(u)}{4}\left(\frac{k}{2}+1\right),$$

where $\lambda(u)\in\mathrm{U}(1)$ is given by the action of t on $L_+|_{m_u}$. For $u=(1,1)$, this phase factor is $\lambda(u)=1$ as above. The total fixed-point contribution is obtained by summing over all $u=(u_1,u_2)$:

$$\mathcal{Q}(D(\mathrm{SO}(3))(q^{k/2+1})=\left(\frac{k}{2}+1\right)\sum_u\frac{\lambda(u)}{4}.$$

Let $\chi\in R_k(\mathrm{SU}(2))$ be defined by

$$\chi=\sum_{j=0}^{k/2}(-1)^j\tau_{2j}=\tau_0-\tau_2+\tau_4+\cdots+(-1)^{k/2}\tau_k. \tag{17}$$

Using the orthogonality relations for level-k characters, one finds that

$$\chi(q^{k/2+1})=\frac{k}{2}+1,\quad \chi(q^s)=0\quad \text{for } s\neq k/2+1.$$

From the localization contributions, we see that

$$\mathcal{Q}(D(\mathrm{SO}(3)))=\frac{1}{4}\left(\mathcal{Q}(D(\mathrm{SU}(2)))+\sum_{u\neq(1,1)}\lambda(u)\,\chi\right).$$

It remains to understand the sum $\sum_{u\neq(1,1)}\lambda(u)$.

[6] The computation is similar to that in Section 11.1. In particular, the symplectic volume of the 2-torus F_u may be computed by working out ω_{F_u} in coordinates; one obtains $\mathrm{vol}(F_u)=1/2$. See [2] for more general calculations along these lines.

Lemma 11.5. *For every even k and any $\phi \in \mathrm{Hom}(\mathbb{Z}_2 \times \mathbb{Z}_2, \mathrm{U}(1))$, the space $D(\mathrm{SO}(3))$ admits a unique prequantization at level-k with the property that*

$$\lambda(u) = (-1)^{k/2}\phi(u)$$

for all $u \neq (1,1)$.

Proof. Changing the prequantization by $\phi \in \mathrm{Hom}(\mathbb{Z}_2 \times \mathbb{Z}_2, \mathrm{U}(1))$ changes $\lambda(u)$ to $\tilde{\lambda}(u) = \lambda(u)\phi(u)$. This shows uniqueness. For existence, we have to find a prequantization with $\lambda(u) = (-1)^{k/2}$ for $u \neq (1,1)$. In fact, it is enough to find such a prequantization for $k = 2$. (The general case will then follow by taking the $k/2$th power of the prequantization at level 2.)

For $k = 2$, and any of the four possible prequantizations, write

$$\mathcal{Q}(D(\mathrm{SO}(3))) = \sum_{l=0}^{2} N(l)\tau_l.$$

The localization formulas for q, q^2, q^3 give equations

$$N(0) + \sqrt{2}N(1) + N(2) = 1,$$
$$N(0) - N(2) = \frac{1}{2} + \frac{1}{2}\sum_{u\neq(1,1)} \lambda(u),$$
$$N(0) - \sqrt{2}N(1) + N(2) = 1.$$

The first and third equations give $N(1) = 0$ and $N(0) + N(2) = 1$. In particular, $N(0) - N(2)$ is an odd integer. The second equation shows that $\sum_{u\neq(1,1)} \lambda(u)$ is a real number. A change of prequantization produces a sign change of exactly two of the $\lambda(u)$'s with $u \neq (1,1)$. Since $\sum_{u\neq(1,1)} \tilde{\lambda}(u)$ is again a real number, it follows that all $\lambda(u)$ are real, and hence equal to ± 1. The number of $\lambda(u)$'s equal to -1 must be odd, or else the second equation would give that $N(0) + N(2) = 0$ or $= 2$, contradicting that $N(0) - N(2)$ is odd. Hence, either all three $\lambda(u)$'s with $u \neq (1,1)$ are equal to -1, or exactly one of them equals -1 and the other two are equal to $+1$. The resulting four cases must correspond to the four prequantizations. In particular, there is a unique level-2 prequantization such that $\lambda(u) = -1$ for all $u \neq (-1,-1)$.

Let $\delta_{\phi,1}$ be equal to 1 if $\phi = 1$, equal to 0 otherwise. Then $\sum_u \phi(u) = 4\delta_{\phi,1}$, i.e., $\sum_{u\neq(1,1)} \phi(u) = -1 + 4\delta_{\phi,1}$. It follows that

$$\mathcal{Q}(D(\mathrm{SO}(3))) = \frac{1}{4}(\mathcal{Q}(D(\mathrm{SU}(2))) + (-1)^{k/2}(-1 + 4\delta_{\phi,1})\,\chi).$$

From the known expansions of $\mathcal{Q}(D(\mathrm{SU}(2)))$ (Proposition 11.1) and χ (equation (17)) in the basis τ_j, we finally obtain the following theorem:

Theorem 11.6. *For k even, let $D(\mathrm{SO}(3))$ carry the level-k prequantization labeled by $\phi \in \mathrm{Hom}(\mathbb{Z}_2 \times \mathbb{Z}_2, \mathrm{U}(1))$. Then*

$$\mathcal{Q}(D(\mathrm{SO}(3))) = \frac{1}{4}\sum_{j=0}^{k/2}(k+1-2j+(-1)^{j+k/2}(-1+4\delta_{\phi,1}))\tau_{2j}.$$

Equivalently, for $s = 1, \ldots, k+1$,

$$\mathcal{Q}(D(\mathrm{SO}(3)))(j(q^s)) = \begin{cases} \frac{1}{4}\sin^{-2}\left(\frac{\pi s}{k+2}\right)\left(\frac{k}{2}+1\right), & s \neq \frac{k}{2}+1, \\ \frac{1}{4}(1+(-1)^{k/2}(-1+4\delta_{\phi,1}))\left(\frac{k}{2}+1\right), & s = \frac{k}{2}+1. \end{cases}$$

Dividing into the various subcases, the formula reads

$$\mathcal{Q}(D(\mathrm{SO}(3))) = \begin{cases} \left(\frac{k}{4}+1\right)\tau_0 + \left(\frac{k}{4}-1\right)\tau_2 + \frac{k}{4}\tau_4 + \left(\frac{k}{4}-2\right)\tau_6 + \cdots, \\ \quad \phi = 1,\ k = 0 \mod 4, \\ \frac{k}{4}\tau_0 + \frac{k}{4}\tau_2 + \left(\frac{k}{4}-1\right)\tau_4 + \left(\frac{k}{4}-1\right)\tau_6 + \cdots, \\ \quad \phi \neq 1,\ k = 0 \mod 4, \\ \left(\frac{k-2}{4}\right)\tau_0 + \left(\frac{k-2}{4}+1\right)\tau_2 + \left(\frac{k-2}{4}-1\right)\tau_4 + \left(\frac{k-2}{4}\right)\tau_6 + \cdots, \\ \quad \phi = 1,\ k = 2 \mod 4, \\ \left(\frac{k-2}{4}+1\right)\tau_0 + \left(\frac{k-2}{4}\right)\tau_2 + \left(\frac{k-2}{4}\right)\tau_4 + \left(\frac{k-2}{4}-1\right)\tau_6 + \cdots, \\ \quad \phi \neq 1,\ k = 2 \mod 4. \end{cases}$$

Using this result, in combination with "quantization commutes with reduction," it is now straightforward to compute the quantizations (Verlinde numbers) for the moduli spaces (16). Note that there are many different prequantizations, since one can choose a different ϕ for each factor. The case with boundary (markings) is still more complicated, and will be discussed elsewhere.

Remark 11.7. For $k = 0 \mod 4$, the result above was proved about eight years ago in joint work [2] with Anton Alekseev and Chris Woodward. Pantev [34] and Beauville [10] had earlier obtained similar results using techniques from algebraic geometry.

References

1. A. Alekseev, A. Malkin, and E. Meinrenken, *Lie group valued moment maps*, J. Differential Geom. **48** (1998), no. 3, 445–495.
2. A. Alekseev, E. Meinrenken, and C. Woodward, *Formulas of Verlinde type for non simply-connected groups*, 2000, Unfinished manuscript.
3. ———, *Group-valued equivariant localization*, Invent. Math. **140** (2000), no. 2, 327–350.
4. A. Alekseev, E. Meinrenken, and C. Woodward, *The Verlinde formulas as fixed point formulas*, J. Symplectic Geom. **1** (2001), no. 1, 1–46.
5. ———, *Duistermaat–Heckman measures and moduli spaces of flat bundles over surfaces*, Geom. and Funct. Anal. **12** (2002), 1–31.
6. M. Atiyah and G. Segal, *Twisted K-theory*, Ukr. Mat. Visn. **1** (2004), no. 3, 287–330.
7. M. F. Atiyah and G. B. Segal, *The index of elliptic operators. II*, Ann. of Math. (2) **87** (1968), 531–545.

8. M. F. Atiyah and I. M. Singer, *The index of elliptic operators I*, Ann. of Math. (2) **87** (1968), 484–530.
9. ———, *The index of elliptic operators III*, Ann. of Math. (2) **87** (1968), 546–604.
10. A. Beauville, *The Verlinde formula for PGL(p).*, The mathematical beauty of physics (Saclay, 1996), Adv. Ser. Math. Phys., vol. 24, World Sci. Publishing, 1997, pp. 141–151.
11. N. Berline, E. Getzler, and M. Vergne, *Heat kernels and Dirac operators*, Grundlehren der mathematischen Wissenschaften, vol. 298, Springer-Verlag, Berlin–Heidelberg–New York, 1992.
12. D. Chatterjee, *On the construction of Abelian gerbs*, Ph.D. thesis, University of Cambridge, 1998.
13. A. Connes, *Noncommutative differential geometry*, Inst. Hautes Études Sci. Publ. Math. (1985), no. 62, 257–360.
14. J. Dixmier and A. Douady, *Champs continus d'espaces hilbertiens et de C^*-algèbres*, Bull. Soc. Math. France **91** (1963), 227–284.
15. J. J. Duistermaat, *Convexity and tightness for restrictions of Hamiltonian functions to fixed point sets of an antisymplectic involution*, Trans. Amer. Math. Soc. **275** (1983), 412–429.
16. ———, *The heat kernel Lefschetz fixed point formula for the Spin-c Dirac operator*, Progress in Nonlinear Differential Equations and Their Applications, vol. 18, Birkhäuser, Boston, 1996.
17. J. J. Duistermaat, V. Guillemin, E. Meinrenken, and S. Wu, *Symplectic reduction and Riemann–Roch for circle actions*, Math. Res. Letters **2** (1995), 259–266.
18. J. J. Duistermaat and G. J. Heckman, *On the variation in the cohomology of the symplectic form of the reduced phase space*, Invent. Math. **69** (1982), 259–268.
19. D. Freed, *The Verlinde algebra is twisted equivariant K-theory*, Turkish J. Math. **25** (2001), no. 1, 159–167.
20. D. Freed, M. Hopkins, and C. Teleman, *Loop Groups and Twisted K-Theory I*, arXiv: 0711.1906.
21. V. Guillemin, *Reduced phase spaces and Riemann–Roch*, Lie Groups and Geometry in Honor of B. Kostant (Massachusetts Institute of Technology, 1994) (R. Brylinski et al., eds.), Progress in Mathematics, vol. 123, Birkhäuser, Boston, 1995, pp. 305–334.
22. V. Guillemin and S. Sternberg, *Geometric quantization and multiplicities of group representations*, Invent. Math. **67** (1982), 515–538.
23. N. Higson and J. Roe, *Analytic K-homology*, Oxford Mathematical Monographs, Oxford University Press, Oxford, 2000, Oxford Science Publications.
24. N. Hitchin, *Lectures on special Lagrangian submanifolds*, Winter School on Mirror Symmetry, Vector Bundles and Lagrangian Submanifolds (Cambridge, MA, 1999), Amer. Math. Soc., Providence, RI, 2001, pp. 151–182.
25. J. Hurtubise, L. Jeffrey, and R. Sjamaar, *Group-valued implosion and parabolic structures*, Amer. J. Math. **128** (2006), no. 1, 167–214.
26. J. C. Hurtubise and L. C. Jeffrey, *Representations with weighted frames and framed parabolic bundles*, Canad. J. Math. **52** (2000), no. 6, 1235–1268.
27. L. Jeffrey and J. Weitsman, *Bohr–Sommerfeld orbits in the moduli space of flat connections and the Verlinde dimension formula*, Comm. Math. Phys. **150** (1992), no. 3, 593–630.
28. G. G. Kasparov, *K-theory, group C^*-algebras, and higher signatures (conspectus)*, Novikov conjectures, index theorems and rigidity, Vol. 1 (Oberwolfach, 1993), London Math. Soc. Lecture Note Ser., vol. 226, Cambridge Univ. Press, Cambridge, 1995, pp. 101–146.
29. D. Krepski, *Pre-quantization of the moduli space of flat G-bundles over a surface*, J. Geom. Phys. **58** (2008), no. 11, 1624–1637.
30. E. Meinrenken, *Symplectic surgery and the Spin^c-Dirac operator*, Adv. in Math. **134** (1998), 240–277.
31. E. Meinrenken, *Twisted K-homology and group-valued moment maps*, Preprint, August 2010, arXiv:1008.1261.
32. E. Meinrenken and R. Sjamaar, *Singular reduction and quantization*, Topology **38** (1999), 699–763.
33. L. O'Shea and R. Sjamaar, *Moment maps and Riemannian symmetric pairs*, Math. Ann. **317** (2000), no. 3, 415–457.

34. T. Pantev, *Comparison of generalized theta functions*, Duke Math. J. **76** (1994), no. 2, 509–539.
35. R. J. Plymen, *Strong Morita equivalence, spinors and symplectic spinors*, J. Operator Theory **16** (1986), no. 2, 305–324.
36. A. Pressley and G. Segal, *Loop groups*, Oxford University Press, Oxford, 1988.
37. J. Rosenberg, *Continuous-trace algebras from the bundle theoretic point of view*, J. Austral. Math. Soc. Ser. A **47** (1989), no. 3, 368–381.
38. F. Schaffhauser, *A real convexity theorem for quasi-Hamiltonian actions*, arXiv:0705.0858.
39. R. Sjamaar and E. Lerman, *Stratified symplectic spaces and reduction*, Ann. of Math. (2) **134** (1991), 375–422.
40. A. Szenes, *Verification of Verlinde's formulas for* SU(2), Internat. Math. Res. Notices (1991), no. 7, 93–98.

Wall-crossing formulas in Hamiltonian geometry

Paul-Emile Paradan

Dedicated to Hans Duistermaat on the occasion of his 65th birthday

Abstract In this article, we study the local invariants associated to the Hamiltonian action of a compact torus. Our main results are wall-crossing formulas between invariants attached to adjacent connected components of regular values of the moment map.

Key words: Symplectic, moment map, geometric quantization, transversally elliptic, partition function

Mathematics Subject Classification (2010): 58J20, 53D20, 53D50

1 Introduction

Let (M, Ω) be a compact connected symplectic manifold with the Hamiltonian action of a compact torus T and moment map $\Phi : M \to \mathfrak{t}^*$. Let us assume that the action is *effective*. We are interested here in two global invariants:

1. the Duistermaat–Heckman measure $\mathrm{DH}(M)$, which is the pushforward by Φ of the Liouville volume form,
2. the Riemann–Roch characters $\mathrm{RR}(M, L^{\otimes k}), k \geq 1$, which are virtual representations of T. Here the data (M, Ω, Φ) is prequantized by a Kostant–Souriau line bundle L.

Let $\Lambda^* \subset \mathfrak{t}^*$ be the weight lattice of T. For every couple $(\mu, k) \in \Lambda^* \times \mathbb{Z}^{>0}$, we denote by $\mathrm{m}(\mu, k) \in \mathbb{Z}$ the multiplicity of the weight μ in $\mathrm{RR}(M, L^{\otimes k})$.

Institut de Mathématiques et de Modèlisation de Montpellier (I3M)
CNRS:UMR5149, Université Montpellier II, France
e-mail: paradan@math.univ-montp2.fr

One striking property of the moment map is that its image $\Phi(M)$ is a convex polytope in $\mathfrak{t}^*$. In fact, as noted for example in [17] or [20], each component of the set of regular values of Φ is either an open convex polytope contained in $\Phi(M)$ or the open subset $\mathfrak{c}_{\text{ext}} = \mathfrak{t}^* \setminus \Phi(M)$.

Let us fix a connected component $\mathfrak{c}$ of regular values of Φ. A celebrated theorem of Duistermaat and Heckman [15] tells us that the measure DH(M) is equal to a *polynomial* $\mathrm{DH}_\mathfrak{c}$ times a Lebesgue measure on the open subset $\mathfrak{c}$. Note that $\mathrm{DH}_{\mathfrak{c}_{\text{ext}}}$ is the zero polynomial.

The "quantization commutes with reduction" theorem [28, 29] shows that there exists a *periodic polynomial*[1] $\mathrm{m}_\mathfrak{c} : \Lambda^* \times \mathbb{Z} \to \mathbb{Z}$ which coincides with the multiplicity map $\mathrm{m} : \Lambda^* \times \mathbb{Z}^{>0} \to \mathbb{Z}$ on the cone of $\mathfrak{t}^* \times \mathbb{R}$ generated by $\mathfrak{c} \times \{1\}$. The periodic polynomial $\mathrm{m}_\mathfrak{c}$ is defined by a Kawasaki–Riemann–Roch formula on a symplectic quotient $\mathcal{M}_a = \Phi^{-1}(a)/T$, where $a \in \mathfrak{c}$. As a corollary, we get that $\mathrm{DH}_\mathfrak{c}$ is the semiclassical limit of $\mathrm{m}_\mathfrak{c}$: one has

$$\lim_{k \to +\infty} \frac{\mathrm{m}_\mathfrak{c}(k\mu, k)}{k^d} = \frac{1}{(2\pi)^d} \mathrm{DH}_\mathfrak{c}(\mu) \tag{1.1}$$

for every $\mu \in \Lambda^*$. Here $d = \frac{1}{2} \dim \mathcal{M}_a$.

We have seen that the *global* invariants DH(M), RR$(M, L^{\otimes k})$, $k \geq 1$, give rise to a family of *local* invariants $\mathrm{DH}_\mathfrak{c}$, $\mathrm{m}_\mathfrak{c}$, where $\mathfrak{c}$ runs over the connected component of regular values of Φ.

This paper is concerned with the differences $\mathrm{DH}_{\mathfrak{c}_+} - \mathrm{DH}_{\mathfrak{c}_-}$ and $\mathrm{m}_{\mathfrak{c}_+} - \mathrm{m}_{\mathfrak{c}_-}$ in the case that $\mathfrak{c}_\pm$ are two *adjacent* connected components of regular values of Φ. Let $\Delta \subset \mathfrak{t}^*$ be the hyperplane that separates $\mathfrak{c}_\pm$. Some continuity properties are known:

1. the polynomial $\mathrm{DH}_{\mathfrak{c}_+} - \mathrm{DH}_{\mathfrak{c}_-}$ is divisible by a certain power of the equation of the hyperplane Δ (see [17] and [12]);
2. the periodic polynomial $\mathrm{m}_{\mathfrak{c}_+} - \mathrm{m}_{\mathfrak{c}_-}$ vanishes on

$$\{(\mu, k) \in \Lambda^* \times \mathbb{Z} \mid \mu \in k\Delta\}. \tag{1.2}$$

See [29].

In this paper, we compute explicitly the difference $\mathrm{DH}_{\mathfrak{c}_+} - \mathrm{DH}_{\mathfrak{c}_-}$, and we show that $\mathrm{m}_{\mathfrak{c}_+} - \mathrm{m}_{\mathfrak{c}_-}$ vanishes also on some translates of (1.2).

Let us introduce some notation. We denote by $T_\Delta \subset T$ the subtorus of dimension 1 that has for Lie algebra the one-dimensional subspace $\mathfrak{t}_\Delta$ which is orthogonal to the hyperplane Δ. Let $\beta \in \mathfrak{t}_\Delta$ be the primitive element of the coweight lattice $\ker(\exp : \mathfrak{t} \to T)$ which is pointing in the direction from $\mathfrak{c}_-$ to $\mathfrak{c}_+$.

We make the choice of a decomposition $T = T/T_\Delta \times T_\Delta$, where T/T_Δ denotes a subtorus of T. At the level of Lie algebras, we have then $\mathfrak{t} = (\mathfrak{t}/\mathfrak{t}_\Delta) \oplus \mathfrak{t}_\Delta$ and $\mathfrak{t}^* = (\mathfrak{t}/\mathfrak{t}_\Delta)^* \oplus \mathfrak{t}^*_\Delta$: hence $\xi + (\mathfrak{t}/\mathfrak{t}_\Delta)^* = \Delta$ for any $\xi \in \Delta$. We denote by $\mathcal{S}(\mathfrak{t})$ the algebra of polynomials on the vector space $\mathfrak{t}^*$. We will consider the polynomial $\mathrm{DH}_{\mathfrak{c}_+} - \mathrm{DH}_{\mathfrak{c}_-} \in \mathcal{S}(\mathfrak{t})$ relatively to the decomposition

[1] See Definition 3.1 for the notion of "periodic polynomial."

$$\mathcal{S}(\mathfrak{t}) = \bigoplus_{j\in\mathbb{N}} \mathcal{S}(\mathfrak{t}/\mathfrak{t}_\Delta)\beta^j.$$

Let us choose (and fix once and for all) $\xi \in \Delta$ in the relative interior of $\overline{\mathfrak{c}_+} \cap \overline{\mathfrak{c}_-}$ in Δ. We consider the family $\mathcal{F}$ of connected components $Z \subset M^{T_\Delta}$ such that $\xi \in \Phi(Z) \subset \Delta$. It is easy to see that $\mathcal{F}$ does not depend of the choice of ξ: we have $\overline{\mathfrak{c}_+} \cap \overline{\mathfrak{c}_-} \subset \Phi(Z)$ for all $Z \in \mathcal{F}$. For each $Z \in \mathcal{F}$, we denote by

$$\Phi_Z : Z \to (\mathfrak{t}/\mathfrak{t}_\Delta)^*$$

the restriction of the map $\Phi - \xi$ to the symplectic submanifold Z. The map Φ_Z is a moment map relative to the Hamiltonian action of T/T_Δ on Z. Let $\mathrm{DH}(Z)$ be Duistermaat–Heckman measure on $(\mathfrak{t}/\mathfrak{t}_\Delta)^*$ associated to the moment map Φ_Z. Since 0 is a regular value of Φ_Z, we may consider the Duistermaat–Heckman polynomial

$$\mathrm{DH}_0(Z) \in \mathcal{S}(\mathfrak{t}/\mathfrak{t}_\Delta)$$

such that $\mathrm{DH}(Z)(a') = \mathrm{DH}_0(Z)(a')Pa'$ in a neighborhood of $0 \in (\mathfrak{t}/\mathfrak{t}_\Delta)^*$. Here da' is the natural Lebesgue measure on $(\mathfrak{t}/\mathfrak{t}_\Delta)^*$, derived from the Lebesgue measure on $\mathfrak{t}^*$ and $d\beta$ on $\mathfrak{t}^*_\Delta$.

For $Z \in \mathcal{F}$, we consider the symplectic reduction

$$\mathcal{Z}_\xi = \Phi_Z^{-1}(0)/(T/T_\Delta)$$

and the normal bundle N_Z of Z in M. Note that N_Z inherits the structure of a symplectic vector bundle, together with a Hamiltonian vector bundle action of the circle T_Δ on it. Let $2d_Z$ be the dimension of $\mathcal{Z}_\xi$ and let $2r_Z$ be the (real) rank of N_Z. We prove in Section 2 the following.

Theorem A. *We have*

$$(\mathrm{DH}_{\mathfrak{c}_+} - \mathrm{DH}_{\mathfrak{c}_-})(a) = \sum_{Z\in\mathcal{F}} \mathbf{D}_Z(a-\xi), \quad a \in \mathfrak{t}^*,$$

where each polynomial $\mathbf{D}_Z \in \mathcal{S}(\mathfrak{t})$ *admits the following decomposition:*

$$\mathbf{D}_Z = \frac{\beta^{r_Z-1}}{\det_Z^{1/2}\left(\frac{-\mathcal{L}_\beta}{2\pi}\right)} \left(\frac{\mathrm{DH}_0(Z)}{(r_Z-1)!} + \sum_{k=1}^{d_Z} \beta^k \mathrm{Q}_{Z,k}\right).$$

Each polynomial $\mathrm{Q}_{Z,k}$ *belongs to* $\mathcal{S}(\mathfrak{t}/\mathfrak{t}_\Delta)$ *and is of degree less than* $d_Z - k$. *The term* $\det_Z^{1/2}\left(\frac{-\mathcal{L}_\beta}{2\pi}\right) \in \mathbb{Z}$ *is the Pfaffian of the infinitesimal action of* $\frac{-\beta}{2\pi}$ *on the fibers of the normal bundle* N_Z.

Theorem **A** generalizes previous results of Guillemin–Lerman–Sternberg [17] and Brion–Procesi [12]. In Section 2.4 we give the precise definition of the polynomials $\mathrm{Q}_{Z,k}$.

Suppose now that M is prequantized by a Kostant–Souriau line bundle L. The hyperplane Δ is defined by the equation

$$\frac{\langle a, \beta\rangle}{2\pi} - r_\Delta = 0, \quad a \in \mathfrak{t}^*, \tag{1.3}$$

for some $r_\Delta \in \mathbb{Z}$. The bundle N_Z decomposes as the sum of two polarized subbundles $N_Z^{\pm,\beta}$. Let $s_Z^\pm \in \mathbb{N}$ be the absolute value of the trace of $\frac{1}{2\pi}\mathcal{L}_\beta$ on $N_Z^{\pm,\beta}$. Note that the integer $s_Z^+ + s_Z^-$ is larger than half of the codimension of Z in M.

We prove in Section 3.5 the following theorem.

Theorem B. *Let* $s^\pm := \inf_{Z\in\mathcal{F}} s_Z^\pm$. *We have* $\mathrm{m}_{\mathfrak{c}_+}(\mu, k) = \mathrm{m}_{\mathfrak{c}_-}(\mu, k)$ *when*

$$-s^- < \frac{\langle \mu, \beta\rangle}{2\pi} - kr_\Delta < s^+. \tag{1.4}$$

Note that the symplectic orbifolds $\mathcal{Z}_\xi$, $Z \in \mathcal{F}$ form the connected component of the symplectic reduction

$$\mathcal{M}_\xi^\Delta := (\Phi^{-1}(\xi) \cap M^{T_\Delta})/(T/T_\Delta).$$

We have the following refinement of Theorem **B**.

Theorem C. *If* $\mathcal{M}_\xi^\Delta$ *is connected, the inequalities (1.4) are optimal, i.e., there exist* (μ, k) *such that* $\frac{\langle\mu,\beta\rangle}{2\pi} - kr_\Delta = \pm s^\pm$ *and* $\mathrm{m}_{\mathfrak{c}_+}(\mu, k) \neq \mathrm{m}_{\mathfrak{c}_-}(\mu, k)$.

In Section 4 we apply Theorem **B** to the particular cases in which M is an integral coadjoint orbit of a compact Lie group G. In Section 4.4, we study more precisely the case $G = \mathrm{SU}(n)$: here our result specifies some results of Billey–Guillemin–Rassart [10].

In Section 5, we obtain a strong version of Theorem **B** in the case of an action of a torus T on a complex vector space $\mathbb{C}^d$. The quantization of this action is in some sense the vector space $\mathrm{Pol}(\mathbb{C}^d)$ of complex polynomials on $\mathbb{C}^d$. The T-multiplicities of $\mathrm{Pol}(\mathbb{C}^d)$ are given by a partition function $N_R : \Lambda^* \to \mathbb{N}$. It was observed in [13, 35] that there exists a finite decomposition of the vector space $\mathfrak{t}^*$ in *conic chambers* such that N_R is a periodic polynomial on each piece.

Let $\mathfrak{c}_\pm$ be two adjacent chambers, and let $P_{\mathfrak{c}\pm}$ be the corresponding periodic polynomials computing N_R on each chamber. The main result of Section 5 is the formula (5.6), which depicts the periodic polynomial $P_{\mathfrak{c}_+} - P_{\mathfrak{c}_-}$ as a convolution of distributions. Recently,[2] Boyal–Vergne [11] and De Concini–Procesi–Vergne [14] proposed different proofs of this formula.

[2] Our present paper is a revised version of the preprint math.SG/0411306.

Acknowledgments

I am grateful to Michèle Vergne for bringing reference [10] to my attention, and for explaining to me her work with András Szenes [36].

Notation

Throughout the paper T will denote a compact, connected abelian Lie group, and $\mathfrak{t}$ its Lie algebra. The integral lattice $\Lambda \subset \mathfrak{t}$ is defined as the kernel of $\exp : \mathfrak{t} \to T$, and the real weight lattice $\Lambda^* \subset \mathfrak{t}^*$ is defined by $\Lambda^* := \hom(\Lambda, 2\pi\mathbb{Z})$. Every $\mu \in \Lambda^*$ defines a 1-dimensional T-representation, denoted by $\mathbb{C}_\mu$, where $t = \exp X$ acts by $t^\mu := e^{i\langle\mu, X\rangle}$. We denote by $R(T)$ the ring of characters of finite-dimensional T-representations. We denote by $R^{-\infty}(T)$ the set of generalized characters of T. An element $\chi \in R^{-\infty}(T)$ is of the form $\chi = \sum_{\mu\in\Lambda^*} a_\mu \mathbb{C}_\mu$, where $\mu \mapsto a_\mu$, $\Lambda^* \to \mathbb{Z}$ has at most polynomial growth.

The symplectic manifolds are oriented by their Liouville volume forms. If (Z, o_Z) is an oriented submanifold of an oriented manifold (M, o_M), we take on the fibers of the normal bundle N of Z in M, the orientation o_N satisfying $o_M = o_Z \cdot o_N$.

2 Duistermaat–Heckman measures

Let (M, Ω) be a connected symplectic manifold of dimension $2n$ equipped with a Hamiltonian action of a torus T, with Lie algebra $\mathfrak{t}$. The moment map $\Phi : M \to \mathfrak{t}^*$ satisfies the relations $\Omega(X_M, -) + d\langle\Phi, X\rangle = 0$, $X \in \mathfrak{t}$, where $X_M(m) := \frac{d}{dt}|_{t=0} e^{-tX} \cdot m$ is the vector field generated by the infinitesimal action of $X \in \mathfrak{t}$.

We assume in this section that Φ is a *proper* map, and that the generic stabilizer Γ_M of T on M is *finite*.

The Duistermaat–Heckman measure $\mathrm{DH}(M)$ is defined as the pushforward by Φ of the Liouville volume form $\frac{\Omega^n}{n!}$ on M. For every $f \in \mathcal{C}^\infty(\mathfrak{t}^*)$ with compact support one has $\int_{\mathfrak{t}^*} \mathrm{DH}(M)(a) f(a) = \int_M f(\Phi)\frac{\Omega^n}{n!}$. Equivalently,

$$\mathrm{DH}(M)(a) = \int_M \delta(a - \Phi)\frac{\Omega^n}{n!}, \quad a \in \mathfrak{t}^*.$$

We can define $\mathrm{DH}(M)$ in terms of equivariant forms as follows. Let $\mathcal{A}(M)$ be the space of differential forms on M with complex coefficients. We denote by $\mathcal{A}^{-\infty}_{\mathrm{temp}}(\mathfrak{t}, M)$ the space of tempered generalized functions over $\mathfrak{t}$ with values in $\mathcal{A}(M)$, and by $\mathcal{M}^{-\infty}_{\mathrm{temp}}(\mathfrak{t}^*, M)$ the space of tempered distributions over $\mathfrak{t}^*$ with values in $\mathcal{A}(M)$. Let $\mathcal{F} : \mathcal{A}^{-\infty}_{\mathrm{temp}}(\mathfrak{t}, M) \to \mathcal{M}^{-\infty}_{\mathrm{temp}}(\mathfrak{t}^*, M)$ be the Fourier transform normalized by the condition that $\mathcal{F}(X \mapsto e^{i\langle \xi, X\rangle})$ is equal to the Dirac distribution $a \mapsto \delta(a-\xi)$.

Let $\Omega_{\mathfrak{t}}(X) = \Omega - \langle \Phi, X\rangle$ be the equivariant symplectic form. We have then $\mathcal{F}(e^{-i\Omega_{\mathfrak{t}}}) = e^{-i\Omega}\delta(a-\Phi)$ and so

$$\mathrm{DH}(M) = (i)^n \int_M \mathcal{F}(e^{-i\Omega_{\mathfrak{t}}}). \tag{2.1}$$

2.1 Equivariant cohomology and localization

We first recall the Cartan model of equivariant cohomology with polynomial coefficients and the extension to generalized coefficients defined by Kumar and Vergne [26]. We then give a brief account of the method of localization developed in [30, 31],

Let M be a manifold provided with an action of a compact connected Lie group K with Lie algebra $\mathfrak{k}$. Let $d : \mathcal{A}(M) \to \mathcal{A}(M)$ be the exterior differentiation. Let $\mathcal{A}_c(M)$ be the subalgebra of compactly supported differential forms. If V is a vector field on M we denote by $c(V) : \mathcal{A}(M) \to \mathcal{A}(M)$ the contraction by V. The action of K on M gives a morphism $X \to X_M$ from $\mathfrak{k}$ to the Lie algebra of vector fields on M.

We consider the space of K-equivariant maps $\mathfrak{k} \to \mathcal{A}(M)$, $X \mapsto \eta(X)$, equipped with the derivation $(D\eta)(X) := (d - c(X_M))(\eta(X))$, $X \in \mathfrak{k}$. Since $D^2 = 0$, one can define the cohomology space $\ker D/\mathrm{Im} D$. The Cartan model [7, 21] considers *polynomial* maps, and the associated cohomology is denoted by $\mathcal{H}^*_K(M)$. Kumar and Vergne [26] studied the cohomology spaces $\mathcal{H}^{\pm\infty}_K(M)$ obtained by taking $\mathcal{C}^{\pm\infty}$ maps. Recall the construction $\mathcal{H}^{-\infty}_K(M)$.

The space $\mathcal{C}^{-\infty}(\mathfrak{k}, \mathcal{A}(M))$ of generalized functions on $\mathfrak{k}$ with values in the space $\mathcal{A}(M)$ is, by definition, the space $\mathrm{Hom}(m_c(\mathfrak{k}), \mathcal{A}(M))$ of continuous $\mathbb{C}$-linear maps from the space $m_c(\mathfrak{k})$ of smooth compactly supported densities on $\mathfrak{k}$ to the space $\mathcal{A}(M)$, both endowed with the $\mathcal{C}^\infty$-topologies. We define $\mathcal{A}^{-\infty}_K(M) := \mathcal{C}^{-\infty}(\mathfrak{k}, \mathcal{A}(M))^K$ as the space of K-equivariant $\mathcal{C}^{-\infty}$-maps from $\mathfrak{k}$ to $\mathcal{A}(M)$.

The differential D defined on $\mathcal{C}^\infty(\mathfrak{k}, \mathcal{A}(M))$ admits a natural extension to $\mathcal{C}^{-\infty}(\mathfrak{k}, \mathcal{A}(M))$ and $D^2 = 0$ on $\mathcal{A}^{-\infty}_K(M)$ [26]. The cohomology associated to $(\mathcal{A}^{-\infty}_K(M), D)$ is called the K-equivariant cohomology with generalized coefficients and is denoted by $\mathcal{H}^{-\infty}_K(M)$. The subspace $\mathcal{A}^{-\infty}_{K,c}(M) := \mathcal{C}^{-\infty}(\mathfrak{k}, \mathcal{A}_c(M))^K$ is stable under the differential D, and we denote by $\mathcal{H}^{-\infty}_{K,c}(M)$ the associated cohomology. When M is oriented, the integration over M gives rise to a map $\int_M : \mathcal{H}^{-\infty}_{K,c}(M) \to \mathcal{C}^{-\infty}(\mathfrak{k})^K$.

Localization procedure. Let λ be a K-invariant 1-form on M and let

$$\Phi_\lambda : M \to \mathfrak{k}^* \tag{2.2}$$

be the K-equivariant map defined by $\langle \Phi_\lambda(m), X\rangle = \lambda(X_M)_m$. Then $D\lambda(X) = d\lambda - \langle \Phi_\lambda, X\rangle$. The localization procedure developed in [30, 31] is based on the existence of an inverse $[D\lambda]^{-1}$ of the K-equivariant form $D\lambda$. It is an equivariantly closed element of $\mathcal{A}_K^{-\infty}(M\backslash\Phi_\lambda^{-1}(0))$ defined by the integral

$$[D\lambda]^{-1}(X) = i\int_0^\infty e^{-i\,t\,D\lambda(X)}dt. \tag{2.3}$$

An open subset $\mathcal{U} \subset M$ is called *adapted to* λ if $\mathcal{U}$ is K-invariant and if $(\partial\mathcal{U}) \cap \Phi_\lambda^{-1}(0) = \emptyset$. In [31], we associate to an open subset $\mathcal{U}$ adapted to λ the following equivariantly closed form with generalized coefficients:

$$\mathrm{P}_\lambda^{\mathcal{U}} = \chi^{\mathcal{U}} + d\chi^{\mathcal{U}}[D\lambda]^{-1}\lambda. \tag{2.4}$$

Here $\chi^{\mathcal{U}} \in \mathcal{C}^\infty(M)$ is a K-invariant function supported in $\mathcal{U}$ which is equal to 1 in a neighborhood of $\mathcal{U} \cap \Phi_\lambda^{-1}(0)$. The cohomology class defined by $\mathrm{P}_\lambda^{\mathcal{U}}$ in $\mathcal{H}_K^{-\infty}(M)$ does not depend of $\chi^{\mathcal{U}}$. In particular, $\mathrm{P}_\lambda^{\mathcal{U}} = 0$ in $\mathcal{H}_K^{-\infty}(M)$ if $\mathcal{U} \cap \Phi_\lambda^{-1}(0) = \emptyset$. If $\mathcal{U} \cap \Phi_\lambda^{-1}(0)$ is compact, we take $\chi^{\mathcal{U}}$ with compact support. Then $\mathrm{P}_\lambda^{\mathcal{U}}$ defines a cohomology class in $\mathcal{H}_{K,c}^{-\infty}(M)$.

2.2 *Localization of* DH(M)

We return to the situation of a Hamiltonian action of a torus T on a symplectic manifold (M,ω). We need two auxiliary data: a T-invariant Riemannian metric on M, denoted by $(\cdot,\cdot)_M$, and a scalar product $(\cdot,\cdot)$ on $\mathfrak{t}^*$ which induces an identification $\mathfrak{t}^* \simeq \mathfrak{t}$.

Let $\mathcal{H}$ be the Hamiltonian vector field of the function $\frac{-1}{2}\|\Phi\|^2 : M \to \mathbb{R}$. For $m \in M$ we have $\mathcal{H}_m = (\Phi(m))_M|_m$. Then for every $\xi \in \mathfrak{t}^*$, the Hamiltonian vector field of $\frac{-1}{2}\|\Phi - \xi\|^2$ is $\mathcal{H} - \xi_M$, and we consider the T-invariant 1-form

$$\lambda_\xi = (\mathcal{H} - \xi_M, \cdot)_M \tag{2.5}$$

with corresponding map $\Phi_{\lambda_\xi} : M \to \mathfrak{t}^*$ (see (2.2)). Here $\Phi_{\lambda_\xi}^{-1}(0)$ coincides with the subset $\mathrm{Cr}(\|\Phi - \xi\|^2) \subset M$ of critical points of the function $\|\Phi - \xi\|^2$, and $m \in \mathrm{Cr}(\|\Phi - \xi\|^2)$ if and only if $(\Phi(m) - \xi)_M$ vanishes at m [30, 31].

Definition 2.1. Let $\mathrm{P}_\xi \in \mathcal{H}_{T,c}^{-\infty}(M)$ be the cohomology class defined by $\mathrm{P}_{\lambda_\xi}^{\mathcal{U}}$, where $\mathcal{U}$ is a T-invariant relatively compact neighborhood of $\Phi^{-1}(\xi)$ such that $\overline{\mathcal{U}} \cap \mathrm{Cr}(\|\Phi - \xi\|^2) = \Phi^{-1}(\xi)$.

The cohomology class P_ξ will be used to localize the Duistermaat–Heckman measure. For every $\xi \in \mathfrak{t}^*$, we define the distribution $\mathrm{DH}_\xi(M)$ by

$$\mathrm{DH}_\xi(M) = (i)^n \mathcal{F}\left(\int_M \mathrm{P}_\xi e^{-i\Omega_{\mathfrak{t}}}\right). \tag{2.6}$$

Here we can put the Fourier transform outside the integral because P_ξ is compactly supported on M. For any $\xi \in \mathfrak{t}^*$, let $r_\xi > 0$ be the smallest nonzero critical value of the function $\|\Phi - \xi\|^2$. As a particular case of Proposition 3.8 in [31], we have

Proposition 2.2. *Let ξ be any point in $\mathfrak{t}^*$. The equality*

$$\mathrm{DH}(M) = \mathrm{DH}_\xi(M)$$

of distributions on $\mathfrak{t}^*$ holds in the open ball $B(\xi, r_\xi) \subset \mathfrak{t}^*$.

We will now use this proposition, first to recover the classical result of Duistermaat and Heckman [15] concerning the polynomial behavior of $\mathrm{DH}(M)$ on the open subset of regular values of Φ and then to determine the difference taken by $\mathrm{DH}(M)$ between two adjacent regions of regular values.

2.3 Polynomial behavior

We recall now the computation of the cohomology class P_ξ when ξ is a regular value of Φ, which is given in [30, Section 6] for the torus case (see [31, Section 3.1] for the case of the Hamiltonian action of a compact Lie group). First recall the following basic result, which shows that $\xi \mapsto \mathrm{DH}_\xi(M)$ is locally constant on the open subset of regular values of Φ.

Lemma 2.3 ([33]). *If ξ and ξ' belong to the same connected component of regular values of Φ, we have $\mathrm{P}_\xi = \mathrm{P}_{\xi'}$ in $\mathcal{H}^{-\infty}_{T,c}(M)$.*

If we combine Lemma 2.3 with Proposition 2.2, we see that for any connected component $\mathfrak{c}$ of regular values of Φ, we have

$$\mathrm{DH}(M)(a) = \mathrm{DH}_\xi(M)(a), \quad a \in \mathfrak{c},$$

for any $\xi \in \mathfrak{c}$. We have to compute $\mathrm{DH}_\xi(M)$ when ξ is a regular value of Φ.

We consider the T-principal bundle $\Phi^{-1}(\xi) \to \mathcal{M}_\xi := \Phi^{-1}(\xi)/T$ with curvature form $\omega_\xi \in \mathcal{H}^2(\mathcal{M}_\xi) \otimes \mathfrak{t}$. The orbifold $\mathcal{M}_\xi$ carries a canonical symplectic 2-form Ω_ξ. We denote by

$$\mathbf{Kir}_\xi : \mathcal{H}^\infty_T(M) \to \mathcal{H}^*(\mathcal{M}_\xi)$$

the Kirwan morphism. For any $\psi \in \mathcal{C}^\infty(\mathfrak{t})$ and $\eta \in \mathcal{H}^\infty_T(M)$ we have $\mathbf{Kir}_\xi(\eta\psi) = \mathbf{Kir}_\xi(\eta)\psi(\omega_\xi)$, where the characteristic class $\psi(\omega_\xi)$ is the value of the differential operator $e^{\omega_\xi(\frac{\partial}{\partial X}|0)}$ against ψ. By [31, Proposition 3.11], we know that the integral

$$\int_\mathfrak{t}\int_M \mathrm{P}_\xi(X)\eta(X)\psi(X)dX$$

is equal to

$$\frac{(-2i\pi)^{\dim T}\mathrm{vol}(T,dX)}{|\Gamma_M|}\int_{\mathcal{M}_\xi}\mathbf{Kir}_\xi(\eta)\psi(\omega_\xi) \tag{2.7}$$

for every equivariant class $\eta \in \mathcal{H}_T^\infty(M)$. Here $\mathrm{vol}(T,dX)$ is the volume of T for the Haar measure compatible with dX, and $|\Gamma_M|$ is the cardinal of Γ_M (note that the generic stabilizer of T on $\Phi^{-1}(\xi)$ is Γ_M). In other words, for every $\eta \in \mathcal{H}_T^\infty(M)$ we have the following equality of generalized functions on $\mathfrak{t}^*$ supported at 0:

$$\int_M \mathrm{P}_\xi(X)\eta(X) = \frac{(-2i\pi)^{\dim T}}{|\Gamma_M|}\int_{\mathcal{M}_\xi}\mathbf{Kir}_\xi(\eta)e^{\omega_\xi(\frac{\partial}{\partial X}|0)}\mathrm{vol}(T,-). \tag{2.8}$$

For $\eta = e^{-i\Omega_\mathfrak{t}}$ we have $\mathbf{Kir}_\xi(\eta) = e^{-i(\Omega_\xi-\langle\xi,\omega_\xi\rangle)}$, and a small computation shows that

$$\mathcal{F}\big(e^{\omega_\xi(\frac{\partial}{\partial X}|0)}\mathrm{vol}(T,-)\big)(a) = e^{-i\langle a,\omega_\xi\rangle}\frac{da}{(2\pi)^{\dim T}}, \quad a\in\mathfrak{t}^*. \tag{2.9}$$

Here da is the Lebesgue measure on $\mathfrak{t}^*$ normalized by the condition $\mathrm{vol}(T,dX)=1$ for the Lebesgue measure dX on $\mathfrak{t}$, which is dual to da.

Finally (2.6), (2.8), and (2.9) give

$$\begin{aligned}\mathrm{DH}_\xi(M)(a) &= \frac{(i)^d}{|\Gamma_M|}\int_{\mathcal{M}_\xi} e^{-i(\Omega_\xi+\langle a-\xi,\omega_\xi\rangle)}\,da\\ &= \frac{1}{|\Gamma_M|}\int_{\mathcal{M}_\xi}\frac{(\Omega_\xi+\langle a-\xi,\omega_\xi\rangle)^d}{d!}\,da,\end{aligned} \tag{2.10}$$

where $2d = \dim\mathcal{M}_\xi$.

Definition 2.4. For any connected component $\mathfrak{c}$ of regular values of Φ we denote by $\mathrm{DH}_\mathfrak{c}$ the polynomial function $a \mapsto \frac{1}{|\Gamma_M|}\int_{\mathcal{M}_\xi}\frac{(\Omega_\xi+\langle a-\xi,\omega_\xi\rangle)^d}{d!}$, where ξ is any point of $\mathfrak{c}$.

With the help of Proposition 2.2 we recover the classical result of Duistermaat and Heckman [15] stating that the measure $\mathrm{DH}(M)$ is locally polynomial[3] on the open subset of regular values of Φ, and its value at a regular element ξ is equal to the symplectic volume of the reduced space $\mathcal{M}_\xi$ (times $|\Gamma_M|^{-1}$). More precisely, we have shown that for a connected component $\mathfrak{c}$ of regular values of Φ we have

$$\mathrm{DH}(M)(a) = \mathrm{DH}_\mathfrak{c}(a)da, \quad a\in\mathfrak{c}. \tag{2.11}$$

[3] It is a polynomial times a Lebesgue measure on $\mathfrak{t}^*$.

2.4 Wall-crossing formulas

Consider now two connected regions $\mathfrak{c}_\pm$ of regular values of Φ separated by a hyperplane $\Delta \subset \mathfrak{t}^*$. In this section we compute the polynomial $\mathrm{DH}_{\mathfrak{c}_+} - \mathrm{DH}_{\mathfrak{c}_-}$. It generalizes previous results of Guillemin–Lerman–Sternberg [17] and Brion–Procesi [12].

Let ξ_+, ξ_- be respectively two elements of $\mathfrak{c}_+$ and $\mathfrak{c}_-$. We know from (2.2), (2.10), and Definition 2.4 that

$$(\mathrm{DH}_{\mathfrak{c}_+} - \mathrm{DH}_{\mathfrak{c}_-})(a)da = (i)^n \mathcal{F}\left(\int_M (\mathrm{P}_{\xi_+} - \mathrm{P}_{\xi_-})e^{-i\Omega_{\mathfrak{t}}}\right)(a), \quad a \in \mathfrak{t}^*. \tag{2.12}$$

We recall now the computation of the cohomology class $\mathrm{P}_{\xi_+} - \mathrm{P}_{\xi_-} \in \mathcal{H}^{-\infty}_{T,c}(M)$ done in [33]. We use the notation defined in the introduction.

Definition 2.5. We denote by M^Δ the union of the connected component Z of the fixed point set M^{T_Δ}, for which we have $\Phi(Z) \subset \Delta$. Let M^Δ_o be the T-invariant open subset of M^Δ where T/T_Δ acts locally freely.

For a connected component $Z \subset M^\Delta$, one has either $\overline{\mathfrak{c}_+} \cap \overline{\mathfrak{c}_-} \subset \Phi(Z)$ or $\overline{\mathfrak{c}_+} \cap \overline{\mathfrak{c}_-} \cap \Phi(Z) = \emptyset$, due to the fact that for any ξ in relative interior of $\overline{\mathfrak{c}_+} \cap \overline{\mathfrak{c}_-}$ in Δ, and any $m \in \Phi^{-1}(\xi)$, the stabilizer $\mathfrak{t}_m \subset \mathfrak{t}$ is either equal to $\mathfrak{t}_\Delta$ or reduced to $\{0\}$.

The symplectic manifold M^Δ carries a Hamiltonian action of T/T_Δ with moment map $\Phi|_{M^\Delta} : M^\Delta \to \Delta$ equal to the restriction of Φ on M^Δ.

Let ξ be a point in the relative interior of $\overline{\mathfrak{c}_+} \cap \overline{\mathfrak{c}_-}$ in Δ. From the previous discussion, we know that ξ is a regular value of $\Phi|_{M^\Delta}$, i.e., $\Phi^{-1}(\xi) \cap M^{T_\Delta}$ is a submanifold of M^Δ_o. Following Definition 2.1 we associate to ξ the cohomology class

$$\mathrm{P}^\Delta_\xi \in \mathcal{H}^{-\infty}_{T/T_\Delta,c}(M^\Delta_o).$$

Let $\mathcal{H}^*(M^\Delta_o)^{\mathrm{bas}}$ be the subalgebra of $\mathcal{H}^*(M^\Delta_o)$ formed by the T-basic elements. Since the T_Δ-action on M^Δ_o is trivial, we have a canonical product operation

$$\mathcal{H}^{-\infty}_{T/T_\Delta,c}(M^\Delta_o) \times \mathcal{C}^{-\infty}(\mathfrak{t}_\Delta, \mathcal{H}^*(M^\Delta_o)^{\mathrm{bas}}) \xrightarrow{\wedge} \mathcal{H}^{-\infty}_{T,c}(M^\Delta_o). \tag{2.13}$$

Proposition 2.6 ([33]). *There exists a generalized function supported at 0,* $\delta^\Delta \in \mathcal{C}^{-\infty}(\mathfrak{t}_\Delta, \mathcal{H}^*(M^\Delta_o)^{\mathrm{bas}})$, *such that*

$$\mathrm{P}_{\xi+} - \mathrm{P}_{\xi-} = (i_\Delta)_*(\mathrm{P}^\Delta_\xi \wedge \delta^\Delta) \quad \text{in} \quad \mathcal{H}^{-\infty}_{T,c}(M).$$

Here $(i_\Delta)_* : \mathcal{H}^{-\infty}_{T,c}(M^\Delta_o) \to \mathcal{H}^{-\infty}_{T,c}(M)$ *is the direct image map relative to the inclusion* $i_\Delta : M^\Delta_o \hookrightarrow M$.

We will now give the precise definition of δ^Δ. The decomposition $T = T_\Delta \times T/T_\Delta$ and the trivial action of T_Δ on M^Δ_o give a canonical isomorphism

$$j_\Delta : \mathcal{H}^*_T(M^\Delta_o) \xrightarrow{\sim} \mathcal{S}(\mathfrak{t}^*_\Delta) \otimes \mathcal{H}^*_{T/T_\Delta}(M^\Delta_o),$$

where $\mathcal{S}(\mathfrak{t}_\Delta^*)$ is the algebra of complex polynomial functions on $\mathfrak{t}_\Delta$. Since the T/T_Δ-action on M_o^Δ is locally free, we have the Chern–Weil isomorphism

$$\mathbf{cv}_\Delta : \mathcal{H}^*_{T/T_\Delta}(M_o^\Delta) \xrightarrow{\sim} \mathcal{H}^*(M_o^\Delta)^{\mathrm{bas}}.$$

Let N_Δ be the T-equivariant normal bundle of M^Δ in M, and let

$$\mathrm{Eul}(N_\Delta) \in \mathcal{H}^*_T(M^\Delta)$$

be the T-equivariant Euler class of N_Δ. Now we consider the restriction of $\mathrm{Eul}(N_\Delta)$ on the open subset $M_o^\Delta \subset M^\Delta$, which we see through the isomorphism $\mathbf{cv}_\Delta \circ j_\Delta$ as an element of $\mathcal{S}(\mathfrak{t}_\Delta^*) \otimes \mathcal{H}^*(M_o^\Delta)^{\mathrm{bas}}$ (for simplicity we keep the same notations $\mathrm{Eul}(N_\Delta)$ for this element). Following [30], we define inverses $\mathrm{Eul}^{-1}_{\pm\beta}(N_\Delta) \in \mathcal{C}^{-\infty}(\mathfrak{t}_\Delta, \mathcal{H}^*(M_o^\Delta)^{\mathrm{bas}})$ by

$$\mathrm{Eul}^{-1}_{\pm\beta}(N_\Delta)(X) = \lim_{s\to+\infty} \frac{1}{\mathrm{Eul}(N_\Delta)(X \pm is\beta)}. \tag{2.14}$$

Here $\beta \in \mathfrak{t}_\Delta$ is chosen so that $\langle \xi^+ - \xi^-, \beta\rangle > 0$.

Definition 2.7. The generalized function $\delta^\Delta \in \mathcal{C}^{-\infty}(\mathfrak{t}_\Delta, \mathcal{H}^*(M_o^\Delta)^{\mathrm{bas}})$ is defined by

$$\delta^\Delta := \mathrm{Eul}^{-1}_{\beta}(N_\Delta) - \mathrm{Eul}^{-1}_{-\beta}(N_\Delta). \tag{2.15}$$

Since the polynomial $\mathrm{Eul}(N_\Delta)$ is invertible in a smooth manner on $\mathfrak{t}_\Delta\backslash\{0\}$, the generalized function δ^Δ is supported at 0.

Let ξ be a point in the relative interior of $\overline{\mathfrak{c}_+} \cap \overline{\mathfrak{c}_-}$ in Δ. We consider the symplectic reduction

$$\mathcal{M}_\xi^\Delta := (M^\Delta \cap \Phi^{-1}(\xi))/(T/T_\Delta).$$

If we restrict δ^Δ to the submanifold $M^\Delta \cap \Phi^{-1}(\xi)$ we get the generalized function

$$\delta_\xi^\Delta \in \mathcal{C}^{-\infty}(\mathfrak{t}_\Delta, \mathcal{H}^*(\mathcal{M}_\xi^\Delta)).$$

Now we are able to compute the right-hand side of (2.12). Let $\omega_\xi^\Delta \in \mathcal{H}^2(\mathcal{M}_\xi^\Delta) \otimes \mathfrak{t}/\mathfrak{t}_\Delta$ be the curvature of the T/T_Δ-principal bundle $M^\Delta \cap \Phi^{-1}(\xi) \to \mathcal{M}_\xi^\Delta$. Let $|S_\xi^\Delta|$ be locally constant function on $M^\Delta \cap \Phi^{-1}(\xi)$, which is equal to the cardinal of the generic stabilizer of T/T_Δ. Let $l = \dim T$. From (2.8) and Proposition 2.6 we have

$$\begin{aligned}
&\int_M (\mathrm{P}_{\xi_+} - \mathrm{P}_{\xi_-})(X) e^{-i\Omega_{\mathfrak{t}}(X)} \\
&\quad = \int_{M_o^\Delta} \mathrm{P}_\xi^\Delta(X')\delta^\Delta(X'') e^{-i\Omega_{\mathfrak{t}}(X'+X'')} \\
&\quad = \frac{(-2i\pi)^{l-1}}{|S_\xi^\Delta|} \int_{\mathcal{M}_\xi^\Delta} e^{\omega_\xi^\Delta(\frac{\partial}{\partial X'}|0)} \mathrm{vol}(T/T_\Delta, -)\mathbf{Kir}_\xi^\Delta(e^{-i\Omega_{\mathfrak{t}}})(X'')\delta_\xi^\Delta(X''). \quad (2.16)
\end{aligned}$$

In the last equation the notation is the following:

1. $X = X' + X''$ with $X' \in \mathfrak{t}/\mathfrak{t}_\Delta$ and $X'' \in \mathfrak{t}_\Delta$,
2. the Kirwan map $\mathbf{Kir}_\xi^\Delta : \mathcal{H}_T^\infty(M) \to \mathcal{C}^\infty(\mathfrak{t}_\Delta, \mathcal{H}^*(\mathcal{M}_\xi^\Delta))$ is the composition of the restriction $\mathcal{H}_T^\infty(M) \to \mathcal{H}_T^\infty(M^\Delta \cap \Phi^{-1}(\xi))$ with the Chern–Weil isomorphism $\mathcal{H}_T^\infty(M^\Delta \cap \Phi^{-1}(\xi)) \xrightarrow{\sim} \mathcal{C}^\infty(\mathfrak{t}_\Delta, \mathcal{H}^*(\mathcal{M}_\xi^\Delta))$.

A direct computation gives that $\mathbf{Kir}_\xi^\Delta(\Omega_\mathfrak{t})(X'') = \Omega_\xi^\Delta - \langle \xi, \omega_\xi^\Delta + X'' \rangle$, where Ω_ξ^Δ is the induced symplectic form on the reduced space $\mathcal{M}_\xi^\Delta$. If we take the Fourier transform in (2.16) we get

$$
\begin{aligned}
&(\mathrm{DH}_{\mathfrak{c}_+} - \mathrm{DH}_{\mathfrak{c}_-})(a)da \\
&= \frac{(i)^{n+1-l}}{|S_\xi^\Delta|} \left(\int_{\mathcal{M}_\xi^\Delta} e^{-i(\Omega_\xi^\Delta + \langle a', \omega_\xi^\Delta \rangle)} \, da' \, \mathcal{F}_{\mathfrak{t}_\Delta}(\delta_\xi^\Delta)(a'') \right)(a - \xi) \\
&= \sum_{Z \in \mathcal{F}} \frac{(i)^{n+1-l}}{|S_\xi^Z|} \left(\int_{\mathcal{Z}_\xi} e^{-i(\Omega_\xi^Z + \langle a', \omega_\xi^Z \rangle)} \, da' \, \mathcal{F}_{\mathfrak{t}_\Delta}(\delta_\xi^Z)(a'') \right)(a - \xi),
\end{aligned}
\tag{2.17}
$$

where $a = a' + a''$ with $a' \in (\mathfrak{t}/\mathfrak{t}_\Delta)^*$ and $a'' \in (\mathfrak{t}_\Delta)^*$. In (2.17), we write $\int_{\mathcal{M}_\xi^\Delta} = \sum_{Z \in \mathcal{F}} \int_{\mathcal{Z}_\xi}$, where the sum is taken over the set $\mathcal{F}$ of connected components Z of M^Δ that intersect $\Phi^{-1}(\xi)$. We take then

$$\mathcal{Z}_\xi = (Z \cap \Phi^{-1}(\xi))/(T/T_\Delta).$$

The 2-forms Ω_ξ^Δ, ω_ξ^Δ, the generic stabilizer S_ξ^Δ, the vector bundle N_Δ, and the generalized function δ_ξ^Δ restrict to each component Z: we denote them respectively by $\Omega_\xi^Z, \omega_\xi^Z, S_\xi^Z, N_Z, \delta_\xi^Z$.

We recall now the computation of the Fourier transform of the inverses $\mathrm{Eul}_{\pm\beta}^{-1}(N_Z) := \mathrm{Eul}_{\pm\beta}^{-1}(N_\Delta)|_Z$ that is given in [30, Proposition 4.8]. We consider a T-invariant scalar product on the fibers of the bundle N_Δ. Let $R \in \mathcal{A}^2(M_o^\Delta, \mathrm{so}(N_\Delta))^{\mathrm{bas}}$ be the curvature of a T-invariant and T/T_Δ-horizontal Euclidean connexion on N_Δ: we denote by $R^Z \in \mathcal{A}^2(Z, \mathrm{so}(N_Z))^{\mathrm{bas}}$ the restriction of R to a component $Z \in \mathcal{F}$. The curvature commutes with the infinitesimal action $\mathcal{L}_X$ of $X \in \mathfrak{t}_\Delta$ and with the complex structure $J_\beta = \mathcal{L}_\beta(-\mathcal{L}_\beta^2)^{1/2}$ on N_Δ defined by $\beta \in \mathfrak{t}_\Delta$.

We denote by $S^\bullet$ the symmetric algebra of the complex vector bundle (N_Δ, J_β). We keep the same notation for the restriction of $S^\bullet$ on the submanifolds $Z, \Phi^{-1}(\xi) \cap M^\Delta$, and for the induced orbifold vector bundle on the reduced spaces $\mathcal{Z}_\xi$ and $\mathcal{M}_\xi^\Delta$. For each $k \in \mathbb{N}$, we denote by $\mathbf{Tr}_{S^k}$ the trace operator defined on the complex endomorphisms of S^k. For a complex endomorphism A of N_Δ, we denote by $A^{\otimes k}$ the induced endomorphism on S^k. For any $X \in \mathfrak{t}_\Delta$, the complex endomorphism $\mathcal{L}_X^{-1} R^Z$ is symmetric. Hence the trace $\mathbf{Tr}_{S^k}((\mathcal{L}_X^{-1} R^Z)^{\otimes k})$ is a basic *real* differential form of degree $2k$ on Z which does not depend of the choice of complex structures (J_β or $J_{-\beta}$).

Let $\beta^* \in \mathfrak{t}_\Delta^*$ the dual of $\beta \in \mathfrak{t}_\Delta$.

Proposition 2.8 ([30]). *For a smooth function f on $\mathfrak{t}_\Delta^*$ with compact support we have $\int_{\mathfrak{t}_\Delta^*} \mathcal{F}_{\mathfrak{t}_\Delta}(\mathrm{Eul}_\beta^{-1}(N_Z))(a'')f(a'') = \int_0^\infty P_Z(t)f(t\beta^*)dt$, where P_Z is the polynomial on $\mathbb{R}$ defined by*

$$P_Z(t) = \frac{(2\pi i)^{r_Z}}{\det_Z^{1/2}(\mathcal{L}_\beta)} \left(\frac{t^{r_Z-1}}{(r_Z-1)!} + \sum_{k=1}^{\dim(Z)/2} (i)^k \alpha_k \frac{t^{r_Z-1+k}}{(r_Z-1+k)!} \right). \tag{2.18}$$

Here $\alpha_k = \mathbf{Tr}_{S^k}((\mathcal{L}_\beta^{-1} R^Z)^{\otimes k}) \in \mathcal{A}^{2k}(Z)^{\mathrm{bas}}$, $\det_Z^{1/2}(\mathcal{L}_\beta)$ is the Pfaffian of $\mathcal{L}_\beta$ on N_Z, and $r_Z = \mathrm{rk}_{\mathbb{C}}(N_Z)$.

One checks then that

$$\begin{aligned}\int_{\mathfrak{t}_\Delta^*} \mathcal{F}_{\mathfrak{t}_\Delta}(\mathrm{Eul}_{-\beta}^{-1}(N_Z))(a'')f(a'') &= \int_0^\infty -P_Z(-t)f(-t\beta^*)dt \\ &= -\int_{-\infty}^0 P_Z(t)f(t\beta^*)dt.\end{aligned}$$

Hence the distribution $\mathcal{F}_{\mathfrak{t}_\Delta}(\delta^Z)$ is equal to $P_Z(\beta)d\beta$. From now on we fix β as the primitive element of $\mathfrak{t}_\Delta \cap \Lambda$ which is pointing out $\mathfrak{c}_-$. Then $d\beta$ and $d\beta^*$ are (dual) Lebesgue measures on $\mathfrak{t}^*$ and $\mathfrak{t}$. We have $\mathrm{vol}(T_\Delta, d\beta^*) = 1$.

Let R_ξ^Z be the restriction of the curvature R^Z to the submanifold $Z \cap \Phi^{-1}(\xi)$. Since R^Z is T/T_Δ-basic, $\mathbf{Tr}_{S^k}((\mathcal{L}_\beta^{-1} R_\xi^Z)^{\otimes k})$ can be seen as a real differential form of degree $2k$ on the orbifold $\mathcal{Z}_\xi = (Z \cap \Phi^{-1}(\xi))/(T/T_\Delta)$.

Each connected component Z of M^Δ is a T/T_Δ Hamiltonian manifold: we take for moment map $\Phi_Z : Z \to (\mathfrak{t}/\mathfrak{t}_\Delta)^*$ the restriction of $\Phi - \xi$ to Z. Hence 0 is a regular value of Φ_Z. Let $\mathrm{DH}_0(Z)$ be the polynomial function on $(\mathfrak{t}/\mathfrak{t}_\Delta)^* = \{a \in \mathfrak{t}^* | \langle \beta, a\rangle = 0\}$ such that $\mathrm{DH}(Z)(a') = \mathrm{DH}_0(Z)(a')da'$ near 0. Finally, (2.17) together with the Proposition 2.8 gives the following theorem.

Theorem 2.9. *We have $(\mathrm{DH}_{\mathfrak{c}_+} - \mathrm{DH}_{\mathfrak{c}_-})(a) = \sum_{Z\in\mathcal{F}} \mathbf{D}_Z(a-\xi)$, $a \in \mathfrak{t}^*$, where each polynomial $\mathbf{D}_Z \in \mathcal{S}(\mathfrak{t})$ admits the following decomposition:*

$$\mathbf{D}_Z = \frac{\beta^{r_Z-1}}{\det_Z^{1/2}\left(\frac{-\mathcal{L}_\beta}{2\pi}\right)} \left(\frac{\mathrm{DH}_0(Z)}{(r_Z-1)!} + \sum_{k=1}^{d_Z} \beta^k \mathrm{Q}_{Z,k} \right). \tag{2.19}$$

The polynomials $\mathrm{Q}_{Z,k} \in \mathcal{S}(\mathfrak{t}/\mathfrak{t}_\Delta)$ are defined by

$$\mathrm{Q}_{Z,k}(a') = \frac{(-1)^k}{(r_Z-1+k)!|S_\xi^Z|} \int_{\mathcal{Z}_\xi} \frac{(\Omega_\xi^Z + \langle a', \omega_\xi^Z\rangle)^{d_Z-k}}{(d_Z-k)!} \mathbf{Tr}_{S^k}((\mathcal{L}_\beta^{-1} R_\xi^Z)^{\otimes k}).$$

Here $2d_Z = \dim \mathcal{Z}_\xi$ and $2r_Z = \dim M - \dim Z$.

Remark 2.10.

- The polynomial $\mathrm{DH}_{\mathfrak{c}_+} - \mathrm{DH}_{\mathfrak{c}_-}$ is divisible by the factor $a \mapsto \langle a - \xi, \beta \rangle^{r-1}$ with $r = \inf_{Z \in \mathcal{F}} r_Z$. If $\Delta \cap \Phi(M)$ is not a facet of the polytope $\Phi(M)$ we have $r_Z \geq 2$ for all connected components $Z \in \mathcal{F}$; hence $r - 1 \geq 1$.
- Suppose now that $\mathfrak{c}_+$ is a connected component of regular values of Φ at the edge of the polytope $\Phi(M)$. Then $\Phi(M) \cap \Delta$ is a facet of the polytope $\Phi(M)$. Here $Z = \Phi^{-1}(\Delta)$ is a connected component of the fixed-point set M^{T_Δ}. In this situation we have $\mathrm{DH}_{\mathfrak{c}_+} = \mathbf{D}_Z$, where the polynomial $\mathbf{D}_Z$ is defined by (2.19).

3 Quantum version of Duistermaat–Heckman measures

We suppose here that the Hamiltonian T-manifold (M, ω, Φ) is prequantized by a T-equivariant Hermitian line bundle L over M, which is equipped with a Hermitian connection ∇ satisfying the Kostant formula

$$\mathcal{L}(X) - \nabla_{X_M} = i \langle \Phi, X \rangle, \qquad X \in \mathfrak{t}. \tag{3.1}$$

The former equation implies that the first Chern class of L is equal to $\left[\frac{\Omega}{2\pi}\right]$. In this section we suppose that M is *compact* and we still assume that the generic stabilizer Γ_M of T on M is *finite*. The quantization of (M, Ω) is defined as the Riemann–Roch character $\mathrm{RR}(M, L) \in R(T)$, which is computed with a T-equivariant almost complex structure on M *compatible* with Ω [32]. For $k \geq 1$, we consider the tensor product $L^{\otimes k}$. Its Riemann–Roch character $\mathrm{RR}(M, L^{\otimes k})$ decomposes as

$$\mathrm{RR}(M, L^{\otimes k}) = \sum_{\mu \in \Lambda^*} \mathrm{m}(\mu, k)\, \mathbb{C}_\mu. \tag{3.2}$$

Let us recall the well-known properties of the map $\mathrm{m}\colon \Lambda^* \times \mathbb{Z}^{>0} \to \mathbb{Z}$. When $\frac{\mu}{k}$ is a regular value of Φ, the "quantization commutes with reduction" theorem [28, 29] tells us that

$$\mathrm{m}(\mu, k) = \mathrm{RR}(\mathcal{M}_{\frac{\mu}{k}}, \mathcal{L}^{\mu,k}), \tag{3.3}$$

where $\mathcal{L}^{\mu,k} = (L^{\otimes k}|_{\Phi^{-1}(\frac{\mu}{k})} \otimes \mathbb{C}_{-\mu})/T$ is an orbifold line bundle over the symplectic orbifold $\mathcal{M}_{\frac{\mu}{k}} = \Phi^{-1}\left(\frac{\mu}{k}\right)/T$. In particular, if $\frac{\mu}{k}$ does not belong to $\Phi(M)$ we have $\mathrm{m}(\mu, k) = 0$. When $\frac{\mu}{k} \in \Phi(M)$ is not necessarily a regular value of Φ, one proceeds by shift desingularization. If $\xi \in \Phi(M)$ is a regular value of Φ close enough to $\frac{\mu}{k}$ then (3.3) becomes

$$\mathrm{m}(\mu, k) = \mathrm{RR}(\mathcal{M}_\xi, \mathcal{L}_\xi^{\mu,k}), \tag{3.4}$$

where $\mathcal{L}_\xi^{\mu,k} = (L^{\otimes k}|_{\Phi^{-1}(\xi)} \otimes \mathbb{C}_{-\mu})/T$ (for a proof see [29, 32]).

Definition 3.1. A function $f : \Xi \to \mathbb{Z}$ defined over a lattice $\Xi \simeq \mathbb{Z}^r$ is called *periodic polynomial* if

$$f(x) = \sum_{i=1}^{p} e^{i\frac{\langle \alpha_j, x\rangle}{N}} P_j(x), \quad x \in \Xi,$$

where $\alpha_1, \ldots, \alpha_p \in \Xi^*$, $N \geq 1$, and the functions $P_1, \ldots, P_p$ are polynomials with complex coefficients.

Remark 3.2. Let $\mathcal{C}$ be a cone with *nonempty* interior in the real vector space $\Xi \otimes_{\mathbb{Z}} \mathbb{R}$. Any *periodic-polynomial* function $f : \Xi \to \mathbb{Z}$ is completely determined by its restriction on $\mathcal{C} \cap \Xi$.

Let $\mathfrak{c} \subset \mathfrak{t}^*$ be a connected component of regular values of Φ. In [29] Meinrenken and Sjamaar proved that there exists a periodic polynomial function $\mathrm{m}_{\mathfrak{c}} : \Lambda^* \times \mathbb{Z} \to \mathbb{Z}$ such that $\mathrm{m}_{\mathfrak{c}}(\mu, k) = \mathrm{m}(\mu, k)$ for every (μ, k) in the cone

$$\mathrm{Cone}(\mathfrak{c}) = \{(\xi, s) \in \mathfrak{t}^* \times \mathbb{R}^{>0} \,|\, \xi \in s \cdot \mathfrak{c}\}. \tag{3.5}$$

Consider now two adjacent connected regions $\mathfrak{c}_\pm$ of regular values of Φ separated by a hyperplane $\Delta \subset \mathfrak{t}^*$. When Δ does not contain a facet of the polytope $\Phi(M)$, Meinrenken and Sjamaar proved also that

$$\mathrm{m}_{\mathfrak{c}_+}(\mu, k) = \mathrm{m}_{\mathfrak{c}_-}(\mu, k) = \mathrm{m}(\mu, k) \tag{3.6}$$

for every $(\mu, k) \in \overline{\mathrm{Cone}(\mathfrak{c}_+)} \cap \overline{\mathrm{Cone}(\mathfrak{c}_-)} = \mathrm{Cone}(\overline{\mathfrak{c}_+} \cap \overline{\mathfrak{c}_-}) \subset \mathrm{Cone}(\Delta)$.

The main purpose of this section is to prove that (3.6) extends to a "strip" containing $\mathrm{Cone}(\Delta)$.

Let $\beta \in \Lambda$ be the primitive orthogonal vector to the hyperplane $\Delta \subset \mathfrak{t}^*$ which is pointing out of $\mathfrak{c}_-$. Then $\Delta = \{\xi \in \mathfrak{t}^* | \frac{\langle \xi, \beta\rangle}{2\pi} = r_\Delta\}$ for some $r_\Delta \in \mathbb{Z}$, $\mathrm{Cone}(\Delta) = \{(\xi, s) \in \mathfrak{t}^* \times \mathbb{R}^{\geq 0} | \frac{\langle \xi, \beta\rangle}{2\pi} - s r_\Delta = 0\}$, and $\mathfrak{c}_- \subset \{\xi \in \mathfrak{t}^* | \frac{\langle \xi, \beta\rangle}{2\pi} < r_\Delta\}$.

Let T_Δ be the subtorus of T generated by β. Let N_Δ be the normal vector bundle of M^{T_Δ} in M. The almost complex structure on M induces a complex structure J on the fibers of N_Δ. We have a decomposition $N_\Delta = \sum_s N_\Delta^s$, where $N_\Delta^s = \{v \in N_\Delta | \mathcal{L}_\beta v = s\, J v\}$. We write $N_\Delta = N_\Delta^{+,\beta} \oplus N_\Delta^{-,\beta}$, where

$$N_\Delta^{\pm,\beta} = \sum_{\pm s > 0} N_\Delta^s. \tag{3.7}$$

Definition 3.3. For every connected component $Z \subset M^{T_\Delta}$ we define $s_Z^\pm \in \mathbb{N}$ respectively as the absolute value of the trace of $\frac{1}{2\pi}\mathcal{L}_\beta$ on $N_\Delta^{\pm,\beta}|_Z$.

Note that $s_Z^+ + s_Z^-$ is larger than half of the codimension of Z in M. We prove in Section 3.5 the following theorem.

Theorem 3.4. *We have* $\mathrm{m}_{\mathfrak{c}_+}(\mu, k) = \mathrm{m}_{\mathfrak{c}_-}(\mu, k)$ *for all* $(\mu, k) \in \Lambda^* \times \mathbb{Z}$ *such that*

$$-s^- < \frac{\langle \mu, \beta\rangle}{2\pi} - k r_\Delta < s^+ . \tag{3.8}$$

The numbers $s^-, s^+ \in \mathbb{N}$ are defined as follows. We take $s^\pm = \inf_Z s_Z^\pm$, where the minimum is taken over the connected components Z of M^{T_Δ} for which $\overline{\mathfrak{c}_+} \cap \overline{\mathfrak{c}_-} \subset \Phi(Z)$.

Similar results were obtained by Billey–Guillemin–Rassart [10] in the case of M a coadjoint orbit of $\mathrm{SU}(n)$, and by Szenes–Vergne [36] in the case of M a complex vector space. See Sections 4.4 and 5, where we study these two particular cases in detail. In Proposition 3.25, we give also a criterion which says when the inequalities in (3.8) are optimal. This criterion is satisfied when there is *only one* component Z of M^{T_Δ} such that $\overline{\mathfrak{c}_+} \cap \overline{\mathfrak{c}_-} \subset \Phi(Z)$. Then (3.8) is *optimal* and $s^+ + s^-$ is larger than half of the codimension of Z in M.

The following easy lemma (see Lemma 7.3. of [32]) gives some basic information about the integer $s_Z^\pm$.

Lemma 3.5. *Let (M, Ω, Φ) be a compact Hamiltonian T-manifold equipped with a T-invariant almost complex structure compatible with Ω. Consider a nonzero vector $\gamma \in \mathfrak{t}$ and let Z be a connected component of the fixed-point set M^γ. Let N be the normal vector of Z in M and let $N^{-,\gamma}$ be the negative polarized normal bundle (see (3.7)). Then $N^{-,\gamma} = 0$ if and only if the function $\langle \Phi, \gamma \rangle : M \to \mathbb{R}$ takes its maximal value on Z.*

This lemma ensures that $s^\pm \geq 1$ in Theorem 3.4 when $\Delta \cap \Phi(M)$ is not a facet of the polytope $\Phi(M)$.

Consider the situation in which $\Delta \cap \Phi(M)$ is a facet of the polytope $\Phi(M)$, so that $\mathfrak{c}_+ \cap \Phi(M) = \emptyset$; hence $\mathrm{m}_{\mathfrak{c}_+} = 0$. If we apply Lemma 3.5 with $\gamma = \beta$, we obtain $N^{-,\beta} = 0$, and so $s^- = 0$. In this situation we get the following corollary.

Corollary 3.6. *Let $\mathfrak{c}_-$ be a connected component of regular values at the edge of the polytope $\Phi(M)$. Then $\Phi(M) \cap \Delta$ is a facet of $\Phi(M)$. Let $\beta \in \Lambda$ be the primitive orthogonal vector to the hyperplane $\Delta \subset \mathfrak{t}^*$ which is pointing out of $\mathfrak{c}_-$. Here $Z = \Phi^{-1}(\Delta)$ is a connected component of the fixed-point set M^{T_Δ}. We have $\mathrm{m}_{\mathfrak{c}_-}(\mu, k) = 0$ for all $(\mu, k) \in \Lambda^* \times \mathbb{Z}$ such that*

$$0 < \frac{\langle \mu, \beta \rangle}{2\pi} - k r_\Delta < s_Z^+. \tag{3.9}$$

Here $s_Z^+ \in \mathbb{N}$ is larger than half of the codimension of Z in M, and the inequalities (3.9) are optimal.

The rest of this section is dedicated to the proof of Theorem 3.4. We start by reviewing some of the results of [32].

3.1 Elliptic and transversally elliptic symbols

We work in the setting of a compact manifold M equipped with a smooth action of a torus T.

Let $p : \mathbf{T}M \to M$ be the projection, and let $(\cdot,\cdot)_M$ be a T-invariant Riemannian metric. If E^0, E^1 are T-equivariant vector bundles over M, a T-equivariant morphism $\sigma \in \Gamma(\mathbf{T}M, \hom(p^*E^0, p^*E^1))$ is called a *symbol*. The subset of all $(m, v) \in \mathbf{T}M$ where $\sigma(m, v) : E^0_m \to E^1_m$ is not invertible is called the *characteristic set* of σ, and is denoted by $\mathrm{Char}(\sigma)$.

Let $\mathbf{T}_T M$ be the following subset of $\mathbf{T}M$:

$$\mathbf{T}_T M = \left\{(m, v) \in \mathbf{T}M,\ (v, X_M(m))_M = 0 \quad \text{for all } X \in \mathfrak{t}\right\}.$$

A symbol σ is *elliptic* if σ is invertible outside a compact subset of $\mathbf{T}M$ ($\mathrm{Char}(\sigma)$ is compact), and is *transversally elliptic* if the restriction of σ to $\mathbf{T}_T M$ is invertible outside a compact subset of $\mathbf{T}_T M$ ($\mathrm{Char}(\sigma) \cap \mathbf{T}_T M$ is compact). An elliptic symbol σ defines an element in the equivariant K-theory of $\mathbf{T}M$ with compact support, which is denoted by $\mathbf{K}_T(\mathbf{T}M)$, and the index of σ is a virtual finite-dimensional representation of T [3, 4, 5, 6].

A *transversally elliptic* symbol σ defines an element of $\mathbf{K}_T(\mathbf{T}_T M)$, and the index of σ is defined as a trace class virtual representation of T (see [1] for the analytic index and [8, 9] for the cohomological one). Observe that any elliptic symbol of $\mathbf{T}M$ is transversally elliptic; hence we have a restriction map $\mathbf{K}_T(\mathbf{T}M) \to \mathbf{K}_T(\mathbf{T}_T M)$, and a commutative diagram

$$\begin{array}{ccc} \mathbf{K}_T(\mathbf{T}M) & \longrightarrow & \mathbf{K}_T(\mathbf{T}_T M) \\ \downarrow{\scriptstyle \mathrm{Index}^T_M} & & \downarrow{\scriptstyle \mathrm{Index}^T_M} \\ R(T) & \longrightarrow & R^{-\infty}(T)\,. \end{array} \tag{3.10}$$

Using the *excision property*, one can easily show that the index map $\mathrm{Index}^T_{\mathcal{U}} : \mathbf{K}_T(\mathbf{T}_T\mathcal{U}) \to R^{-\infty}(T)$ is still defined when $\mathcal{U}$ is a T-invariant relatively compact open subset of a T-manifold (see [32, Section 3.1]).

3.2 Localization of the Riemann–Roch character

We suppose now that the compact T-manifold M is equipped with a T-invariant almost complex structure J. Let us recall the definitions of the Thom symbol $\mathrm{Thom}(M, J)$ and of the Riemann–Roch character [32].

Consider a T-invariant Riemannian metric q on M such that J is orthogonal relative to q, and let h be the Hermitian structure on $\mathbf{T}M$ defined by $h(v, w) = q(v, w) - \imath q(Jv, w)$ for $v, w \in \mathbf{T}M$. The symbol

$$\mathrm{Thom}(M, J) \in \Gamma(\mathbf{T}M, \hom(p^*(\wedge^{\mathrm{even}}_{\mathbb{C}}\mathbf{T}M),\ p^*(\wedge^{\mathrm{odd}}_{\mathbb{C}}\mathbf{T}M)))$$

at $(m, v) \in \mathbf{T}M$ is equal to the Clifford map

$$\mathrm{Cl}_m(v) : \wedge^{\mathrm{even}}_{\mathbb{C}}\mathbf{T}_m M \longrightarrow \wedge^{\mathrm{odd}}_{\mathbb{C}}\mathbf{T}_m M, \tag{3.11}$$

where $\mathrm{Cl}_m(v)\cdot w = v\wedge w - c_h(v)\cdot w$ for $w\in \wedge_{\mathbb{C}}^{\bullet}\mathbf{T}_x M$. Here $c_h(v) : \wedge_{\mathbb{C}}^{\bullet}\mathbf{T}_m M \to \wedge^{\bullet-1}\mathbf{T}_m M$ denotes the contraction map relative to h. Since the map $\mathrm{Cl}_m(v)$ is invertible for all $v\neq 0$, the symbol $\mathrm{Thom}(M,J)$ is elliptic.

The Riemann–Roch character $\mathrm{RR}(M,-) : \mathbf{K}_T(M)\to R(T)$ is defined by the relation

$$\mathrm{RR}(M,E) = \mathrm{Index}_M^T(\mathrm{Thom}(M,J)\otimes p^*E). \tag{3.12}$$

The important point is that for any T-vector bundle E, $\mathrm{Thom}(M,J)\otimes p^*E$ corresponds to the *principal symbol* of the twisted Spin^c Dirac operator $\mathcal{D}_E$ [16]; hence $\mathrm{RR}(M,E)\in R(T)$ is also defined as the (analytical) index of the elliptic operator $\mathcal{D}_E$.

Consider now the case of a *compact* Hamiltonian T-manifold (M,ω,Φ). Here J is a T-invariant almost complex structure compatible with Ω: $(v,w)\mapsto \Omega(v,Jw)$ that defines a Riemannian metric on M. As in Section 2.2, we make the choice of a scalar product $(\cdot,\cdot)$ on $\mathfrak{t}^*$ (which induces an identification $\mathfrak{t}^*\simeq\mathfrak{t}$) and we consider for any $\xi\in\mathfrak{t}^*$ the function $\frac{-1}{2}\|\Phi-\xi\|^2 : M\to\mathbb{R}$ and its Hamiltonian vector field $\mathcal{H}-\xi_M$.

Definition 3.7. For any $\xi\in\mathfrak{t}^*$ and any T-invariant open subset $\mathcal{U}\subset M$ we define the symbol $\mathrm{Thom}_\xi(\mathcal{U})$ by the relation

$$\mathrm{Thom}_\xi(\mathcal{U})(m,v) := \mathrm{Thom}(M,J)(m, v-(\mathcal{H}-\xi_M)(m)),\quad (m,v)\in\mathbf{T}\mathcal{U}.$$

The characteristic set of $\mathrm{Thom}_\xi(\mathcal{U})$ corresponds to $\{(m,v)\in\mathbf{T}\mathcal{U}, v=(\mathcal{H}-\xi_M)(m)\}$, the graph of the vector field $\mathcal{H}-\xi_M$ over $\mathcal{U}$. Since $\mathcal{H}-\xi_M$ belongs to the set of tangent vectors to the T-orbits, we have

$$\begin{aligned}\mathrm{Char}(\mathrm{Thom}_\xi(\mathcal{U}))\cap\mathbf{T}_T\mathcal{U} &= \{(m,0)\in\mathbf{T}\mathcal{U}\mid(\mathcal{H}-\xi_M)(m)=0\}\\ &\cong\{m\in\mathcal{U},\ d\|\Phi-\xi\|^2_m=0\}.\end{aligned}$$

Therefore the symbol $\mathrm{Thom}_\xi(\mathcal{U})$ is transversally elliptic if and only if

$$\mathrm{Cr}(\|\Phi-\xi\|^2)\cup\partial\mathcal{U}=\emptyset. \tag{3.13}$$

Definition 3.8. When (3.13) holds we say that the couple $(\mathcal{U},\xi)$ is *good*.

Definition 3.9. Let $(\mathcal{U},\xi)$ be a good couple. For any T-vector bundle $E\to M$, the tensor product $\mathrm{Thom}_\xi(\mathcal{U})\otimes p^*E$ belongs to $\mathbf{K}_T(\mathbf{T}_T\mathcal{U})$ and we denote by

$$\mathrm{RR}^\xi_{\mathcal{U}}(M,E)\in R^{-\infty}(T)$$

its index.

Proposition 3.10. *Let $(\mathcal{U},\xi)$ be a good couple.*

- *If $\mathcal{U}$ possesses two T-invariant open subsets $\mathcal{U}^1,\mathcal{U}^2$ such that $\overline{\mathcal{U}^1}\cap\overline{\mathcal{U}^2}\cap\mathrm{Cr}(\|\Phi-\xi\|^2)=\emptyset$ and $(\mathcal{U}^1\cup\mathcal{U}^2)\cap\mathrm{Cr}(\|\Phi-\xi\|^2)=\mathcal{U}\cap\mathrm{Cr}(\|\Phi-\xi\|^2)$, then the couples $(\mathcal{U}^1,\xi)$ and $(\mathcal{U}^2,\xi)$ are good and*

$$\mathrm{RR}^{\xi}_{\mathcal{U}}(M,-) = \mathrm{RR}^{\xi}_{\mathcal{U}^1}(M,-) + \mathrm{RR}^{\xi}_{\mathcal{U}^2}(M,-).$$

In particular, $\mathrm{RR}^{\xi}_{\mathcal{U}}(M,-) = \mathrm{RR}^{\xi}_{\mathcal{U}^1}(M,-)$ *if* $\mathcal{U}^1$ *is an open subset of* $\mathcal{U}$ *such that* $\mathcal{U}^1 \cap \mathrm{Cr}(\|\Phi-\xi\|^2) = \mathcal{U} \cap \mathrm{Cr}(\|\Phi-\xi\|^2)$.

- *If* $\xi' \in \mathfrak{t}^*$ *is close enough to* ξ, *then* $(\mathcal{U},\xi')$ *is good and*

$$\mathrm{RR}^{\xi}_{\mathcal{U}}(M,-) = \mathrm{RR}^{\xi'}_{\mathcal{U}}(M,-).$$

Proof. The first point is a direct consequence of the excision property (see Proposition 4.1. in [32]).

Let us prove the second point. Consider now the scalar product

$$\phi(s) := (\mathcal{H} - \xi^s_M, \mathcal{H} - \xi_M)_M,$$

where $\xi^s = s\xi' + (1-s)\xi$, $s \in [0,1]$. Each $\phi(s)$ is a smooth function on M. We have $\phi(s) = \|\mathcal{H}-\xi_M\|^2 + s((\xi-\xi')_M, \mathcal{H}-\xi_M)$ and then the following inequality holds on M:

$$\phi(s) \geq \|\mathcal{H}-\xi_M\|^2(\|\mathcal{H}-\xi_M\| - s\|\xi_M - \xi'_M\|). \tag{3.14}$$

Since $\partial\mathcal{U}$ is compact, we have the following inequalities on it: $\|\mathcal{H}-\xi_M\| \geq c_1 > 0$ and $\|X_M\| \leq c_2\|X\|$ for any $a \in \mathfrak{t}$. So (3.14) implies the following inequality on $\partial\mathcal{U}$:

$$\phi(s) \geq c_1(c_1 - s\|\xi-\xi'\|) \quad \text{for} \quad s \in [0,1].$$

So if ξ' is close enough to ξ, we have $\|\mathcal{H}-\xi^s_M\| \geq c_3 > 0$ on $\partial\mathcal{U}$ for any $s \in [0,1]$. We have first to prove that the couple $(\mathcal{U},\xi^s)$ is good for any $s \in [0,1]$. We see then that the family of transversally elliptic symbols $\mathrm{Thom}_{\xi^s}(\mathcal{U})$, $s \in [0,1]$, defines a homotopy between $\mathrm{Thom}_{\xi}(\mathcal{U})$ and $\mathrm{Thom}_{\xi'}(\mathcal{U})$. Hence $\mathrm{Thom}_{\xi}(\mathcal{U}) = \mathrm{Thom}_{\xi'}(\mathcal{U})$ in $\mathbf{K}_T(\mathbf{T}_T\mathcal{U})$.

The first point of Proposition 3.10 shows that $\mathrm{RR}^{\xi}_{\mathcal{U}}(M,-)$ depends closely on the intersection $\mathcal{U} \cap \mathrm{Cr}(\|\Phi-\xi\|^2)$. In particular, $\mathrm{RR}^{\xi}_{\mathcal{U}}(M,-) = 0$ when $\mathcal{U} \cap \mathrm{Cr}(\|\Phi-\xi\|^2) = \emptyset$. Recall that

$$\mathrm{Cr}(\|\Phi-\xi\|^2) = \bigcup_{\gamma\in\mathcal{B}_\xi} M^\gamma \cap \Phi^{-1}(\gamma+\xi), \tag{3.15}$$

where $\mathcal{B}_\xi \subset \mathfrak{t}^*$ is a finite set [24].

Definition 3.11. For any $\xi \in \mathfrak{t}^*$ and $\gamma \in \mathcal{B}_\xi$, we denote simply by

$$\mathrm{RR}^{\xi}_{\gamma}(M,-) : \mathbf{K}_T(M) \to R^{-\infty}(T)$$

the map $\mathrm{RR}^{\xi}_{\mathcal{U}}(M,-)$, where $\mathcal{U}$ is a T-invariant open neighborhood of $M^\gamma \cap \Phi^{-1}(\gamma+\xi)$ such that $\mathrm{Cr}(\|\Phi-\xi\|^2) \cap \overline{\mathcal{U}} = M^\gamma \cap \Phi^{-1}(\gamma+\xi)$.

Proposition 3.10 ensures that the maps $\mathrm{RR}^{\xi}_{\gamma}(M,-)$ are well defined, and for any good couple $(\mathcal{U},\xi)$ we have

$$\mathrm{RR}^{\xi}_{\mathcal{U}}(M,-) = \sum_{\gamma\in\mathcal{B}_\xi\cap\Phi(\mathcal{U})} \mathrm{RR}^{\xi}_{\gamma}(M,-). \tag{3.16}$$

If we take $\mathcal{U} = M$, we have $\mathrm{RR}^{\xi}_{\mathcal{U}}(M,-) = \mathrm{RR}(M,-) = \sum_{\gamma\in\mathcal{B}_\xi} \mathrm{RR}^{\xi}_{\gamma}(M,-)$ (see [32, Section 4]).

3.3 Periodic polynomial behavior of the multiplicities

We suppose here that the Hamiltonian T-manifold (M,Ω,Φ) is prequantized by a T-complex line bundle L satisfying (3.1) for a suitable invariant connection. In this section we will characterize the periodic polynomial behavior of the multiplicities $\mathrm{m}(\mu,k)$ with the help of the localized Riemann–Roch character $\mathrm{RR}^{\xi}_{0}(M,-)$.

Let us introduce some vocabulary. We say that two generalized characters $\chi^{\pm} = \sum_{\mu\in\Lambda^*} a^{\pm}_{\mu}\,\mathbb{C}_\mu$ *coincide* on a region $D\subset\mathfrak{t}^*$ if $a^{+}_{\mu} = a^{-}_{\mu}$ for every $\mu\in D\cap\Lambda^*$. A generalized character $\chi = \sum_\mu a_\mu\,\mathbb{C}_\mu$ is *supported* on a region $D\subset\mathfrak{t}^*$ if $a_\mu = 0$ for $\mu\notin D$. A weight $\mu\in\Lambda^*$ *occurs* in $\chi = \sum_\mu a_\mu\,\mathbb{C}_\mu$ if $a_\mu\neq 0$.

For $\xi\in\mathfrak{t}^*$, we define $r_\xi > 0$ as the smallest nonzero critical value of the function $\|\Phi-\xi\|$, and we denote by $B(\xi,r_\xi)$ the open ball of center ξ and radius r_ξ.

Theorem 3.12 ([32]). *For any $\xi\in\mathfrak{t}^*$, the generalized character $\mathrm{RR}^{\xi}_{0}(M,L^{\otimes k})$ coincides with $\mathrm{RR}(M,L^{\otimes k})$ on the open ball $k\cdot B(\xi,r_\xi)$.*

The arguments of [32] for the proof of this theorem will be needed later, so we recall them. Let $\xi\in\mathfrak{t}^*$. We start with the decomposition

$$\mathrm{RR}(M,L^{\otimes k}) = \sum_{\gamma\in\mathcal{B}_\xi} \mathrm{RR}^{\xi}_{\gamma}(M,L^{\otimes k}). \tag{3.17}$$

We recall now, for a nonzero $\gamma\in\mathcal{B}_\xi$, the localization of the map $\mathrm{RR}^{\xi}_{\gamma}$ on the fixed-point set M^γ [32].

Let N be the normal bundle of M^γ in M. The almost complex structure on M induces an almost complex structure on M^γ and a complex structure on the bundles N and $N_{\mathbb{C}} := N\otimes\mathbb{C}$. Following (3.7), we define the γ-polarized complex vector bundles $N^{+,\gamma}$ and $(N_{\mathbb{C}})^{+,\gamma}$.

The manifold M^γ is a symplectic submanifold of M equipped with an induced Hamiltonian action of T; its moment map is the restriction of Φ on M^γ. Following Definition 3.11, we have on M^γ a localized Riemann–Roch character $\mathrm{RR}^{\xi}_{\gamma}(M^\gamma,-)$. On M^γ, the Hamiltonian vector fields of the functions $\|\Phi-\xi\|^2$ and $|\Phi-(\xi+\gamma)\|^2$ coincide; hence

$$\mathrm{RR}^{\xi}_{\gamma}(M^\gamma,-) = \mathrm{RR}^{\xi+\gamma}_{0}(M^\gamma,-). \tag{3.18}$$

We prove in [32, Theorem 5.8] that

$$\mathrm{RR}^{\xi}_{\gamma}(M, E) = \sum_{k\in\mathbb{N}} (-1)^{l}\mathrm{RR}^{\xi}_{\gamma}(M^{\gamma}, E|_{M^{\gamma}} \otimes \det(N^{+,\gamma}) \otimes S^{k}(N^{+,\gamma}_{\mathbb{C}})) \tag{3.19}$$

for every T-vector bundle E. Here l is the locally constant function on M^{γ} equal to the complex rank of $N^{+,\gamma}$.

Proposition 3.13 ([32], Section 5). *Let $\overline{N}$ be the T-vector bundle N with the opposite complex structure on the fibers. The sum $(-1)^{l}\sum_{k\in\mathbb{N}}\det(N^{+,\gamma})\otimes S^{k}(N^{+,\gamma}_{\mathbb{C}})$ is an inverse of $\wedge^{\bullet}_{\mathbb{C}}\overline{N}$, which we denote by $[\wedge^{\bullet}_{\mathbb{C}}\overline{N}\,]^{-1}_{\gamma}$.*

If we use the notation of Proposition 3.13 and (3.18), the localization (3.19) can be rewritten as

$$\mathrm{RR}^{\xi}_{\gamma}(M, E) = \mathrm{RR}^{\xi+\gamma}_{0}(M^{\gamma}, E|_{M^{\gamma}} \otimes [\wedge^{\bullet}_{\mathbb{C}}\overline{N}\,]^{-1}_{\gamma}). \tag{3.20}$$

Let $i : T_{\gamma} \hookrightarrow T$ be the inclusion of the subtorus generated by γ. Let F be a T-vector bundle on M^{γ}.

Lemma 3.14 ([32], Lemma 9.4.). *A weight $\mu \in \Lambda^{*}$ occurs in $\mathrm{RR}^{\xi}_{\gamma}(M^{\gamma}, F)$ only if $i^{*}(\mu)$ occurs as a weight for the T_{γ}-action on the fibers of $F \otimes [\wedge^{\bullet}_{\mathbb{C}}\overline{N}\,]^{-1}_{\gamma}$.*

Since the T_{γ} weights on the bundles $N^{+,\gamma}_{\mathbb{C}}$ and $N^{+,\gamma}$ are polarized by γ, the localization (3.19) gives the following:

Corollary 3.15. *For a nonzero $\gamma \in \mathcal{B}_{\xi}$, the generalized character $\mathrm{RR}^{\xi}_{\gamma}(M, L^{\otimes k})$ is supported on the half-space $\{a \in \mathfrak{t}^{*} | (\gamma, a - k(\xi+\gamma)) \geq 0\}$.*

Since the condition $(\gamma, a - k(\xi+\gamma)) \geq 0$ implies that $\|a - k\xi\| \geq k\|\gamma\| \geq kr_{\xi}$, the last proposition shows that every weight of the open ball $k \cdot B(\xi, r_{\xi})$ does not occur in $\mathrm{RR}^{\xi}_{\gamma}(M, L^{\otimes k})$. This last remark together with (3.17) proves Theorem 3.12.

For the localized Riemann–Roch character $\mathrm{RR}^{\xi}_{0}(M, -)$ we have the following lemma, which is very similar to Lemma 2.3.

Lemma 3.16. *Let $\mathfrak{c} \subset \mathfrak{t}^{*}$ be a connected component of regular values of Φ. For every $\xi, \xi' \in \mathfrak{c}$, we have $\mathrm{RR}^{\xi}_{0}(M, -) = \mathrm{RR}^{\xi'}_{0}(M, -)$.*

Proof. We have to show that the map $\xi \mapsto \mathrm{RR}^{\xi}_{0}(M, -)$ is locally constant on $\mathfrak{c}$. Let $\xi \in \mathfrak{c}$ and take an open neighborhood $\mathcal{U}$ of $\Phi^{-1}(\xi)$ small enough that the stabilizer $T_m = \{t \in T \,|\, t \cdot m = m\}$ is finite for every $m \in \overline{\mathcal{U}}$. We see then that $\overline{\mathcal{U}} \cap \mathrm{Cr}(\|\Phi - \xi'\|^2) = \Phi^{-1}(\xi')$ and $\partial\mathcal{U} \cap \mathrm{Cr}(\|\Phi - \xi'\|^2) = \emptyset$ if ξ' is close enough to ξ: hence $\mathrm{RR}^{\xi'}_{0}(M, -) = \mathrm{RR}^{\xi'}_{\mathcal{U}}(M, -)$ for ξ' close enough to ξ. The second point of Proposition 3.10 finishes the proof.

When ξ is a regular value of Φ, the localized Riemann–Roch character $\mathrm{RR}_0^{\xi}(M,-)$ has been computed in [32] as follows. Let $\mathrm{RR}(\mathcal{M}_\xi,-)$ be the Riemann–Roch map defined on the orbifold $\mathcal{M}_\xi = \Phi^{-1}(\xi)/T$ by means of an almost complex structure compatible with the induced symplectic structure. For every T-vector bundle $E \to M$ we define the following family of orbifold vector bundles over $\mathcal{M}_\xi$:

$$\mathcal{E}_\xi^\mu := (E|_{\Phi^{-1}(\xi)} \otimes \mathbb{C}_{-\mu})/T, \quad \mu \in \Lambda^*. \tag{3.21}$$

For every T-vector bundle E on M, we proved in [32, Section 6.2] the following equality in $R^{-\infty}(T)$:

$$\mathrm{RR}_0^{\xi}(M,E) = \sum_{\mu\in\Lambda^*} \mathrm{RR}(\mathcal{M}_\xi, \mathcal{E}_\xi^\mu)\, \mathbb{C}_\mu. \tag{3.22}$$

This decomposition was first obtained by Vergne [37] when T is the circle group and when M is Spin. The number $\mathrm{RR}(\mathcal{M}_\xi, \mathcal{E}_\xi^\mu) \in \mathbb{Z}$ is then equal to the T-invariant part of the index $\mathrm{RR}_0^{\xi}(M,E) \otimes \mathbb{C}_{-\mu}$.

Remark 3.17. Let $t \to t^\lambda$ be a character of T. Suppose that a subgroup $H \subset T$ acts trivially on M and with the character $t \in H \to t^\lambda$ on the fibers of the T-vector bundle E. Then H acts with the character $t \in H \to t^{\lambda-\mu}$ on $\mathrm{RR}_0^{\xi}(M,E) \otimes \mathbb{C}_{-\mu}$, and then $\mathrm{RR}(\mathcal{M}_\xi, \mathcal{E}_\xi^\mu) \neq 0$ only if $t^{\lambda-\mu} = 1$ for every $t \in H$. So the sum in (3.22) can be restricted to $\lambda + \Lambda_H^*$, where Λ_H^* is the sublattice of Λ^* formed by the element $\alpha \in \Lambda^*$ satisfying $t^\alpha = 1$, $\forall\, t \in H$.

This remark applies also to the usual character $\mathrm{RR}(M,E) = \sum_{\mu\in\Lambda^*} m_\mu \mathbb{C}_\mu$. The multiplicity $m_\mu \in \mathbb{Z}$ is equal to the (virtual) dimension of the T-invariant part of $\mathrm{RR}(M,E) \otimes \mathbb{C}_{-\mu}$. With the same hypothesis as above we see that $m_\mu \neq 0$ only if $\mu \in \lambda + \Lambda_H^*$.

Let Γ_M be the generic stabilizer for the action of T on M. Consider a weight α_o such that Γ_M acts on the fibers of L with the character $t \mapsto t^{\alpha_o}$. We define the sublattice $\Xi(M,L) \subset \Lambda^* \times \mathbb{Z}$ by

$$\Xi(M,L) := \{(\mu,k) \in \Lambda^* \times \mathbb{Z} | k\alpha_o - \mu \in \Lambda_{\Gamma_M}^*\}. \tag{3.23}$$

We know then that $\mathrm{m}(\mu,k) = 0$ if $(\mu,k) \notin \Xi(M,L)$.

Proposition 3.18. *Let $\mathfrak{c}$ be a connected component of regular values of Φ and let* $\mathrm{Cone}(\mathfrak{c})$ *be the corresponding cone in $\mathfrak{t}^* \times \mathbb{R}^{>0}$ (see (3.5)). Let $\xi \in \mathfrak{c}$. For any $(\mu,k) \in \mathrm{Cone}(\mathfrak{c}) \cap \Xi(M,L)$ we have*

$$\mathrm{m}(\mu,k) = \mathrm{RR}(\mathcal{M}_\xi, \mathcal{L}_\xi^{\mu,k}), \tag{3.24}$$

where

$$\mathcal{L}_\xi^{\mu,k} = (L^{\otimes k}|_{\Phi^{-1}(\xi)} \otimes \mathbb{C}_{-\mu})/T. \tag{3.25}$$

Proof. Let $(\mu, k) \in \mathrm{Cone}(\mathfrak{c})$ and let $\xi' = \frac{\mu}{k} \in \mathfrak{c}$. We know from Theorem 3.12 that the generalized character $\mathrm{RR}_0^{\xi'}(M, L^{\otimes k})$ coincides with $\mathrm{RR}(M, L^{\otimes k})$ on the open ball $k \cdot B(\xi', r_{\xi'}) = B(\mu, kr_{\xi'})$. So $\mathrm{m}(\mu, k)$ is equal to the μ-multiplicity in $\mathrm{RR}_0^{\xi'}(M, L^{\otimes k})$. Take now any $\xi \in \mathfrak{c}$. We know from Lemma 3.16 that $\mathrm{RR}_0^{\xi}(M, -) = \mathrm{RR}_0^{\xi'}(M, -)$, and (3.22) shows that the μ-multiplicity in $\mathrm{RR}_0^{\xi}(M, L^{\otimes k})$ is equal to $\mathrm{RR}(\mathcal{M}_\xi, \mathcal{L}_\xi^{\mu,k})$.

Definition 3.19. Take $\xi \in \mathfrak{c}$. The map $\mathrm{m}_\mathfrak{c} : \Lambda^* \times \mathbb{Z} \to \mathbb{Z}$ is defined by the equation

$$\mathrm{m}_\mathfrak{c}(\mu, k) = \mathrm{RR}(\mathcal{M}_\xi, \mathcal{L}_\xi^{\mu,k}), \tag{3.26}$$

where $\mathcal{L}_\xi^{\mu,k}$ is the orbifold line bundle defined by (3.25).

In other words, the map $\mathrm{m}_\mathfrak{c}$ is defined by the following equality in $R^{-\infty}(T)$:

$$\sum_{\mu \in \Lambda^*} \mathrm{m}_\mathfrak{c}(\mu, k)\, \mathbb{C}_\mu = \mathrm{RR}_0^{\xi}(M, L^{\otimes k}),$$

for all $k \in \mathbb{Z}$. From Remark 3.17, we know that $\mathrm{m}_\mathfrak{c}$ is supported on the sublattice $\Xi(M, L)$ defined in (3.23).

We will now exploit the Riemann–Roch theorem for orbifolds due to Kawasaki [23] to show that the map $\mathrm{m}_\mathfrak{c}$ is a periodic polynomial.

3.4 Riemann–Roch–Kawasaki theorem

First we recall how the Riemann–Roch character $\mathrm{RR}(\mathcal{M}_\xi, \mathcal{E}_\xi)$ is defined when ξ is a regular value of Φ and $\mathcal{E}_\xi = E_{|\Phi^{-1}(\xi)}/T$ is the reduction of a complex T-vector bundle E over M. The number $\mathrm{RR}(\mathcal{M}_\xi, \mathcal{E}_\xi) \in \mathbb{Z}$ is defined as the T-invariant part of the index of a transversally elliptic operator D_E on $\Phi^{-1}(\xi)$. Since the index of D_E depends only of the class of its symbol $\sigma(D_E)$ in $K_T(\mathbf{T}_T\Phi^{-1}(\xi))$, it is enough to define the transversally elliptic symbol $\sigma(D_E)$. Since the action of T on $\Phi^{-1}(\xi)$ is locally free, $V := \mathbf{T}_T\Phi^{-1}(\xi)$ is a vector bundle. It carries a canonical symplectic structure on the fibers, and we choose any compatible complex structure making V into a Hermitian vector bundle. At $(m, v) \in \mathbf{T}\Phi^{-1}(\xi)$, the map $\sigma(D_E)(m, v)$ is the Clifford action

$$\mathrm{Cl}_m(v_1) \otimes \mathrm{Id}_{E_m} : (\wedge_\mathbb{C}^{\mathrm{even}} V_m) \otimes E_m \longrightarrow (\wedge_\mathbb{C}^{\mathrm{odd}} V_m) \otimes E_m.$$

where $v_1 \in V_m$ is the V-component of the vector $v \in \mathbf{T}_m\Phi^{-1}(\xi)$. We explain now the formula of Kawasaki for $\mathrm{RR}(\mathcal{M}_\xi, \mathcal{E}_\xi)$ when $\xi \in \Phi(M)$ is a regular value of Φ [23].

Let $\mathbb{F}$ be the collection of the finite subgroups of T that are stabilizers of points in M. Consider the orbit type stratification of $\Phi^{-1}(\xi)$ and denote by $\mathbb{S}_\xi$ the set of its orbit type strata. Each stratum S is a connected component of the smooth submanifold

$$\Phi^{-1}(\xi)_{H_S} := \{m \in \Phi^{-1}(\xi) | \mathrm{Stab}_T(m) = H_S\} \tag{3.27}$$

for a unique $H_S \in \mathbb{F}$. The orbifold $\mathcal{M}_\xi$ decomposes as a disjoint union $\cup_{S \in \mathbb{S}_\xi} S/T$ of smooth components, and each quotient $\overline{S}/T$ is a suborbifold of $\mathcal{M}_\xi$. The generic stabilizer Γ_M of T on M is also the generic stabilizer of T on the fiber[4] $\Phi^{-1}(\xi)$, and is associated to an open and dense stratum $S_{\max}$.

Suppose that $E \to M$ is a Hermitian T-vector bundle. On each suborbifold $\overline{S}/T$, we get the orbifold complex vector bundle

$$\mathcal{E}_S := E_{|\overline{S}}/T. \tag{3.28}$$

We define twisted characteristic classes $\mathrm{Ch}^-(\mathcal{E}_S)$ and $D^-(\mathcal{E}_S)$ by

$$\mathrm{Ch}^\gamma(\mathcal{E}_S) := \mathbf{Tr}\big(\gamma^{\mathcal{E}_S} \cdot e^{\frac{i}{2\pi} R(\mathcal{E}_S)}\big), \quad \gamma \in H_S, \tag{3.29}$$

and

$$D^\gamma(\mathcal{E}_S) := \det\big(1 - (\gamma^{\mathcal{E}_S})^{-1} \cdot e^{-\frac{i}{2\pi} R(\mathcal{E}_S)}\big), \quad \gamma \in H_S. \tag{3.30}$$

Here $R(\mathcal{E}_S) \in \mathcal{A}^2(\overline{S}/T, \mathrm{End}(\mathcal{E}_S))$ is the curvature of a horizontal Hermitian connection on $E_{|\overline{S}}$, and $\gamma \mapsto \gamma^{\mathcal{E}_S}$ is the linear action of H_S on the fibers of $E_{|\overline{S}}$.

Let N_S be the normal bundle of $\overline{S}$ in $\Phi^{-1}(\xi)$. The symplectic structure on M induces a symplectic form Ω_S on each suborbifold $\overline{S}/T$, and a symplectic structure on the fibers of the bundle N_S. Choose a compatible almost complex structure on $\overline{S}/T$, and a compatible complex structure on the fibers of N_S making the tangent bundle of $\overline{S}/T$ and $\mathcal{N}_S := N_S/T$ into Hermitian vector bundles. Consider a Hermitian connexion on $\mathbf{T}(\overline{S}/T)$, with curvature $R(\overline{S}/T)$, and let

$$\mathrm{Todd}(\overline{S}/T) = \det\left(\frac{(i/2\pi)R(\overline{S}/T)}{1 - e^{-(i/2\pi)R(\overline{S}/T)}}\right) \tag{3.31}$$

be the corresponding Todd forms. As in (3.30), we associate to the complex orbifold vector bundle $\mathcal{N}_S$, the twisted form $D^-(\mathcal{N}_S)$, which is a map form H_S to $\mathcal{A}^{\mathrm{even}}(\overline{S}/T)$. The 0-degree part of $D^\gamma(\mathcal{N}_S)$ is equal to $\det(1 - (\gamma^{\mathcal{N}_S})^{-1})$; hence $D^\gamma(\mathcal{N}_S)$ is invertible in $\mathcal{A}^{\mathrm{even}}(\overline{S}/T)$ when γ belongs to

$$H_S^o = \{\gamma \in H_S | \det(1 - (\gamma^{\mathcal{N}_S})^{-1}) \neq 0\}. \tag{3.32}$$

Note that H_S^o corresponds to the set of $\gamma \in H_S$ for which $\overline{S}$ is a connected component of $(\Phi^{-1}(\xi))^\gamma$.

[4] Since a neighborhood of $\Phi^{-1}(\xi)$ in M is T-equivariantly diffeomorphic to $\Phi^{-1}(\xi) \times \mathfrak{t}^*$.

Theorem 3.20 (Kawasaki). *The number* $\mathrm{RR}(\mathcal{M}_\xi, \mathcal{E}_\xi) \in \mathbb{Z}$ *is given by the formula*

$$\mathrm{RR}(\mathcal{M}_\xi, \mathcal{E}_\xi) = \sum_{S \in \mathbb{S}_\xi} \frac{1}{|H_S|} \sum_{\gamma \in H_S^o} \int_{\overline{S}/T} \frac{\mathrm{Todd}(\overline{S}/T)\mathrm{Ch}^\gamma(\mathcal{E}_S)}{D^\gamma(\mathcal{N}_S)}. \tag{3.33}$$

We now exploit Theorem 3.20 to show that the map $\mathrm{m}_\mathfrak{c} : \Lambda^* \times \mathbb{Z} \to \mathbb{Z}$ defined by (3.26) is a periodic polynomial. We need the classical computation of the first Chern class of the line bundle

$$\mathcal{L}_S^{\mu,k} = (L^{\otimes k} \otimes \mathbb{C}_{-\mu})_{|\overline{S}}/T. \tag{3.34}$$

The curvature form $\omega_\xi \in H^2(\mathcal{M}_\xi) \otimes \mathfrak{t}$ of the principal T-bundle $\Phi^{-1}(\xi) \to \mathcal{M}_\xi$ restricts to a curvature form $\omega_S \in H^2(\overline{S}/T) \otimes \mathfrak{t}$ on each stratum.

Lemma 3.21. *The first Chern class of the line bundle* $\mathcal{L}_S^{\mu,k}$ *is given by*

$$c_1(\mathcal{L}_S^{\mu,k}) = \frac{1}{2\pi}(k\Omega_S - \langle k\xi - \mu, \omega_S \rangle).$$

For a stratum S, we consider $\alpha_S \in \Lambda^*$ such that $\gamma \in H_S \mapsto \gamma^{\alpha_S}$ corresponds to the action of H_S on the fibers of $L_{|\overline{S}}$. Finally, we have the decomposition

$$\mathrm{m}_\mathfrak{c}(\mu, k) = \sum_{S \in \mathbb{S}_\xi} P_S(\mu, k), \tag{3.35}$$

where

$$P_S(\mu, k) = \frac{1}{|H_S|} \sum_{\gamma \in H_S^o} \gamma^{k\alpha_S - \mu} \int_{\overline{S}/T} \frac{\mathrm{Todd}(\overline{S}/T)}{D^\gamma(\mathcal{N}_S)} e^{\frac{1}{2\pi}(k\Omega_S - \langle k\xi - \mu, \omega_S \rangle)}. \tag{3.36}$$

When S is the principal open dense stratum $S_{\max}$, the map P_S is

$$P_{\max}(\mu, k) = \frac{\sum_{\gamma \in \Gamma_M} \gamma^{k\alpha_o - \mu}}{|\Gamma_M|} \int_{\mathcal{M}_\xi} \mathrm{Todd}(\mathcal{M}_\xi) e^{\frac{1}{2\pi}(k\Omega_\xi - \langle k\xi - \mu, \omega_\xi \rangle)}. \tag{3.37}$$

The term $\frac{\sum_{\gamma \in \Gamma_M} \gamma^{k\alpha_o - \mu}}{|\Gamma_M|}$ is equal to 1 when (μ, k) belongs to the lattice $\Xi(M, L)$ (see (3.23)), and is equal to 0 in the other cases. From (3.36) we see that P_S is a periodic polynomial of degree less than $\frac{\dim(\overline{S}/T)}{2}$, and for $S = S_{\max}$ we have on $\Xi(M, L)$,

$$P_{\max}(\mu, k) = \frac{1}{(2\pi)^d} \int_{\mathcal{M}_\xi} \frac{(k\Omega_\xi - \langle k\xi - \mu, \omega_\xi \rangle)^d}{d!} + O(d - 1), \tag{3.38}$$

where $d = \frac{\dim \mathcal{M}_\xi}{2}$ and $O(d-1)$ denotes a polynomial of degree less than $d - 1$. If we use the polynomial $\mathrm{DH}_\mathfrak{c}$ defined in Section 2 we can conclude our computations with the following:

Proposition 3.22. *The map* $\mathrm{m}_{\mathfrak{c}}$ *is a periodic polynomial of degree* $d = \frac{\dim \mathcal{M}_\xi}{2}$ *supported on* $\Xi(M, L)$. *For* $(\mu, k) \in \Xi(M, L)$ *we have*

$$\mathrm{m}_{\mathfrak{c}}(\mu, k) = |\Gamma_M| \frac{k^d}{(2\pi)^d} \mathrm{DH}_{\mathfrak{c}}\left(\frac{\mu}{k}\right) + O(d-1),$$

where $O(d-1)$ *means a periodic polynomial of degree less than* $d-1$.

3.5 Wall-crossing formulas for the $\mathrm{m}_{\mathfrak{c}}$

Let $\mathfrak{c}_+$ and $\mathfrak{c}_-$ be two adjacent connected components of regular values of Φ separated by a hyperplane Δ. The aim of this section is to compute the periodic polynomial $\mathrm{m}_{\mathfrak{c}_+} - \mathrm{m}_{\mathfrak{c}_-}$.

We consider two points $\xi_\pm \in \mathfrak{c}_\pm$ such that $\xi = \frac{1}{2}(\xi_+ + \xi_-)$ belongs to the relative interior of $\overline{\mathfrak{c}_+} \cap \overline{\mathfrak{c}_-}$ in Δ. We suppose furthermore that $\xi^+ - \xi^-$ is orthogonal to Δ. Using the identification $\mathfrak{t}^* \simeq \mathfrak{t}$ given by the scalar product, the vector $\gamma = \frac{1}{2}(\xi_+ - \xi_-)$, seen as a vector of $\mathfrak{t}_\Delta$, belongs[5] to $\mathbb{R}^{>0}\beta$. We noticed in Section 2.4 that for all $m \in \Phi^{-1}(\xi)$ the stabilizer $\mathfrak{t}_m$ is equal either to $\mathfrak{t}_\Delta$ or to $\{0\}$. Then there exists an open T-invariant neighborhood $\mathcal{U}$ of $\Phi^{-1}(\xi)$ in M such that for all $m \in \overline{\mathcal{U}}$ either $\mathfrak{t}_m := \{0\}$, or $\mathfrak{t}_m = \mathfrak{t}_\Delta$ and $\Phi(m) \in \Delta$.

One can see easily that the couple $(\mathcal{U}, \xi)$ is good and the second point of Proposition 3.10 tells us that

$$\mathrm{RR}^{\xi}_{\mathcal{U}}(M, -) = \mathrm{RR}^{\xi_-}_{\mathcal{U}}(M, -) = \mathrm{RR}^{\xi_+}_{\mathcal{U}}(M, -) \tag{3.39}$$

when $\xi_\pm$ are close enough to ξ. Since $\mathcal{U} \cap \mathrm{Cr}(\|\Phi - \xi\|^2) = \Phi^{-1}(\xi)$, we have $\mathrm{RR}^{\xi}_{\mathcal{U}}(M, -) = \mathrm{RR}^{\xi}_{0}(M, -)$. If $\xi_\pm$ are close enough to ξ we have

$$\mathcal{U} \cap \mathrm{Cr}(\|\Phi - \xi_\pm\|^2) = \Phi^{-1}(\xi_\pm) \bigcup M^\gamma \cap \Phi^{-1}(\xi). \tag{3.40}$$

The former decomposition is due to (3.15) and to the fact that the stabilizer of $\mathfrak{t}$ on $\mathcal{U}$ is equal either to $\mathfrak{t}_\Delta$ or to $\{0\}$. Notice that $\xi_- + \gamma = \xi_+ + \gamma = \xi$. The decomposition (3.40) gives

$$\mathrm{RR}^{\xi_\pm}_{\mathcal{U}}(M, -) = \mathrm{RR}^{\xi_\pm}_{0}(M, -) + \mathrm{RR}^{\xi_\pm}_{\mp\gamma}(M, -), \tag{3.41}$$

where $\mathrm{RR}^{\xi_-}_{\gamma}(M, -)$ (respectively $\mathrm{RR}^{\xi_+}_{-\gamma}(M, -)$) is the Riemann–Roch character localized on $M^\gamma \cap \Phi^{-1}(\xi)$ by the vector field $\mathcal{H} - (\xi_-)_M$ (respectively $\mathcal{H} - (\xi_+)_M$). Now (3.39) and (3.41) prove the following result.

Proposition 3.23. *If* $\xi_\pm$ *are close enough to* Δ, *we have*

$$\mathrm{RR}^{\xi_+}_{0}(M, -) - \mathrm{RR}^{\xi_-}_{0}(M, -) = \mathrm{RR}^{\xi_-}_{\gamma}(M, -) - \mathrm{RR}^{\xi_+}_{-\gamma}(M, -).$$

---

[5] β is the primitive vector of $\mathfrak{t}_\Delta \cap \Lambda$ pointing out of $\mathfrak{c}_-$.

We know from Proposition 3.18 that $\mathrm{m}_{\mathfrak{c}_\pm}(\mu,k)$ is equal to the μ-multiplicity of T in $\mathrm{RR}_0^{\xi_\pm}(M, L^{\otimes k})$. Hence $\mathrm{m}_{\mathfrak{c}_+}(\mu,k) - \mathrm{m}_{\mathfrak{c}_-}(\mu,k)$ is equal to the μ-multiplicity of T in $\mathrm{RR}_\gamma^{\xi_-}(M, L^{\otimes k}) - \mathrm{RR}_{-\gamma}^{\xi_+}(M, L^{\otimes k})$.

Let N_Δ be the normal bundle of M^{T_Δ} in M, and let $[\wedge_{\mathbb{C}}^\bullet \overline{N_\Delta}\,]^{-1}_{\pm\beta}$ be the polarized inverses of $\wedge_{\mathbb{C}}^\bullet \overline{N_\Delta}$ (see Proposition 3.13). Since $\xi = \xi_+ - \gamma = \xi_- + \gamma$ and $\gamma \in \mathbb{R}^{>0}\beta$, the localization (3.20) gives

$$\mathrm{RR}_\gamma^{\xi_-}(M, L^{\otimes k}) = \sum_{Z\in\mathcal{F}} \mathrm{RR}_0^{\xi}(Z, L^{\otimes k}|_Z \otimes [\wedge_{\mathbb{C}}^\bullet \overline{N_Z}\,]^{-1}_{\beta}),$$

$$\mathrm{RR}_{-\gamma}^{\xi_+}(M, L^{\otimes k}) = \sum_{Z\in\mathcal{F}} \mathrm{RR}_0^{\xi}(Z, L^{\otimes k}|_Z \otimes [\wedge_{\mathbb{C}}^\bullet \overline{N_Z}\,]^{-1}_{-\beta}).$$

Finally, $\mathrm{m}_{\mathfrak{c}_+}(\mu,k) - \mathrm{m}_{\mathfrak{c}_-}(\mu,k) = \sum_{Z\in\mathcal{F}} \mathcal{A}_Z(\mu,k)$, where $\mathcal{A}_Z(\mu,k)$ is equal to the μ-multiplicity of T in

$$\mathrm{RR}_0^{\xi}(Z, L^{\otimes k}|_Z \otimes [\wedge_{\mathbb{C}}^\bullet \overline{N_Z}\,]^{-1}_{-\beta}) - \mathrm{RR}_0^{\xi}(Z, L^{\otimes k}|_Z \otimes [\wedge_{\mathbb{C}}^\bullet \overline{N_Z}\,]^{-1}_{\beta}). \tag{3.42}$$

Let $\beta' \in \mathfrak{t}_\Delta^* \cap \Lambda^*$, which is defined by the relation $\langle \beta', \beta\rangle = 2\pi$, so that $\Lambda^*_{\mathfrak{t}_\Delta} = \mathbb{Z}\beta'$. Concerning the T_Δ-weights we have

1. The T_Δ-weight on $L^{\otimes k}|_Z$ is equal to $kr_\Delta\beta'$.
2. The T_Δ-weight on $\det(N_Z^{+,\pm\beta})$ is $\pm s_Z^\pm \beta'$, where $s_Z^\pm \in \mathbb{N}$ is the absolute value of the trace of $\frac{1}{2\pi}\mathcal{L}_\beta$ on $N_Z^{+,\pm\gamma}$.
3. The T_Δ-weights on $S^{>0}(N_Z^{+,\beta})$ (respectively $S^{>0}(N_Z^{+,-\beta})$) are of the form $p\beta'$ with $p > 0$ (respectively $p < 0$).

Now Lemma 3.14 shows that if a weight μ occurs in $\mathrm{RR}_0^{\xi}(Z, L^{\otimes k}|_Z \otimes [\wedge_{\mathbb{C}}^\bullet \overline{N_Z}\,]^{-1}_{\beta})$, we have $i^*(\mu) = (kr_\Delta + s_Z^+ + p)\beta'$ with $p \geq 0$, and then

$$\frac{\langle \mu, \beta\rangle}{2\pi} - kr_\Delta \geq s_Z^+.$$

Similarly, if a weight μ occurs in $\mathrm{RR}_0^{\xi}(Z, L^{\otimes k}|_Z \otimes [\wedge_{\mathbb{C}}^\bullet \overline{N_Z}\,]^{-1}_{-\beta})$, we have

$$\frac{\langle \mu, \beta\rangle}{2\pi} - kr_\Delta \leq -s_Z^-.$$

Finally, $\mathcal{A}_Z(\mu,k) = 0$ when $-s_Z^- < \frac{\langle\mu,\beta\rangle}{2\pi} - kr_\Delta < s_Z^+$. We have proved the following theorem.

Theorem 3.24. *Let $s^\pm = \inf_Z s_Z^\pm$, where the infimum is taken over the connected components Z of M^{T_Δ} for which $\overline{\mathfrak{c}_+} \cap \overline{\mathfrak{c}_-} \subset \Phi(Z)$. For every $(\mu,k) \in \Lambda^* \times \mathbb{Z}$, we have $\mathrm{m}_{\mathfrak{c}_+}(\mu,k) = \mathrm{m}_{\mathfrak{c}_-}(\mu,k)$ if*

$$-s^- < \frac{\langle \mu, \beta\rangle}{2\pi} - kr_\Delta < s^+. \tag{3.43}$$

Sometimes the inequalities (3.43) are optimal.

Proposition 3.25.

- *Consider the connected components $Z \in \mathcal{F}$ for which s_Z^+ is minimal. Among them consider the subset $\mathcal{F}_+$ where* $\dim(Z)$ *is maximal. If the integers* $\mathrm{rk}_{\mathbb{C}}(N_Z^{+,\beta})$, *$Z \in \mathcal{F}_+$ have the same parity, then the condition $\frac{\langle \mu,\beta\rangle}{2\pi} - kr_\Delta < s^+$ is optimal in (3.43).*
- *In the same way, consider the connected components $Z \in \mathcal{F}$ for which s_Z^- is minimal. Among them consider the subset $\mathcal{F}_-$ where* $\dim(Z)$ *is maximal. If the integers* $\mathrm{rk}_{\mathbb{C}}(N_Z^{+,\beta})$*, $Z \in \mathcal{F}_+$, have the same parity, then the condition $-s^- < \frac{\langle \mu,\beta\rangle}{2\pi} - kr_\Delta$ is optimal in (3.43).*

Remark 3.26. The last proposition applies when there is a *unique* connected component Z of M^{T_Δ} for which $\overline{\mathfrak{c}_+} \cap \overline{\mathfrak{c}_-} \subset \Phi(Z)$.

Proof. We consider only the first point, since the other point works similarly. We restrict our attention to the couples (μ, k) such that $\frac{\langle \mu,\beta\rangle}{2\pi} - kr_\Delta = s^+$. They are of the form

$$\mu = (kr_\Delta + s^+)\beta' + \mu_2 \tag{3.44}$$

with $\mu_2 \in \Lambda^*_{\mathfrak{t}/\mathfrak{t}_\Delta}$. Let us denote by $D(\mu_2, k)$ the restriction of $\mathrm{m}_{\mathfrak{c}_+}(\mu,k) - \mathrm{m}_{\mathfrak{c}_-}(\mu,k)$ to the set of couples (μ, k) parametrized by (3.44). We want to prove that $D(\mu_2, k)$ is not identically equal to zero.

From the previous discussion one knows that

$$D(\mu_2, k) = \sum_{Z,\, s_Z^+ = s^+} (-1)^{\mathrm{rk}_{\mathbb{C}}(N_Z^{+,\beta})} D_Z(\mu_2, k), \tag{3.45}$$

where $D_Z(\mu_2, k)$ is the μ-multiplicity of T in

$$\mathrm{RR}_0^{\xi}(Z, L^{\otimes k}|_Z \otimes \det(N_Z^{+,\beta})).$$

Let us make a few remarks concerning the maps $\mathrm{RR}_0^{\xi}(Z,-) : \mathbf{K}_T(Z) \to R^{-\infty}(T)$. Since T_Δ acts trivially on Z, the decomposition $T = T/T_\Delta \times T_\Delta$ induces a canonical isomorphism $\mathbf{K}_T(Z) \simeq \mathbf{K}_{T/T_\Delta}(Z) \otimes R(T_\Delta)$; i.e., every T-equivariant vector bundle $E \to Z$ decomposes as

$$E = \sum_{\mu_1 \in \mathbb{Z}\beta'} E^{\mu_1} \otimes \mathbb{C}_{\mu_1}. \tag{3.46}$$

Here each E^{μ_1} is a T/T_Δ-equivariant vector bundle on Z, and $\mathbb{C}_{\mu_1}$ denotes the one-dimensional T_Δ-representation associated to $\mu_1 \in \Lambda^*_{\mathfrak{t}_\Delta}$.

For every T-equivariant vector bundle $E \to Z$, the character $\mathrm{RR}_0^{\xi}(Z, E)$ is equal to the T-equivariant index of the T-transversally elliptic symbol $\mathrm{Thom}_\xi(\mathcal{V}) \otimes p^*(E)$, where $\mathcal{V}$ is a small neighborhood of $\Phi^{-1}(\xi) \cap Z$ in Z (see Definition 3.9).

Since the T_Δ action is trivial on Z, the symbol $\mathrm{Thom}_\xi(\mathcal{V})$ is also T/T_Δ transversally elliptic and the action of T_Δ is trivial on it. We have then

$$\mathrm{RR}_0^\xi(Z, E) = \sum_{\mu_1 \in \mathbb{Z}\beta'} \mathrm{RR}_0^\xi(Z, E^{\mu_1}) \otimes \mathbb{C}_{\mu_1}, \tag{3.47}$$

where the character $\mathrm{RR}_0^\xi(Z, E^{\mu_1}) \in R^{-\infty}(T/T_\Delta)$ is computed by Theorem 3.22 applied to the Hamiltonian T/T_Δ-manifold Z. For every T-vector bundle $E \to Z$ we define the family $\mathcal{E}_\xi^{\mu_1,\mu_2}$, $\mu_1 \in \mathbb{Z}\beta'$, $\mu_2 \in \Lambda^*_{\mathfrak{t}/\mathfrak{t}_\Delta}$, of orbifold vector bundles over the reduced space $\mathcal{Z}_\xi = Z \cap \Phi^{-1}(\xi)/(T/T_\Delta)$ by

$$\mathcal{E}_\xi^{\mu_1,\mu_2} := (E^{\mu_1} \otimes \mathbb{C}_{-\mu_2})|_{\Phi^{-1}(\xi)\cap Z}/(T/T_\Delta). \tag{3.48}$$

Finally, (3.22) and (3.47) give the following:

$$\begin{aligned}\mathrm{RR}_0^\xi(Z, E) &= \sum_{\mu_1 \in \mathbb{Z}\beta'} \sum_{\mu_2 \in \Lambda^*_{\mathfrak{t}/\mathfrak{t}_\Delta}} \mathrm{RR}(\mathcal{Z}_\xi, \mathcal{E}_\xi^{\mu_1,\mu_2}) \otimes \underbrace{\mathbb{C}_{\mu_1}}_{\in R(T_\Delta)} \otimes \underbrace{\mathbb{C}_{\mu_2}}_{\in R(T/T_\Delta)} \\ &= \sum_{\mu \in \Lambda^*} \mathrm{RR}(\mathcal{Z}_\xi, \mathcal{E}_\xi^{\mu_1,\mu_2})\, \mathbb{C}_\mu. \end{aligned} \tag{3.49}$$

In (3.49) we write $\mu \in \Lambda^*$ as a sum of $\mu_1 \in \mathbb{Z}\beta'$ with $\mu_2 \in \Lambda^*_{\mathfrak{t}/\mathfrak{t}_\Delta}$, so that $\mathbb{C}_\mu \in R(T)$ is equal to the tensor product $\mathbb{C}_{\mu_1} \otimes \mathbb{C}_{\mu_1}$.

When the vector bundle $E \to Z$ is the line bundle $\mathbb{L} := L^{\otimes k}|_Z \otimes \det(N_Z^{+,\beta})$ we have $\mathbb{L} = \mathbb{L}^{j\beta'} \otimes \mathbb{C}_{j\beta'}$ for $j = kr_\Delta + s^+$. Finally, we have

$$D_Z(\mu_2, k) = \mathrm{RR}(\mathcal{Z}_\xi, \mathbb{L}_\xi^{(kr_\Delta+s^+)\beta',\mu_2}).$$

Now we use the results of Section 3.4 to study the map

$$D_Z : \Lambda^*_{\mathfrak{t}/\mathfrak{t}_\Delta} \times \mathbb{Z} \longrightarrow \mathbb{Z}. \tag{3.50}$$

Let $\Gamma_Z \subset T/T_\Delta$ be the generic stabiliser of T/T_Δ on a component Z. Let $\alpha_Z, \delta_Z \in \Lambda^*_{\mathfrak{t}/\mathfrak{t}_\Delta}$ be such that the action of Γ_Z on the fibers of $L_{|Z}$ and $\det(N_Z^{+,\beta})$ are respectively $t \to t^{\alpha_Z}$ and $t \to t^{\delta_Z}$. From Remark 3.17 we know that the map (3.50) is supported on the subset

$$\Xi_Z := \{(\mu_2, k) \in \Lambda^*_{\mathfrak{t}/\mathfrak{t}_\Delta} \times \mathbb{Z} \mid t^{k\alpha_Z+\delta_Z+\mu_2} = 1,\ \forall\, t \in \Gamma_Z\}. \tag{3.51}$$

The only difference with the computations done in Section 3.4 is the line bundle $\det(N_Z^{+,\beta})$. But this does not change the global behaviour of the map (3.50) on Ξ_Z: it is a periodic polynomial map of degree $d_Z = \dim(\mathcal{Z}_\xi)/2$, and we have

$$D_Z(\mu_2, k) = \frac{1}{(2\pi)^{d_Z}} \int_{\mathcal{Z}_\xi} \frac{(k\Omega_{\mathcal{Z}_\xi} - \langle k\xi - \mu_2, \omega_{\mathcal{Z}_\xi}\rangle)^{d_Z}}{d_Z!} + O(d_Z - 1) \tag{3.52}$$

for all $(\mu_2, k) \in \Xi_Z$.

Suppose now that all the signs $(-1)^{\mathrm{rk}_{\mathbb{C}}(N_Z^{+,\beta})}$ coincide when $Z \in \mathcal{F}_+$. From (3.45), we get that $D(\mu_2, k)$ does not vanish for large values of (μ_2, k).

4 Multiplicities of group representations

Let K be a semisimple compact Lie group with Lie algebra $\mathfrak{k}$, and let T be a maximal torus in K with Lie algebra $\mathfrak{t}$. In this section we denote by $(-,-)$ the scalar product on $\mathfrak{k}$ induced by the Killing form, and we keep the same notation for the induced scalar products on $\mathfrak{t}^*$ and on $\mathfrak{t}$.

Let $\Lambda^* \subset \mathfrak{t}^*$ be the weight lattice, and let $\mathfrak{R} \subset \Lambda^*$ be the set of roots for the action of T on $\mathfrak{k} \otimes \mathbb{C}$: we denote by $\Lambda^*_{\mathfrak{R}}$ the sublattice of Λ^* generated by $\mathfrak{R}$. We choose a system of positive roots $\mathfrak{R}^+ \subset \mathfrak{R}$, and we denote by $\mathfrak{t}^*_+$ the corresponding Weyl chamber.

The irreducible representations of K are parametrized by the set $\Lambda^*_+ = \Lambda^* \cap \mathfrak{t}^*_+$. For $\lambda \in \Lambda^*_+$ we denote by V_λ the irreducible representation of K with highest weight λ. Here we are interested in the T-multiplicities in $V_\lambda|_T$. Let $\mathrm{m} : \Lambda^* \times \Lambda^*_+ \to \mathbb{N}$ be the map defined by

$$V_\lambda|_T = \sum_{\mu \in \Lambda^*} \mathrm{m}(\mu, \lambda)\, \mathbb{C}_\mu \tag{4.1}$$

for every $\lambda \in \Lambda^*_+$.

Definition 4.1. For every $\lambda \in \Lambda^*_+$, we denote by $\mathrm{m}^\lambda : \Lambda^* \times \mathbb{Z}^{>0} \to \mathbb{N}$ the map defined by $\mathrm{m}^\lambda(\mu, k) = \mathrm{m}(\mu, k\lambda)$. So $\mathrm{m}^\lambda(\mu, k)$ is equal to the multiplicity of $\mathbb{C}_\mu$ in $V_{k\lambda}|_T$.

4.1 Borel–Weil Theorem

First we recall the realization of the K-representation V_λ given by the Borel–Weil theorem. The coadjoint orbit $K \cdot \lambda$ is equipped with the Kirillov–Kostant–Souriau symplectic form Ω, which is defined by

$$\Omega(X_M, Y_M)_m = \langle m, [X, Y] \rangle, \quad \text{for} \quad m \in K \cdot \lambda \quad \text{and} \quad X, Y \in \mathfrak{k}. \tag{4.2}$$

The action of K on $K \cdot \lambda$ is Hamiltonian with moment map $K \cdot \lambda \hookrightarrow \mathfrak{k}^*$ equal to the inclusion. The action of T on $K \cdot \lambda$ is also Hamiltonian with moment map $\Phi : K \cdot \lambda \to \mathfrak{t}^*$ equal to the composition of the inclusion $K \cdot \lambda \hookrightarrow \mathfrak{k}^*$ with the projection map $\mathfrak{k}^* \to \mathfrak{t}^*$.

There exists a unique K-invariant complex structure on $K \cdot \lambda$ compatible with the symplectic form. In this situation the Kostant–Souriau prequantum line bundle over $K \cdot \lambda$ is

$$\mathbb{C}_{[\lambda]} = K \times_{K_\lambda} \mathbb{C}_\lambda.$$

Here we use the canonical identification $K/K_\lambda \simeq K \cdot \lambda$, $[k] \mapsto k \cdot \lambda$, where K_λ is the stabilizer of λ in K. The line bundle $\mathbb{C}_{[\lambda]}$ over the complex manifold $K \cdot \lambda$ carries a canonical holomorphic structure. If one works with the symplectic form $k\Omega$, for an integer $k \geq 1$, the corresponding Kostant–Souriau prequantum line bundle is $\mathbb{C}_{[\lambda]}^{\otimes k} = K \times_{K_\lambda} \mathbb{C}_{k\lambda} = \mathbb{C}_{[k\lambda]}$.

Let $\mathcal{H}^q(K \cdot \lambda, \mathbb{C}_{[\lambda]}^{\otimes k})$ be the qth cohomology group of the sheaves of holomorphic sections of $\mathbb{C}_{[\lambda]}^{\otimes k}$ over $K \cdot \lambda$. The Borel–Weil theorem tells us that

$$\mathcal{H}^0(K \cdot \lambda, \mathbb{C}_{[\lambda]}^{\otimes k}) = V_{k\lambda} \tag{4.3}$$

and

$$\mathcal{H}^q(K \cdot \lambda, \mathbb{C}_{[\lambda]}^{\otimes k}) = 0 \quad \text{for} \quad q \geq 1. \tag{4.4}$$

If $\mathrm{RR}^K(K \cdot \lambda, -) : \mathbf{K}_K(K \cdot \lambda) \to R(K)$ is the K-equivariant Riemann–Roch character defined by the compatible complex structure, (4.3) and (4.4) give

$$\mathrm{RR}^K(K \cdot \lambda, \mathbb{C}_{[\lambda]}^{\otimes k}) = V_{k\lambda} \quad \text{in} \quad R(K). \tag{4.5}$$

Now if we denote by $\mathrm{RR}(K \cdot \lambda, -) : \mathbf{K}_T(K \cdot \lambda) \to R(T)$ the T-equivariant Riemann–Roch character, we have $V_{k\lambda}|_T = \mathrm{RR}(K \cdot \lambda, \mathbb{C}_{[\lambda]}^{\otimes k})$. The multiplicity function $\mathrm{m}^\lambda : \Lambda^*_+ \times \mathbb{N}^* \to \mathbb{N}$ is characterized by the relation

$$\mathrm{RR}(K \cdot \lambda, \mathbb{C}_{[\lambda]}^{\otimes k}) = \sum_{\mu \in \Lambda^*} \mathrm{m}^\lambda(\mu, k)\, \mathbb{C}_\mu, \quad \text{in } R(T), \tag{4.6}$$

for $k \geq 1$.

The sublattice $\Lambda^*_{\mathfrak{R}}$ of Λ^* generated by the roots is characterized by the (finite) center $Z(K)$ of K as follows. For $\alpha \in \Lambda^*$ we have

$$\lambda \in \Lambda^*_{\mathfrak{R}} \Longleftrightarrow t^\lambda = 1, \quad \forall t \in Z(K), \tag{4.7}$$

and for $t \in T$ we have $t \in Z(K) \Longleftrightarrow t^\lambda = 1, \forall \lambda \in \Lambda^*_{\mathfrak{R}}$. The finite abelian group $\Lambda^*/\Lambda^*_{\mathfrak{R}}$ is then naturally identified with the dual of $Z(K)$. We have the following well-known fact.

Lemma 4.2. *The multiplicity map* m^λ *is supported on the sublattice* $\Xi_\lambda = \{(\mu, k) \in \Lambda^* \times \mathbb{Z} | \mu - k\lambda \in \Lambda^*_{\mathfrak{R}}\}$.

Proof. The center $Z(K)$ of K acts trivially on $K \cdot \lambda$ and with the character $t \in Z(K) \mapsto t^{k\lambda}$ on the fibers of the line bundle $\mathbb{C}_{[\lambda]}^{\otimes k}$. Since $\mathrm{m}^\lambda(\mu, k)$ is equal to the dimension of the T-invariant subspace of $\mathrm{RR}(K \cdot \lambda, \mathbb{C}_{[\lambda]}^{\otimes k}) \otimes \mathbb{C}_{-\mu}$, we have following Lemma 3.17 that $\mathrm{m}^\lambda(\mu, k) \neq 0$ only if $t^{\mu - k\lambda} = 1, \forall t \in Z(K)$. We conclude then with (4.7).

In this section we study the periodic polynomials

$$\mathrm{m}^\lambda_{\mathfrak{c}} : \Lambda^* \times \mathbb{Z} \longrightarrow \mathbb{Z}. \tag{4.8}$$

defined for every connected component $\mathfrak{c} \subset \mathfrak{t}^*$ of regular values of the moment map $\Phi : K \cdot \lambda \to \mathfrak{t}^*$. We know that $\mathrm{m}_{\mathfrak{c}}^{\lambda}$ is also supported on the sublattice Ξ_λ (see Section 3.3).

In order to apply Theorem 3.24 to the periodic polynomials $\mathrm{m}_{\mathfrak{c}}^{\lambda}$, we have to compute the critical values of the moment map $\Phi : K \cdot \lambda \to \mathfrak{t}^*$.

4.2 Critical points of $\Phi : K \cdot \lambda \to \mathfrak{t}^*$

Let $\{\alpha_1, \ldots, \alpha_{\dim T}\}$ be the simple roots of the set $\mathfrak{R}_+$ of positive weights. The fundamental weights ϖ_k, $1 \le k \le \dim T$, are defined by the conditions

$$2\frac{(\varpi_i, \alpha_j)}{|\alpha_j|^2} = \delta_{i,j} \quad \text{for all} \quad 1 \le i, j \le \dim T. \tag{4.9}$$

Recall that the fundamental weights generate the lattice Λ^*_{alg} of *algebraic* integral elements of $\mathfrak{t}^*$. We have $\Lambda^* \subset \Lambda^*_{\mathrm{alg}}$ and equality holds only if K is simply connected.

Let W be the Weyl group of (K, T). We will look at

$$\mathcal{G} = \{\sigma \cdot \varpi_i | \sigma \in W,\ 1 \le i \le \dim T\} \tag{4.10}$$

as a subset of $\mathfrak{t}$ modulo the identification $\mathfrak{t} \simeq \mathfrak{t}^*$ given by the scalar product. The singular points of Φ have the following nice description. This result first appeared in Heckman's thesis [22].

Proposition 4.3 ([22, 17]). *The set of critical points of $\Phi : K \cdot \lambda \to \mathfrak{t}^*$ is the union of the fixed-point set $(K \cdot \lambda)^\beta$, $\beta \in \mathcal{G}$. For each $\beta \in \mathcal{G}$ we have*

$$(K \cdot \lambda)^\beta = \bigcup_{\sigma \in W} K^\beta \cdot \sigma\lambda,$$

where K^β is the stabilizer subgroup of β in K.

The fixed points of the action of T on $K \cdot \lambda$ characterize the image of Φ completely: $\Phi(K \cdot \lambda)$ is the convex polytope

$$\mathrm{conv}(W \cdot \lambda) := \text{ convex hull of } W \cdot \lambda. \tag{4.11}$$

This result was first proved by Kostant [25]. This is a particular case of the convexity theorem of Atiyah, Guillemin, and Sternberg [2, 18]. From Proposition 4.3, we know that the singular values of $\Phi : K \cdot \lambda \to \mathfrak{t}^*$ are the convex polytopes

$$\mathrm{conv}(W^\beta \cdot \sigma\lambda), \quad \beta \in \mathcal{F},\ \sigma \in W/W^\beta, \tag{4.12}$$

where W^β is the stabilizer[6] of β in W, i.e., W^β is the Weyl group of (K^β, T). Each convex polytope $\mathrm{conv}(W^\beta \cdot \sigma\lambda)$ lies in the hyperplane

$$\Delta_{\beta,\sigma} = \{\xi \in \mathfrak{t}^* | (\xi - \sigma\lambda, \beta) = 0\}. \tag{4.13}$$

We will use the following lemma.

Lemma 4.4.

- $K^\beta \cdot \sigma\lambda = K^\beta \cdot \sigma'\lambda$ *if and only if* $\sigma\lambda \in W^\beta\sigma'\lambda$,
- $\mathrm{conv}(W^\beta \cdot \sigma\lambda) \cap \mathrm{conv}(W^\beta \cdot \sigma'\lambda) \neq \emptyset$ *if and only if* $\Delta_{\beta,\sigma} = \Delta_{\beta,\sigma'}$.

Proof. The first point follows from the fact that the intersection of a coadjoint orbit $K^\beta \cdot \mu$, $\mu \in \mathfrak{t}^*$, with $\mathfrak{t}^*$ is equal to $W^\beta \cdot \mu$.

It is sufficient to prove the second point for $\beta = \varpi_i$. The half-line $\mathbb{R}^{>0}\varpi_i$ is an edge of the Weyl chamber. It is well known that the following vector subspaces coincide:

- the line $\mathbb{R}\varpi_i$,
- the vector subspace of K^{ϖ_i}-invariant elements of $\mathfrak{k}^*$,
- the vector subspace of W^i-invariant elements of $\mathfrak{t}^*$.

Each convex polytope $\mathrm{conv}(W^i \cdot \sigma\lambda)$ contains the W^i-invariant element

$$\frac{1}{|W^i|} \sum_{\tau \in W^i} \tau \cdot \sigma\lambda,$$

which is equal to the intersection of the hyperplane $\Delta_{\beta,\sigma}$ with the line $\mathbb{R}\varpi_i$. Hence, if $\Delta_{\beta,\sigma} = \Delta_{\beta,\sigma'}$, the intersection $\mathrm{conv}(W^\beta \cdot \sigma\lambda) \cap \mathrm{conv}(W^\beta \cdot \sigma'\lambda)$ contains a W^i-invariant element, and then is not empty.

Definition 4.5. An element $\lambda \in \Lambda^*_+$ is *generic* if for every fundamental root ϖ_i and any $\sigma, \sigma' \in W$, we have

$$\Delta_{\beta,\sigma} \neq \Delta_{\beta,\sigma} \tag{4.14}$$

whenever the submanifolds $K^\beta \cdot \sigma\lambda$ and $K^\beta \cdot \sigma'\lambda$ are not equal.

This condition of genericity imposes that $(\sigma\lambda, \varpi_i) \neq (\sigma'\lambda, \varpi_i)$ when $\sigma\lambda \notin W^i\sigma'\lambda$.

Example 4.6. Consider the case of SU(4). Take the coadjoint orbit trough $\lambda = (2, 1, -1, -2)$, and σ, σ' such that $\sigma\lambda = (2, -2, 1, -1)$ and $\sigma'\lambda = (1, -1, 2, -2)$. Take the fundamental weight $\varpi_2 = \frac{1}{2}(1, 1, -1, -1)$. In this case λ is not "generic," since $\sigma\lambda \notin W^i\sigma'\lambda$ but $(\sigma\lambda, \varpi_2) = 0 = (\sigma'\lambda, \varpi_2)$.

[6] When $\beta = \varpi_i$, we denote by W^i the stabilizer of ϖ_i in W.

4.3 Main theorems

Let $\mathfrak{c}_+$ and $\mathfrak{c}_-$ be two adjacent connected components of regular values of $\Phi : K \cdot \lambda \to \mathfrak{t}^*$. The intersection $\overline{\mathfrak{c}_+} \cap \overline{\mathfrak{c}_+}$ is contained in a hyperplane orthogonal to $\beta \in \mathcal{F}$.

Definition 4.7. Let $\mathcal{A}(\mathfrak{c}_+, \mathfrak{c}_-)$ be the set of all $[\sigma] \in W/W^\beta$ such that the convex polytope $\mathrm{conv}(W^\beta \cdot \sigma\lambda)$ contains $\overline{\mathfrak{c}_+} \cap \overline{\mathfrak{c}_+}$.

Then

$$\bigcup_{[\sigma]\in\mathcal{A}(\mathfrak{c}_+,\mathfrak{c}_-)} K^\beta \cdot \sigma\lambda$$

is the union of the connected components of $(K \cdot \lambda)^\beta$ that intersect $\Phi^{-1}(\xi)$ when $\xi \in \overline{\mathfrak{c}_+} \cap \overline{\mathfrak{c}_+}$.

Remark 4.8. When λ is a regular element of $\mathfrak{t}^*$, all polytopes $\mathrm{conv}(W^\beta \cdot \sigma\lambda)$ are of codimension 1. When λ is "generic" (see Definition 4.5), the set $\mathcal{A}(\mathfrak{c}_+, \mathfrak{c}_-)$ is reduced to one element.

The multiplicity function $\mathrm{m}^\lambda : \Lambda^* \times \mathbb{Z}^{>0} \to \mathbb{N}$ is invariant under the action of the Weyl group: $\mathrm{m}^\lambda(\sigma\mu, k) = \mathrm{m}^\lambda(\mu, k)$ for every $\sigma \in W$. The set of connected components of regular values of Φ is also invariant under the action of W.

So, for the rest of this section we restrict our attention to case that $\mathfrak{c}_+$ and $\mathfrak{c}_-$ are separated by a hyperplane orthogonal to a fundamental weight $\beta = \varpi_i$: the vector ϖ_i is pointing out of $\mathfrak{c}_-$. We denote by K^i the stabilizer of ϖ_i in K.

Consider $[\sigma] \in \mathcal{A}(\mathfrak{c}_+, \mathfrak{c}_-)$ and let $K^i \cdot \sigma\lambda$ be the corresponding connected component of $(K \cdot \lambda)^\beta$. The tangent space of $K \cdot \lambda$ at $\sigma\lambda$ is the following $K^{\sigma\lambda}$-module:

$$\mathbf{T}_{\sigma\lambda}(K \cdot \lambda) = \sum_{(\alpha,\sigma\lambda)>0} \mathfrak{k}_\alpha, \tag{4.15}$$

where $\mathfrak{k}_\alpha \subset \mathfrak{k}\otimes\mathbb{C}$ is the one-dimensional complex subspace associated to the weight $\alpha \in \mathfrak{R}$. In the same way, the tangent space of $K^i \cdot \sigma\lambda$ at $\sigma\lambda$ is the $K^i \cap K^{\sigma\lambda}$-module defined by

$$\mathbf{T}_{\sigma\lambda}(K^i \cdot \sigma\lambda) = \sum_{\substack{(\alpha,\sigma\lambda)>0 \\ (\alpha,\varpi_i)=0}} \mathfrak{k}_\alpha. \tag{4.16}$$

Finally, the normal bundle of $K^i \cdot \sigma\lambda \simeq K^i/(K^i \cap K^{\sigma\lambda})$ in $K \cdot \lambda$ is $\mathcal{N}_{\sigma,i} = K^i \times_{K^i \cap K^{\sigma\lambda}} N_{\sigma,i}$, where

$$N_{\sigma,i} = \sum_{\substack{(\alpha,\sigma\lambda)>0 \\ (\alpha,\varpi_i)\neq 0}} \mathfrak{k}_\alpha. \tag{4.17}$$

For an element $\mu \in \mathfrak{t}^*$, we have $\mu = \sum_{i=1}^{\dim T} [\mu]_k \, \alpha_k$, where

$$[\mu]_k = 2\frac{(\varpi_k, \mu)}{|\alpha_k|^2} \in \mathbb{R}.$$

Note that $[\mu]_k \in \mathbb{Z}$ when μ belongs to the lattice $\Lambda^*_{\mathfrak{R}}$.

Definition 4.9. For $[\sigma] \in \mathcal{A}(\mathfrak{c}_+, \mathfrak{c}_-)$ we define the positive integers

$$s^{\pm}_{\sigma,i} = \pm \sum_{\substack{(\alpha,\sigma\lambda)>0 \\ \pm(\alpha,\varpi_i)>0}} [\alpha]_i .$$

Note that $s^{+}_{\sigma,i} + s^{-}_{\sigma,i}$ is larger than half of the codimension of $K^i \cdot \sigma\lambda$ in $K \cdot \lambda$.

Theorem 4.10. *Let $\mathfrak{c}_+$ and $\mathfrak{c}_-$ be two adjacent connected components of regular values of $\Phi : K \cdot \lambda \to \mathfrak{t}^*$ separated by a hyperplane orthogonal to a fundamental weight ϖ_i: we denote by r_i the common value $[\xi]_i$ for all ξ in this hyperplane. Let $\mathrm{m}^{\lambda}_{\mathfrak{c}_\pm} : \Lambda^* \times \mathbb{Z} \longrightarrow \mathbb{Z}$ be the corresponding periodic polynomials which are supported on the sublattice $\Xi_\lambda := \{(\mu, k) | \mu \in k\lambda + \Lambda^*_{\mathfrak{R}}\}$.*

For all $(\mu, k) \in \Xi_\lambda$, we have $\mathrm{m}^{\lambda}_{\mathfrak{c}_+}(\mu, k) = \mathrm{m}^{\lambda}_{\mathfrak{c}_-}(\mu, k)$ when the integer $[\mu]_i - kr_i$ satisfies

$$-s^{-}_{i} < [\mu]_i - kr_i < s^{+}_{i} . \tag{4.18}$$

Here the positive integers $s^{\pm}_i$ are defined by

$$s^{\pm}_i = \inf_{[\sigma]\in\mathcal{A}(\mathfrak{c}_+,\mathfrak{c}_-)} s^{\pm}_{\sigma,i} . \tag{4.19}$$

When $\mathcal{A}(\mathfrak{c}_+, \mathfrak{c}_-)$ is reduced to one element σ, for example if λ is "generic," the integer $s^{+}_i + s^{-}_i$ is larger than half of the codimension of $K^i \cdot \sigma\lambda$ in $K \cdot \lambda$.

Another way to express the result of Theorem 4.10 is to introduce, as in [36] the convex polytope

$$\square(\mathfrak{c}_+, \mathfrak{c}_-) = \bigcap_{\sigma\in\mathcal{A}(\mathfrak{c}_+,\mathfrak{c}_-)} \left(\sum_{(\alpha,\sigma\lambda)>0} [0, 1[\, \alpha \right) . \tag{4.20}$$

Let Δ be the hyperplane which separates $\mathfrak{c}_+$ and $\mathfrak{c}_-$. Equation (4.18) is equivalent to saying that

$$\mathrm{m}^{\lambda}_{\mathfrak{c}_+}(\mu, k) = \mathrm{m}^{\lambda}_{\mathfrak{c}_-}(\mu, k) \quad \text{if} \quad \mu \in k\Delta + \square(\mathfrak{c}_+, \mathfrak{c}_-). \tag{4.21}$$

Proof. Theorem 4.10 is a direct consequence of Theorem 3.24. The main difference between them is the decomposition of the lattice supporting the periodic polynomials. In the former we use the decomposition $\Lambda^* = \Lambda^*_{\mathfrak{t}_\Delta} \oplus \Lambda^*_{\mathfrak{t}/\mathfrak{t}_\Delta}$ associated to the choice of a subtorus T/T_Δ. Here we use the decomposition $\Lambda^*_{\mathfrak{R}} = \mathbb{Z}\alpha_i \oplus \sum_{k\neq i} \mathbb{Z}\alpha_k$. Note first that for $(\mu, k) \in \Xi_\lambda$, we have $\mu - \sigma\lambda \in \Lambda^*_{\mathfrak{R}}$ and then $[\mu - \sigma\lambda]_i = [\mu]_i - kr_i$ is an integer.

We start as after Proposition 3.23: $\mathrm{m}^{\lambda}_{\mathfrak{c}_+}(\mu, k) - \mathrm{m}^{\lambda}_{\mathfrak{c}_-}(\mu, k)$ is equal to the μ-multiplicity in $\sum_{\sigma\in\mathcal{A}(\mathfrak{c}_+,\mathfrak{c}_-)} A^{-}_{\sigma} - A^{+}_{\sigma}$, where

$$A^{\pm}_{\sigma} = \mathrm{RR}^{\xi}_{0}(K^i \cdot \sigma\lambda, \mathbb{C}^{\otimes k}_{[\lambda]} \otimes [\wedge^{\bullet}_{\mathbb{C}} \overline{\mathcal{N}_{\sigma,i}}\,]^{-1}_{\mp\varpi_i}). \tag{4.22}$$

Here ξ belongs to the relative interior of $\overline{\mathfrak{c}_+} \cap \overline{\mathfrak{c}_+}$, the line bundle $\mathbb{C}_{[\lambda]}^{\otimes k}$ is equal to $K^i \times_{K^i \cap K^{\sigma\lambda}} \mathbb{C}_{k\sigma\lambda}$, and $[\wedge_{\mathbb{C}}^{\bullet} \overline{\mathcal{N}_{\sigma,i}}\,]^{-1}_{\pm\varpi_i}$ corresponds to $(-1)^{\mathrm{rk}_{\mathbb{C}}(N^{\pm}_{\sigma,i})}$ times

$$K^i \times_{K^i \cap K^{\sigma\lambda}} (\det(N^{\pm}_{\sigma,i}) \otimes S^{\bullet}((N_{\sigma,i} \otimes \mathbb{C})^{\pm})),$$

with

$$N^{\pm}_{\sigma,i} = \sum_{\substack{(\alpha,\sigma\lambda)>0\\ \pm(\alpha,\varpi_i)>0}} \mathfrak{k}_\alpha$$

and

$$(N_{\sigma,i} \otimes \mathbb{C})^{\pm} = \sum_{\substack{(\alpha,\sigma\lambda)\neq 0\\ \pm(\alpha,\varpi_i)>0}} \mathfrak{k}_\alpha .$$

Now we can apply Remark 3.17 with the subgroup $H \subset T$ equal to the center $Z(K^i)$ of K^i: an element $\gamma \in \Lambda^*$ belong to $\sum_{k\neq i} \mathbb{Z}\alpha_k$ if and only if $t^\gamma = 1$ for all $t \in Z(K^i)$.

The group $Z(K^i)$ acts trivially on the manifolds $K^i \cdot \sigma\lambda$, and with the characters associated to the weights

$$k\sigma\lambda + \sum_{\substack{(\alpha,\sigma\lambda)>0\\ (\alpha,\varpi_i)>0}} \alpha + \delta \quad \text{with} \quad (\delta, \varpi_i) \geq 0$$

on the bundle $\mathbb{C}_{[\lambda]}^{\otimes k} \otimes [\wedge_{\mathbb{C}}^{\bullet} \overline{\mathcal{N}_{\sigma,i}}\,]^{-1}_{\varpi_i}$, and with the characters associated to the weights

$$k\sigma\lambda + \sum_{\substack{(\alpha,\sigma\lambda)>0\\ (\alpha,\varpi_i)<0}} \alpha + \delta \quad \text{with} \quad (\delta, \varpi_i) \leq 0$$

on the bundle $\mathbb{C}_{[\lambda]}^{\otimes k} \otimes [\wedge_{\mathbb{C}}^{\bullet} \overline{\mathcal{N}_{\sigma,i}}\,]^{-1}_{-\varpi_i}$. Now, the μ-multiplicity in $A^{\pm}_\sigma$ is not equal to 0 only if

$$k\sigma\lambda + \sum_{\substack{(\alpha,\sigma\lambda)>0\\ \pm(\alpha,\varpi_i)>0}} \alpha + \delta - \mu \in \sum_{k\neq i} \mathbb{Z}\alpha_k \quad \text{with} \quad \pm(\delta, \varpi_i) \geq 0. \tag{4.23}$$

Condition (4.23) implies that $[\mu]_i \geq k[\sigma\lambda]_i + s^+_{\sigma,i}$ or $[\mu]_i \leq k[\sigma\lambda]_i - s^-_{\sigma,i}$. Finally, we have to prove that $\mathrm{m}^{\lambda}_{\mathfrak{c}_+}(\mu, k) = \mathrm{m}^{\lambda}_{\mathfrak{c}_-}(\mu, k)$ if

$$-s^-_{\sigma,i} < [\mu]_i - k[\sigma\lambda]_i < s^+_{\sigma,i}$$

for all $\sigma \in \mathcal{A}(\mathfrak{c}_+, \mathfrak{c}_-)$.

4.4 The case of SU(n)

Let T be the maximal torus of SU(n) consisting of the diagonal matrices. The dual $\mathfrak{t}^*$ can be identified with the subspace $x_1 + \cdots + x_n = 0$ of $\mathbb{R}^n$. The roots are $\mathfrak{R} = \{e_i - e_j \,|\, 1 \leq i \neq j \leq n\}$, and we will choose the positive ones to be $\mathfrak{R}^+ = \{e_i - e_j | 1 \leq i < j \leq n\}$. The simple roots are then $\alpha_i = e_i - e_{i+1}$, for $1 \leq i \leq n-1$, and for these simple roots, the fundamental weights are

$$\varpi_k = \frac{1}{n}(\underbrace{n-k, n-k, \ldots, n-k}_{k \text{ times}}, \underbrace{-k, -k, \ldots, -k}_{n-k \text{ times}}), \quad 1 \leq k \leq n-1. \quad (4.24)$$

Consider now the coadjoint orbit $\mathcal{O}_\lambda$ for $\lambda \in \mathfrak{t}^*$. Let $\Phi : \mathcal{O}_\lambda \to \mathfrak{t}^*$ be the moment map associated to the Hamiltonian action of T on $\mathcal{O}_\lambda$. The center of SU(n), which we denote by $\mathbb{Z}_n$ corresponds to the set of matrices zI with $z^n = 1$. Recall the following well-known fact.

Lemma 4.11. *Let ξ be a regular value of $\Phi : \mathcal{O}_\lambda \to \mathfrak{t}^*$. Then for every $m \in \Phi^{-1}(\xi)$ the stabilizer subgroup $T_m := \{t \in T \mid t \cdot m = m\}$ is equal to $\mathbb{Z}_n$.*

Proof. Since ξ is a regular value, we know that T_m is finite for every $m \in \Phi^{-1}(\xi)$. The dual of the Lie algebra $\mathfrak{su}(n)$ decomposes as $\mathfrak{su}(n)^* = \mathfrak{t}^* \oplus \sum_{\alpha \in \mathfrak{R}^+} \mathfrak{su}(n)^*_\alpha$, where $\mathfrak{su}(n)^*_\alpha \simeq \mathbb{C}_{-\alpha}$ as a T-module. For $m \in \Phi^{-1}(\xi)$, we have $m = m_0 + \sum_{\alpha \in \mathfrak{R}^+} m_\alpha$ with $m_\alpha \in \mathfrak{su}(n)^*_\alpha$, and then $T_m = \bigcap_{m_\alpha \neq 0} \ker(t \mapsto t^\alpha)$. So the lattice Λ^*_m generated by the set $\{\alpha \in \mathfrak{R}^+ \mid m_\alpha \neq 0\}$ is a subgroup of $\Lambda^*_{\mathfrak{R}}$ with $\Lambda^*_{\mathfrak{R}}/\Lambda^*_m$ finite. We have to show that $\Lambda^*_m = \Lambda^*_{\mathfrak{R}}$. For this purpose we introduce the following equivalence relation on $\{1, \ldots, n\}$:

$$i \sim j \Longleftrightarrow e_i - e_j \in \Lambda^*_m.$$

Suppose that $\{1, \ldots, n\}/\sim$ is not reduced to a point. Let C_1 and C_2 be two distinct equivalent classes and let $\beta = (\beta_1, \ldots, \beta_n)$ be the element of $\mathfrak{t}^*$ defined by $\beta_i = \frac{1}{|C_1|}$ if $i \in C_1$, $\beta_i = \frac{-1}{|C_2|}$ if $i \in C_2$, and $\beta_i = 0$ in the other cases. We see then that $(\beta, \alpha) = 0$ for all $\alpha \in \Lambda^*_m$, which is in contradiction with the fact that $\Lambda^*_{\mathfrak{R}}/\Lambda^*_m$ is finite. We have proved that $e_i - e_j \in \Lambda^*_m$ for all $i, j \in \{1, \ldots, n\}$.

We are in the particularly nice situation in which the symplectic reduction $(\mathcal{O}_\lambda)_\xi = \Phi^{-1}(\xi)/T$ is a smooth manifold for every regular value ξ.

Suppose now that λ is a positive weight, and let $\mathfrak{c}$ be a connected component of regular values of $\Phi : \mathcal{O}_\lambda \to \mathfrak{t}^*$. We know that $\mathrm{m}^\lambda_{\mathfrak{c}} : \Lambda^* \times \mathbb{Z} \longrightarrow \mathbb{Z}$ is supported on the sublattice $\Xi_\lambda := \{(\mu, k) | \mu \in k\lambda + \Lambda^*_{\mathfrak{R}}\}$.

Corollary 4.12. *The map $\mathrm{m}^\lambda_{\mathfrak{c}} : \Xi_\lambda \longrightarrow \mathbb{Z}$ is a* polynomial *of degree $\frac{(n-1)(n-2)}{2} - d_\lambda$, where d_λ is the number of positive roots orthogonal to λ.*

Proof. Take $\xi \in \mathfrak{c}$. Following Proposition 3.18, the periodic polynomial $\mathrm{m}^\lambda_{\mathfrak{c}}$ is defined by $\mathrm{m}^\lambda_{\mathfrak{c}}(\mu, k) = \mathrm{RR}((\mathcal{O}_\lambda)_\xi, \mathcal{L}^k_{\xi,\mu})$ for all $(\mu, k) \in \Xi_\lambda$. Here $(\mathcal{O}_\lambda)_\xi =$

$\Phi^{-1}(\xi)/T$ is a *smooth* manifold, and the line bundle $\mathcal{L}^k_{\xi,\mu} = (L^{\otimes k}|_{\Phi^{-1}(\xi)} \otimes \mathbb{C}_{-\mu})/T$ is also *smooth*, since the center $\mathbb{Z}_n$ acts trivially on $L^{\otimes k}|_{\Phi^{-1}(\xi)} \otimes \mathbb{C}_{-\mu}$. Now the Atiyah–Singer integral formula for the Riemann–Roch number $\mathrm{RR}((\mathcal{O}_\lambda)_\xi, \mathcal{L}^k_{\xi,\mu})$ shows that $\mathrm{m}^\lambda_{\mathfrak{c}}$ is a polynomial of degree $\frac{\dim(\mathcal{O}_\lambda)_\xi}{2} = \frac{\dim \mathcal{O}_\lambda}{2} - (n-1) = \frac{(n-1)(n-2)}{2} - d_\lambda$.

Now we rewrite Theorem 4.10 for the group $\mathrm{SU}(n)$. Let $\lambda = (\lambda_1 \geq \cdots \geq \lambda_n)$ be a positive weight and let $\mathfrak{c}_+$ and $\mathfrak{c}_-$ be two adjacent connected components of regular values of $\Phi : \mathcal{O}_\lambda \to \mathfrak{t}^*$ separated by a hyperplane orthogonal to a fundamental weight ϖ_i. The vector ϖ_i is pointing out of $\mathfrak{c}_-$, and let $(\varpi_i, \xi) - r_i = 0$ be the equation of this hyperplane. We consider the linear map

$$Q(\xi, t) := (\varpi_i, \xi) - t r_i.$$

The hyperplane $\{Q = 0\} \subset \mathfrak{t}^* \times \mathbb{R}$ separates $\mathrm{Cone}(\mathfrak{c}_+)$ and $\mathrm{Cone}(\mathfrak{c}_-)$.

The conditions $(e_k - e_l, \sigma\lambda) > 0$ and $(e_k - e_l, \varpi_i) > 0$ are respectively equivalent to $\lambda_{\sigma(k)} > \lambda_{\sigma(l)}$ and $k \leq i < l$. For $\mathrm{SU}(n)$, the number $[\alpha]_i$ is equal to 0, 1, or -1 for any roots α and any $i = 1, \ldots, n-1$. Hence for every $\sigma \in \mathcal{A}(\mathfrak{c}_+, \mathfrak{c}_-)$, the integers $s^-_{\sigma,i}, s^+_{\sigma,i} \geq 0$ introduced in Definition 4.9 are equal to

$$s^+_{\sigma,i} = \mathrm{rk}_{\mathbb{C}}(N^+_{\sigma,i}) = \sharp\{k \leq i < l \text{ such that } \lambda_{\sigma(k)} > \lambda_{\sigma(l)}\}, \tag{4.25}$$

$$s^-_{\sigma,i} = \mathrm{rk}_{\mathbb{C}}(N^-_{\sigma,i}) = \sharp\{k \leq i < l \text{ such that } \lambda_{\sigma(k)} < \lambda_{\sigma(l)}\}, \tag{4.26}$$

and the sum $s^+_{\sigma,i} + s^-_{\sigma,i}$ is equal to half of the codimension of $K^i \cdot \sigma\lambda$ in $K \cdot \lambda$, that is, $s^+_{\sigma,i} + s^-_{\sigma,i} = i(n-i) - \dim(K^{\sigma\lambda}/K^i \cap K^{\sigma\lambda})/2$.

Now we make precise the results of [10].

Theorem 4.13.

- *The polynomial* $\mathrm{m}^\lambda_{\mathfrak{c}_-} - \mathrm{m}^\lambda_{\mathfrak{c}_+} : \Xi_\lambda \to \mathbb{Z}$ *is divisible by the linear factors*

$$(Q - s^-_i + 1), (Q - s^-_i + 2), \ldots, Q, \ldots, (Q + s^+_i - 2), (Q + s^+_i - 1),$$

where $s^\pm_i = \inf_{[\sigma] \in \mathcal{A}(\mathfrak{c}_+, \mathfrak{c}_-)} s^\pm_{\sigma,i}$.
- *The linear factors* $(Q - s^-_i)$ *and* $(Q - s^+_i)$ do not divide $\mathrm{m}^\lambda_{\mathfrak{c}_-} - \mathrm{m}^\lambda_{\mathfrak{c}_+}$.

Proof. The first part is the translation of Theorem 4.10. We have just to prove that the linear factors $(Q - s^-_i)$ and $(Q - s^+_i)$ do not divide $\mathrm{m}^\lambda_{\mathfrak{c}_-} - \mathrm{m}^\lambda_{\mathfrak{c}_+}$. This point is a direct application of Proposition 3.25. The only fact we use here is that $\mathrm{rk}_{\mathbb{C}}(N^\pm_{\sigma,i}) = s^\pm_{\sigma,i}$. So the number $\mathrm{rk}_{\mathbb{C}}(N^\pm_{\sigma,i})$ is constant for all $\sigma \in \mathcal{A}(\mathfrak{c}_+, \mathfrak{c}_-)$ for which $s^\pm_{\sigma,i} = s^\pm_i$.

We now rewrite Theorem 4.13 in the particular case that $\mathcal{A}(\mathfrak{c}_+, \mathfrak{c}_-)$ contains just one element. This happens when λ is a "generic" positive weight (see Definition 4.5), or when $\mathfrak{c}_+$ does not intersect $\Phi(\mathcal{O}_\lambda)$. Here a positive weight $\lambda = (\lambda_1 \geq \cdots \geq \lambda_n)$ is "generic" if for every pair of permutations σ, σ' and any $k = 1, \ldots, n-1$, we have

$$\sum_{i=1}^{k} \lambda_{\sigma(i)} \neq \sum_{i=1}^{k} \lambda_{\sigma'(i)}$$

when $(\lambda_{\sigma(1)}, \ldots, \lambda_{\sigma(n)}) \notin \mathfrak{S}_k \times \mathfrak{S}_{n-k}(\lambda_{\sigma'(1)}, \ldots, \lambda_{\sigma'(n)})$.

Corollary 4.14. *Let λ be a* regular *weight. Let $\mathfrak{c}_+$ and $\mathfrak{c}_-$ be two adjacent connected components of regular values of $\Phi : \mathcal{O}_\lambda \to \mathfrak{t}^*$ and suppose that $\mathcal{A}(\mathfrak{c}_+, \mathfrak{c}_-)$ contains just one element σ. Then the polynomial $\mathrm{m}^\lambda_{\mathfrak{c}_-} - \mathrm{m}^\lambda_{\mathfrak{c}_+} : \Xi_\lambda \to \mathbb{Z}$ is divisible by the $i(n-i)$ linear factors*

$$(Q - s_i^- + 1), (Q - s_i^- + 2), \ldots, q, \ldots, (Q + s_i^+ - 2), (Q + s_i^+ - 1),$$

where $s_i^\pm = s_{\sigma,i}^\pm$ are defined by (4.25) and (4.26). Moreover, the linear factors $(Q - s_i^-)$ and $(Q - s_i^+)$ do not divide $\mathrm{m}^\lambda_{\mathfrak{c}_-} - \mathrm{m}^\lambda_{\mathfrak{c}_+}$.

5 Vector partition functions

Let T be a torus with Lie algebra $\mathfrak{t}$ and let $\Lambda^* \subset \mathfrak{t}^*$ be the weight lattice. Let $R = \{\alpha_1, \ldots, \alpha_d\}$ be a subset of not necessarily distinct elements of Λ^* which are in an open half-space of $\mathfrak{t}^*$. We associate with the collection R a function

$$N_R : \Lambda^* \longrightarrow \mathbb{N}$$

called the *vector partition function* associated to R. By definition, for a weight μ, the value $N_R(\mu)$ is the number of solutions of the equation

$$\sum_{j=1}^{d} k_j \alpha_j = \mu, \quad k_j \in \mathbb{Z}^{\geq 0}, \quad j = 1, \ldots, d. \tag{5.1}$$

Let $C(R) \subset \mathfrak{t}^*$ be the closed convex cone generated by the elements of R, and denote by $\Lambda_R^* \subset \Lambda^*$ the sublattice generated by R. Obviously, $N_R(\mu)$ vanishes if μ does not belong to $C(R) \cap \Lambda_R^*$.

Suppose now that R generates the vector space $\mathfrak{t}^*$. Following [36], we will call a vector *singular* with respect to R if it is in a cone $C(\nu)$ generated by a subset $\nu \subset R$ of cardinality strictly less than $\dim T$. The connected components of $\mathfrak{t}^* \backslash \{\text{singular vectors}\}$ are called *conic chambers*. The periodic polynomial behavior of N_R on closures of conic chambers of the cone $C(R)$ is proved in [35]. We have the following refinement due to Szenes and Vergne [36]. Let us introduce the convex polytope

$$\square(R) = \sum_{j=1}^{d} [0, 1]\alpha_j. \tag{5.2}$$

We observe that $\mathfrak{c} - \square(R)$ is a neighborhood of $\overline{\mathfrak{c}}$ for any conic chamber $\mathfrak{c}$ of the cone $C(R)$. We have the following qualitative result.

Theorem 5.1 ([36]). *Let $\mathfrak{c}$ be a conic chamber of the cone $C(R)$. There exists a periodic polynomial $P_\mathfrak{c}$ on Λ^* such that for each $\mu \in \mathfrak{c} - \square(R)$, we have*

$$N_R(\mu) = P_\mathfrak{c}(\mu).$$

In Section 5.4 we will give another proof of Theorem 5.1.

Let $\mathfrak{c}_\pm \subset \mathfrak{t}^*$ be two adjacent conic chambers separated by the hyperplane $\Delta = \{\xi \in \mathfrak{t}^* | \langle \xi, \beta \rangle = 0\}$. Here $\beta \in \mathfrak{t}$ is chosen so that $\mathfrak{c}_\pm \subset \{\xi \in \mathfrak{t}^* | \pm \langle \xi, \beta \rangle > 0\}$. The aim of this section is to give a wall-crossing formula for the periodic polynomial $P_{\mathfrak{c}_+} - P_{\mathfrak{c}_-}$.

Note that the vector space Δ is generated by $R \cap \Delta$. We polarize the elements of R that are outside Δ. We define

$$R' = \{\epsilon_j \alpha_j | \langle \alpha_j, \beta \rangle \neq 0 \text{ and } \epsilon_j = \text{sign } \langle \alpha_j, \beta \rangle\}, \tag{5.3}$$

$$\delta^\pm = \sum_{\pm\langle \alpha_j, \beta \rangle > 0} \alpha_j, \tag{5.4}$$

and

$$r^\pm = \sharp\{j | \pm \langle \alpha_j, \beta \rangle > 0\}. \tag{5.5}$$

We now look at the vector space Δ equipped with the subset $R \cap \Delta \subset \Lambda^* \cap \Delta$, which lies entirely in an open half-space. Let $N_{R\cap\Delta} : \Lambda^* \cap \Delta \to \mathbb{N}$ be the corresponding vector partition function. It is easy to see that $\overline{\mathfrak{c}_+} \cap \overline{\mathfrak{c}_-}$ is contained in the closure of a conic chamber $\mathfrak{c}' \subset \Delta$ relative to $R \cap \Delta$. Following Proposition 5.1 there exists a periodic polynomial $P_{\mathfrak{c}'}$ on $\Lambda^* \cap \Delta$ such that for each $\mu \in \overline{\mathfrak{c}'} \cap \Lambda^*$, we have

$$N_{R\cap\Delta}(\gamma) = P_{\mathfrak{c}'}(\gamma).$$

Let $N_{R'} : \Lambda^* \to \mathbb{N}$ be the vector partition function associated to the polarized set of weight R' (see (5.3)). The main result of this section is the following.

Theorem 5.2. *The periodic polynomial $P_{\mathfrak{c}_+} - P_{\mathfrak{c}_-} : \Lambda^* \to \mathbb{Z}$ satisfies*

$$P_{\mathfrak{c}_+}(\mu) - P_{\mathfrak{c}_-}(\mu) = \sum_{\gamma \in \Lambda^* \cap \Delta} D(\mu - \gamma) P_{\mathfrak{c}'}(\gamma), \quad \mu \in \Lambda^*, \tag{5.6}$$

where $D : \Lambda^* \to \mathbb{Z}$ is defined by

$$D(\mu) = (-1)^{r^-} N_{R'}(\mu + \delta^-) - (-1)^{r^+} N_{R'}(-\mu - \delta^+).$$

The proof of Theorem 5.2 will be given in Section 5.5.

Corollary 5.3. $P_{\mathfrak{c}_+}(\mu) = P_{\mathfrak{c}_-}(\mu)$ *for all the weights $\mu \in \Lambda^*$ satisfying the condition*

$$-\langle\delta^+,\beta\rangle < \langle\mu,\beta\rangle < -\langle\delta^-,\beta\rangle.$$

These inequalities are optimal, since

$$(P_{\mathfrak{c}_+} - P_{\mathfrak{c}_-})(-\delta^- + \gamma) = (-1)^{r^-} P_{\mathfrak{c}'}(\gamma)$$

and

$$(P_{\mathfrak{c}_+} - P_{\mathfrak{c}_-})(-\delta^+ + \gamma) = (-1)^{1+r^+} P_{\mathfrak{c}'}(\gamma)$$

for all $\gamma \in \Lambda^* \cap \Delta$.

Proof. In (5.6), the term $D(\mu - \gamma)P_{\mathfrak{c}'}(\gamma)$ does not vanish only if $\mu - \gamma \in -\delta^- + C(R')$ or $-(\mu - \gamma) \in \delta^+ + C(R')$ for some $\gamma \in C(R \cap \Delta)$. These two conditions impose respectively that $\langle\mu,\beta\rangle \geq -\langle\delta^-,\beta\rangle$ and $\langle\mu,\beta\rangle \leq -\langle\delta^+,\beta\rangle$. If one takes $\mu = -\delta^- + \gamma$ with $\gamma \in \Lambda^* \cap \Delta$, (5.6) becomes $(P_{\mathfrak{c}_+} - P_{\mathfrak{c}_-})(-\delta^- + \gamma) = \sum_{\gamma' \in \Lambda^* \cap \Delta} D(-\delta^- + \gamma - \gamma')P_{\mathfrak{c}'}(\gamma')$ with

$$D(-\delta^- + \gamma - \gamma') = (-1)^{r^-} N_{R'}(\gamma - \gamma') - (-1)^{r^+} N_{R'}(\delta^- - \delta^+ - \gamma + \gamma').$$

Since the cone $C(R')$ intersects Δ only at $\{0\}$, $N_{R'}(\gamma - \gamma') = 0$ if $\gamma \neq \gamma'$. Since $\langle\delta^- - \delta^+,\beta\rangle < 0$, we always have $N_{R'}(\delta^- - \delta^+ - \gamma + \gamma') = 0$. We get finally that $(P_{\mathfrak{c}_+} - P_{\mathfrak{c}_-})(-\delta^- + \gamma) = (-1)^{r^-} P_{\mathfrak{c}'}(\gamma)$. One can show in the same way that $(P_{\mathfrak{c}_+} - P_{\mathfrak{c}_-})(-\delta^+ + \gamma) = -(-1)^{r^+} P_{\mathfrak{c}'}(\gamma)$.

5.1 Quantization of $\mathbb{C}^d$

We consider the complex vector space $\mathbb{C}^d$ equipped with the canonical symplectic form $\Omega = \frac{i}{2}\sum_{i=1}^d dz_j \wedge d\overline{z}_j$. The standard complex structure J on $\mathbb{C}^d$ is compatible with Ω. Let T be a torus, let $\alpha_j \in \mathfrak{t}^*$, $j = 1, \ldots, d$, be weights of T, and let T act on $\mathbb{C}^d$ as

$$t \cdot (z_1, \ldots, z_d) = (t^{-\alpha_1} z_1, \ldots, t^{-\alpha_d} z_d). \tag{5.7}$$

The action of T preserves the symplectic form Ω, and the moment map associated with this action is

$$\Phi(z) = \frac{1}{2}\sum_{i=1}^d |z_j|^2 \alpha_j. \tag{5.8}$$

The prequantization data $(L, \langle,\rangle, \nabla)$ on the Hamiltonian T-manifold $(\mathbb{C}^d, \Omega, \Phi)$ is a trivial line bundle L with a trivial action of T equipped with the Hermitian structure $\langle s, s'\rangle_z = e^{\frac{-|z|^2}{2}} s\overline{s'}$ and the Hermitian connexion $\nabla = d - \theta$, where $\theta = \frac{1}{2}\sum_{i=1}^d \overline{z}_j dz_j$.

The quantization of the Hamiltonian T-manifold $(\mathbb{C}^d, \Omega)$, which we denote by $\mathcal{Q}^T(\mathbb{C}^d)$, is the *Bargman space* of entire holomorphic functions on $\mathbb{C}^d$ which are $\mathcal{L}^2$ integrable with respect to the Gaussian measure $e^{\frac{-|z|^2}{2}} \Omega^d$.

We suppose now that the set of weights $R = \{\alpha_1, \ldots, \alpha_d\}$ is *polarized* by $\eta \in \mathfrak{t}$, which means that $\langle \alpha_j, \eta \rangle > 0$ for all j. The T-representation $\mathcal{Q}^T(\mathbb{C}^d)$ is then admissible, and we have the following equality in $R^{-\infty}(T)$:

$$\mathcal{Q}^T(\mathbb{C}^d) = \sum_{\mu \in \Lambda^*} N_R(\mu)\, \mathbb{C}_\mu, \tag{5.9}$$

where $N_R : \Lambda^* \to \mathbb{N}$ is the vector partition function associated to R. In other words, the generalized character of $\mathcal{Q}^T(\mathbb{C}^d)$ coincides with the generalized character of the symmetric algebra $S^\bullet(\overline{\mathbb{C}^d})$, where $\overline{\mathbb{C}^d}$ means $\mathbb{C}^d$ with the opposite complex structure.

For the remaining part of Section 5, we assume that the set of weights $R = \{\alpha_1, \ldots, \alpha_d\}$ is *polarized*, and *generates* the vector space $\mathfrak{t}^*$. The first assumption is equivalent to the fact that the moment map $\Phi : \mathbb{C}^d \to \mathfrak{t}^*$ is *proper*, and the second assumption is equivalent to the fact that the generic stabiliser of T on $\mathbb{C}^d$ is *finite*. Notice that the vectors of $\mathfrak{t}^*$ which are *singular* with respect to R correspond to the *singular values* of Φ.

In the next section we will show that $\mathcal{Q}^T(\mathbb{C}^d)$, viewed as an element of $R^{-\infty}(T)$, can be realized as the index of transversally elliptic symbols on $\mathbb{C}^d$. Then we will apply the techniques developed in Section 3. The main difference here is that we work with the *noncompact* manifold $\mathbb{C}^d$.

5.2 Transversally elliptic symbols on $\mathbb{C}^d$

Let $p : \mathbf{T}\mathbb{C}^d \to \mathbb{C}^d$ be the canonical projection. We consider the Thom symbol

$$\mathrm{Thom}(\mathbb{C}^d) \in \Gamma(\mathbf{T}\mathbb{C}^d, \hom(p^*(\wedge_{\mathbb{C}}^{\mathrm{even}} \mathbf{T}\mathbb{C}^d), p^*(\wedge_{\mathbb{C}}^{\mathrm{odd}} \mathbf{T}\mathbb{C}^d)))$$

associated to the standard Hermitian structure on $\mathbb{C}^d$. Obviously the symbol $\mathrm{Thom}(\mathbb{C}^d)$ is not elliptic, since its characteristic set is equal to the zero section in $\mathbf{T}\mathbb{C}^d$ (hence is not compact).

Now we deform the symbol $\mathrm{Thom}(\mathbb{C}^d)$ in order to obtain transversally elliptic symbols. Since $\mathbb{C}^d$ can be realized as an open subset of a compact T-manifold, we have a well-defined index map

$$\mathrm{Index}^T_{\mathbb{C}^d} : \mathbf{K}_T(\mathbf{T}_T\mathbb{C}^d) \longrightarrow R^{-\infty}(T).$$

Definition 5.4. For any $\eta \in \mathfrak{t}$, we define the symbol $\mathrm{Thom}^\eta(\mathbb{C}^d)$ by

$$\mathrm{Thom}^\eta(\mathbb{C}^d)(z, v) = \mathrm{Thom}(\mathbb{C}^d)(z, v - \eta_{\mathbb{C}^d}(z)), \quad (z, v) \in \mathbf{T}\mathbb{C}^d,$$

where $\eta_{\mathbb{C}^d}$ is the vector field on $\mathbb{C}^d$ generated by η.

The symbols $\mathrm{Thom}^\eta(\mathbb{C}^d)$ were studied in [32]. It is easy to see that $\mathrm{Thom}^\eta(\mathbb{C}^d)$ is transversally elliptic if and only if the vector subspace $(\mathbb{C}^d)^\eta$ is reduced to $\{0\}$, i.e., if $\langle\alpha_j,\eta\rangle \neq 0$ for all $j = 1,\ldots,d$. We prove in Proposition 5.4 of [32] that

$$\mathrm{Index}^T_{\mathbb{C}^d}(\mathrm{Thom}^\eta(\mathbb{C}^d)) = S^\bullet(\overline{\mathbb{C}^d}) \quad \text{in} \quad R^{-\infty}(T), \tag{5.10}$$

when $\langle\alpha_j,\eta\rangle > 0$ for all $j = 1,\ldots,d$.

In order to compute the multiplicities $N_R(\mu)$ of $\mathcal{Q}^T(\mathbb{C}^d)$ we introduce the following transversally elliptic symbols. Take a scalar product $b(\cdot,\cdot)$ on $\mathfrak{t}^*$, and denote by $\xi \mapsto \xi^b$, $\mathfrak{t}^* \simeq \mathfrak{t}$ the induced isomorphism. For each $\xi \in \mathfrak{t}^*$, the Hamiltonian vector field of the function $\frac{-1}{2}\|\Phi - \xi\|^2_b$ is the vector field

$$z \mapsto ((\Phi(z) - \xi)^b)_{\mathbb{C}^d}(z),$$

which we denote $\mathcal{H}^b - \xi^b_{\mathbb{C}^d}$.

Definition 5.5. For any $\xi \in \mathfrak{t}^*$ and any scalar product $b(\cdot,\cdot)$ on $\mathfrak{t}^*$, we define the symbol $\mathrm{Thom}_{\xi,b}(\mathbb{C}^d)$ by

$$\mathrm{Thom}_{\xi,b}(\mathbb{C}^d)(z,v) = \mathrm{Thom}(\mathbb{C}^d)(z, v - (\mathcal{H}^b - \xi^b_{\mathbb{C}^d})(z)), \quad (z,v) \in \mathbf{T}\mathbb{C}^d.$$

Let $\mathrm{Char}(\mathrm{Thom}_{\xi,b}(\mathbb{C}^d)) \subset \mathbf{T}\mathbb{C}^d$ be the characteristic set of $\mathrm{Thom}_{\xi,b}(\mathbb{C}^d)$. We know that $\mathrm{Char}(\mathrm{Thom}_{\xi,b}(\mathbb{C}^d)) \cap \mathbf{T}_T\mathbb{C}^d$ is equal to the critical set $\mathrm{Cr}(\|\Phi - \xi\|^2_b)$ of the function $\|\Phi - \xi\|^2_b : \mathbb{C}^d \to \mathbb{R}$ (see Section 3.2). A straightforward computation gives that $z \in \mathrm{Cr}(\|\Phi - \xi\|^2_b)$ if and only if

$$b(\Phi(z) - \xi, \alpha_j)\, z_j = 0 \quad \text{for all} \quad j = 1,\ldots,d. \tag{5.11}$$

The former relation implies in particular that $b(\Phi(z) - \xi, \Phi(z)) = \frac{1}{2}\sum_j b(\Phi(z) - \xi, \alpha_j)\,|z_j|^2 = 0$. Hence $\|\Phi(z)\|^2_b = b(\Phi(z),\xi)$, which implies

$$\|\Phi(z)\|_b \leq \|\xi\|_b. \tag{5.12}$$

Take now $\eta \in \mathfrak{t}$ such that $\langle\alpha_j,\eta\rangle > 0$ for all j, and let $\eta_b \in \mathfrak{t}^*$ be such that $(\eta_b)^b = \eta$. We have then

$$C_\eta\|z\|^2 \leq \langle\Phi(z),\eta\rangle = b(\Phi(z),\eta_b) \leq \|\Phi(z)\|_b\|\eta_b\|_b, \tag{5.13}$$

where $C_\eta = \frac{1}{2}\inf_j\langle\alpha_j,\eta\rangle$, and $z \mapsto \|z\|^2$ is the usual Hermitian form on $\mathbb{C}^d$. With (5.11) and (5.13) we get the following.

Lemma 5.6. *The critical set* $\mathrm{Cr}(\|\Phi - \xi\|^2_b) \subset \mathbb{C}^d$ *is contained in the ball of radius*

$$\frac{\|\xi\|_b\,\|\eta_b\|_b}{C_\eta},$$

where $\eta \in \mathfrak{t}$ *is such that* $C_\eta = \frac{1}{2}\inf_j\langle\alpha_j,\eta\rangle > 0$.

We have then proved that the symbols $\mathrm{Thom}_{\xi,b}(\mathbb{C}^d)$ are transversally elliptic.

Proposition 5.7. *The class of the transversally elliptic symbol* $\mathrm{Thom}_{\xi,b}(\mathbb{C}^d)$ *in* $\mathbf{K}_T(\mathbf{T}_T\mathbb{C}^d)$ *does not depend of the data* ξ, b*, and is equal to the class defined by* $\mathrm{Thom}^\eta(\mathbb{C}^d)$*, where* $\eta \in \mathfrak{t}$ *is chosen so that* $\langle \alpha_j, \eta\rangle > 0$ *for all* j.

Proof. By Lemma 5.6, we know that for any scalar product $b(\cdot,\cdot)$ on $\mathfrak{t}^*$, the characteristic set of $\mathrm{Thom}_{0,b}(\mathbb{C}^d)$ intersects $\mathbf{T}_T\mathbb{C}^d$ at $\{0\}$. If b_0 and b_1 are two scalar products on $\mathfrak{t}^*$, we consider the family $b_t = tb_1 + (1-t)b_0$, $0 \le t \le 1$, of scalar products on $\mathfrak{t}^*$. Hence $\mathrm{Thom}_{0,b_t}(\mathbb{C}^d)$, $t \in [0,1]$, defines a homotopy of transversally elliptic symbols. We have proved that $\mathrm{Thom}_{0,b_0}(\mathbb{C}^d) = \mathrm{Thom}_{0,b_1}(\mathbb{C}^d)$ in $\mathbf{K}_T(\mathbf{T}_T\mathbb{C}^d)$ for any $\xi \in \mathfrak{t}^*$.

Fix now the scalar product b and an element $\xi \in \mathfrak{t}^*$. For any $t \in [0,1]$ the characteristic set of $\mathrm{Thom}_{t\xi,b}(\mathbb{C}^d)$ intersects $\mathbf{T}_T\mathbb{C}^d$ in the ball of radius

$$\frac{\|\xi\|_b\, \|\eta_b\|_b}{C_\eta}.$$

Hence $\mathrm{Thom}_{t\xi,b}(\mathbb{C}^d)$, $t \in [0,1]$, defines a homotopy of transversally elliptic symbols: $\mathrm{Thom}_{\xi,b}(\mathbb{C}^d) = \mathrm{Thom}_{0,b}(\mathbb{C}^d)$ in $\mathbf{K}_T(\mathbf{T}_T\mathbb{C}^d)$. We have proved that the class of the transversally elliptic symbol $\mathrm{Thom}_{\xi,b}(\mathbb{C}^d)$ in $\mathbf{K}_T(\mathbf{T}_T\mathbb{C}^d)$ does not depend on the data ξ, b.

Since the weights α_j lie entirely in an open half-space of $\mathfrak{t}^*$, there exists a scalar product $b_+(\cdot,\cdot)$ on $\mathfrak{t}^*$ for which we have

$$b_+(\alpha_i, \alpha_j) > 0$$

for all $i, j = 1,\ldots,d$. Let $\mathcal{H}^{b_+}$ be the Hamiltonian vector field of the function $\frac{-1}{2}\|\Phi\|^2_{b_+}$, and let $\eta_{\mathbb{C}^d}$ be the vector field on $\mathbb{C}^d$ generated by $\eta \in \mathfrak{t}$ such that $\langle \alpha_j, \eta\rangle > 0$ for all j. A straightforward computation gives that

$$(\mathcal{H}^{b_+}(z), \eta_{\mathbb{C}^d}(z)) > 0 \tag{5.14}$$

for all nonzero $z \in \mathbb{C}^d$. Consider now the following family of symbols on $\mathbb{C}^d$:

$$\sigma_t(z,v) = \mathrm{Thom}(\mathbb{C}^d)(z, v - (t\mathcal{H}^{b_+} + (1-t)\eta_{\mathbb{C}^d})(z)), \quad (z,v) \in \mathbf{T}\mathbb{C}^d,$$

so that $\sigma_0 = \mathrm{Thom}^\eta(\mathbb{C}^d)$ and $\sigma_1 = \mathrm{Thom}_{0,b_+}(\mathbb{C}^d)$. The inequality (5.14) shows that $\mathrm{Char}(\sigma_t) \cap \mathbf{T}_T\mathbb{C}^d = \{0\}$ for all $t \in [0,1]$. Hence σ_t, $t \in [0,1]$, defines a homotopy of transversally elliptic symbols: $\mathrm{Thom}^\eta(\mathbb{C}^d) = \mathrm{Thom}_{0,b_+}(\mathbb{C}^d)$ in $\mathbf{K}_T(\mathbf{T}_T\mathbb{C}^d)$.

For the remaining part of this paper we fix a scalar product on $\mathfrak{t}^*$, and we consider the family of transversally elliptic symbols $\mathrm{Thom}_\xi(\mathbb{C}^d)$, $\xi \in \mathfrak{t}^*$ (to simplify, we do not mention the scalar product in the notation). Proposition 5.7 and (5.10) imply the following.

Proposition 5.8. *For every* $\xi \in \mathfrak{t}^*$, $\mathcal{Q}^T(\mathbb{C}^d)$ *is equal to the generalized character* $\mathrm{Index}^T_{\mathbb{C}^d}(\mathrm{Thom}_\xi(\mathbb{C}^d))$.

Now we apply the techniques developed in Section 3 in order to compute the multiplicities of $\mathrm{Index}^T_{\mathbb{C}^d}(\mathrm{Thom}_\xi(\mathbb{C}^d))$.

5.3 Localization in a noncompact setting

As in Section 3.2 we start with a definition.

Definition 5.9. For any $\xi \in \mathfrak{t}^*$ and any T-invariant *relatively compact* open subset $\mathcal{U} \subset \mathbb{C}^d$ we define the symbol $\mathrm{Thom}_\xi(\mathcal{U})$ by the relation

$$\mathrm{Thom}_\xi(\mathcal{U})(z,v) := \mathrm{Thom}(\mathbb{C}^d)(z, v - (\mathcal{H} - \xi_{\mathbb{C}^d})(z)) \quad (z,v) \in \mathbf{T}\mathcal{U}.$$

The symbol $\mathrm{Thom}_\xi(\mathcal{U})$ is transversally elliptic when $\mathrm{Cr}(\|\Phi - \xi\|^2) \cap \partial\mathcal{U} = \emptyset$ (the couple $(\mathcal{U}, \xi)$ is called *good*) and we denote by

$$\mathrm{RR}^\xi_{\mathcal{U}}(\mathbb{C}^d) \in R^{-\infty}(T)$$

its index. Proposition 3.10 is still valid here. In particular, for a good couple $(\mathcal{U}, \xi)$, we have $\mathrm{RR}^{\xi'}_{\mathcal{U}}(\mathbb{C}^d) = \mathrm{RR}^\xi_{\mathcal{U}}(\mathbb{C}^d)$ if ξ' is close enough to ξ. Consider now the decomposition

$$\mathrm{Cr}(\|\Phi - \xi\|^2) = \bigcup_{\gamma \in \mathcal{B}_\xi} (\mathbb{C}^d)^\gamma \cap \Phi^{-1}(\gamma + \xi).$$

Here $\mathcal{B}_\xi \subset \mathfrak{t}^*$ is finite set, since $\mathbb{C}^d$ has a finite number of stabilizers. Since $0 \in (\mathbb{C}^d)^\gamma$ and $z \mapsto \langle \Phi(z), \gamma \rangle$ is constant on $(\mathbb{C}^d)^\gamma$, we have

$$(\gamma + \xi, \gamma) = 0 \tag{5.15}$$

for all $\gamma \in \mathcal{B}_\xi$.

Definition 5.10. For any $\xi \in \mathfrak{t}^*$ and $\gamma \in \mathcal{B}_\xi$, we denote simply by

$$\mathrm{RR}^\xi_\gamma(\mathbb{C}^d) \in R^{-\infty}(T)$$

the generalized character $\mathrm{RR}^\xi_{\mathcal{U}}(\mathbb{C}^d)$, where $\mathcal{U}$ is a T-invariant relatively compact open neighborhood of $(\mathbb{C}^d)^\gamma \cap \Phi^{-1}(\gamma + \xi)$ such that $\mathrm{Cr}(\|\Phi - \xi\|^2) \cap \overline{\mathcal{U}} = (\mathbb{C}^d)^\gamma \cap \Phi^{-1}(\gamma + \xi)$.

Since $\mathrm{RR}^\xi_{\mathbb{C}^d}(\mathbb{C}^d)$ is equal to $\mathcal{Q}^T(\mathbb{C}^d)$ (see Proposition 5.8), part (*a*) of Proposition 3.10 ensures that we have the decomposition

$$\mathcal{Q}^T(\mathbb{C}^d) = \sum_{\gamma \in \mathcal{B}_\xi} \mathrm{RR}^\xi_\gamma(\mathbb{C}^d).$$

Let $\mathfrak{c} \subset \mathfrak{t}^*$ be a conic chamber of the cone $C(R)$, and take ξ in $\mathfrak{c}$. Then ξ is a regular value of the moment map $\Phi : \mathbb{C}^d \to \mathfrak{t}^*$ defined in (5.8). Let Ω_ξ be the symplectic structure on the orbifold $(\mathbb{C}^d)_\xi = \Phi^{-1}(\xi)/T$ that is induced from Ω. The orbifold $(\mathbb{C}^d)_\xi$ is also equipped with a *complex structure* J_ξ that is induced from the standard complex structure on $\mathbb{C}^d$, in such a way that the orbifold $((\mathbb{C}^d)_\xi, \Omega_\xi, J_\xi))$ is a Kähler orbifold. If ξ belongs to the lattice Λ^*, the reduced space $(\mathbb{C}^d)_\xi$ is the

Kähler toric variety corresponding to the polytope $\{s \in (\mathbb{R}^{\geq 0})^d | \sum s_j \alpha_j = \xi\}$ of $\mathbb{R}^d$. For every $\mu \in \Lambda$ we consider the holomorphic orbifold line bundle

$$\mathcal{L}_{\xi,\mu} = (\Phi^{-1}(\xi) \times \mathbb{C}_{-\mu})/T$$

on $(\mathbb{C}^d)_\xi$.

Definition 5.11. The periodic polynomial $P_{\mathfrak{c}} : \Lambda^* \to \mathbb{Z}$ associated to the conic chamber $\mathfrak{c}$ is given by

$$P_{\mathfrak{c}}(\mu) = \mathrm{RR}((\mathbb{C}^d)_\xi, \mathcal{L}_{\xi,\mu}), \tag{5.16}$$

where the right-hand side is the Riemann–Roch number associated to the holomorphic orbifold line bundle $\mathcal{L}_{\xi,\mu}$.

Another way to define the periodic polynomial $P_{\mathfrak{c}}$ is to consider the generalized character $\mathrm{RR}_0^\xi(\mathbb{C}^d)$ for $\xi \in \mathfrak{c}$: here $\gamma = 0$ parametrizes the component $\Phi^{-1}(\xi)$ of $\mathrm{Cr}(\|\Phi - \xi\|^2)$. By (3.22) we have

$$\mathrm{RR}_0^\xi(\mathbb{C}^d) = \sum_{\mu \in \Lambda^*} P_{\mathfrak{c}}(\mu)\, \mathbb{C}_\mu \quad \text{in} \quad R^{-\infty}(T). \tag{5.17}$$

By Lemma 3.16, we know that $\mathrm{RR}_0^{\xi'}(\mathbb{C}^d) = \mathrm{RR}_0^\xi(\mathbb{C}^d)$ when ξ, ξ' are two elements of $\mathfrak{c}$: hence the polynomial $P_{\mathfrak{c}}$ does not depend on the choice of ξ in $\mathfrak{c}$.

5.4 *Proof of Theorem 5.1*

Consider a weight $\mu \in (\mathfrak{c} - \square(R)) \cap \Lambda^*$ of the form $\mu = \xi' - \sum_j t_j \alpha_j$ with $\xi' \in \mathfrak{c}$ and $t_j \in [0, 1]$. We start with the decomposition

$$\mathcal{Q}^T(\mathbb{C}^d) = \sum_{\gamma \in \mathcal{B}_{\xi'}} \mathrm{RR}_\gamma^{\xi'}(\mathbb{C}^d).$$

Since $N_R(\mu)$ and $P_{\mathfrak{c}}(\mu)$ are respectively the multiplicities of $\mathbb{C}_\mu$ in $\mathcal{Q}^T(\mathbb{C}^d)$ and in $\mathrm{RR}_0^{\xi'}(\mathbb{C}^d)$, the proof will be complete if we show that the multiplicity of $\mathbb{C}_\mu$ in $\mathrm{RR}_\gamma^{\xi'}(\mathbb{C}^d)$ is equal to zero when $\gamma \neq 0$.

Consider a *nonzero* element γ in $\mathcal{B}_{\xi'}$. For the character $\mathrm{RR}_\gamma^{\xi'}(\mathbb{C}^d)$ the localization (3.20) gives

$$\mathrm{RR}_\gamma^{\xi'}(\mathbb{C}^d) = \mathrm{RR}_0^{\xi'+\gamma}((\mathbb{C}^d)^\gamma) \otimes [\wedge_{\mathbb{C}}^\bullet \overline{N}]_\gamma^{-1}, \tag{5.18}$$

where $N = \sum_{(\alpha_j,\gamma)\neq 0} \mathbb{C}_{-\alpha_j}$ corresponds to the normal bundle of $(\mathbb{C}^d)^\gamma$ in $\mathbb{C}^d$. The inverse $[\wedge_{\mathbb{C}}^\bullet \overline{N}]_\gamma^{-1}$ is equal to $(-1)^l \mathbb{C}_{\delta(\gamma)} \otimes S^\bullet(N_{\mathbb{C}}^{+,\gamma})$, where

$$\delta(\gamma) = -\sum_{(\alpha_j,\gamma)<0} \alpha_j.$$

Since γ acts trivially on $(\mathbb{C}^d)^\gamma$, all the weights $\mu' \in \Lambda^*$ that appear in $RR_0^{\xi'+\gamma}((\mathbb{C}^d)^\gamma)$ satisfy $(\mu', \gamma) = 0$. Since the weights of $N_{\mathbb{C}}^{+,\gamma}$ are polarized by γ, we see from (5.18) that all the weights $\mu' \in \Lambda^*$ that appear in $RR_\gamma^{\xi'}(\mathbb{C}^d)$ must satisfy

$$(\mu', \gamma) \geq (\delta(\gamma), \gamma). \tag{5.19}$$

Consider now the weight $\mu = \xi' - \sum_j t_j \alpha_j$. Since $\xi' \in \mathfrak{c}$, the equality (5.15) implies $(\xi', \gamma) < 0$ and then

$$(\mu, \gamma) = \underbrace{(\xi', \gamma)}_{<0} + \underbrace{\sum_{(\alpha_j, \gamma)>0} -t_j(\alpha_j, \gamma) - \sum_{(\alpha_j,\gamma)<0} t_j(\alpha_j, \gamma)}_{\leq 0} < - \sum_{(\alpha_j,\gamma)<0} (\alpha_j, \gamma).$$

So we have proved that $(\mu, \gamma) < (\delta(\gamma), \gamma)$; hence the multiplicity of $\mathbb{C}_\mu$ in $RR_\gamma^{\xi'}(\mathbb{C}^d)$ is equal to zero.

5.5 *Proof of Theorem 5.2*

Let $\mathfrak{c}_\pm$ be two adjacent conic chambers separated by the hyperplane $\Delta = \{\xi \in \mathfrak{t}^* | \langle \xi, \beta \rangle = 0\}$. Here β is pointing out of $\mathfrak{c}_-$.

We consider two points $\xi_\pm \in \mathfrak{c}_\pm$ such that $\xi = \frac{1}{2}(\xi^+ + \xi^-) \in \Delta$ belongs to the conic chamber $\mathfrak{c}'$. We suppose also that the orthogonal projections of $\xi_\pm$ on Δ are equal to ξ. We know that $P_{\mathfrak{c}_+}(\mu) - P_{\mathfrak{c}_-}(\mu)$ is equal to the μ-multiplicity of $RR_0^{\xi_+}(\mathbb{C}^d) - RR_0^{\xi_-}(\mathbb{C}^d)$. Proposition 3.23 tells us that

$$RR_0^{\xi_+}(\mathbb{C}^d) - RR_0^{\xi_-}(\mathbb{C}^d) = RR_\gamma^{\xi_-}(\mathbb{C}^d) - RR_{-\gamma}^{\xi_+}(\mathbb{C}^d),$$

where $\gamma \in \mathbb{R}^{>0}\beta$ is such that $\xi_- + \gamma = \xi_+ - \gamma = \xi$. The localization (3.20) gives then

$$RR_\gamma^{\xi_-}(\mathbb{C}^d) - RR_{-\gamma}^{\xi_+}(\mathbb{C}^d) = RR_0^{\xi}((\mathbb{C}^d)^\beta) \otimes ([\wedge_{\mathbb{C}}^\bullet \overline{N}]_\beta^{-1} - [\wedge_{\mathbb{C}}^\bullet \overline{N}]_{-\beta}^{-1}). \tag{5.20}$$

Let $P_{\mathfrak{c}'} : \Lambda^* \cap \Delta \to \mathbb{Z}$ be the periodic polynomial map which coincides with the vector partition function $N_{R\cap\Delta}$ on $\overline{\mathfrak{c}'} \cap \Lambda^*$. If we work with the vector space $(\mathbb{C}^d)^\beta$ equipped with the Hamiltonian action of T/T_Δ, (3.22) gives the following equality in $R^{-\infty}(T/T_\Delta)$:

$$RR_0^{\xi}((\mathbb{C}^d)^\beta) = \sum_{\gamma \in \Lambda^* \cap \Delta} P_{\mathfrak{c}'}(\gamma)\mathbb{C}_\gamma. \tag{5.21}$$

A straightforward computation gives

$$[\wedge_{\mathbb{C}}^\bullet \overline{N}]_\beta^{-1} = (-1)^{r^-} \sum_{\mu \in \Lambda^*} N_{R'}(\mu + \delta^-)\mathbb{C}_\mu \tag{5.22}$$

and

$$[\wedge_{\mathbb{C}}^{\bullet}\overline{N}]_{-\beta}^{-1} = (-1)^{r^+} \sum_{\mu\in\Lambda^*} N_{-R'}(\mu+\delta^+)\mathbb{C}_\mu, \tag{5.23}$$

where $r^{\pm}$, $\delta^{\pm}$, R' are defined in (5.3), (5.4), and (5.5). Since $N_{-R'}(\mu) = N_{R'}(-\mu)$, the equations (5.21), (5.22), and (5.23) show that the right-hand side of (5.20) is equal to

$$\sum_{\mu\in\Lambda^*}\sum_{\gamma\in\Lambda^*\cap\Delta} D(\mu)P_{\mathfrak{c}'}(\gamma)\mathbb{C}_{\mu+\gamma}$$

with $D(\mu) = (-1)^{r^-}N_{R'}(\mu+\delta^-)-(-1)^{r^+}N_{R'}(-\mu-\delta^+)$. Finally, we have proved that $P_{\mathfrak{c}_+}(\mu) - P_{\mathfrak{c}_-}(\mu) = \sum_{\gamma\in\Lambda^*\cap\Delta} D(\mu-\gamma)P_{\mathfrak{c}'}(\gamma)$.

References

1. M.F. ATIYAH, Elliptic operators and compact groups, Springer, 1974. Lecture notes in Mathematics, **401**.
2. M.F. ATIYAH, Convexity and commuting Hamiltonians, *Bull. London Math. Soc.* **14**, 1982, pp. 1–15.
3. M.F. ATIYAH and G.B. SEGAL, The index of elliptic operators II, *Ann. Math.* **87**, 1968, pp. 531–545.
4. M.F. ATIYAH and I.M. SINGER, The index of elliptic operators I, *Ann. Math.* **87**, 1968, pp. 484–530.
5. M.F. ATIYAH and I.M. SINGER, The index of elliptic operators III, *Ann. Math.* **87**, 1968, pp. 546–604.
6. M.F. ATIYAH and I.M. SINGER, The index of elliptic operators IV, *Ann. Math.* **93**, 1971, pp. 139–141.
7. N. BERLINE, E. GETZLER, and M. VERGNE, *Heat kernels and Dirac operators*, Grundlehren, vol. 298, Springer, Berlin, 1991.
8. N. BERLINE and M. VERGNE, The Chern character of a transversally elliptic symbol and the equivariant index, *Invent. Math.* **124**, 1996, pp. 11–49.
9. N. BERLINE and M. VERGNE, L'indice équivariant des opérateurs transversalement elliptiques, *Invent. Math.* **124**, 1996, pp. 51–101.
10. S. BILLEY, V. GUILLEMIN, and E. RASSART, A vector partition function for the multiplicities of $\mathrm{sl}_k(\mathbb{C})$, *J. Algebra* **278**, 2004, pp. 251–293.
11. A. BOYSAL and M. VERGNE, Paradan's wall crossing formula for partition functions and Khovanski–Pukhlikov differential operator, arXiv:math, CO/0803.2810.
12. M. BRION and C. PROCESI, *Action d'un tore dans une variété projective*, Operator algebras, unitary representations, enveloping algebras, and invariant theory (Paris, 1989), pp. 509–539, Progr. Math., 92, Birkhäuser Boston, Boston, MA, 1990.
13. W. DAHMEN and C.A. MICHELLI, The number of solutions to linear Diophantine equations and multivariate splines, *Trans. Amer. Math. Soc.* **308**, 1988, pp. 509–532.
14. C. DE CONCINI, C. PROCESI, and M. VERGNE, Partition function and generalized Dahmen–Micchelli spaces, arXiv:math, CO/0805.2907.
15. J.J. DUISTERMAAT and G.J. HECKMAN, On the variation in the cohomology in the symplectic form of the reduced phase space, *Invent. Math.* **69**, 1982, pp. 259–268; addendum, *ibid.* **72**, 1983; pp. 153–158.
16. J.J. DUISTERMAAT, *The heat Lefschetz fixed point formula for the Spin^c-Dirac operator*, Progress in Nonlinear Differential Equations and Their Applications, vol. 18, Birkhäuser, Boston, 1996.

17. V. GUILLEMIN, E. LERMAN, and S. STERNBERG, *Symplectic fibrations and multiplicity diagrams*. Cambridge University Press 1996.
18. V. GUILLEMIN and S. STERNBERG, Convexity properties of the moment mapping, *Invent. Math.* **67**, 1982, pp. 491–513.
19. V. GUILLEMIN and S. STERNBERG, Geometric quantization and multiplicities of group representations, *Invent. Math.* **67**, 1982, pp. 515–538.
20. V. GUILLEMIN and S. STERNBERG, Birational equivalence in the symplectic category, *Invent. Math.* **97**, 1989, pp. 485–522.
21. V. GUILLEMIN and S. STERNBERG, *Supersymmetry and equivariant de Rham theory*. With an appendix containing two reprints by Henri Cartan. Mathematics Past and Present. Springer-Verlag, Berlin, 1999.
22. G. J. HECKMAN, *Projections of orbits and asymptotic behavior of multiplicities for compact Lie groups*, PhD thesis, University of Leiden, 1980.
23. T. KAWASAKI, The Riemann–Roch theorem for complex V-manifolds, *Osaka J. Math.* **16**, 1979, pp. 151–157.
24. F. KIRWAN, *Cohomology of quotients in symplectic and algebraic geometry*, Princeton Univ. Press, Princeton, 1984.
25. B. KOSTANT, On convexity, the Weyl group and the Iwasawa decomposition, *Annales Scientifiques de l'E. N. S.* **6**, 1973, pp. 413–455.
26. S. KUMAR and M. VERGNE, Equivariant cohomology with generalized coefficients, *Astérisque*, 215, 1993, pp. 109–204.
27. E. MEINRENKEN, On Riemann–Roch formulas for multiplicities, *J. Amer. Math. Soc.* **9**, 1996, pp. 373–389.
28. E. MEINRENKEN, Symplectic surgery and the Spin^c-Dirac operator, *Adv. in Math.* **134**, 1998, pp. 240–277.
29. E. MEINRENKEN and S. SJAMAAR, Singular reduction and quantization, *Topology* **38**, 1999, pp. 699–762.
30. P.-E. PARADAN, Formules de localisation en cohomologie équivariante, *Compositio Mathematica* **117**, 1999, pp. 243–293.
31. P.-E. PARADAN, The moment map and equivariant cohomology with generalized coefficient, *Topology* **39**, 2000, pp. 401–444.
32. P.-E. PARADAN, Localization of the Riemann–Roch character, *J. Funct. Anal.* **187**, 2001, pp. 442–509.
33. P.-E. PARADAN, Note sur les formules de saut de Guillemin–Kalkman, *C. R. Acad. Sci. Paris*, Ser. I **339**, 2004, pp. 467–472.
34. R. SJAMAAR, Symplectic reduction and Riemann–Roch formulas for multiplicities, *Bull. Amer. Math. Soc.* **33**, 1996, pp. 327–338.
35. B. STURMFELS, On vector partition functions, *J. Combinatorial Theory* **71**, 1995, pp. 302–309.
36. A. SZENES and M. VERGNE, Residue formulae for vector partitions and Euler–Maclaurin sums, *Adv. in Applied Math.* **30**, 2003, pp. 295–342.
37. M. VERGNE, Multiplicity formula for geometric quantization, Part I, Part II, and Part III, *Duke Math. Journal* **82**, 1996, pp. 143–179, pp. 181–194, pp. 637–652.
38. M. VERGNE, Quantification géométrique et réduction symplectique, *Séminaire Bourbaki* **888**, 2001.

Eigenvalue distributions and Weyl laws for semiclassical non-self-adjoint operators in 2 dimensions

Johannes Sjöstrand

Dedicated to Hans Duistermaat

Abstract In this note we compare two recent results about the distribution of eigenvalues for semiclassical pseudodifferential operators in two dimensions. For classes of analytic operators A. Melin and the author [6] obtained a complex Bohr–Sommerfeld rule, showing that the eigenvalues are situated on a distorted lattice. On the other hand, with M. Hager [4] we showed in any dimension that Weyl asymptotics holds with probability close to 1 for small random perturbations of the operator. In both cases the eigenvalues distribute to leading order according to smooth densities, and we show here that the two densities are in general different.

Key words: Weyl law, random

Mathematics Subject Classification (2010): 35P20, 35R60

1 Introduction

In a classical paper by J.J. Duistermaat and L. Hörmander [1], one very interesting application is about (pseudo)differential operators with principal symbol p such that the Poisson bracket $\{p, \overline{p}\}$ vanishes on the zero set $p^{-1}(0)$ and the differentials of the real and imaginary part of p are independent there, so that the zero set is a codimension-2 submanifold of the cotangent space. The authors gave interesting existence results under noncompactness assumptions on the bicharacteristic foliation. In my thesis under the direction of L. Hörmander my task was to study the

CMLS, École Polytechnique, FR-91128 Palaiseau, France, UMR 7640, CNRS
Current address: Institut de Mathématiques de Bourgogne, Université de Bourgogne, 9 avenue Alain Savary - BP 47870, FR-21078 Dijon cedex, UMR 5584 du CNRS
e-mail: johannes.sjostrand@u.bourgogne.fr

case $\{p, \overline{p}\} \neq 0$ on the zero set of the symbol, and in a subsequent paper with Duistermaat [2] we introduced and studied certain microlocal projections onto the kernel and the cokernel of the operator. The full history of this subject can be traced back to the famous counterexample of Hans Lewy to local solvability and subsequent work by Hörmander and others, and there is also quite a rich recent history.

There has been a renewed interest in non-self-adjoint operators and the related notion of pseudospectrum, promoted by L.N. Trefethen, E.B. Davies, M. Zworski, and others. Again the Poisson bracket $i^{-1}\{p, \overline{p}\}$ plays an important role as a source of pseudospectral behaviour, including spectral instability. (We observe here that the above Poisson bracket is equal to the principal symbol of the commutator of the corresponding (pseudo)differential operator and its adjoint.) We refer to the surveys [8, 9], where further references can be found.

Possibly, as a reaction to these developments, the author participated in two projects:

- With A. Melin [6] we discovered for a fairly wide and stable class of non-self-adjoint semiclassical pseudodifferential operators in dimension 2 with analytic symbols that the individual eigenvalues in certain regions can be determined by a Bohr–Sommerfeld quantization rule defined in terms of certain complex Lagrangian tori (close to the real domain). The underlying idea is here to change the Hilbert space norm by means of exponential weights in such a way that the operator becomes (more) normal.
- M. Hager [3] considered certain non-self-adjoint h-pseudodifferential operators in dimension 1 with small multiplicative random perturbations and showed that with probability tending to 1 when $h \to 0$, the eigenvalues distribute according to the classical Weyl law, well known in the context of self-adjoint operators for almost a century. The same type of result was subsequently obtained in any dimension by Hager and the author [4] for a certain class of nonmultiplicative random perturbations and recently also for multiplicative random perturbations in any dimension by the author [10].

In the present note we shall compare the resulting distributions of eigenvalues in dimension 2. More precisely, let P have leading symbol p and satisfy the assumptions of [6] in the slightly strengthened form of Theorem 1 below. Then for generic p the density of eigenvalues of P according to [6] is different from the one for small random perturbations of P as in [4]. This means that the random perturbations will change radically the asymptotic distribution of eigenvalues. The intuitive explanation of this phenomenon is that the result of [6] depends on the geometry in the complex domain, while the random perturbation destroys analyticity, and hence the eigenvalue distribution should be given in terms of the real phase space, where the Weyl law is the natural candidate.

The corresponding phenomenon in one dimension is similar and easier. Here the (expected and sometimes well established) general situation is that the eigenvalues of an analytic pseudodifferential operator are confined to curves possibly with branch points, determined by complex Bohr–Sommerfeld rules, while the Weyl law for random perturbations will distribute the eigenvalues in the image by p of the real phase space. For example, if $P = (hD_x)^2 + ix^2$, $p(x,\xi) = \xi^2 + ix^2$, the eigenvalues will be confined to $e^{i\pi/4}[0, +\infty[$, while those of random perturbations of P will spread inside the first quadrant $= p(T^*\mathbf{R})$.

We next describe the main result of [6]. Let $p(x,\xi)$ be bounded and holomorphic in a tubular neighborhood of $\mathbf{R}^4$ in $\mathbf{C}^4 = \mathbf{C}^2_x \times \mathbf{C}^2_\xi$. (The assumptions near ∞ can be varied in many ways and we can let p belong to some more general symbol space as long as we have the appropriate form of ellipticity near infinity; cf. (2) below.) Assume that

$$\mathbf{R}^4 \cap p^{-1}(0) \neq \emptyset \text{ is connected,} \tag{1}$$

for simplicity. Also assume that

$$\text{on } \mathbf{R}^4 \text{ we have } |p(x,\xi)| \geq 1/C, \text{ for } |(x,\xi)| \geq C, \tag{2}$$

for some $C > 0$, and

$$d\Re p(x,\xi), d\Im p(x,\xi) \text{ are linearly independent for all } (x,\xi) \in p^{-1}(0) \cap \mathbf{R}^4. \tag{3}$$

It follows that $p^{-1}(0) \cap \mathbf{R}^4$ is a compact (2-dimensional) surface.

Also assume that

$$|\{\Re p, \Im p\}| \text{ is sufficiently small on } p^{-1}(0) \cap \mathbf{R}^4. \tag{4}$$

By "sufficiently small" we mean that $|\{\Re p, \Im p\}| < \delta$ for some $\delta > 0$ that will depend on all constants (implicit or explicit) that are required to express the other conditions above uniformly.

In [6] we showed that $p^{-1}(z) \cap \mathbf{R}^4$ is a real torus for $z \in \text{neigh}\,(0, \mathbf{C})$ (i.e., some neighborhood of 0 in $\mathbf{C}$) and that there exists a smooth 2-dimensional torus $\Gamma(z) \subset p^{-1}(z) \cap \mathbf{C}^4$ close to $p^{-1}(z) \cap \mathbf{R}^4$ such that $\sigma_{|\Gamma(z)} = 0$ and $I_j(z) \in \mathbf{R}$, $j = 1, 2$. Here $I_j(z) := \int_{\gamma_j(z)} \xi \cdot dx$ are the actions along two fundamental cycles $\gamma_1(z), \gamma_2(z) \subset \Gamma(z)$ and $\sigma = \sum_1^2 d\xi_j \wedge dx_j$ is the complex symplectic (2,0)-form. Moreover, $\Gamma(z)$, $I_j(z)$ depend smoothly on $z \in \text{neigh}\,(0)$.

The main result of [6], valid under slightly more general assumptions than the ones above, is then the following

Theorem 1. *Under the above assumptions, there exist a neighborhood V of $0 \in \mathbf{C}$, $\theta_0 \in (\frac{1}{2}\mathbf{Z})^2$, $\theta_j \in C^\infty(V; \mathbf{R}^2)$, and $\theta(z; h) \sim \theta_0 + \theta_1(z)h + \theta_2(z)h^2 + \cdots$ in $C^\infty(V; \mathbf{R}^2)$ such that for $z \in V$ and for $h > 0$ sufficiently small, z is an eigenvalue of $P = p^w(x, hD_x)$ iff*

$$\frac{(I_1(z), I_2(z))}{2\pi h} = k - \theta(z; h), \text{ for some } k \in \mathbf{Z}^2. \tag{BS}$$

Here $p^w(x, hD)$ denotes the Weyl quantization of the symbol $p(x, h\xi)$.

Let us also assume that

$$\text{the map } z \mapsto I(z) := (I_1(z), I_2(z)) \text{ is a diffeomorphism from } V \text{ to } I(V). \tag{5}$$

This assumption is satisfied if we strengthen (4) by assuming that $|\{\Re p, \Im p\}|$ is sufficiently small on $p^{-1}(z)$ for all $z \in \mathrm{neigh}(0, \mathbf{C})$ and choose V small enough. The eigenvalues near 0 will then form a distorted lattice, and we introduce the leading spectral density function $0 < \omega(z) \in C^\infty(V)$ by

$$dI_1(z) \wedge dI_2(z) = \pm\omega(z) d\Re z \wedge d\Im z, \tag{6}$$

where the sign is chosen so that ω becomes positive. Then from Theorem 1 it follows that for every $W \Subset V$ with smooth boundary, the number of eigenvalues in W satisfies

$$N(W; h) = \frac{1}{(2\pi h)^2}\left(\int_W \omega(z) L(dz) + o(1)\right), \quad h \to 0. \tag{7}$$

Here $L(dz) = d\Re z\, d\Im z$ denotes the Lebesgue measure.

Now we turn to the results in [3, 4, 10]. Again the unperturbed operator is of the form $P = p^w(x, hD_x)$, where the complex-valued smooth symbol should belong to a suitable symbol class and satisfy an ellipticity condition at infinity which guarantees that the spectrum of P in a given open set $\Omega \Subset \mathbf{C}$ is discrete. The perturbed operator is of the form $P_\delta = P + \delta Q_\omega$, where the parameter δ is small, say bounded from above by some positive power of h and from below by $e^{-h^{-\alpha}}$ for some suitable value $\alpha \in]0, 1]$. Under some additional assumptions on the type of random perturbation and about nonconstancy of the symbol p, it is shown in the cited works that with a probability that tends to 1 when $h \to 0$, the number of eigenvalues of P_δ in $W \Subset \Omega$ obeys

$$N_\delta(W; h) = \frac{1}{(2\pi h)^n}(\mathrm{vol}(p^{-1}(W)) + o(1)) \tag{8}$$

uniformly for W in a class of subsets of Ω with uniformly smooth boundary. (In the case of multiplicative perturbations, an additional symmetry assumption on the symbol is imposed which cannot be completely eliminated.)

Notice that this result can be formulated as in (7) with the density ω replaced by the Weyl density $w(z)L(dz)$, defined to be the direct image of the symplectic volume element under the map p, so that

$$\int f(z) w(z) L(dz) = \iint f(p(x,\xi)) dx\, d\xi, \quad f \in C_0^\infty(V). \tag{9}$$

In the two-dimensional case there are situations (for instance in the case of the symbol $p(x,\xi) = \frac{1}{2}((x_1^2 + \xi_1^2) + i(x_2^2 + \xi_2^2))$ − const and small perturbations of that symbol) where Theorem 1 applies to P and the results of [4, 7] apply to small random perturbations, and it is then of interest to compare the spectral densities. We shall see that $\omega(z) = w(z)$ in the integrable case when $\{\Re p, \Im p\} \equiv 0$ but that these quantities are different in general.

Theorem 2. *Under the assumptions (1)–(5) we have generically that* $w \not\equiv \omega$.

In other words, if $w \equiv \omega$, then there are arbitrarily small perturbations of P within the class of operators as in the theorem for which $w \not\equiv \omega$.

2 The integrable case

In this section, we strengthen the assumption (4) to

$$\{\Re p, \Im p\} \equiv 0. \tag{10}$$

It is then well known by the Liouville–Mineur–Arnold theorem (see [11]) that there exists a real symplectic diffeomorphism $\kappa : \text{neigh}\,(\eta = 0, T^*\mathbf{T}^2) \to \text{neigh}\,(p^{-1}(0) \cap \mathbf{R}^4, \mathbf{R}^4)$ (i.e., from a neighborhood of $\{\eta = 0\}$ in $T^*\mathbf{T}^2$ to a neighborhood of $p^{-1}(0) \cap \mathbf{R}^4$ in $\mathbf{R}^4$) such that

$$p \circ \kappa = \widetilde{p}(\eta) \tag{11}$$

is independent of y, where $\mathbf{T}^2 = (\mathbf{R}/2\pi\mathbf{Z})^2$ and $T^*\mathbf{T}^2 \simeq \mathbf{T}^2_y \times \mathbf{R}^2_\eta$.

In this case $\Gamma(z)$ is simply the real Lagrangian torus $p^{-1}(z) \cap \mathbf{R}^4$ and

$$I_j(z) = 2\pi\eta_j + I_j(0), \quad \widetilde{p}(\eta) = z. \tag{12}$$

It follows that up to composition with κ, we get the same quantities $\omega(z)$, $w(z)$ if we do the computation directly on $T^*\mathbf{T}^2$ for $\widetilde{p}(\eta)$, and we restrict the attention to that case and drop the tilde.

From (12) and (6) we get

$$\frac{\omega(z)}{(2\pi)^2} = \left|\det \frac{\partial(\eta_1, \eta_2)}{\partial(\Re p, \Im p)}\right|, \quad p(\eta) = z. \tag{13}$$

From (9), we get for $f \in C_0^\infty(V)$:

$$\begin{aligned}\int f(z)w(z)L(dz) &= \iint f(p(y,\eta))dy\,d\eta \\ &= (2\pi)^2 \int f(p(\eta))d\eta \\ &= (2\pi)^2 \int f(z) \left|\det \frac{\partial(\eta_1, \eta_2)}{\partial(\Re p, \Im p)}\right| L(dz),\end{aligned}$$

which shows that $w(z)$ also satisfies (13), so

$$w(z) = \omega(z), \quad z \in V, \tag{14}$$

in the completely integrable case (10).

3 The general case

In this section we shall prove Theorem 2 by means of calculations similar to those in [5]. Let p satisfy the assumptions (1)–(5). Let $G_t(x,\xi)$ for $t \in \mathrm{neigh}\,(0, \mathbf{R})$ be a smooth family of functions that are holomorphic and uniformly bounded in a fixed tubular neighborhood of $\mathbf{R}^4$. Possibly after decreasing the neighborhood of $t = 0$ we get a smooth family of canonical transformations κ_t from a fixed tubular neighborhood of $\mathbf{R}^4$ onto a neighborhood of $\mathbf{R}^4$, by solving the Cauchy problem

$$\frac{d}{dt}\kappa_t(\rho) = (\kappa_t)_*(\widehat{iH_{G_t}})(\rho), \quad \kappa_0(\rho) = \rho, \tag{15}$$

where

$$H_{G_t} = \frac{\partial G_t}{\partial \xi}\frac{\partial}{\partial x} - \frac{\partial G_t}{\partial x}\frac{\partial}{\partial \xi}$$

is the holomorphic Hamilton field (of type $(1,0)$) and we identify iH_{G_t} with the corresponding real vector field $\widehat{iH_{G_t}} := iH_{G_t} + \overline{iH_{G_t}}$.

Put $p_t = p \circ \kappa_t$. Then (possibly after further shrinking the neighborhood of $t = 0$) p_t will satisfy the assumptions (1)–(5), and since κ_t are complex canonical transformations, we also know that

$$\omega_t = \omega \text{ is independent of } t. \tag{16}$$

In order to prove Theorem 2, it suffices to prove the following result.

Theorem 3. *For every neighborhood V of $0 \in \mathbf{C}$, we can find a family G_t as above such that every neighborhood of $t = 0$ will contain a t for which $w_t \not\equiv w$ in V. Here w_t denotes the Weyl density of p_t, defined as in (9).*

Remark 1. Actually, we shall prove the theorem in all dimensions (replacing 2 by any $0 < n \in \mathbf{N}$) for any $p \not\equiv 0$ that is bounded and holomorphic in a tubular neighborhood of $\mathbf{R}^{2n}$ in $\mathbf{C}^{2n}$ that satisfies (2) and for which $p^{-1}(0) \cap \mathbf{R}^{2n} \neq \emptyset$. Let $w_t(z)L(dz)$ be the measure defined as in (9) with $p = p_0$ replaced by $p_t = p \circ \kappa_t$.

Proof. For $f \in C_0^\infty(V; \mathbf{R})$ we get

$$\begin{aligned}\int f(z)\frac{\partial w_t(z)}{\partial t}L(dz) &= \frac{d}{dt}\int f(z)w_t(z)L(dz)\\ &= \frac{d}{dt}\iint f(p_t(x,\xi))dx\,d\xi\\ &= \iint \left(\frac{\partial f}{\partial z}(p_t)\frac{\partial p_t}{\partial t} + \frac{\partial f}{\partial \overline{z}}(p_t)\frac{\partial \overline{p}_t}{\partial t}\right)dx\,d\xi.\end{aligned}$$

Here, we have

$$\frac{\partial p_t}{\partial t} = iH_{G_t}p_t,$$

and using that f is real,

$$\begin{aligned}
\int f(z)\frac{\partial w_t(z)}{\partial t}L(dz) &= 2\Re\left(i\iint \frac{\partial f}{\partial z}(p_t)H_{G_t}p_t dx\,d\xi\right)\\
&= -2\Re\left(i\iint \frac{\partial f}{\partial z}(p_t)H_{p_t}(G_t)dx\,d\xi\right)\\
&= 2\Re\left(i\iint H_{p_t}\left(\frac{\partial f}{\partial z}(p_t)\right)G_t\,dx\,d\xi\right)\\
&= 2\Re\left(i\iint\left(\frac{\partial^2 f}{\partial z^2}H_{p_t}(p_t)+\frac{\partial^2 f}{\partial\overline{z}\partial z}H_{p_t}(\overline{p}_t)\right)G_t\,dx\,d\xi\right)\\
&= 2\Re\left(\iint \frac{\partial^2 f}{\partial\overline{z}\partial z}(p_t)i\{p_t,\overline{p}_t\}G_t\,dx\,d\xi\right)\\
&= \iint(\Delta f)(p_t)\{\Re p_t,\Im p_t\}\Re G_t\,dx\,d\xi. \qquad (17)
\end{aligned}$$

If $\{\Re p,\Im p\}=\frac{i}{2}\{p,\overline{p}\}$ does not vanish identically, there are points arbitrarily close to $p^{-1}(0)$ where it does not vanish and we can choose $f\in C_0^\infty(V;\mathbf{R})$ (where V is any fixed neighborhood of $0\in\mathbf{C}$) such that $(\Delta f)(p)\{\Re p,\Im p\}$ does not vanish identically. We can then choose $G=G_0$ independent of t with the properties above so that

$$\int f(z)\left(\frac{\partial}{\partial t}\right)_{t=0}w_t(z)L(dz)=\iint(\Delta f)(p)\{\Re p,\Im p\}\Re G\,dx\,d\xi\neq 0.$$

We get the conclusion of Theorem 3 in this case.

If $\{\Re p,\Im p\}\equiv 0$, we choose G real and independent of t in (17) and differentiate that identity once with respect to t at $t=0$ to get

$$\begin{aligned}
\int f(z)\left(\frac{\partial^2 w_t}{\partial t^2}\right)_{t=0}L(dz) &= \iint(\Delta f)(p)\left(\frac{\partial}{\partial t}\right)_{t=0}\left(\frac{i}{2}\{p_t,\overline{p_t}\}\right)G\,dx\,d\xi\\
&= \iint(\Delta f)(p)\frac{i}{2}(\{iH_G p,\overline{p}\}+\{p,\overline{iH_G p}\})G\,dx\,d\xi\\
&= -\frac{1}{2}\iint(\Delta f)(p)(H_{\overline{p}}H_pG+H_pH_{\overline{p}}G)G\,dx\,d\xi.
\end{aligned}$$

Here we integrate by parts and use that $H_pp=0$, $H_{\overline{p}}p=0$, to get

$$\int f(z)\left(\frac{\partial^2 w_t}{\partial t^2}\right)_{t=0}L(dz)=\iint(\Delta f)(p)|H_pG|^2dx\,d\xi.$$

Again we see that we can find $f \in C_0^\infty(V; \mathbf{R})$ and $G = G_0$ as above, so that the last integral is $\neq 0$. The conclusion in the theorem follows in this case also.

References

1. J.J. Duistermaat, L. Hörmander, *Fourier integral operators. II,* Acta Math. (3–4)128(1972), 183–269.
2. J.J. Duistermaat, J. Sjöstrand, *A global construction for pseudo-differential operators with non-involutive characteristics,* Invent. Math. 20(1973), 209–225.
3. M. Hager, *Instabilité spectrale semiclassique d'opérateurs non-autoadjoints. II,* Ann. Henri Poincaré 7(6)(2006), 1035–1064.
4. M. Hager, J. Sjöstrand, *Eigenvalue asymptotics for randomly perturbed non-selfadjoint operators,* Math. Annalen, 342(1)(2008), 177–243.
5. A. Melin, J. Sjöstrand, *Determinants of pseudodifferential operators and complex deformations of phase space,* Methods and Applications of Analysis, 9(2)(2002), 177–238. Preprint: http://xxx.lanl.gov/abs/math.SP/0111292
6. A. Melin, J. Sjöstrand, *Bohr–Sommerfeld quantization condition for non-selfadjoint operators in dimension 2,* Astérique 284(2003), 181–244. Preprint: http://xxx.lanl.gov/abs/math.SP/0111293
7. J. Sjöstrand, *Function spaces associated to global I-Lagrangian manifolds,* pages 369–423 in Structure of solutions of differential equations, Katata/Kyoto, 1995, World Scientific 1996.
8. J. Sjöstrand, *Spectral instability for non-selfadjoint operators,* Algebraic analysis of differential equations, Springer 2008, 265–273.
9. J. Sjöstrand, *Some results on non-self-adjoint operators, a survey,* Further progress in analysis, 45–74, World Sci. Publ., Hackensack, NJ, 2009. Preprint: http://arxiv.org/abs/0804.3701
10. J. Sjöstrand, *Eigenvalue distribution for non-self-adjoint operators with small multiplicative random perturbations,* Annales Fac. Sci. Toulouse 18(4)(2009), 739–795. Preprint: http://arxiv.org/abs/0802.3584
11. S. Vũ Ngọc, *Systèmes intégrables semi-classiques: du local au global,* Panoramas et Synthèses, 22. Société Mathématique de France, Paris, 2006.

Symplectic inverse spectral theory for pseudodifferential operators

San Vũ Ngọc

Hans Duistermaat's enthusiasm and taste for good mathematics and mathematical writings were contagious, and have been a great source of inspiration for me.
This article is dedicated to his memory.

Abstract We prove, under some generic assumptions, that the semiclassical spectrum modulo $\mathcal{O}(\hbar^2)$ of a one-dimensional pseudodifferential operator completely determines the symplectic geometry of the underlying classical system. In particular, the spectrum determines the Hamiltonian dynamics of the principal symbol.

Key words: Inverse spectral theory, semiclassical spectral asymptotics, symplectic classification, microlocal analysis, pseudodifferential operators, Bohr-Sommerfeld rules

Mathematics Subject Classification (2010): 47G30, 81Q20, 35P20, 58F05

1 Introduction

In this article I would like to advocate an inverse spectral theory for pseudodifferential operators. What does this mean? One of the most famous inverse spectral problems, made fashionable by Kac's very entertaining article [12], with a mind-catching title "Can one hear the shape of a drum?"[1] was about the Laplace operator on a bounded domain $\Omega \subset \mathbb{R}^n$. Frequencies ν, solutions to the eigenvalue problem

$$\frac{1}{2}\Delta u = \nu^2 u, \qquad u = 0 \text{ on } \partial\Omega,$$

IRMAR (UMR 6625), Université de Rennes 1, Campus de Beaulieu, 35042 Rennes cedex (France)
e-mail: san.vu-ngoc@univ-rennes1.fr

[1] Kac attributes the problem to Bochner and the title to Bers.

may be viewed as harmonics that can be heard when the interior of the "membrane" Ω vibrates freely. The question was whether the knowledge of *all* frequencies completely determines Ω (up to isometry, of course). As Kac mentioned, this question appears naturally in the context of quantum mechanics, for a particle trapped in a hard potential well. An important observation in Kac's paper was the relevance of the *Weyl law*, which let us find the *volume* (or area when $n = 2$) of Ω from the *asymptotic behaviour of large eigenvalues*.

Counterexamples are now known: there are nonisometric shapes in $\mathbb{R}^2$ that produce different frequencies [10]. Nonetheless, this fact should not let us think that the problem has become obsolete. The seemingly simple case of a convex, bounded domain $\Omega \in \mathbb{R}^2$ with analytic boundary is still open; the closest result, in the presence of symmetry, is due to Zelditch, see [24, 23, 27].

Understanding this problem requires putting it in a wider perspective. A natural variant of Kac's problem is whether the spectrum of the Laplace operator Δ_g on a compact Riemannian manifold (M, g) determines the metric g. Although here again, counterexamples have been known for a long time [15], our understanding remains relatively poor. Recent work by Zelditch and Guillemin suggests that microlocal tools are quite relevant for all these questions. This, in turn, is a hint that more general operators than the Laplacian could be dealt with similarly.

From a quantum-mechanical viewpoint, Kac's situation is quite extreme. A more natural setting would involve a particle "trapped" by a smooth potential well. No more boundary problems, but instead a Schrödinger operator on $\mathbb{R}^n$,

$$P = -\frac{\hbar^2}{2}\Delta + V(x).$$

Of course now the *potential function* V should be recovered from the spectrum of P. This inverse spectral problem has been much studied, but only very recently have microlocal tools similar to those used by Guillemin and Zelditch been applied to it [11, 6, 3].

Here, I would like to shift again the initial problem one step further away. Instead of the Laplacian, or the Schrödinger operator, why not consider *any* (elliptic) differential operator, or even, while we're at it, any *pseudodifferential* operator? Of course, since there is no longer a domain Ω, no potential function V, the sensible question is, *what* should we try to recover from the spectrum?

The inverse spectral problems I have mentioned here can all be understood as *semiclassical limits*. From a quantum object, the spectrum, one wants to recover classical observables such as the metric g, or the potential V. These quantities, in turn, fully determine the *classical dynamics* of the system. For general pseudodifferential operators, semiclassical analysis still shows the strong relationship between the classical dynamics and the quantum spectrum, so I believe that the most natural "object" that we should try to recover from the spectrum is precisely this classical dynamics. This, precisely, amounts to determining the *principal symbol* of the operator. In fact, if we keep in mind Weyl's asymptotics, this sounds fairly natural, for it is well known that Weyl's asymptotics extend to arbitrary pseudodifferential operators, provided that we compute phase space volumes defined by energy ranges given by the principal symbol [22, 16].

As in the Riemannian case, one should take into account a symmetry group acting on the classical data. For general pseudodifferential operators, there is only one available: the group of symplectomorphisms, acting on the phase space M. This is a much bigger group than the group of Riemannian isometries, in accordance with the fact that the space of principal symbols $C^\infty(M)$ is much bigger than the space of Riemannian metrics, or potential functions.

Acknowledgements

I would like to thank the referee for suggesting several simplifications of the proofs. I am also grateful to Michael Van Valkenburgh who pointed out a mistake in the proof of Theorem 4.2.

2 The setting

Since we aim at recovering the classical dynamics from the spectrum, we are going to work in the setting of semiclassical pseudodifferential operators, which we recall here. Throughout this work, we consider only the *one-dimensional* theory. It would be very interesting to have higher-dimensional results, but it is not expected that such precise results would persist. However, a reasonable challenge would be to undertake a similar study for the *completely integrable* case.

The classes $\Psi^d(m)$ of semiclassical pseudodifferential operators that we use are standard. Let $M = T^*\mathbb{R} = \mathbb{R}^2_{(x,\xi)}$. Let d and m be real numbers. Let $S^d(m)$ be the set of all families $(p(\cdot;\hbar))_{\hbar\in(0,1]}$ of functions in $C^\infty(M)$ such that

$$\forall \alpha \in \mathbb{N}^2, \quad |\partial^\alpha_{(x,\xi)} p(x,\xi;\hbar)| \leq C_\alpha \hbar^d (1+|x|^2+|\xi|^2)^{\frac{m}{2}}, \tag{1}$$

for some constant $C_\alpha > 0$, uniformly in $\hbar$. Then $\Psi^d(m)$ is the set of all (unbounded) linear operators P on $L^2(\mathbb{R})$ that are $\hbar$-Weyl quantisations of symbols $p \in S^d(m)$:

$$(Pu)(x) = (Op^w_\hbar(p)u)(x) = \frac{1}{2\pi\hbar}\int_{\mathbb{R}^2} e^{\frac{i}{\hbar}\langle x-y,\xi\rangle} p\left(\frac{x+y}{2},\xi;\hbar\right) u(y)|dy\,d\xi|.$$

The number d in (1) is called the $\hbar$-order of the operator. Unless specified, it will always be zero here. In this work all symbols are assumed to admit a "classical" asymptotic expansion in integral powers of $\hbar$ (that is to say, in the ladder $(S^d(m))_{d\in\mathbb{Z},d\geq d_0}$ for some $d_0 \in \mathbb{Z}$). The leading term in this expansion is called the principal symbol of the operator.

Thus, the Schrödinger operator $P = -\frac{\hbar^2}{2}\Delta + V$ on $\mathbb{R}$ is a good candidate, of $\hbar$-order zero, whenever V has at most polynomial growth.

We use in this article the standard properties of such pseudodifferential operators. In particular, the composition sends $\Psi^d(m) \times \Psi^{d'}(m')$ to $\Psi^{d+d'}(m+m')$. Moreover, all $P \in \Psi^0(0)$ are bounded: $L^2(\mathbb{R}) \to L^2(\mathbb{R})$, uniformly for $0 < \hbar \leq 1$.

An operator $P \in \Psi(m)$ is said to be *elliptic at infinity* if there exists a constant $C > 0$ such that the principal symbol p satisfies

$$|p(x,\xi)| \geq \frac{1}{C}(|x|^2 + |\xi|^2)^{m/2}$$

for $|x|^2 + |\xi|^2 \geq C$.

If P has a real-valued Weyl symbol, then it is a symmetric operator on L^2 with domain $C_0^\infty(\mathbb{R})$. If furthermore the principal symbol is elliptic at infinity, then P is essentially self-adjoint (see for instance [8, Proposition 8.5]).

Finally, when $P \in \Psi^0(m)$ is self-adjoint and elliptic at infinity, then for any $f \in C_0^\infty(\mathbb{R})$, the operator $f(P)$ defined by functional calculus satisfies $f(P) \in \bigcap_{k\in\mathbb{N}}(\Psi^0(-km))$. See for instance [8] or [17] for details.

The advantage of the semiclassical theory is that it allows us to use richer versions of Weyl's asymptotics. Instead of considering the limit of large eigenvalues, we fix a bounded *spectral window* $I = [E_0, E_1] \subset \mathbb{R}$ and study the asymptotics of all eigenvalues in I, as $\hbar \to 0$.

Definition 2.1. We say that Assumption $\mathcal{A}(P, \mathcal{J}, I)$ holds whenever:

1. P is a self-adjoint pseudodifferential operator in $\Psi^0(m)$ with principal symbol p, elliptic at infinity.
2. $\mathcal{J} \subset [0, 1]$ is an infinite subset with zero as an accumulation point.
3. There exists a neighbourhood J of I such that $p^{-1}(J)$ is compact in M.

When $m > 0$, the properness condition 3 is implied by condition 1. If Assumption $\mathcal{A}(P, \mathcal{J}, I)$ holds, we denote by $\Sigma_\hbar(P, I)$ the spectrum of $P = P(\hbar)$ in I (including multiplicities). We denote by $\Sigma(P, \mathcal{J}, I)$ the family of all $\{\Sigma_\hbar(P, I); \hbar \in \mathcal{J}\}$. It is well known that $\Sigma_\hbar(P, I)$ is discrete for $\hbar$ small enough (see, e.g., [17, Théorème 3.13], in a slightly different setup).

Proposition 2.2. *Let P be a self-adjoint pseudodifferential operator in $\Psi(m)$, with principal symbol p, elliptic at infinity. Let $J \subset \mathbb{R}$ be a closed interval such that $p^{-1}(J)$ is compact. Then for any open interval $I \subset J$ there exists $\hbar_0 > 0$ such that the spectrum of P in I is discrete for $\hbar \leq \hbar_0$.*

Proof. Let $f \in C_0^\infty(J)$ be equal to 1 on I. Then by pseudodifferential functional calculus, $f(P)$ is compact for $\hbar$ small enough (see for instance [8, p. 115]). Therefore, denoting by Π_I the spectral projector on I, we have that $\Pi_I = \Pi_I f(P)$ is compact. This implies that Π_I has finite rank: the spectrum in I is discrete.

The goal of this article is to recover the dynamics of the Hamiltonian p in the region $p^{-1}(I)$ for any operator P for which Assumption $\mathcal{A}(P, \mathcal{J}, I)$ holds, for some subset $\mathcal{J} \subset [0, 1]$. Of course if we can do it for an arbitrary compact interval $I \subset \mathbb{R}$, we recover the full dynamics of p.

It turns out that under some genericity conditions, these inverse spectral problems are fairly easy, compared to the general multidimensional problems alluded to in the introduction, in the sense that they require only a few terms in the asymptotics of the spectrum. Having this in mind, for $\alpha \in \mathbb{R}$ we denote by $\Sigma(P, \mathcal{J}, I) + \mathcal{O}(\hbar^{\alpha})$ the equivalence class of all $\Sigma_{\hbar}(P, I)$ modulo $\hbar^{\alpha}$. Our main result is Theorem 5.2, but we also state several intermediate results that require weaker hypotheses. An informal statement of Theorem 5.2 is as follows.

Theorem 2.3 (Theorem 5.2). *Let Assumption $\mathcal{A}(P, \mathcal{J}, I)$ hold, and set $M = p^{-1}(I)$. Suppose that $p_{\restriction M}$ is a Morse function. Assume that the graphs of the periods of all trajectories of the Hamiltonian flow defined by $p_{\restriction M}$, as functions of the energy, intersect generically.*

Then the knowledge of $\Sigma(P, \mathcal{J}, I) + \mathcal{O}(\hbar^2)$ determines the dynamics of the Hamiltonian system $p_{\restriction M}$.

In fact, we determine completely the Hamiltonian p up to symplectic equivalence. Perhaps the most difficult step, for which Weyl's asymptotics are not enough, is the seemingly simple problem of counting the number of connected components of $p^{-1}(E)$, for a regular energy $E \in I$ (Theorem 4.2).

Although we state everything for pseudodifferential operators defined on $\mathbb{R}$, it is most probable that all results extend to the case of pseudodifferential operators defined on a one-dimensional compact manifold equipped with a smooth density, and to the case of Toeplitz operators on two-dimensional symplectic manifolds.

The plan of the paper follows a fairly logical progression. Since we always work modulo symplectomorphisms, it is not reasonable to look for a formula that would give the principal symbol p. Instead we will try to recover as many *symplectic invariants* as possible from the spectrum, so that given two spectra, we should be able to tell whether they come from isomorphic systems.

Thus, the geometric object under study is a proper map $p : M \to \mathbb{R}$, where M is a symplectic 2-manifold. The simplest symplectic invariants of this map are in fact topological invariants, and are dealt with in Sections 3 and 4. Indeed, it follows from the action-angle theorem that as soon as $E \in \mathbb{R}$ is a regular value of p, then the fibres of p consist of a finite number of closed loops, each one diffeomorphic to a circle. Therefore, we need to be able to detect

1. whether an energy $E \in \mathbb{R}$ is a regular or critical value of p; this is done in Section 3 (Theorem 3.1);
2. when E is a regular value, the number of connected components of the fibre $p^{-1}(E)$; Section 4.1 discusses this point (Theorem 4.2).

Putting these results together, we are able to recover the topological type of the singular fibration (Theorem 4.5). Then in Section 5, relying on the classification result of Dufour–Molino–Toulet [9, 18] (and some additional argument) we finally manage to recover the symplectic geometry of the system (Theorem 5.2).

3 Singularities

In order to detect whether a given energy $E_0 \in \mathbb{R}$ is a critical value of p, it is enough to know the spectrum of P in a small ball around E_0, at least under some nondegeneracy conditions.

Recall that a function $f : M \to \mathbb{R}$ is said to have a nondegenerate critical point $m \in M$ when $df(m) = 0$ and the Hessian $f''(m)$ is a nondegenerate quadratic form. Since M has dimension 2, there are only two cases:

1. **Elliptic case:** there are local symplectic coordinates (x, ξ) in T_mM such that $f''(m)(x, \xi) = C(x^2 + \xi^2)$, for some constant $C \neq 0$.
2. **Hyperbolic case:** there are local symplectic coordinates (x, ξ) in T_mM such that $f''(m)(x, \xi) = Cx\xi$, for some constant $C \neq 0$.

We refer to each of these two cases as the *type* of the singularity m.

Theorem 3.1. *Let I be an interval containing E_0 in its interior, and let Assumption $\mathcal{A}(P, \mathcal{J}, I)$ hold. Assume also that p has only nondegenerate critical values in I, and that any two critical points with the same singularity type cannot have the same image by p. Then from the knowledge of $\Sigma(P, \mathcal{J}, I) + \mathcal{O}(\hbar^2)$ one can infer*

1. whether E_0 is a critical value of p;
2. in case E_0 is a critical value, the type of the singularity.

Proof. We use Weyl's asymptotics, which can be obtained from a semiclassical trace formula as in [7, 3]: of $E_1 \in I$, then the number of eigenvalues of P in $[E_0, E_1]$ is equivalent, as $\hbar \to 0$, to

$$\frac{1}{2\pi\hbar} \int_{p^{-1}([E_0, E_1])} dx\, d\xi.$$

We fix E_0 and consider the behaviour of the "action" function

$$A(E_1) := \int_{p^{-1}([E_0, E_1])} dx\, d\xi.$$

Since $p^{-1}(E_0)$ is compact, it follows from the action-angle theorem that if E_0 is a regular value of p, then A is smooth in a neighbourhood of E_0. If E_0 is an elliptic critical value, then one of the Liouville tori in $p^{-1}(E)$, for E close to E_0 (on one side of E_0) degenerates to a point in the way that a harmonic oscillator energy level does at the origin. Then the function A admits a discontinuous derivative at E_0. All of this follows from the local normal form of elliptic singularities. Finally, if E_0 is a hyperbolic singularity, we can also use the local normal form at the singularity, which tells us that two Liouville tori coalesce into a unique one. In this case, $A(E)$ behaves like $(E - E_0) \ln |E - E_0|$, and thus its derivative diverges to $\pm\infty$.

Another approach to this theorem is to consider the *density of states* in small regions around E_0, which can be proved using Bohr–Sommerfeld rules. In fact,

Bohr–Sommerfeld rules are arguably too strong for the following proposition, but we use them here because we will have to refer to them later.

Proposition 3.2. *Let* $\gamma \in (0,1)$*, and for* $E \in I$*,*

$$\rho_\hbar(E) = \hbar^{1-\gamma}\#(\Sigma_\hbar(P, B(E, \hbar^\gamma))).$$

Then for any $E \in I$*, the limit* $\rho(E) = \lim_{\hbar\to 0}\rho_\hbar(E)$ *exists (in* $[0, +\infty]$*), and*

1. *if E is a regular value of p, then* ρ *is smooth at E;*
2. *if E is an elliptic critical value of p, then* ρ *is discontinuous;*
3. *if E is a hyperbolic critical value of p, then* $\rho(E) = +\infty$*.*

Proof. In case E is a regular value, the result follows directly from Weyl's asymptotics as before, or from the semiclassical Bohr–Sommerfeld rules as in [19]. Let us recall the Bohr–Sommerfeld approach. There exists an $\epsilon > 0$ such that the eigenvalues of P inside $[E-\epsilon, E+\epsilon]$ modulo $O(\hbar^\infty)$ are the union (with multiplicities) of a finite number of spectra σ_k, $k = 1, \dots, N$, where N is the number of connected components of $p^{-1}(E)$, and each σ_k is determined by quasi-modes microlocalised on the corresponding component. Precisely, the elements of σ_k are given by the solutions λ to the equation

$$g^{(k)}(\lambda;\hbar) \in 2\pi\hbar\mathbb{Z}, \tag{2}$$

where the function $g^{(k)}$ admits an asymptotic expansion of the form

$$g^{(k)}(\lambda;\hbar) \sim g_0^{(k)}(\lambda) + \hbar g_1^{(k)}(\lambda) + \hbar^2 g_2^{(k)}(\lambda) + \cdots \tag{3}$$

with smooth coefficients g_j. Moreover, if we denote by $\mathcal{C}_k(\lambda)$ the kth connected component of $p^{-1}(\lambda)$, in such a way that the family $(\mathcal{C}_k(\lambda))$ is smooth in the variable λ, then $g_0^{(k)}$ is the *action integral*:

$$g_0^{(k)}(\lambda) = \int_{\mathcal{C}_k(\lambda)} \xi dx. \tag{4}$$

From (2) it follows that for $\hbar$ small enough,

$$\#(\sigma_k \cap B(E,\epsilon)) = (2\pi\hbar)^{-1}|g^{(k)}(E+\epsilon;\hbar) - g^{(k)}(E-\epsilon;\hbar)| + \delta,$$

where $\delta \in [-1, 1]$ is here to take care of the appropriate integer part of the right-hand side. Hence

$$\#(\sigma_k \cap B(E,\epsilon)) = (2\pi\hbar)^{-1}\left|2\epsilon\frac{\partial g_0^{(k)}(E)}{\partial E} + \mathcal{O}(\epsilon^2) + \mathcal{O}(\hbar)\right| + \delta.$$

With $\epsilon = \hbar^\gamma$, this gives

$$\#(\sigma_k \cap B(E, \hbar^\gamma)) = \frac{\hbar^{\gamma-1}}{\pi} \left| \frac{\partial g_0^{(k)}(E)}{\partial E} \right| + \mathcal{O}(\hbar^{2\gamma-1}) + \mathcal{O}(1).$$

Summing up all contributions for $k = 1, \dots, N$, we get the first claim of the theorem, with

$$\rho(E) = \frac{1}{\pi} \left| \frac{\partial g_0^{(k)}(E)}{\partial E} \right|.$$

The second claim can be proved in a similar way, using Bohr–Sommerfeld rules for elliptic singularities [21]. For our purposes, a Birkhoff normal form as in [2] would even be enough, since we deal with energy intervals of size $\mathcal{O}(\hbar^\gamma)$. Here again there exists an $\epsilon > 0$ such that the eigenvalues of P inside $[E-\epsilon, E+\epsilon]$ modulo $O(\hbar^\infty)$ are the union (with multiplicities) of a finite number of spectra σ_k corresponding to the various connected components of $p^{-1}(E)$. The difference is that not all components need have critical points. In fact, by assumption only one component may have an elliptic critical point. Let us call σ_k the corresponding spectrum, and $\mathcal{C}_k(\lambda)$ the corresponding family of connected components. Since an elliptic critical point is a local extremum for p, the sets $\mathcal{C}_k(\lambda)$ are empty for all λ in one of the halves of the interval $[E-\epsilon, E+\epsilon]$. Without loss of generality, one can assume that $\mathcal{C}_k(\lambda) = \emptyset$, $\forall \lambda \in [E-\epsilon, E[$. Then $\mathcal{C}_k(E)$ is just a point, while $\mathcal{C}_k(\lambda)$ is a circle for all $\lambda \in]E, E+\epsilon]$.

The Bohr–Sommerfeld rules for elliptic singularities say that the elements of σ_k are the solutions λ to an equation of the form

$$e^{(k)}(\lambda; \hbar) \in 2\pi\hbar\mathbb{N}, \tag{5}$$

where the function $e^{(k)}$ admits an asymptotic expansion exactly as $g^{(k)}$ above (3). What is more, it is equally true that the principal term is an action integral:

$$e^{(k)}(E) = 0, \qquad e_0^{(k)}(\lambda) = \int_{\mathcal{C}_k(\lambda)} \xi\, dx, \quad \forall \lambda \in [E, E+\epsilon].$$

Calculating along the same lines as above, we find, for the quantity

$$\rho_\hbar^{(k)}(\lambda) := \hbar^{1-\gamma} \#(\sigma_k \cap B(\lambda, \hbar^\gamma))),$$

the following limits:

1. when $\lambda \in [E-\epsilon, E[$, $\lim_{\hbar\to 0} \rho_\hbar^{(k)}(\lambda) = 0$;
2. when $\lambda \in]E, E+\epsilon, E]$, $\lim_{\hbar\to 0} \rho_\hbar^{(k)}(\lambda) = \frac{1}{\pi}\left|\frac{\partial e_0^{(k)}(\lambda)}{\partial \lambda}\right|$;
3. $\lim_{\hbar\to 0} \rho_\hbar^{(k)}(E) = \frac{1}{2\pi}\left|\frac{\partial e_0^{(k)}(E)}{\partial E}\right|$;

Finally, let E be a hyperbolic critical value for p. Weyl asymptotics for such a situation have been worked out in [1], and the singular Bohr–Sommerfeld rules have been established in [4]. Using the latter result it can be proven as in [13] that the number of semiclassical eigenvalues generated by a hyperbolic fixed point, in a neighbourhood of size $\epsilon = \hbar^\gamma$ of the critical value, is of order $\epsilon |\ln \hbar|/\hbar$. Therefore, since there may be only one hyperbolic point in $p^{-1}(0)$, it follows from this estimate and the results we just proved above for the regular and the elliptic cases that

$$\rho_\hbar(E) \geq C|\ln \hbar|,$$

for some constant $C > 0$. This gives $\rho(E) = +\infty$.

Remark 3.3. It is probable that the nondegeneracy condition can be avoided. It is known quite generally that Weyl asymptotics hold for critical energies [25]. Thus, in all cases, we recover the action integral as the integrated density of states. It would remain to show that the behaviour of the action integral completely determines the singularities of p. This is easy in the Schrödinger case $p = \xi^2 + V(x)$.

4 Topology

As we already mentioned above, once the singular fibres of p have been excluded, the topology is easy to understand. The map p becomes a locally trivial fibration whose fibres are disjoint unions of circles.

Thus, if E_0 is a regular value of p, the semiglobal problem around E_0 just amounts to counting the number of connected components of $p^{-1}(E_0)$.

The topology of singular fibres strongly depends on the type of singularity. Under the nondegeneracy assumption, the topology of the singular foliation in a neighbourhood of a singular fibre is essentially determined by the type of the singularity, and thus by Theorem 3.1.

4.1 Connected components

Let I be a compact interval of regular values of p. As above, we denote by $\mathcal{C}_k(\lambda)$, for $k = 1, \ldots, N$ and $\lambda \in I$, the smooth families of connected components of $p^{-1}(\lambda)$. Each $\mathcal{C}_k(\lambda)$ is globally invariant by the Hamiltonian flow generated by p. Thus, this flow is periodic on $\mathcal{C}_k(\lambda)$. Let $|\tau_k(\lambda)| \neq 0$ be its primitive period (the sign is determined by the formula below). It follows from the action-angle theorem that τ_k is a smooth function of λ. In fact, it is well known that the period is the derivative of the action, and we have already met this quantity in the proof of Proposition 3.2. Using the action integral (4), we get

$$\tau_k(\lambda) = \frac{\partial g_0^{(k)}(\lambda)}{\partial \lambda}.$$

Notice again that τ_k never vanishes on I.

Definition 4.1. We say that a point $(\lambda, t) \in (I \times \mathbb{R}^*)$ is **resonant** whenever there exist (k, j) and (k', j') in $\{1, \ldots, N\} \times \mathbb{Z}^*$, with $k \neq k'$, such that

$$j\tau_k(\lambda) = j'\tau_{k'}(\lambda) = -t.$$

Theorem 4.2. *Let I be an interval of regular values of p, and let Assumption $\mathcal{A}(P, \mathcal{J}, I)$ hold. Assume also that the set of resonant points in $I \times \mathbb{R}$ is discrete. Then the number N of connected components of $p^{-1}(\lambda)$, $\lambda \in I$, is determined by the spectrum $\Sigma(P, \mathcal{J}, I) + \mathcal{O}(\hbar^2)$.*

Before proving the theorem, let us just remark that the leading term of Weyl's asymptotics is not sharp enough for this. Indeed, it gives only the density ρ (Proposition 3.2):

$$\rho(\lambda) = \frac{1}{\pi}\sum_{k=1}^{N} |\tau_k(\lambda)|. \tag{6}$$

From this one cannot distinguish, for example, one component with period τ from two components with periods $|\tau_1| + |\tau_2| = |\tau|$.

Observe also that the condition on resonant points is not adapted to systems with symmetries. For instance, a Schrödinger operator with a symmetric double well has two components with equal periods.

Proof of Theorem 4.1. We introduce the period lattice $\mathcal{L}_k(I)$:

$$\begin{aligned}\mathcal{L}_k(I) :&= \{(\lambda, t) \in I \times \mathbb{R};\ \exp(t\mathcal{X}_p) \text{is periodic on } \mathcal{C}_k(\lambda)\}\\ &= \{(\lambda, j\tau_k(\lambda));\ \lambda \in I, j \in \mathbb{Z}\}\end{aligned}$$

and $\mathcal{L}(I) = \bigcup_{k=1}^{N} \mathcal{L}_k(I)$. The set $\mathcal{L}(I)$ is a union of smooth graphs that may intersect. The intersection points for $t \neq 0$ are precisely the resonant points.

In order to prove the theorem, we split the argument into two steps. The first one is to prove that $\Sigma(P, \mathcal{J}, I) + \mathcal{O}(\hbar^2)$ determines $\mathcal{L}(I)$. The second step consists in showing why the knowledge of $\mathcal{L}(I)$—and the assumption on the set of resonant points—allows us to count the number N of connected components.

Step 1. Returning to the Bohr–Sommerfeld rules discussed in the proof of Proposition 3.2, we recall that the spectrum of P modulo $O(\hbar^\infty)$ is the superposition of the spectra σ_k generated by $\mathcal{C}_k$, for $k = 2, \ldots, N$. For each k, σ_k has a periodic structure that makes it close to an arithmetic progression. Thus, a simple and naive idea to distinguish between the different periodic structures is to perform a *frequency analysis*, via a Fourier transform. Because we have at our disposal only a truncated sequence of eigenvalues (those that belong to I), we need to introduce a cut-off. Let $I' \Subset I$ and let $\chi \in C^\infty(\mathbb{R})$ have compact support in the interior of I and be equal to 1 on I'. We introduce the spectral measure

$$D_0(\lambda; \hbar) = \sum_{E \in \Sigma_\hbar(P, I)} \chi(E)\delta_E(\lambda),$$

where δ_E is the Dirac distribution at E. The quantity we want to investigate is its Fourier transform. Since the mean spacing between consecutive eigenvalues is of order $\hbar$, we use a corresponding scale for the time variable t, and thus introduce

$$Z(t;\hbar) = \sum_{E\in\Sigma_\hbar(P,I)} \chi(E)e^{-itE/\hbar}.$$

The function Z is called the *partition function*. In fact, the idea we have just described is very well known in the semiclassical context, and is part of the general formalism of *trace formulas*. We can consider the Schrödinger group $U(t;\hbar) = \exp(-itP/\hbar)$, and then $Z(t;\hbar) = \mathrm{Trace}(\chi(P)U(t;\hbar))$. It is well known that $\chi(P)U(t;\hbar)$ is a Fourier integral operator, whose canonical transformation is the classical flow of p. Moreover, its trace is a Lagrangian (or WKB) distribution associated with the Lagrangian manifold of periods

$$\Lambda_p = \{(E,\tau)\in\mathbb{R}^2;\ \exists z\in p^{-1}(E), \exp(\tau\mathcal{X}_p)(z) = z\} = \mathcal{L}(I').$$

Such a result would almost finish the proof of Step 1. In fact, this statement exists in many versions, depending on various possible situations and hypotheses. For this reason we are not using it here as is, but instead resort once again to the Bohr–Sommerfeld rules, which is arguably the easiest way to go.

We can split the partition function as

$$Z(t;\hbar) = \sum_{k=1}^{N}\sum_{E\in\sigma_k} \chi(E)e^{-itE/\hbar}.$$

Then from (2) one can introduce $c\mapsto f^{(k)}(c;\hbar)$ as the inverse of $\lambda\mapsto g^{(k)}(\lambda;\hbar)$, which exists for $\hbar$ small enough, and write

$$Z(t;\hbar) = \sum_{j\in\mathbb{Z}} \varphi_t(2\pi\hbar j;\hbar) \tag{7}$$

(which, as before, is a finite sum) with

$$\varphi_t(c;\hbar) := \sum_{k=1}^{N} \chi(f^{(k)}(c;\hbar))e^{-itf^{(k)}(c;\hbar)/\hbar}. \tag{8}$$

Note that $\varphi_t(\cdot;\hbar)\in C_0^\infty(\mathbb{R})$. By the Poisson summation formula,

$$Z(t;\hbar) = \frac{1}{2\pi\hbar}\sum_{j\in\mathbb{Z}} \hat{\varphi}_t(j/\hbar) \tag{9}$$

(which, in contrast to (7), is a truly infinite sum) with

$$\frac{1}{2\pi\hbar}\hat{\varphi}_t(j/\hbar) = \frac{1}{2\pi\hbar}\int_{\mathbb{R}} e^{-icj/\hbar}\varphi_t(c)dc = \sum_{k=1}^{N} Z_k(t;\, j,\hbar) \tag{10}$$

and

$$Z_k(t;\, j,\hbar) = \frac{1}{2\pi\hbar}\int e^{-i\hbar^{-1}(cj+tf^{(k)}(c;\hbar))}\chi(f^{(k)}(c;\hbar))dc.$$

Let us now fix j and consider Z_k as a function of t. Returning to the spectral variable $\lambda = f^{(k)}(c;\hbar)$ we can write

$$\begin{aligned} Z_k(t;\, j,\hbar) &= \frac{1}{2\pi\hbar}\int e^{-it\lambda/\hbar}e^{-ijg^{(k)}(\lambda;\hbar)/\hbar}\chi(\lambda)\frac{\partial g^{(k)}(\lambda;\hbar)}{\partial\lambda}d\lambda. \\ &= \mathcal{F}_{\hbar,\lambda\mapsto t}\left(e^{-ijg^{(k)}(\lambda;\hbar)/\hbar}\chi(\lambda)\frac{\partial g^{(k)}(\lambda;\hbar)}{\partial\lambda}\right). \end{aligned}$$

Thus Z_k is the semiclassical Fourier transform of a compactly supported WKB function with phase function $\lambda \mapsto -jg_0^{(k)}(\lambda)$. Since $\frac{\partial g^{(k)}(\lambda;\hbar)}{\partial\lambda} = \tau_k(\lambda) + \mathcal{O}(\hbar)$, its associated lagrangian submanifold is defined by the equation

$$t = -j\tau_k(\lambda). \tag{11}$$

More precisely, since the amplitude of this WKB function is $\tau_k(\lambda)\chi(\lambda)+\mathcal{O}(\hbar)$, and τ_k does not vanish in the support of χ, the semiclassical wavefront-set of $Z_k(\cdot;\, j,\hbar)$ is, for fixed $j \in \mathbb{Z}$,

$$\mathrm{WF}_{\hbar}(Z_k) = \{(\lambda,t) \in \mathbb{R}^2; \quad t = -j\tau_k(\lambda), \quad \chi(\lambda) \neq 0\}.$$

Let us now turn to the behaviour of $Z(t,\hbar)$ for positive times. For this we consider the localisation of Z modulo $O(\hbar^\infty)$. Let $t_0 > 0$, $\epsilon > 0$, and let $\rho \in C_0^\infty(B(t_0,\epsilon))$. There is no solution to (11) in the support of ρ for $|j|$ outside the interval

$$I_k(\epsilon) := \left(\frac{t_0-\epsilon}{\sup_J |\tau_k|}, \frac{t_0+\epsilon}{\inf_J |\tau_k|}\right).$$

Making explicit the nonstationary phase argument, we can write, for any $\ell \in \mathbb{N}$,

$$Z_k(t;\, j,\hbar) = \left(\frac{\hbar}{ji}\right)^{\ell}\int e^{-i\hbar^{-1}(cj+tf_0^{(k)}(c;\hbar))}L^{\ell}(a(c;\hbar))dc,$$

where L is the linear differential operator defined by

$$(Lu)(c) = \frac{d}{dc}\left(\frac{u(c)}{1+\frac{t}{j}\frac{\partial f_0^{(k)}}{\partial c}}\right)$$

and $a(\cdot;\hbar) \in C_0^\infty(I)$ admits an asymptotic expansion in nonnegative powers of $\hbar$, in the C^∞ topology. Let $b(c) = \left(1 + \frac{t}{j}\frac{\partial f_0^{(k)}}{\partial c}\right)^{-1}$. Then b is uniformly bounded on I for $|j| > (t_0 + \epsilon)/\inf_J |\tau_k|$, and for any $\ell \in \mathbb{N}^*$, there exists a positive constant C_ℓ, independent of j and $\hbar$, such that $\left|\frac{d^\ell b}{dc^\ell}\right| \leq C_\ell / j$. Therefore, there exist constants $\tilde{C}_\ell > 0$ such that

$$|L^\ell(a)| \leq \tilde{C}_\ell,$$

and we get, again when $|j| > (t_0 + \epsilon)/\inf_J |\tau_k|$,

$$|\rho(t) Z_k(t; j, \hbar)| \leq \tilde{C}_\ell \left(\frac{\hbar}{j}\right)^\ell.$$

Thus, for $\ell \geq 2$,

$$\sum_{|j| > \frac{t_0+\epsilon}{\inf_J |\tau_k|}} |\rho(t) Z_k(t; j, \hbar)| \leq \tilde{C}_\ell \hbar^\ell.$$

This shows that only a finite (independent of $\hbar$) number of terms contributes to $\rho(t) Z(t;\hbar)$ modulo $O(\hbar^\infty)$. Thus the (non)stationary phase approximations are jointly valid. Therefore $Z(t;\hbar)$ microlocally vanishes at any point that does not belong to $\mathcal{L}(I)$; this can be written

$$\mathrm{WF}_\hbar(Z(\cdot;\hbar)) \subset \mathcal{L}(I).$$

More precisely,

$$\mathrm{WF}_\hbar(\rho Z(\cdot;\hbar)) \subset \{(\lambda, j\tau_k(\lambda));\ \lambda \in I, |j| \in I_k(\epsilon), k = 1, \ldots, N\}.$$

Moreover, at a non-resonant point $(\lambda, j\tau_k(\lambda))$, no other period $j'\tau_{k'}$ can contribute, and thus $Z(\cdot;\hbar)$ is a lagrangian distribution microlocally equal to $Z_k(\cdot; j, \hbar)$. Since the set of resonant points is discrete, and $\mathrm{WF}_\hbar(Z)$ is closed in T^*I, we must have $\mathrm{WF}_\hbar(Z(\cdot;\hbar)) = \mathcal{L}(I)$, which finishes the proof of the first step.

Step 2. We are now left with a simple geometric inverse problem: given the set of periods $\mathcal{L}(I)$, how can one recover the number N of connected components?

Our strategy is to recover the fundamental periods $|\tau_1|, \ldots, |\tau_N|$. First of all, by Weyl's asymptotics (6), one obtains the a priori bound $|\tau_k(\lambda)| \leq \pi\rho(\lambda)$. Let $R := \max_J \pi\rho$. Then by assumption, the set of resonant points inside $I \times]0, R]$ is finite; therefore, one can always find a smaller, nonempty interval $\tilde{I} \subset I$ such that there is *no* resonant point at all in $\tilde{I} \times]0, R]$.

We extract the periods τ_k from $\mathcal{L}_1 := \mathcal{L}(\tilde{I}) \cap (\tilde{I} \times]0, R])$ inductively, as follows.

1. Consider a point $(\lambda_1, \tau_1) \in \mathcal{L}_1$ with "minimal height" $\tau_1 : \forall (\lambda, \tau) \in \mathcal{L}_1, \tau_1 \leq \tau$.
2. By the nonresonance assumption, the connected component of (λ_1, τ_1) in $\mathcal{L}_1$ is the graph of a smooth function of the interval $\tilde{I}$. We denote this function by $\lambda \mapsto \tau_1(\lambda)$.

3. Consider the set

$$\mathcal{L}_2 := \mathcal{L}_1 \backslash \{(\lambda, j\tau_1(\lambda));\ \lambda \in \tilde{I}, j \in \mathbb{Z}^*\}.$$

Again by the nonresonance assumption, $\mathcal{L}_1$ remains a union of nonintersecting smooth graphs.
4. If $\mathcal{L}_1$ is empty, then $N = 1$. Otherwise, start again by replacing $\mathcal{L}_0$ by $\mathcal{L}_1$, and so on. If $\mathcal{L}_k$ is empty, then $N = k - 1$.

Remark 4.3. If we disregard symmetry issues, our assumption on the resonant set is quite weak. For instance, one can allow the crossing of two periods to be flat (all derivatives are equal at a point λ), simply because we put ourselves in a region with no crossing at all. However, it is easy to prove Step 2 with even weaker assumptions. For instance, it may work even if there are some open intervals of values of λ which admits resonant pairs. It would be interesting to know whether Step 1 could hold in this case as well. It would then involve subprincipal terms in the Bohr–Sommerfeld expansion.

4.2 Singular fibres

As we already mentioned, the following result comes for free.

Theorem 4.4. *Let Assumption $\mathcal{A}(P, \mathcal{J}, I)$ hold, and let $E_0 \in I$ be a nondegenerate critical value of p. Assume also that $p^{-1}(E_0)$ contains only one critical point. Then from the knowledge of $\Sigma(P, \mathcal{J}, I) + \mathcal{O}(\hbar^2)$ one can determine the topology of the singular foliation induced by p, in a saturated neighbourhood of $p^{-1}(E_0)$.*

Proof. Under these assumptions, the topology of the singular foliation induced by p in a saturated neighbourhood of $p^{-1}(E_0)$ is known to be completely characterised by the type of the singularity [9, 26], which is determined by Theorem 3.1. For the convenience of the reader, we briefly recall the two possible cases.

1. *The elliptic case.* The singular fibre $p^{-1}(E_0)$ is just a point and the foliation is homeomorphic to the one given by the Hamiltonian $H(x, \xi) = x^2 + \xi^2$.
2. *The hyperbolic case.* The singular fibre is a circle with a transversal self-intersection (the figure eight). It separates a saturated neighbourhood into three connected parts: two on one side, and one on the other side. It is homeomorphic to the foliation given by the Hamiltonian $H(x, \xi) = \xi^2 + x^4 - x^2$, in a neighbourhood of $H^{-1}(0)$.

4.3 Global topology

We say that a Hamiltonian system p on the symplectic 2-manifold M is topologically equivalent to the Hamiltonian system $\tilde{p}$ on $\tilde{M}$ if there is a homeomorphism

$\varphi : M \to \tilde{M}$ such that

$$p = \tilde{p} \circ \varphi.$$

Notice that this implies that φ respects the foliation, fibre by fibre. In particular, p and $\tilde{p}$ have the same set of regular values and the same set of critical values. If I is an open interval, then two Hamiltonian systems p and $\tilde{p}$ are called topologically equivalent over I when they are topologically equivalent when restricted to the symplectic manifolds $p^{-1}(I)$, $(\tilde{p})^{-1}(I)$.

We call the *topological type* of a Hamiltonian system the equivalence class of topologically equivalent systems.

Theorem 4.5. *Let Assumption $\mathcal{A}(P, \mathcal{J}, I)$ hold, and assume that p has only non-degenerate critical values in some neighbourhood of I, such that any two critical points with the same singularity type cannot have the same image by p. Let $c_1 < \cdots < c_n$ be the critical values of p in I. Suppose that in each interval (c_i, c_{i+1}), $i = 1, \ldots, n-1$, there exists a nonempty subinterval I_i such that the set of resonant points in $I_i \times \mathbb{R}$ is discrete in $\mathbb{R}^2$. Then the knowledge of $\Sigma(P, \mathcal{J}, I) + \mathcal{O}(\hbar^2)$ determines the topological type of the Hamiltonian system p over I.*

Proof. Upon a possible enlargement of I, one may assume that $I = (E_0, E_1)$ for regular values E_0, E_1. Using symplectic cutting [14] or surgery [26], one may replace the phase space $\mathbb{R}^2$ by a compact symplectic manifold where $p^{-1}(I)$ is embedded. Then we apply the result of [9] that says that the topological type of p on M is determined by its Reeb graph: the set of leaves of the foliation, as a topological 1-complex. This graph is characterised by the relative positions of critical values, and the number of fibres between two consecutive critical values. The former is determined by the spectrum in I thanks to Theorem 3.1, while the latter is determined for each $i = 1, \ldots, n-1$ by the spectrum in I_i, thanks to Theorem 4.2. This gives the topological type of p, up to some homeomorphism of the Reeb graph itself. But since we know the precise values of p at singularities, we can in fact assume that this homeomorphism is the identity.

5 Symplectic geometry

The tools we have used so far give us the *periods* of the classical Hamiltonian system, which is of course much more than mere topological information. We show here that it is indeed sufficient to recover the full dynamics of the systems.

We say that a Hamiltonian system p on the symplectic 2-manifold M is symplectically equivalent to the Hamiltonian system $\tilde{p}$ on $\tilde{M}$ if there is a smooth symplectomorphism $\varphi : M \to \tilde{M}$ such that

$$p = \tilde{p} \circ \varphi.$$

Thus, the dynamics of p on the level set $\{p = E\}$ is transported via φ to the dynamics of $\tilde{p}$ on the level set $\{\tilde{p} = E\}$.

We call the *symplectomorphism type* of a Hamiltonian system the equivalence class of symplectically equivalent systems. As before, one may restrict this equivalence to an interval I of values of p and $\tilde{p}$.

Definition 5.1. Let $(\lambda, t) \in I \times \mathbb{R}$ be a resonant point for p. Thus

$$j\tau_k(\lambda) = j'\tau_{k'}(\lambda) = -t$$

for some $j, j', k \neq k'$. We say that this resonance is *weakly transversal* if there exists an integer $n \in \mathbb{N}^*$ such that the nth derivatives of the periods are not equal:

$$j\tau_k^{(n)}(\lambda) \neq j'\tau_{k'}^{(n)}(\lambda).$$

Theorem 5.2. *Let Assumption $\mathcal{A}(P, \mathcal{J}, I)$ hold, and suppose that p has only non-degenerate critical values in some neighbourhood of I, such that any two critical points with the same singularity type cannot have the same image by p. Let $c_1 < \cdots < c_n$ be the critical values of p in I. Suppose that for each interval $J_i := (c_i, c_{i+1})$, $i = 1, \ldots, n-1$, the set of resonant points in $J_i \times \mathbb{R}$ is discrete. Finally, assume that all such resonant points are weakly transversal.*

Then the knowledge of $\Sigma(P, \mathcal{J}, I) + \mathcal{O}(\hbar^2)$ determines the symplectic type of the Hamiltonian system p over I.

Proof. We use the symplectic classification of [9, 18] using *weighted* Reeb graphs. Under our assumptions, the Reeb graph has vertices of degrees 1 and 3. A vertex of degree 1, a *bout*, corresponds to an elliptic critical value, while a vertex of degree 3, called a *bifurcation point*, corresponds to a hyperbolic critical value. At a bifurcation point we can distinguish one particular edge, called the *trunk*, corresponding to the side of the figure 8 with only one connected component. The two other edges are called the *branches*. A weighted Reeb graph is a Reeb graph each of whose edges is associated with a positive real number, its *length*, and such that each of the two branches of each bifurcation point is associated with a formal Taylor series (i.e., a sequence of real numbers). The hypotheses of the theorem allow for determining the topological Reeb graph via Theorem 4.5. Thus, the next step of the proof is to show how the numbers that constitute the weighted Reeb graph can be recovered from the spectrum. The final step is to obtain the symplectic equivalence in the sense that we have just defined above.

The lengths

Let $\mathcal{C}_k(J_i)$, for $k = 1, \ldots, N$, be the connected components of $p^{-1}(J_i)$. Let $K_{k,i} \in C^\infty(\mathcal{C}_k(J_i))$ be an action variable for the regular Lagrangian fibration $p_{\restriction \mathcal{C}_k(J_i)}$; it is unique up to a sign and an additive constant. By definition the *length* of the edge corresponding to the set of leaves in $\mathcal{C}_k(J_i)$ is

$$\ell_{k,i} := \left| \lim_{c \to c_i} K_{k,i}(c) - \lim_{c \to c_{i+1}} K_{k,i}(c) \right|. \tag{12}$$

In learned terminology, this is the Duistermaat–Heckman measure of J_i for the S^1-action defined by $K_{k,i}$, or equivalently, it is the affine length of J_i endowed with its natural integral affine structure given by $p_{\restriction \mathcal{C}_k(J_i)}$.

It follows from the local models for elliptic and hyperbolic singularities that this length is always finite. This is obvious at elliptic singularities, where the action has the form $x^2 + \xi^2$. At a hyperbolic singularity m, one can introduce a foliation function q such that in some local symplectic coordinates around m, $q = x\xi$, and $q > 0$ on the branches, while $q < 0$ on the trunk. Then the Duistermaat–Heckman measure has the form

$$\begin{cases} d\mu_j(q) = (\ln q + g_j(q))dq & \text{on each branch } (j = 1, 2), \\ d\mu(q) = (2\ln|q| + g(q))dq & \text{on the trunk,} \end{cases} \tag{13}$$

with some smooth functions g, g_1, g_2 satisfying

$$\forall p, \qquad g^{(p)}(0) = g_1^{(p)}(0) + g_2^{(p)}(0).$$

In this form, the Taylor series of the functions g, g_1, g_2 at the origin are uniquely defined [18, 20].

Using the proof of Theorem 4.2, from the spectrum in I we can recover the periods $\tau_k(\lambda)$, $k = 1, \ldots, N$, for λ in any interval in J_i where the graphs of the periods τ_k don't cross. At a crossing the difficulty is to put the labels k correctly, so that the connected components $\mathcal{C}_k(\lambda)$ remain in the same $\mathcal{C}_k(J_i)$ when λ varies. This can be overcome precisely thanks to the weak resonant assumption at each crossing, because each τ_k is C^∞ in J_i. This was the main issue. Now, fixing a point $\lambda_i \in J_i$, the action variable $K_{k,i}$ we can compute by the formula

$$K_{k,i}(\lambda) := \int_{\lambda_i}^{\lambda} \tau_k(\lambda) d\lambda, \quad \lambda \in J_i.$$

This gives the length of $\mathcal{C}_k(J_i)$ via equation (12).

The Taylor series at the bifurcation points

By definition, the sequences of numbers associated with a bifurcation point in the Reeb graph are the Taylor series of the functions g_1, g_2 (defined in equation (13)) at the origin.

Let us show how to recover the Taylor series of g from the spectrum. The procedure is completely analogous for g_1 and g_2.

Thus, we consider a hyperbolic critical value c_{i+1}. We want to express the Duistermaat–Heckman measure on the trunk in terms of the principal symbol

p. By a theorem of Colin de Verdire and Vey [5], there exist local symplectic coordinates (x, ξ) at the hyperbolic point, and a smooth, locally invertible function $f : (\mathbb{R}, c_{i+1}) \to (\mathbb{R}, 0)$ such that

$$f(p) = x\xi = q.$$

For notational purposes, one may assume that $f'(c_{i+1}) > 0$, which amounts to saying that the trunk is sent by p to $\lambda < c_{i+1}$. Then from (13), for λ close to c_{i+1}, $\lambda < c_{i+1}$,

$$d\mu(\lambda) = (2 \ln |f(\lambda)| + g \circ f(\lambda)) f'(\lambda) d\lambda.$$

On the other hand, if the connected component corresponding to the trunk is $\mathcal{C}_k(J_i)$, one has by definition of the Duistermaat–Heckman measure $d\mu(\lambda) = \tau_k(\lambda)d\lambda$. Therefore

$$\tau_k(\lambda) = f'(\lambda)(2 \ln |f(\lambda)| + g \circ f(\lambda)) = 2f'(\lambda) \ln |\lambda - c_{i+1}| + h(\lambda),$$

for some smooth function h at $\lambda = c_{i+1}$. There, using Taylor's formula, we have written $f(\lambda) = \alpha(\lambda - c_{i+1}) + (\lambda - c_{i+1})^2 \hat{f}(\lambda)$, with $\alpha > 0$ and $\hat{f}$ smooth at c_{i+1}, and hence

$$h(\lambda) = 2f'(\lambda) \ln |\alpha + (\lambda - c_{i+1}) \hat{f}(\lambda)| + f'(\lambda) g \circ f(\lambda). \tag{14}$$

This shows that h is smooth for λ close to c_{i+1}.

It is easy to see that any smooth function ϕ in a neighbourhood of the origin such that $\phi(t) \ln t$ extends to a smooth function at $t = 0$ must be flat. Hence the knowledge of $\tau_k(\lambda)$ for $\lambda < c_{i+1}$ completely determines the Taylor series of $f'(\lambda)$ (and hence $f(\lambda)$) at $\lambda = c_{i+1}$.

Then one can recover the Taylor series of h using

$$h(\lambda) = \tau_k(\lambda) - 2f'(\lambda) \ln |\lambda - c_{i+1}|, \quad \forall \lambda < c_{i+1}.$$

Finally, from (14) and the fact that f is locally invertible, one can recover the Taylor series of g at the origin.

Symplectic equivalence

We have proved that the weighted Reeb graph is determined by the spectrum. By Toulet's classification [9, 18], if two such systems (M, p) and $(M, \tilde{p})$ have the same weighted Reeb graph, there exists a symplectomorphism $\varphi : M \to \tilde{M}$ such that p and $\tilde{p} \circ \varphi$ define the same singular foliation on M (φ induces a homeomorphism of the leaf space, fixing the vertices). If we assume that the operators P and $\tilde{P}$ have the same spectrum (modulo $\hbar^2$) and satisfy the requirements of the theorem, then we also know that p and $\tilde{p} \circ \varphi$ share the same set of critical values c_i. The fact that p and $\tilde{p} \circ \varphi$ define the same foliation implies that for each connected component

$\mathcal{C}_k(J_i)$, there exists a smooth, invertible function $f : J_i \to J_i$ such that

$$p = f \circ \tilde{p} \circ \varphi \qquad \text{on } \mathcal{C}_k(J_i). \tag{15}$$

Since the singular fibres at the ends of $\mathcal{C}_k(J_i)$ are fixed by φ, f must be increasing, and thus extends to a homeomorphism of $\overline{J_i}$.

As we already saw, the spectrum also determines the periods at a given energy $E = \lambda$. Hence for $\lambda \in J_i$, $\tau_k(\lambda) = \tilde{\tau}_k(\lambda)$. Since τ_k is integrable at c_{i+1}, we can define action integrals for $\lambda < c_{i+1}$ as

$$K_{k,i}(\lambda) := \int_{c_{i+1}}^{\lambda} \tau_k(\lambda) d\lambda, \qquad \tilde{K}_{k,i}(\lambda) := \int_{c_{i+1}}^{\lambda} \tilde{\tau}_k(\lambda) d\lambda.$$

We have $K_{k,i}(\lambda) = \tilde{K}_{k,i}(\lambda)$. On the other hand, the action is a symplectic invariant of the foliation. From (15) on can compute the action on the curve $\varphi(\mathcal{C}_k(f(\lambda))) = \tilde{\mathcal{C}}_k(\lambda) : K_{k,i}(f(\lambda)) = \tilde{K}_{k,i}(\lambda) + \text{const}$. Therefore

$$K_{k,i}(\lambda) = K_{k,i}(f(\lambda)).$$

Since τ_k does not vanish in J_i, $K_{k,i}$ is strictly monotonic on J_i. Therefore

$$f(\lambda) = \lambda, \qquad \forall \lambda \in J_i.$$

Thus $p = \tilde{p} \circ \varphi$ on each $\mathcal{C}_k$, and by continuity

$$p = \tilde{p} \circ \varphi \quad \text{on } M.$$

This finishes the proof of the theorem.

References

1. R. Brummelhuis, T. Paul, and A. Uribe. Spectral estimates around a critical level. *Duke Math. J.*, 78(3):477–530, 1995.
2. L. Charles and S. Vũ Ngọc. Spectral asymptotics via the semiclassical Birkhoff normal form. *Duke Math. J.*, 143(3):463–511, 2008.
3. Y. Colin de Verdière. A semi-classical inverse problem II: reconstruction of the potential. `oai:hal.archives-ouvertes.fr:hal-00251590_v1`.
4. Y. Colin de Verdière and B. Parisse. Singular Bohr–Sommerfeld rules. *Commun. Math. Phys.*, 205:459–500, 1999.
5. Y. Colin de Verdière and J. Vey. Le lemme de Morse isochore. *Topology*, 18:283–293, 1979.
6. Y. Colin de Verdière and V. Guillemin. Semi-classical inverse problem I: Taylor expansions. preprint, `hal-00250568`.
7. M. Combescure, J. Ralston, and D. Robert. A proof of the Gutzwiller semiclassical trace formula using coherent states decomposition. *Commun. Math. Phys.*, 202(2):463–480, 1999.
8. M. Dimassi and J. Sjöstrand. *Spectral asymptotics in the semi-classical limit*, volume 268 of *London Mathematical Society Lecture Note Series*. Cambridge University Press, Cambridge, 1999.

9. J.-P. Dufour, P. Molino, and A. Toulet. Classification des systèmes intégrables en dimension 2 et invariants des modèles de Fomenko. *C. R. Acad. Sci. Paris Sér. I Math.*, 318:949–952, 1994.
10. C. Gordon, D. Webb, and S. Wolpert. Isospectral plane domains and surfaces via Riemannian orbifolds. *Invent. Math.*, 110(1):1–22, 1992.
11. V. Guillemin and A. Uribe. Some inverse spectral results for semi-classical Schrödinger operators. *Math. Res. Lett.*, 14(4):623–632, 2007.
12. M. Kac. Can one hear the shape of a drum? *The American Math. Monthly*, 73(4):1–23, 1966.
13. O. Lablée. Sur le spectre semi-classique d'un système intégrable de dimension 1 autour d'une singularité hyperbolique. *Annales de la Faculté des Sciences de Toulouse Mathématiques*, 19(1):85–123, 2010.
14. E. Lerman. Symplectic cuts. *Math. Res. Lett.*, 2:247–258, 1995.
15. J. Milnor. Eigenvalues of the Laplace operator on certain manifolds. *Proc. Natl. Acad. Sci. USA*, 51:542, 1964.
16. D. Robert. Propriétés spectrales d'opérateurs pseudo-différentiels. *Comm. Partial Differential Equations*, 3:755–826, 1978.
17. D. Robert. *Autour de l'approximation semi-classique*, volume 68 of *Progress in Mathematics*. Birkhäuser, 1987.
18. A. Toulet. *Classification des systèmes intégrables en dimension 2*. PhD thesis, Université de Montpellier II, 1996.
19. S. Vũ Ngọc. Bohr–Sommerfeld conditions for integrable systems with critical manifolds of focus–focus type. *Comm. Pure Appl. Math.*, 53(2):143–217, 2000.
20. S. Vũ Ngọc. On semi-global invariants for focus–focus singularities. *Topology*, 42(2): 365–380, 2003.
21. S. Vũ Ngọc. *Systèmes intégrables semi-classiques: du local au global*. Number 22 in Panoramas et synthèses. SMF, 2006.
22. A. Weinstein. Application des opérateurs intégraux de Fourier aux spectres des variétés riemanniennes. *C. R. Acad. Sci. Paris Sér. A*, 279:229–230, 1974.
23. S. Zelditch. The inverse spectral problem. In *Surveys in differential geometry. Vol. IX*, Surv. Differ. Geom., IX, pages 401–467. Int. Press, Somerville, MA, 2004. With an appendix by Johannes Sjöstrand and Maciej Zworski.
24. S. Zelditch. Inverse spectral problem for analytic domains. I: Balian–Bloch trace formula. *Commun. Math. Phys.*, 248(2):357–407, 2004.
25. L. Zielinski. Sharp semiclassical estimates for the number of eigenvalues below a degenerate critical level. *Asymptotic Analysis*, 53(1–2):97–123, 2007.
26. Nguyên Tiên Zung. Symplectic topology of integrable Hamiltonian systems, I: Arnold–Liouville with singularities. *Compositio Math.*, 101:179–215, 1996.
27. S. Zelditch. Inverse spectral problem for analytic domains. II. $\mathbb{Z}_2$-symmetric domains. *Ann. of Math.*, (2) 170(1):205–269, 2009.